LEE W. JOHNSON

R. DEAN RIESS

*Virginia Polytechnic Institute and State University*

# NUMERICAL ANALYSIS

100606

ADDISION-WESLEY PUBLISHING COMPANY

*Reading, Massachusetts • Menlo Park, California*

*London • Amsterdam • Don Mills, Ontario • Sydney*

Copyright © 1977 by Addison-Wesley Publishing Company, Inc. Philippines copyright 1977 by Addison-Wesley Publishing Company, Inc.

All rights reserved. No part of this publication may be reproduced, stored in a retrieval system, or transmitted, in any form or by any means, electronic, mechanical, photocopying, recording, or otherwise, without the prior written permission of the publisher. Printed in the United States of America. Published simultaneously in Canada. Library of Congress Catalog Card No. 76-14658.

ISBN 0-201-03442-5
ABCDEFGHIJ-MA-7987

QA
297
J63

# CONTENTS

# GLOSSARY OF SYMBOLS

Below is a list of frequently used symbols and the pages where they are first defined. Other symbols do occur, but they are used only in the section where they are defined.

| | |
|---|---|
| $I$ | identity matrix, 17 |
| $A^{-1}$ | matrix inverse, 17 |
| $\mathbf{e}_j$ | canonical unit vectors, 17–18 |
| $\boldsymbol{\theta}$ | zero vector, 18 |
| $A^T$ | matrix transpose, 20 |
| $R^n$ | real $n$-tuples, 38 |
| $\ell_p$ | normed vector space, 38 |
| $\mathcal{O}$ | zero matrix, 41 |
| $\bar{B}$ | conjugate matrix, 107 |
| $B^*$ | conjugate transpose, 107 |
| $g(I) \subseteq I$ | contraction function, 121 |
| $C^k[a, b]$ | functions of $k$th-order continuity, 125 |
| $C[a, b]$ | continuous functions on $[a, b]$, 167 |
| $\mathscr{P}_n$ | polynomials of degree $n$ or less, 167 |
| $\ell_j(x)$ | Lagrange polynomials, 171 |
| $\delta_{ij}$ | Kronecker delta, 171 |
| $\Delta f(x)$ | forward difference, 179 |
| $\Delta^k f(x)$ | $k$th forward difference, 179 |
| $\binom{k}{i}$ | binomial coefficients, 179 |

| | |
|---|---|
| $\mathrm{Spr}\{y_i\}_{i=1}^n$ | spread of $\{y_1, y_2, \ldots, y_n\}$, 187 |
| $T_k(x)$ | Chebyshev polynomials, 190 |
| $\langle f, g \rangle$ | inner-product of functions, 212 |
| $I(f)$ | weighted integral, 235 |
| $Q_n(f)$ | interpolatory quadrature, 236 |
| $T_N(f)$ | composite trapezoidal rule, 245 |
| $S_N(f)$ | composite Simpson's rule, 246 |
| $B_{2j}$ | Bernoulli numbers, 249 |
| $C_{2\pi}$ | continuous periodic functions, 275 |
| $\Phi(x_i, y_i; h)$ | increment function, 290 |
| $\tau_i$ | local truncation error, 297 |
| $\nabla f(x, y)$ | gradient, 337 |

# PREFACE: SUGGESTIONS ON THE USE OF THIS TEXT

In writing this book, our primary goal has been to provide a *textbook* that is well suited for use in an introductory numerical analysis course at the undergraduate level. We view the essential ingredients of such a text to be a proper range of topics, a cohesive and understandable presentation, numerous examples which stress insight into the topics, and large selections of exercises which both reinforce the material of the text and encourage further investigation.

The list of topics is quite extensive, ranging from important classical material to a number of relatively modern concepts. However, the text is by no means an encyclopedia of all possible numerical methods. Rather it presents a selection of commonly used basic procedures together with a comparative analysis of their strengths and weaknesses. Each topic is introduced in its simplest and most understandable form and then developed to a point where the reader has a sound and fundamental background in the subject. With this background, the student may independently delve deeper into the advanced aspects of problems of interest. The examples, for the most part, are kept as simple and direct as possible in order to illustrate a point without obscuring it. There are a few fairly intricate examples, however, which should help the student appreciate the complex nature of typical real-world problems. The exercises include both theoretical and computational problems ranging from the routine to the challenging.

Although this book is intended for use in the classroom, we believe it will serve as a valuable reference book as well since it includes an introduction to several modern topics not usually found in many such texts. These topics include: solutions of over-determined linear systems, spline approximation, the fast Fourier transform, adaptive quadrature, collocation methods for differential equations, and an introduction to some optimization techniques such as quasi-Newton methods, Lagrange multipliers, and linear programming. The text should appeal to engineers, scientists and mathematicians alike, in that it presents computationally efficient and practical methods and also includes sufficient mathematical theory for a thorough understanding of each method presented. The theoretical material is kept at a minimum and is presented in an expository and intuitive fashion, but is still sufficiently comprehensive so that the reader can understand why each technique works, how efficient it is, and, possibly most important, what can cause it to fail.

The text can be used either for a one-term course in introductory numerical methods or for a full-year numerical analysis course at the junior-senior level. This is possible since the more elementary material is placed at the beginning of each chapter, while the more sophisticated material is near the end of the chapter where it can be omitted without loss of continuity. We have given some illustrations at the end of this preface as to how the text may be adapted to fulfill the purposes of either type of course. There are other alternatives to our suggestions, as the chart of chapter dependencies shows. For example, the instructor of a one-term course may wish to omit the material on eigenvalues entirely and concentrate more on interpolation, quadrature, or differential equations. In either type of course, the instructor can present the material in a different order than it appears in the text. For example, interpolation can be covered first, if so desired. The starred sections are more sophisticated than the others and can be either skimmed or omitted. The depth of coverage of these special sections suggests one way of adjusting the level of the course. The broad range of exercises also provides flexibility in adjusting the level of the course.

The prerequisites for either type of course are basic calculus and a familiarity with the ideas of a matrix and a determinant. The fundamental results from calculus that we use frequently are listed in Chapter 1 and most of the fundamentals of matrix theory are reviewed in Chapter 2. We occasionally introduce material that is not always covered in a freshman/sophomore calculus course, such as norms, inner-products and eigenvalues. In these instances, we give a

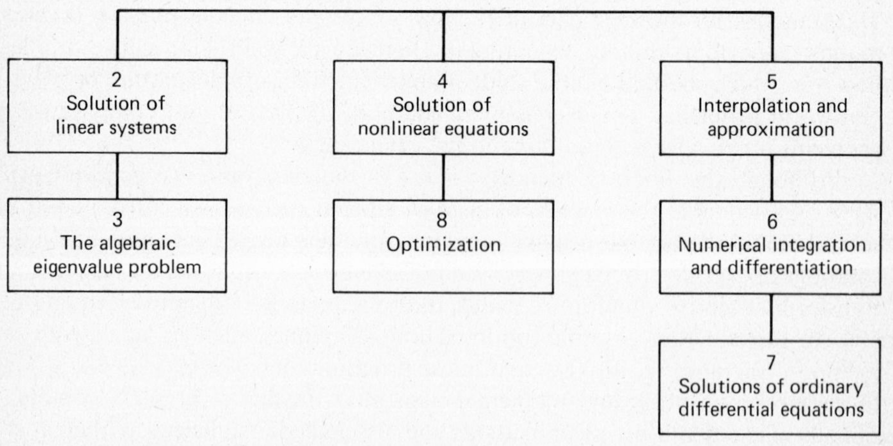


**Fig. 1** Relationship of the chapters.

careful exposition and the reader should have little difficulty understanding the material and its relevance to the particular topic or method being discussed. Whenever possible, material that requires some mathematical maturity is left to the later sections of a chapter or included in the starred (optional) sections.

Although some computational experience can be gained on a desk calculator, it is useful for a student to have a modest programming ability in a language such as FORTRAN. To aid the student in gaining computational experience, we have included a number of fairly simple programs for some of the numerical methods. These programs are in the form of subroutines which are documented and easy to understand since they are written to parallel the statement of the algorithms. In the interest of clarity, we have not tried to include all-purpose, fool-proof codes that handle every possible contingency. Comprehensive, well-tested programs of this sort are available in the literature and in most computer libraries. As we develop techniques to avoid pitfalls that can occur in a particular method the reader should be able to expand the simple programs given in the text to accommodate these techniques.

*Chapter 2:* Sections 1., 2., 2.1, 2.2, 2.3, 2.4
*Chapter 3:* Sections 1., 2.
*Chapter 4:* Sections 1., 2., 3., 3.1, 3.2, 3.3, 3.4, 4., 4.1
*Chapter 5:* Sections 1., 2., 2.1, 2.2, 2.3, 2.4
*Chapter 6:* Sections 1., 2., 2.1, 2.2, 2.3, 2.4
*Chapter 7:* Sections 1., 2., 2.1, 2.2, 2.3, 2.4, 3., 4., 5.

**Fig. 2** Suggested list of sections for a self-contained one-term course in numerical methods.

The first chapter presents basic material on rounding errors and floating-point arithmetic. These topics are further discussed and illustrated as they pertain to particular methods in succeeding chapters. However, since a thorough understanding of the effects of rounding and ill-conditioning requires a considerable theoretical background in mathematics and statistics, we primarily present these sources of error intuitively and illustrate them in the methods where they can cause the greatest difficulties.

The chart in Fig. 1 shows the relationships among the chapters of this book. For example, Chapters 5 and 6 are prerequisite to Chapter 7, but Chapter 7 does not depend on the material in Chapters 2, 3 and 4. (In a few instances, some starred sections and starred problems depend on other chapters.) Figure 2 is a list of sections that are appropriate for a comprehensive and self-contained one-term course on numerical methods. The instructor may wish to modify this suggested set of topics deleting some and including others that we have not listed. Furthermore, there are sufficient topics that the instructor of a full-year sequence can selectively omit or skim some, while treating others more carefully.

*Blacksburg, Virginia* L.W.J
*January 1977* R.D.R

# 1

# COMPUTATIONAL AND MATHEMATICAL PRELIMINARIES

## 1.1 INTRODUCTION

In this text we shall concentrate on that portion of numerical analysis that is concerned with the solution of scientific problems utilizing modern high-speed computers. Even from this possibly narrow point of view, numerical analysis is quite widely interdisciplinary. It involves engineering and physics in converting a physical phenomenon into a mathematical model; it involves mathematics in developing techniques for the solution (or approximate solution) of the mathematical equations describing the model; finally it involves computer science for the implementation of these techniques in an optimal fashion for the particular computer available. These three processes are not independent and we should not lose sight of this dependence as we consider a particular aspect of a problem. For example, it is often the case that certain idealistic restrictions must be placed on a physical system in order to obtain a tractable mathematical model. Furthermore, it is possible that mathematical techniques which are derived to "solve" the equations of the model and which serve very well in theory cannot be practically implemented on a particular computer. The task of the numerical analyst is therefore to synthesize these processes and obtain "acceptable" numerical answers. (It is also part of the task to be able to determine when and why "acceptable" answers for a particular problem on a particular computer *cannot* be attained.)

In this text we shall concentrate on the latter two phases of this three-fold process in order to cover more topics, but we will try not to lose sight of the "real world" origins of the problems. In this chapter we shall discuss some basic concepts such as computer representation of numbers, floating-point arithmetic, and rounding errors. We shall also list some basic mathematical results from the calculus which will be useful in the analysis of the numerical methods.

## 1.2 ERRORS IN COMPUTATIONS

In analyzing the accuracy of numerical results, the numerical analyst should be aware of the possible sources of error in each stage of the computational process and of the extent to which these errors can affect the final answer. We will consider that there are three types of errors which occur in a computation. First, there are errors which we call "initial data" errors. These are errors which arise when the equations of the mathematical model are formed, due to sources such as the idealistic assumptions made to simplify the model, inaccurate measurements of data, miscopying of figures, the inaccurate representation of mathematical constants (for example, if the constant $\pi$ occurs in an equation, we must replace $\pi$ by 3.1416 or 3.141593, etc.). Another class of errors, "truncation" errors, occurs when we are forced to use mathematical techniques which give approximate, rather than exact, answers. For example, suppose we use the Maclaurin's series expansion to represent $e^x$, so that $e^x = 1 + x + x^2/2! + \cdots + x^n/n! + \cdots$. If we want a number that approximates $e^\beta$ for some $\beta$, we must terminate the expansion in order to obtain $e^\beta \approx 1 + \beta + \beta^2/2! + \cdots + \beta^k/k!$. Thus $e^\beta = 1 + \beta + \beta^2/2! + \cdots + \beta^k/k! + E$, where $E$ is the truncation error introduced in the calculation. Truncation errors in numerical analysis usually occur because many numerical methods are iterative in nature, with the approximations theoretically becoming more accurate as we take more iterations. As a practical matter, we must stop the iteration after a finite number of steps, thus introducing a truncation error. The last type of error we shall consider, "round-off" or "rounding" errors, is due to the fact that a computer has a finite word length. Thus most numbers and the results of arithmetic operations on most numbers cannot be represented exactly on a computer. Even though the computer is capable of representing numerical values and performing operations on them, we should be aware of how this is accomplished so that we can understand the error that is produced by inexact representation.

Initial data errors and truncation errors are dependent mostly on the particular problem we are examining, and we shall deal with them as they arise in the context of the different numerical methods we derive throughout the text. The total effect of round-off errors is sometimes dependent on the particular problem, in the sense that the more operations we perform, the more we can probably expect the round-off error to affect the solution. The individual round-off error due to any individual number representation or arithmetic operation is dependent, however, on the particular computer being used, and thus we shall examine the possible sources of this error in the next section, before we introduce any specific numerical methods. We emphasize that it is the effect of total error, from any and all sources, with which we are ultimately concerned. For example, we shall later see problems where a "small" error (no matter from what source) can cause a "large" error in the final solution. Problems of this type are called *ill-conditioned* and must be treated very carefully to obtain an acceptable computed answer.

There are two ways to measure the size of errors. In analyzing the error of a computation, if we let $\tilde{x}$ represent the "computed solution" to the "true solution" $x$, then we define the *absolute error* to be $(x - \tilde{x})$ and the *relative error* to be $(x - \tilde{x})/x$. (If the true solution $x$ is zero, then we say the relative error is undefined.) We shall consider these concepts again in later sections, but we briefly pause to mention here that the relative error is usually more significant than the absolute error, and hence we shall try to establish bounds for the relative error whenever possible. To illustrate this point, suppose that in "Computation A" we have $x = 0.5 \times 10^{-4}$ and $\tilde{x} = 0.4 \times 10^{-4}$ while in "Computation B" we have $x = 5000$ and $\tilde{x} = 4950$. The absolute errors are $0.1 \times 10^{-4}$ and 50, respectively, but the relative errors are 0.2 and 0.01, respectively. Stated differently, Computation A has a 20% error, while Computation B has only a 1% error.

In investigating the effect of the total error in various methods, we shall often mathematically derive an "error bound," which is a limit on how large the error can be. (This applies to both absolute and relative errors.) It is important that the reader realize that the error bound can be much larger than the actual error and that this is often the case in practice. Any mathematically derived error bound must account for the worst possible case that can occur and is often based upon certain simplifying assumptions about the problem which in many particular cases cannot be actually tested. For the error bound to be used in any practical way, the user must have a good understanding of how the error bound was derived in order to know how crude it is, i.e., how likely it is to overestimate the actual error. Of course, whenever possible, our goal is to eliminate or lessen the effects of errors, rather than trying to estimate them after they occur.

## 1.3 ROUNDING ERRORS AND FLOATING POINT ARITHMETIC

The first type of rounding error that we encounter in performing a computation evolves from the fact that most real numbers cannot be represented exactly on a computer. Superficially, readers are probably not surprised by this statement since they are aware that irrational numbers such as $\pi$ or $e$ have an infinite non-repeating decimal expansion. Thus they know that even for a hand computation they must use an approximation such as 3.14159 for $\pi$ or 2.718 for $e$, and they carry as many digits in their approximations as they feel are necessary for a particular computation. Nevertheless, they realize that once these approximations have been used an error has been introduced into the calculation that can never exactly be corrected.

Since only a finite number of digits can be represented in computer memory, each number $x$ must be represented in some fashion that uses only a fixed number of digits. One of the most common forms is the "floating-point" form, where one position is used to identify the sign of $x$, a prescribed number of digits are used to represent the "mantissa" or fractional form of $x$, and an integer is used to represent the "exponent" or "characteristic" of $x$ with respect to the base $b$ of the representation. (Modern, preferred terminology uses "significand" for mantissa and "exrad" for exponent.) Thus each $x$ can be thought of as being represented by a number $\bar{x}$ of the form $\pm(0.a_1a_2 \cdots a_m) \times (b^c)$ where $m$ is the number of digits allowed in the mantissa, $b$ is the base of the representation, $c$ is the exponent. Additionally, there are two machine-dependent constants, $\mu$ and $M$, such that $\mu \leq c \leq M$. We shall consider three bases: (i) $b = 10$ (decimal)—with which the reader is familiar and which is used on some machines; (ii) $b = 16$ (hexadecimal), which is common to the IBM 360 and 370 series; and (iii) $b = 2$ (binary), which is, in a sense, the most fundamental of the three. To be in proper form we require the mantissa, $a \equiv 0.a_1a_2 \cdots a_m$, to satisfy $|a| < 1 = b^0$ and each $a_i$, $1 \leq i \leq m$, to be an integer such that $0 \leq a_i \leq b - 1$, with $a_1 \neq 0$ (unless $\bar{x} = 0$). The floating-point representation, $\bar{x}$, is then said to be "normalized."

The floating-point decimal form ($b = 10$) should be familiar to the reader. For example, the decimal number 150.623 is the same as $0.150623 \times 10^3$ and can also be regarded as

$$(1 \times 10^2) + (5 \times 10^1) + (0 \times 10^0) + (6 \times 10^{-1}) + (2 \times 10^{-2}) + (3 \times 10^{-3}).$$

Likewise, the binary number 110.011 equals

$$(1 \times 2^2) + (1 \times 2^1) + (0 \times 2^0) + (0 \times 2^{-1}) + (1 \times 2^{-2}) + (1 \times 2^{-3}),$$

and the hexadecimal number 15F.A03 equals

$$(1 \times 16^2) + (5 \times 16^1) + (\mathrm{F} \times 16^0) + (\mathrm{A} \times 16^{-1}) + (0 \times 16^{-2}) + (3 \times 16^{-3}).$$

Note that there are only two digits, 0 and 1, in the binary system, and there are sixteen digits in the hexadecimal system; 0, 1, 2, 3, 4, 5, 6, 7, 8, 9, A, B, C, D, E, F; where the decimal equivalents of A, B, C, D, E, F are 10, 11, 12, 13, 14, 15; respectively. Also note that the hexadecimal system is a natural extension of the binary system, since $2^4 = 16$, and hence there is precisely one hexadecimal digit for each group of four binary digits ("bits") and vice versa ($0 = (0000)_2$, $7 = (0111)_2$, $\mathrm{A} = (1010)_2$, $\mathrm{F} = (1111)_2$, etc.).

The conversion of an integer from one system to another is fairly simple and can probably best be presented in terms of an example. Let $k = 275$ in decimal form, that is, $k = (2 \times 10^2) + (7 \times 10^1) + (5 \times 10^0)$. Now $(k/16^2) > 1$, but $(k/16^3) < 1$, so in hexadecimal form $k$ can be written as $k = (\alpha_2 \times 16^2) + (\alpha_1 \times 16^1) + (\alpha_0 \times 16^0)$. Now, $275 = 1(16^2) + 19 = 1(16^2) + 1(16) + 3$, and so the decimal integer, 275, can be written in hexadecimal form as 113, that is, $(275)_{10} = (113)_{16}$. The reverse process is even simpler. For example, $(5\mathrm{C}3)_{16} = 5(16^2) + 12(16) + 3 = 1280 + 192 + 3 = (1475)_{10}$. Conversion of a hexadecimal fraction to a decimal is similar. For example, $(0.2\mathrm{A}8)_{16} = (2/16) + (\mathrm{A}/16^2) + (8/16^3) = (2(16^2) + 10(16) + 8)/16^3 = (680)/4096 = (0.166)_{10}$, (carrying only three digits in the decimal form). Conversion of a decimal fraction to hexadecimal (or binary) proceeds as in the following example. Consider the number $r_1 = 1/10 = 0.1$ (decimal form). Then there exist constants $\{\alpha_k\}_{k=1}^{\infty}$ such that

$$r_1 = 0.1 = \alpha_1/16 + \alpha_2/16^2 + \alpha_3/16^3 + \alpha_4/16^4 + \cdots.$$

Now, $16r_1 = 1.6 = \alpha_1 + \alpha_2/16 + \alpha_3/16^2 + \alpha_4/16^3 + \cdots$. Thus $\alpha_1 = 1$ and $r_2 \equiv 0.6 = \alpha_2/16 + \alpha_3/16^2 + \alpha_4/16^3 + \cdots$. Again, $16r_2 = 9.6 = \alpha_2 + \alpha_3/16 + \alpha_4/16^2 + \cdots$, so $\alpha_2 = 9$ and $r_3 \equiv 0.6 = \alpha_3/16 + \alpha_4/16^2 + \cdots$. From this stage on we see that the process will repeat itself, and so we have $(0.1)_{10}$ equals the infinitely repeating hexadecimal fraction, $(0.1999\cdots)_{16}$. Since $1 = (0001)_2$ and $9 = (1001)_2$ we also have the infinite binary expansion

$$r_1 = (0.1)_{10} = (0.1999\cdots)_{16} = (0.0001\ 1001\ 1001\ 1001\cdots)_2.$$

From the above example we begin to discern one problem of number representation. Not only do we have problems with irrational numbers and infinite repeating decimal expansions such as $1/3 = 0.333\cdots$, but we also see that an $m$-digit terminating fraction with respect to one base may not have an $n$-digit terminating representation in another base. (If we were performing an iteration on a hexadecimal machine, the above example suggests that we should probably choose a step-size of $h = 1/16$ over $h = 1/10$, $h = 1/1024 = 1/2^{10}$ over $h = 1/1000$, etc., if at all possible.) In Table 1.1 we have taken several integers $k$, formed the reciprocals, $1/k$, and added the reciprocal to itself $k$ times. The theoretical result of each computation should, of course, equal 1. The calculations were performed on a six-digit hexadecimal machine.[†]

**Table 1.1**

| $k$ | Sum$(1/k)$ | $k$ | Sum$(1/k)$ | $k$ | Sum$(1/k)$ |
|---|---|---|---|---|---|
| 2 | 0.1000000E 01 | 9 | 0.9999999E 00 | 16 | 0.1000000E 01 |
| 3 | 0.9999999E 00 | 10 | 0.9999996E 00 | ⋮ | |
| 4 | 0.1000000E 01 | 11 | 0.9999997E 00 | 1000 | 0.9999878E 00 |
| 5 | 0.9999999E 00 | 12 | 0.9999998E 00 | 1006 | 0.9999912E 00 |
| 6 | 0.9999998E 00 | 13 | 0.9999999E 00 | 1012 | 0.9999843E 00 |
| 7 | 0.9999999E 00 | 14 | 0.9999995E 00 | 1018 | 0.9999678E 00 |
| 8 | 0.1000000E 01 | 15 | 0.9999999E 00 | 1024 | 0.1000000E 01 |

We must, of course, realize that part of the error in the above table is due to the round-off from addition. We shall discuss this momentarily, but for now the reader should notice the relative accuracies for the values of $k$ which were powers of 2. For example, compare $k = 1000$ to $k = 1024 = 2^{10}$.

We next consider how numbers are represented in an $m$-digit machine, particularly those numbers for which $m$ digits are not sufficient to represent the

[†] Most of the computer programs in this text were run on an IBM 370/158.

number exactly. For example, how might a number like $\frac{1}{3}$ be represented on a 5-digit decimal computer, since $\frac{1}{3} = 0.3333333 \cdots$. As a general example, suppose a real number $x$ is given exactly by

$$x = \pm(0.\tilde{a}_1\tilde{a}_2 \cdots \tilde{a}_m\tilde{a}_{m+1} \cdots) \times 10^c, \qquad \tilde{a}_1 \neq 0$$

and suppose we want to represent $x$ in an $m$-digit decimal computer. There are two common ways of representing $x$. First, we can simply let $a_k = \tilde{a}_k$, $1 \leq k \leq m$, discard the remaining digits, and let the computer representation be

$$\bar{x} = \pm(0.a_1a_2 \cdots a_m) \times 10^c.$$

This is known as "chopping." The other representation is the familiar "symmetric rounding" process which is equivalent to adding $5 \times 10^{c-m-1}$ to $x$ and then chopping (we think of adding 5 to $\tilde{a}_{m+1}$). Of course, if $c < \mu$ or $c > M$, the number lies outside the range of admissable computer representations. If $c < \mu$, (underflow), it is common to regard $x$ as zero, notify the user, and continue further computation. If $c > M$, then overflow results. It is usually deemed not worth the added expense to use symmetric rounding and most machines simply chop. Whether a representation is obtained via chopping or symmetric rounding, we shall hereafter refer to the machine representation, $\bar{x}$, as being "rounded." Note that the above was illustrated in decimal form for simplicity; the analogy can be carried over to other bases. For example, if $x$ has the hexadecimal form

$$x = \pm(0.\tilde{a}_1\tilde{a}_2 \cdots \tilde{a}_m\tilde{a}_{m+1} \cdots) \times 16^c, \qquad \tilde{a}_1 \neq 0,$$

then the "decimal" point is really a "hexadecimal" point, and

$$x = \left(\sum_{k=1}^{\infty} (\tilde{a}_k \times 16^{-k})\right) \times 16^c.$$

The chopped form is still obtained by deleting $\tilde{a}_k$, $k \geq m + 1$. Symmetric rounding is done by adding $8 \times 16^{c-m-1}$ to $x$ and then chopping.

Returning to decimal form for simplicity, we first consider the error in symmetric rounding. If $\tilde{a}_{m+1} \geq 5$, then

$$x - \bar{x} = (\pm 0.\tilde{a}_1 \cdots \tilde{a}_m\tilde{a}_{m+1} \cdots) \times 10^c - [(\pm 0.\tilde{a}_1 \cdots \tilde{a}_m\tilde{a}_{m+1}) + (\pm 0.0 \cdots 05)] \times 10^c.$$

If $\tilde{a}_{m+1} < 5$, then

$$x - \bar{x} = (\pm 0.\tilde{a}_1 \cdots \tilde{a}_m\tilde{a}_{m+1} \cdots) \times 10^c - (\pm 0.\tilde{a}_1 \cdots \tilde{a}_m) \times 10^c.$$

In either case we have that $|x - \bar{x}| < 0.5 \times 10^{c-m}$, and this is a bound on the absolute error. To get a bound on the relative error, we note that since $\tilde{a}_1 \neq 0$, then $|x| \geq 0.1 \times 10^c$. Thus the relative error satisfies

$$\frac{|x - \bar{x}|}{|x|} \leq \frac{0.5 \times 10^{c-m}}{0.1 \times 10^c} = 5 \times 10^{-m} = 0.5 \times 10^{-m+1}.$$

In a similar manner we can show that the relative error from chopping is $1 \times 10^{-m+1}$. It is interesting to note that neither of these relative error bounds depend on the magnitude of $x$, that is, the size of $c$. Rather they depend only on the value of $m$, which is thus said to be the number of "significant digits" of the computer. For example, the IBM System 360 and 370 are hexadecimal with a mantissa of $m = 6$. The relative error from chopping in number representation thus does not exceed $1 \times 16^{-5}$. To compare this with the accuracy of relative errors we expect in the decimal mode, we set $1 \times 16^{-5} = 1 \times 10^{-m+1}$. Solving for $m$ yields $m \approx 7$. Thus we have approximately seven-significant-digit decimal accuracy with respect to the relative error of the representation. This does *not* mean that any seven-digit decimal number can be represented exactly on this machine as the previous example of $x = 1/10$ shows. It does say, however, that $|x - \bar{x}|/|x| \leq 10^{-6}$ (approximately, since $m \approx 7$).

In the discussion above, we analyzed the error made by replacing the true value of $x$ by its machine representation $\bar{x}$. Now we wish to assess the effect of each arithmetic operation $(+,-,\cdot,\div)$ to see how errors propagate. Let us use $\bar{x}$ to denote the machine approximation to the true value $x$, where the error, $e(x) = x - \bar{x}$, includes *all* errors that have been made in going from $x$ to $\bar{x}$, that is, $e(x)$ not only includes the error of representation but also includes errors from previous calculations, initial data errors, etc. In other words, $e(x)$ includes all errors from the beginning of the calculation that have led to any discrepancies between $x$ and $\bar{x}$. With $e(y)$ defined similarly for any quantity $y$, we investigate the error resulting from the "machine addition" of $x$ and $y$. Now,

$$x + y = (\bar{x} + e(x)) + (\bar{y} + e(y)) = (\bar{x} + \bar{y}) + (e(x) + e(y)).$$

At first glance it seems that the error of the addition is merely the sum of the individual errors. However, this is not always true, since even though $\bar{x}$ and $\bar{y}$ can be represented exactly on the machine, it is *not* necessarily true that their sum can also be, i.e., it does not follow that $\bar{x} + \bar{y} = \overline{\bar{x} + \bar{y}}$. For example, consider a four-digit floating-point machine and let $\bar{x} = 0.9621 \times 10^0$ and $\bar{y} = 0.6732 \times 10^0$. Then $\bar{x} + \bar{y} = 1.6353 \times 10^0$. Now it is not uncommon for an $m$-digit machine to perform arithmetic operations in a $2m$-digit accumulator and then to round the answer. Assuming this to be true, we have $\overline{\bar{x} + \bar{y}} = 0.1635 \times 10^1$. Returning to our general analysis we see that the true error, $z - \bar{z}$, where $z = x + y$, is actually $(e(x) + e(y))$ *plus* the error between $\bar{x} + \bar{y}$ and $\overline{\bar{x} + \bar{y}}$.

Before examining the other three arithmetic operations, let us consider addition somewhat further. We can see immediately that addition can lead to overflow, for instance, $\bar{x} = 0.9621 \times 10^M$ and $\bar{y} = 0.6732 \times 10^M$, thus $\bar{x} + \bar{y}$ does not fall within the range of the computer. Another factor that we must consider is that before a machine can perform an addition, it must align the decimal points. For example, again let $m = 4$ with $\bar{x}_1 = 0.5055 \times 10^4$ and $\bar{x}_2 = \bar{x}_3 = \cdots \bar{x}_{11} = 0.4000 \times 10^0$. To perform the addition, $\bar{x}_1 + \bar{x}_2$, the computer must shift the decimal four places to the left in $\bar{x}_2$ and form $\bar{x}_1 + \bar{x}_2 = (0.5055 \times 10^4) +$

$(0.00004 \times 10^4) = (0.50554 \times 10^4)$, which rounds to $\overline{\bar{x}_1 + \bar{x}_2} = 0.5055 \times 10^4 = \bar{x}_1$. Continuing, we see that

$$\overline{(\cdots(((\bar{x}_1 + \bar{x}_2) + \bar{x}_3) + \bar{x}_4) + \cdots + \bar{x}_{11})} = \bar{x}_1,$$

but

$$\overline{(\cdots(((\bar{x}_{11} + \bar{x}_{10}) + \bar{x}_9) + \bar{x}_8) + \cdots + \bar{x}_1)} = 0.5059 \times 10^4$$

which is the correct answer. Thus the machine calculates $\sum_{i=0}^{10} \bar{x}_{11-i}$ correctly, but not the sum $\sum_{i=1}^{11} \bar{x}_i$. This example illustrates the rule of thumb that if we have several numbers of the same sign to add, we should add them in ascending order of magnitude to minimize the propagation of round-off error. The mathematical foundation underlying this statement is the fact that machine addition is not associative, i.e., it can happen that

$$\overline{(\bar{x} + \bar{y}) + \bar{z}} \neq \overline{\bar{x} + (\bar{y} + \bar{z})}.$$

The following numerical examples were run to illustrate this phenomenon. (We used a hexadecimal machine with $m = 6$.) Letting $x = 1048576 = 16^5 = \bar{x}$ and $y = z = 1/2 = 8/16 = \bar{y} = \bar{z}$, our machine results were

$$(\bar{x} + \bar{y}) + \bar{z} = 0.1048576\text{E}07$$

and

$$\bar{x} + (\bar{y} + \bar{z}) = 0.1048577\text{E}07,$$

as our above analysis leads us to expect. Letting $w_0 = 1048576 = 16^5 = \overline{w_0}$ and $w_k = 1/16 = \bar{w}_k$, $1 \leq k \leq 256$, we also obtained the following results (adding in the order indicated by the sum):

$$\sum_{k=0}^{256} w_k = 0.1048576\text{E}07 = w_0$$

but

$$\sum_{k=0}^{256} w_{256-k} = 0.1048592\text{E}07,$$

the latter being the correct result (which we can easily check by hand). As a final example we computed the sum, $S \equiv \sum_{k=1}^{999} (1/k(k+1)) \equiv 0.999$, by adding forwards and backwards and obtained

$$S \approx 0.9989709\text{E}00 \text{ (forwards)}, \qquad S \approx 0.9989992\text{E}00 \text{ (backwards)}.$$

The error analysis for subtraction is much the same as addition, in that

$$x - y = (\bar{x} - \bar{y}) + (e(x) - e(y)),$$

but now we have the problem that subtraction can cause loss of significant digits.

This form of error occurs when $x$ and $y$ are nearly equal. For example, if $x = 0.6532849 \times 10^2$, $y = 0.6531212 \times 10^2$, and $m = 4$, then $\bar{x} = 0.6532 \times 10^2$ and $\bar{y} = 0.6531 \times 10^2$ (rounding by chopping). Now $\bar{x} - \bar{y} = 0.1000 \times 10^{-1}$ which is the machine representation for $\bar{x} - \bar{y}$, that is, $\overline{\bar{x} - \bar{y}} = 0.1000 \times 10^{-1}$. However, $x - y = 0.1637 \times 10^{-1}$. The problem we encounter here is not with this computation itself, but in its effect on later computations, since the zeros in $\overline{\bar{x} - \bar{y}}$ have no significance whatsoever and are, in effect, just filling up spaces. The answer is smaller than $x$ or $y$, and hence errors already present, $e(x)$ and $e(y)$, can have a drastic effect on the relative error of any further calculations, since the zeros will be treated thereafter as though they were significant. This source of error is known as *subtractive cancellation* and leads to serious errors in many computations.

When this type of error occurs we may try to locate the "sensitive" subtraction and eliminate it, if possible, by analytical manipulation. For example, consider the function $f(x) = (1 + x - e^x)/x^2$. Even if we have a very precise way of evaluating $e^x$, if $x$ is "small" the evaluation of $f(x)$ will suffer from subtractive cancellation in the numerator which will be magnified by the division by a number very close to zero. One way to help alleviate this problem is to use the Maclaurin series expansion, $e^x = \sum_{k=0}^{\infty} x^k/k! = 1 + x + x^2/2! + x^3/3! + \cdots$. Then, $1 + x - e^x = -x^2/2! - x^3/3! - x^4/4! - \cdots$, and we see that the factor $[(1 + x) - (1 + x)]$, which was essentially causing the trouble, is eliminated. Furthermore, $f(x) = -1/2! - x/3! - x^2/4! - \cdots$, thus this analytical technique also eliminates the problem of "division by zero." Since we are considering the evaluation of $f(x)$ for small values of $x$, this series should be fairly rapidly convergent, as evidenced in the following computer results. However, for large values of $x$, the series is no longer rapidly convergent and we experience considerable difficulties when the series is truncated after a finite number of terms.

We note from the above convergent series expansion for $f(x)$ that $f(0) = -1/2$. In Table 1.2 the series evaluation was truncated after the seventh term which theoretically should give us at least seven correct decimal place accuracy for the values of $x$ that were used. We see in Table 1.2 that the series evaluations yield good answers whereas direct substitution becomes worse as $x$ decreases.

**Table 1.2**

| $x$ | $f(x)$ (Direct Substitution) | $f(x)$ (Series Evaluation) |
|---|---|---|
| 0.1 | −0.5171781E 00 | −0.5170917E 00 |
| 0.01 | −0.5054476E 00 | −0.5016708E 00 |
| 0.001 | −0.9536749E 00 | −0.5001667E 00 |
| 0.0001 | −0.9536745E 02 | −0.5000166E 00 |
| 0.00001 | 0.0 | −0.5000016E 00 |

Analysis of multiplication yields

$$x \cdot y = (\bar{x} + e(x)) \cdot (\bar{y} + e(y)) = \bar{x}\bar{y} + \bar{x}e(y) + \bar{y}e(x) + e(x)e(y).$$

We can usually assume that the term $e(x)e(y)$ is negligible with respect to the other terms and we omit it. Furthermore, the product of the $m$-digit numbers, $\bar{x}$ and $\bar{y}$, is at most a $2m$-digit number and can be represented exactly in the $2m$-digit mode that computers commonly use for individual operations. The product will be rounded to $m$ digits, of course, to form its machine representation, $\overline{\bar{x}\bar{y}}$. Thus, even though $\bar{x}\bar{y} \neq \overline{\bar{x}\bar{y}}$ (necessarily), the error is merely that of rounding $\bar{x}\bar{y}$ to its machine representation. Thus the relative error of the difference between $\bar{x}\bar{y}$ and $\overline{\bar{x}\bar{y}}$ is bounded by $0.5 \times 10^{-m+1}$ for symmetric rounding and by $1 \times 10^{-m+1}$ for chopping.

It is common, especially in matrix methods, to have to compute a quantity of the form, $\sum_{j=1}^{n} \bar{x}_j\bar{y}_j$, called an "inner-product." The multiplications, $\bar{x}_j\bar{y}_j$, are individually done in the $2m$-digit mode, but often the addition is not. If this is the case, it is often a simple task in a high-level language to program the machine to do the additions of this type in double precision as well. This is called "double precision accumulation of inner-products," and serves to lessen the errors due to the multiplications, $\bar{x}_j\bar{y}_j$, $1 \leq j \leq n$.

Finally, we analyze the division operation

$$\frac{x}{y} = \frac{\bar{x} + e(x)}{\bar{y} + e(y)} = \left(\frac{x + e(x)}{\bar{y}}\right)\left(\frac{1}{1 + e(y)/\bar{y}}\right).$$

Assuming that $e(y)$ is "small" with respect to $\bar{y}$, we recall the geometric series, where for $|r| < 1$,

$$\frac{1}{1 + r} = 1 - r + r^2 - r^3 + \cdots.$$

Letting $r \equiv e(y)/\bar{y}$ in the above expression for $x/y$ we have

$$\frac{x}{y} = \left(\frac{\bar{x} + e(x)}{\bar{y}}\right)(1 - e(y)/\bar{y} + e(y)^2/\bar{y}^2 - e(y)^3/\bar{y}^3 + \cdots) \simeq \frac{\bar{x}}{\bar{y}} + \frac{e(x)}{\bar{y}} - \frac{\bar{x}e(y)}{\bar{y}^2},$$

where we assume the omitted terms in the expansion are negligible with respect to the first three.

Division is much the same as multiplication in that the difference between $\bar{x}/\bar{y}$ and its machine representation, $(\overline{\bar{x}/\bar{y}})$, is merely that of rounding the $2m$-digit representation for $\bar{x}/\bar{y}$. The preceding expression clearly illustrates the disastrous effect on the total error that division by a value of $\bar{y}$ very close to zero can have. As mentioned earlier this problem can sometimes be avoided by alternate approaches which analytically circumvent "zero divisions."

## PROBLEMS

1. a) Convert the following decimal numbers to hexadecimal form:

   i. 1023 ii. 1025

   iii. 278.5 iv. 14.09375

   v. 0.1240234375

   b) Convert the above answers to binary form.

2. Convert the following hexadecimal numbers to decimal form.

   a) 1023, b) 100, c) 1A4.C, d) 6B.1C, e) FFF.118

3. Convert the numbers $\pi$, $e$, and 1/3 into hexadecimal form with a mantissa of six hex-digits using

   a) symmetric rounding.

   b) chopping.

*4. Prove that any proper fraction which has a terminating hexadecimal expansion also has a terminating decimal expansion. Explain in general terms why the converse of this statement is not valid.

5. For the computer example which computed the sums of reciprocals, compute the relative errors for the operations corresponding to $k = 1000, 1006, 1012, 1018$.

6. a) If $x$ is a real number and $\bar{x}$ is its chopped machine representation on a computer with base $b$ and mantissa length $m$, show that the relative error, $|x - \bar{x}|/|x|$, is bounded by $10^{-m+1}$.

   b) If $\delta = 10^{-m+1}$ (in the case of chopping) or $\delta = 0.5 \times 10^{-m+1}$ (in the case of symmetric rounding), show that $\bar{x} = x(1 + \varepsilon)$, where $|\varepsilon| \leq \delta$.

   c) Given a function $F(x)$, $F'$ continuous, we have from the mean-value theorem that in evaluating $F$ at $x$ the relative error due to this source alone is:

$$\frac{F(\bar{x}) - F(x)}{F(x)} = \frac{F(x + \varepsilon x) - F(x)}{F(x)} = \frac{F'(\xi)[(x + \varepsilon x) - x]}{F(x)} \simeq \varepsilon x \frac{F'(x)}{F(x)}.$$

   Analogously, $F(\bar{x}) - F(x) \simeq \varepsilon x F'(x)$. Use these formulas to give estimates of absolute and relative error if

   i. $F(x) = x^k$, for various values of $k$ and $x$;

   ii. $F(x) = e^x$, for large values of $x$;

   iii. $F(x) = \sin(x)$, for small values of $x$;

   iv. $F(x) = x^2 - 1$, for $x$ near 1;

   v. $F(x) = \cos(x)$, for $x$ near $\pi/2$.

7. Write a program to compute the value of $S_N = \sum_{k=1}^{N} 1/k$ for various values of $N$. Adapt the program so that it sums forwards and backwards and compare the results as $N$ increases. Find the value of $N$ such that $S_n = S_N$ for all $n \geq N$ for the forward summation. How is this reconciled with the fact that theoretically $\lim_{N\to\infty} S_N = \infty$? Would there be such a number $N$ for any modern digital computer and would it be the same on all such machines?

8. Consider a computer with mantissa length $m = 3$, base $b = 10$, and exponent constraints $\mu = -2 \leq c \leq 2 = M$. How many real numbers will this computer represent exactly?

9. For any positive integer $N$ and a fixed constant, $r \neq 1$, we recall the formula for the geometric sum,

$$G_N \equiv 1 + r + r^2 + \cdots + r^N = (1 - r^{N+1})/(1 - r) \equiv Q_N.$$

Write a computer program to compute $G_N$ and $Q_N$ for arbitrary values of $r$ and $N$. Let $r$ be chosen close to 1 and note the discrepancy. In which of the two calculations do you have the most faith? Use values of $r$ for input for which you can easily check the exact value of $Q_N$ by hand calculation (for example, $r = 1 - 10^{-k}$ for some integer $k \geq 0$). Calculate the relative and absolute errors of these computations.

## 1.4 REVIEW OF FUNDAMENTAL MATHEMATICAL RESULTS

In this section we shall briefly list some basic mathematical results, mainly from the calculus, that are useful in the development and investigation of numerical procedures in subsequent chapters. We are, of course, assuming familiarity with the rudiments of calculus such as the real number system, functions, and at least an intuitive understanding of the concept of limits (the latter being especially true with respect to sequences, since many numerical methods are iterative in nature and thus generate a sequence of approximations to the true solution). A simple example of this is the classical Newton's method applied to finding the square root of an arbitrary positive real number, $\alpha$. Given an initial guess, $x_0$, for $\sqrt{\alpha}$, Newton's method generates the sequence: $x_{n+1} = \frac{1}{2}(x_n + \alpha/x_n)$, for $n = 0, 1, 2, \ldots$. Three questions which immediately come to mind if we are to be able to successfully use this algorithm in practice are: (1) Is the sequence generated by the algorithm convergent to $\sqrt{\alpha}$ and $\sqrt{\alpha}$ alone? (2) How is the convergence related to the choice of the initial guess $x_0$? (3) How fast does the method converge, or, how large is the error, $x_n - \sqrt{\alpha}$, after the $n$th iteration? These are questions involving the concept of convergence of sequences. There are, of course, other important questions to answer such as: (4) How does rounding error affect the method? (5) Is it competitive with other methods for finding $\sqrt{\alpha}$? (6) How well can the method be implemented on the particular computer available and what cost is necessary to obtain an acceptable answer?

Other basic ideas necessary from calculus are the concepts of continuity, differentiation, and integration. We have assumed in Chapters 2 and 3 that the reader is able to evaluate determinants. We have, however, tried to keep the material involving determinants minimal, and the other necessary material from matrix theory and linear algebra is self-contained. We have also assumed that the reader is aware of the importance of differential equations in modeling and solving modern technological problems, although we assume very little background in the mathematical theory of differential equations in Chapter 7.

We list here some fundamental theorems of calculus and algebra used in later chapters with which the reader should already be familiar. We will present other results only as they are needed to treat particular topics, and they will only be presented in the context of the particular topic being treated. The following results and their proofs can be found in almost any sophomore-level calculus text.

**Theorem 1.1. Rolle's Theorem.** If $f(x)$ is a continuous function in the closed interval $[a, b]$ and is differentiable in $(a, b)$, and if $f(a) = f(b)$, then there exists at least one point $\xi$ in $(a, b)$ such that $f'(\xi) = 0$.

**Theorem 1.2. Mean Value Theorem.** If $f(x)$ is a continuous function in the closed interval $[a, b]$ and is differentiable in $(a, b)$, then there exists at least one number $\xi$ in $(a, b)$ such that

$$f(b) - f(a) = (b - a)f'(\xi).$$

**Theorem 1.3. First Mean Value Theorem for Definite Integrals.** If $f(x)$ is continuous for $a \le x \le b$, then there exists a number $\xi$, $a < \xi < b$, such that

$$\int_a^b f(x)\,dx = f(\xi)(b - a).$$

**Theorem 1.4. Second Mean Value Theorem for Definite Integrals.** Let $g(x)$ be an integrable function which does not change sign on the interval $[a, b]$. If $f(x)$ is continuous on $[a, b]$ then there exists a number $\xi$ in $[a, b]$ such that

$$\int_a^b f(x)g(x)\,dx = f(\xi)\int_a^b g(x)\,dx$$

**Theorem 1.5.** Let $f$ be a continuous function on the closed finite interval $[a, b]$. Then there exist points, $x_m$ and $x_M$, in $[a, b]$ such that $f(x_m) \le f(x) \le f(x_M)$ for any $x$ in $[a, b]$, that is, "a continuous function attains its maximum and minimum on a closed interval."

**Corollary 1.5.1.** Under the hypotheses of Theorems 1.3 and 1.5,

$$\left|\int_a^b f(x)\,dx\right| \le (b - a)\left\{\max_{a \le x \le b} |f(x)|\right\}$$

**Theorem 1.6. Intermediate Value Theorem.** If $f(x)$, $x_m$, and $x_M$ are given as in Theorem 1.5, and if $m \equiv f(x_m)$ and $M \equiv f(x_M)$, then for any value $y^*$ such that $m \le y^* \le M$, there exists at least one number $x^*$ in $[a, b]$ such that $f(x^*) = y^*$.

**Theorem 1.7. Taylor's Theorem.** Let $f(x)$ be a function such that its $(n + 1)$st derivative, $f^{(n+1)}(x)$, is continuous on the interval $(a, b)$. If $x$ and $x^*$ are any two points in $(a, b)$, then

$$f(x) = f(x^*) + \frac{f'(x^*)}{1!}(x - x^*) + \frac{f''(x^*)}{2!}(x - x^*)^2 + \cdots$$
$$+ \frac{f^{(n)}(x^*)}{n!}(x - x^*)^n + R_{n+1}(x^*; x),$$

where there exists a number $\xi$ between $x$ and $x^*$ such that

$$R_{n+1}(x^*; x) = \frac{f^{(n+1)}(\xi)}{(n+1)!}(x - x^*)^{n+1}.$$

(Note that Theorem 1.2 is a special case of this theorem with $n \equiv 0$.)

**Theorem 1.8. Fundamental Theorem of Algebra.** Let $p(x) \equiv a_n x^n + a_{n-1}x^{n-1} + \cdots + a_1 x + a_0$ be an $n$th degree polynomial with $n \geq 1$, $a_n \neq 0$. Then there exist uniquely $n$ constants, $\{r_j\}_{j=1}^{n}$, ("zeros of $p(x)$") such that

$$p(x) = a_n(x - r_1)(x - r_2) \cdots (x - r_n),$$

where the $r_j$'s need not all be distinct or all real. (This is not quite the usual statement of the Fundamental Theorem of Algebra, but this form is more suited to our purposes.)

# 2

# SOLUTION OF LINEAR SYSTEMS OF EQUATIONS

## 2.1 INTRODUCTION

One of the most basic and important problems in science and engineering is the accurate and efficient simultaneous solution of a system of $m$ linear equations in $n$ unknowns. This problem is written in the form

$$\begin{aligned} a_{11}x_1 + a_{12}x_2 + \cdots + a_{1n}x_n &= b_1 \\ a_{21}x_1 + a_{22}x_2 + \cdots + a_{2n}x_n &= b_2 \\ \vdots \qquad\qquad\qquad\quad \vdots \quad &\quad \vdots \\ a_{m1}x_1 + a_{m2}x_2 + \cdots + a_{mn}x_n &= b_m \end{aligned} \tag{2.1}$$

where each $a_{ij}$; $1 \le i \le m$, $1 \le j \le n$, and each $b_i$, $1 \le i \le m$, are known values and $x_j$, $1 \le j \le n$, are the unknowns for which we must solve. Problems of this type arise in almost all disciplines and most students have studied the elimination of variables techniques for the solution of (2.1) which are introduced in elementary algebra courses.

For the study of various numerical methods for solving (2.1) and their implementation on the computer, we will first review some of the basic ideas associated with matrices. In the following we use the common notation that any $(k \times \ell)$ matrix $C$ will be written as "$C = (c_{ij})$," where $c_{ij}$ is the element in the $i$th row and $j$th column of $C$ and where it is understood that $1 \le i \le k$ and $1 \le j \le \ell$. Let $R = (r_{ij})$ be an $(m \times n)$ matrix and let $S = (s_{ij})$ be an $(n \times p)$ matrix. Then the *product* of $R$ and $S$ is defined and is an $(m \times p)$ matrix $T = (t_{ij})$, where $t_{ij} = \sum_{k=1}^{n} r_{ik}s_{kj}$. We write $T = RS$ to denote the product of $R$ and $S$. If $\alpha$ is any real number, then the product $\alpha R$ is formed by multiplying each entry of $R$ by the number $\alpha$. If $Q = (q_{ij})$ is any $(m \times n)$ matrix then the *sum* of $R$ and $Q$ is defined to be the $(m \times n)$ matrix $V = (v_{ij})$, where each element $v_{ij}$ of $V$ is given by $v_{ij} = r_{ij} + q_{ij}$. We write $V = R + Q$ to denote the sum of the two matrices $R$ and $Q$. Notice that the product of $R$ and $Q$ cannot be defined if $m \neq n$ and the sum of $R$ and $S$ cannot be defined unless $R$ and $S$ both have the same number of rows and columns. Also the product $SR$ is not defined unless $m = p$. Moreover, even for $m = p = n$, $SR$ need not equal $RS$. It is true, however, that if $R + Q$ is defined then so is $Q + R$ and also $R + Q = Q + R$. (We say that two $(m \times n)$ matrices are *equal* if and only if each of their corresponding entries are equal.)

**Example 2.1.** Let

$$R = \begin{bmatrix} 1 & 2 & 1 & 1 \\ -1 & 0 & 1 & 1 \\ 1 & 0 & 2 & 3 \end{bmatrix}, \quad S = \begin{bmatrix} 1 & 1 & 1 \\ 1 & 0 & 1 \\ -1 & 1 & -1 \\ 2 & 0 & 3 \end{bmatrix}, \quad \text{and} \quad Q = \begin{bmatrix} 1 & 0 & 1 \\ -1 & 1 & -1 \\ 2 & 1 & 0 \end{bmatrix}.$$

Then

$$RS = \begin{bmatrix} 4 & 2 & 5 \\ 0 & 0 & 1 \\ 5 & 3 & 8 \end{bmatrix}, \quad SR = \begin{bmatrix} 1 & 2 & 4 & 5 \\ 2 & 2 & 3 & 4 \\ -3 & -2 & -2 & -3 \\ 5 & 4 & 8 & 11 \end{bmatrix},$$

but $RQ$ is not defined and $RS \neq SR$. Also,

$$RS + 2Q = \begin{bmatrix} 6 & 2 & 7 \\ -2 & 2 & -1 \\ 9 & 5 & 8 \end{bmatrix}$$

but $R + Q$ and $S + Q$ are not defined.

Using matrix notation, we may write the linear system (2.1) in the compact form

$$A\mathbf{x} = \mathbf{b} \tag{2.2}$$

where $A = (a_{ij})$ represents the $(m \times n)$ *coefficient matrix* for the system and where

$$\mathbf{x} = \begin{bmatrix} x_1 \\ x_2 \\ \vdots \\ x_n \end{bmatrix} \quad \text{and} \quad \mathbf{b} = \begin{bmatrix} b_1 \\ b_2 \\ \vdots \\ b_m \end{bmatrix} \tag{2.3}$$

are $(n \times 1)$ and $(m \times 1)$ column vectors, respectively.

**Example 2.2.** Consider the linear system

$$\begin{aligned} x_1 \quad\quad + x_3 &= 1 \\ -x_1 + x_2 - x_3 &= 0 \\ 2x_1 + x_2 \quad\quad &= 0 \end{aligned} \quad \text{(or in matrix form)} \quad \begin{bmatrix} 1 & 0 & 1 \\ -1 & 1 & -1 \\ 2 & 1 & 0 \end{bmatrix} \begin{bmatrix} x_1 \\ x_2 \\ x_3 \end{bmatrix} = \begin{bmatrix} 1 \\ 0 \\ 0 \end{bmatrix}.$$

This system is easily solved by elimination of variables. Adding the first and second equations yields $x_2 = 1$. Then the third equation gives $x_1 = -1/2$, and the first equation gives $x_3 = 3/2$.

In the following sections we will be primarily concerned with solving (2.1) in the special case where $A$ is $(n \times n)$ and $\mathbf{b}$ is $(n \times 1)$. In this case, the number of equations equals the number of unknowns and $A$ is said to be a *square* matrix. (If $m \neq n$, $A$ is said to be *rectangular*.) Considering (2.1) with $m = n$, the coefficient matrix $A$ is called *non-singular* if there always exists a unique solution, $\mathbf{x}$, to (2.1) for any choice of $b_1, b_2, \ldots, b_n$. It can be shown in this case that there exists a unique $(n \times n)$ matrix, $A^{-1}$, called the *inverse of* $A$ such that $A^{-1}A = AA^{-1} = I$, where $I$ is the $(n \times n)$ *identity matrix*:

$$I = \begin{bmatrix} 1 & 0 & 0 & \cdots & 0 \\ 0 & 1 & 0 & \cdots & 0 \\ \vdots & & & & \\ 0 & 0 & 0 & \cdots & 1 \end{bmatrix}. \tag{2.4}$$

It is easily seen that if $B$ is any $(n \times n)$ matrix, then $IB = BI = B$. Thus if we knew $A^{-1}$, we could multiply both sides of (2.2) by $A^{-1}$ and obtain the solution $\mathbf{x}$ by

$$A^{-1}(A\mathbf{x}) = (A^{-1}A)\mathbf{x} = I\mathbf{x} = \mathbf{x} = A^{-1}\mathbf{b}.$$

It is often very useful to have an alternative way of expressing the product, $A\mathbf{x}$, where $A = (a_{ij})$ is an $(m \times n)$ matrix, and $\mathbf{x}$ is an $(n \times 1)$ column vector as given in (2.3). Let $\mathbf{A}_k$ denote the $k$th column of $A$,

$$\mathbf{A}_k = \begin{bmatrix} a_{1k} \\ a_{2k} \\ \vdots \\ a_{mk} \end{bmatrix}, \qquad \text{for } 1 \leq k \leq n.$$

Then each $\mathbf{A}_k$ is an $(m \times 1)$ column vector. By the rules of matrix multiplication, the $j$th component of the vector $A\mathbf{x}$ is given by $c_j = \sum_{k=1}^{n} a_{jk}x_k$, for each $j$, $1 \leq j \leq m$. We claim that each $c_j$ is also the $j$th component of the $(m \times 1)$ vector

$$x_1\mathbf{A}_1 + x_2\mathbf{A}_2 + \cdots + x_n\mathbf{A}_n.$$

This is easily seen, since the $j$th component of each vector $x_k\mathbf{A}_k$ is equal to $x_k a_{jk}$, and adding these for $1 \leq k \leq n$ yields $\sum_{k=1}^{n} x_k a_{jk} = c_j$. Thus we have shown that $A\mathbf{x}$ may be written in the form $A\mathbf{x} = x_1\mathbf{A}_1 + x_2\mathbf{A}_2 + \cdots + x_n\mathbf{A}_n$. (We will often find it useful to express the matrix $A$ symbolically in column form as $A = [\mathbf{A}_1, \mathbf{A}_2, \ldots, \mathbf{A}_n]$.)

Similarly, from the definition of matrix multiplication, it is easily verified that if $A$ and $B$ are any two $(n \times n)$ matrices, then for any $j$, $1 \leq j \leq n$, the $j$th column of the product $AB$ is equal to $A\mathbf{B}_j$, where $B = [\mathbf{B}_1, \mathbf{B}_2, \ldots, \mathbf{B}_n]$. This observation shows how we can compute $A^{-1}$. To be specific: Let $\mathbf{e}_j$, for $1 \leq j \leq n$, be the

$(n \times 1)$ column vector which has a 1 in the $j$th component and zeros in all other components. Thus $\mathbf{e}_j$ is the $j$th column vector of $I$, or $I = [\mathbf{e}_1, \mathbf{e}_2, \ldots, \mathbf{e}_n]$. If $A$ is non-singular, then for $1 \le j \le n$, the $j$th column of $AA^{-1}$ equals $\mathbf{e}_j$ since $AA^{-1} = I$. We can therefore find the $j$th column of $A^{-1}$, $\mathbf{x}^{(j)}$, by solving the linear system

$$A\mathbf{x}^{(j)} = \mathbf{e}_j.$$

We can use any of the numerical methods described in the following sections to find $A^{-1}$ by solving each of the $n$ systems: $A\mathbf{x}^{(j)} = \mathbf{e}_j$, $1 \le j \le n$, as in Example 2.3.

**Example 2.3.** Consider the two linear systems

$$\begin{aligned} x_1 \qquad\;\; + x_3 &= 0 \\ -x_1 + x_2 - x_3 &= 1 \\ 2x_1 + x_2 \qquad\;\; &= 0 \end{aligned} \qquad \text{and} \qquad \begin{aligned} x_1 \qquad\;\; + x_3 &= 0 \\ -x_1 + x_2 - x_3 &= 0 \\ 2x_1 + x_2 \qquad\;\; &= 1 \end{aligned}$$

Proceeding exactly as in Example 2.2, we obtain $x_1 = -1/2, x_2 = 1, x_3 = 1/2$ as the solution to the first system and $x_1 = 1/2, x_2 = 0, x_3 = -1/2$ as the solution to the second. Now, the coefficient matrix of all three systems in Example 2.2 and Example 2.3 is the matrix $Q$ of Example 2.1. Taking these three solutions as column vectors, we get

$$Q^{-1} = \begin{bmatrix} -\frac{1}{2} & -\frac{1}{2} & \frac{1}{2} \\ 1 & 1 & 0 \\ \frac{3}{2} & \frac{1}{2} & -\frac{1}{2} \end{bmatrix},$$

and we may easily verify that $QQ^{-1} = Q^{-1}Q = I$. Thus, if $\mathbf{b} = \begin{bmatrix} 1 \\ 1 \\ 1 \end{bmatrix}$, for example, then the solution of $Q\mathbf{x} = \mathbf{b}$ is given by

$$\mathbf{x} = Q^{-1}\mathbf{b} = \begin{bmatrix} -\frac{1}{2} & -\frac{1}{2} & \frac{1}{2} \\ 1 & 1 & 0 \\ \frac{3}{2} & \frac{1}{2} & -\frac{1}{2} \end{bmatrix} \begin{bmatrix} 1 \\ 1 \\ 1 \end{bmatrix} = \begin{bmatrix} -\frac{1}{2} \\ 2 \\ \frac{3}{2} \end{bmatrix}.$$

The general problem of solving several linear systems with the same coefficient matrix will be discussed in more detail in Section 2.2.3, with an emphasis on efficient programming for a digital computer.

There are a number of *theoretical* tests from the theory of matrices which can be used to determine whether a square matrix is non-singular or not. Two of the most familiar characterizations of non-singular matrices are:

1) Let $\boldsymbol{\theta}$ denote the $(n \times 1)$ column vector which has all components equal to zero. ($\boldsymbol{\theta}$ is called the *zero vector.*) $A$ is non-singular if and only if $\boldsymbol{\theta}$ is the only solution of $A\mathbf{x} = \boldsymbol{\theta}$. (Note: $\mathbf{x} = \boldsymbol{\theta}$ is always a solution of $A\mathbf{x} = \boldsymbol{\theta}$.)

2) $A$ is non-singular if and only if the determinant of $A$ is nonzero, that is, $\det(A) \ne 0$. (We assume here that the reader is familiar with the rules for calculating determinants of square matrices.)

**Example 2.4.** Let

$$A = \begin{bmatrix} 1 & 2 & 3 & 1 \\ -1 & 0 & 2 & 1 \\ 2 & 1 & -3 & 0 \\ 1 & 1 & 2 & 1 \end{bmatrix}.$$

Recall that the evaluation of det($A$) by cofactor expansion of the first row yields:

$$\det(A) = \begin{vmatrix} 0 & 2 & 1 \\ 1 & -3 & 0 \\ 1 & 2 & 1 \end{vmatrix} - 2\begin{vmatrix} -1 & 2 & 1 \\ 2 & -3 & 0 \\ 1 & 2 & 1 \end{vmatrix} + 3\begin{vmatrix} -1 & 0 & 1 \\ 2 & 1 & 0 \\ 1 & 1 & 1 \end{vmatrix} - \begin{vmatrix} -1 & 0 & 2 \\ 2 & 1 & -3 \\ 1 & 1 & 2 \end{vmatrix}$$

$$= \left\{0\begin{vmatrix} -3 & 0 \\ 2 & 1 \end{vmatrix} - 2\begin{vmatrix} 1 & 0 \\ 1 & 1 \end{vmatrix} + \begin{vmatrix} 1 & -3 \\ 1 & 2 \end{vmatrix}\right\} - 2\left\{-\begin{vmatrix} -3 & 0 \\ 2 & 1 \end{vmatrix} - 2\begin{vmatrix} 2 & 0 \\ 1 & 1 \end{vmatrix} + \begin{vmatrix} 2 & -3 \\ 1 & 2 \end{vmatrix}\right\}$$

$$+ 3\left\{-\begin{vmatrix} 1 & 0 \\ 1 & 1 \end{vmatrix} - 0\begin{vmatrix} 2 & 0 \\ 1 & 1 \end{vmatrix} + \begin{vmatrix} 2 & 1 \\ 1 & 1 \end{vmatrix}\right\} - \left\{-\begin{vmatrix} 1 & -3 \\ 1 & 2 \end{vmatrix} - 0\begin{vmatrix} 2 & -3 \\ 1 & 2 \end{vmatrix} + 2\begin{vmatrix} 2 & 1 \\ 1 & 1 \end{vmatrix}\right\}$$

$$= (0 - 2 + 5) - 2(3 - 4 + 7) + 3(-1 - 0 + 1) - (-5 - 0 + 2) = -6.$$

Thus $A$ is non-singular since $\det(A) \neq 0$. (Note that the above evaluation could be facilitated by elementary row and column operations, but it does not serve our purpose to go into this here.) Similarly, it is easy to verify that if

$$\begin{bmatrix} 1 & 2 & 3 & 1 \\ -1 & 0 & 2 & 1 \\ 2 & 1 & -3 & 0 \\ 1 & 1 & 2 & 1 \end{bmatrix}\begin{bmatrix} x_1 \\ x_2 \\ x_3 \\ x_4 \end{bmatrix} = \begin{bmatrix} 0 \\ 0 \\ 0 \\ 0 \end{bmatrix}$$

then $x_1 = x_2 = x_3 = x_4 = 0$.

We hasten to add, that while determinants and inverses are of great importance to a theoretical analysis of the solution of (2.1), they are rarely recommended for actual use in computational procedures. We will return to this point later and confine ourselves for the moment to saying that procedures of this type, such as the well-known Cramer's Rule for solving (2.1) or computing $A^{-1}$ and forming $\mathbf{x} = A^{-1}\mathbf{b}$ are not competitive methods.

For theoretical purposes it is also important to recall that if $A$ and $B$ are $(n \times n)$ non-singular matrices, then the product $(AB)$ is non-singular and $(AB)^{-1} = B^{-1}A^{-1}$ (Problem 12). In the case that $A$ is singular, then $A$ has no inverse and the equation $A\mathbf{x} = \mathbf{b}$ will either have no solution or infinitely many solutions. It will be demonstrated later that a direct computational method such as Gauss elimination (Section 2.2.1) can be implemented to conveniently display all of the solutions (if any exist) of $A\mathbf{x} = \mathbf{b}$, even if $A$ is rectangular or if $A$ is square but singular. (See Example 2.6.)

In the following sections on numerical procedures for solving the equation $A\mathbf{x} = \mathbf{b}$, we will be working extensively with *triangular* matrices. A square matrix

$U = (u_{ij})$ is said to be *upper triangular* if $u_{ij} = 0$ whenever $i > j$ (that is, all entries below the main diagonal are zero). Similarly, a square matrix $L = (\ell_{ij})$ is called *lower triangular* if all entries above the main diagonal are zero (that is, $\ell_{ij} = 0$ whenever $i < j$). A square matrix $D = (d_{ij})$ is called *diagonal* if $d_{ij} = 0$ whenever $i \neq j$. Thus a diagonal matrix is simultaneously upper triangular and lower triangular. Finally, we wish to recall the definition of the *transpose* of a matrix. If $A = (a_{ij})$ is an $(m \times n)$ matrix, then the transpose of $A$, denoted $A^T$, is an $(n \times m)$ matrix where the $ij$th entry of $A^T$ is $a_{ji}$; $1 \leq j \leq n$, $1 \leq i \leq m$; that is, $A^T$ is formed from $A$ by interchanging the rows and columns of $A$. It is easily verified (Problem 13) that $(AB)^T = B^T A^T$ whenever the product $AB$ is defined and that $(A + B)^T = A^T + B^T$ whenever the sum is defined.

**Example 2.5.** Let

$$A = \begin{bmatrix} 2 & 1 & 3 \\ 6 & 0 & 4 \end{bmatrix}, \qquad \text{then} \qquad A^T = \begin{bmatrix} 2 & 6 \\ 1 & 0 \\ 3 & 4 \end{bmatrix}.$$

For an example of a triangular matrix, let $L$ be given by

$$L = \begin{bmatrix} 1 & 0 & 0 \\ 2 & 4 & 0 \\ 1 & 3 & 5 \end{bmatrix}$$

Then $L$ is lower triangular and, in addition, $L^T$ is upper triangular. Transposes are used frequently in connection with column and row vectors. For example, let

$$\mathbf{x} = \begin{bmatrix} 1 \\ 3 \\ 0 \end{bmatrix} \qquad \text{and} \qquad \mathbf{y} = \begin{bmatrix} 2 \\ 1 \\ 2 \end{bmatrix}.$$

Then $\mathbf{x}$ and $\mathbf{y}$ are $(3 \times 1)$ column vectors (or $(3 \times 1)$ matrices, if we choose to think of them in that fashion). So, $\mathbf{x}^T = (1, 3, 0)$ and $\mathbf{y}^T = (2, 1, 2)$. According to the rules of matrix multiplication, we can form both of the products $\mathbf{x}^T\mathbf{y}$ and $\mathbf{x}\mathbf{y}^T$, obtaining

$$\mathbf{x}^T\mathbf{y} = 5; \qquad \mathbf{x}\mathbf{y}^T = \begin{bmatrix} 2 & 1 & 2 \\ 6 & 3 & 6 \\ 0 & 0 & 0 \end{bmatrix}.$$

There are many other fundamental concepts from matrix theory that could be mentioned here (such as *rank*, *augmented matrix*, *echelon matrix*, etc.). However, since we will be primarily concerned with the computational solution of $A\mathbf{x} = \mathbf{b}$ where $A$ is $(n \times n)$ and non-singular, a review of this additional material is not necessary for our purposes. Some additional concepts are introduced as they become necessary (such as *symmetric matrix*, *positive definite*, *diagonally dominant*, etc.), but they are explained and illustrated in detail so that the reader with a calculus background plus some very rudimentary experience with matrices should have no difficulty with the material.

## PROBLEMS

1. Using the definition of matrix multiplication, write a computer program to form the product of two arbitrary $(n \times n)$ matrices, where $n \leq 10$.
2. In Problem 1, let $A$ and $B$ be the two $(n \times n)$ matrices to be multiplied and suppose that $C = AB$ denotes the product of $A$ and $B$. It is most natural to store $A$ and $B$ as doubly subscripted arrays (so, for example, the $ij$th entry of $A$ is $A(I, J)$). In applications involving large matrices (or small computers), storage requirements can sometimes present problems. In Problem 1, if $A$, $B$, and $C$ are all saved, then $3n^2$ locations are needed. Suppose that $A$ is no longer needed once the product $C$ has been formed. How might the program in Problem 1 be modified so that only $2n^2 + n$ locations are needed to form $C$? (Hint: Once the first row of $C$ is formed, the first row of $A$ is no longer needed.)
3. How many multiplications must be performed to form the product of two $(n \times n)$ matrices?
4. How many multiplications must be performed to form the product $L\mathbf{x}$, when $L$ is an $(n \times n)$ lower triangular matrix and $\mathbf{x}$ is an $(n \times 1)$ column vector? (Do not count multiplications by zero.)

*5. Write a computer program to evaluate the determinant of an arbitrary $(5 \times 5)$ matrix, using a cofactor expansion along the first row. (This is a fairly challenging programming problem.)

6. How many multiplications must be performed to evaluate the determinant of an $(n \times n)$ matrix according to the procedure of Problem 5 when $n = 3$, when $n = 4$, when $n = 5$, and for arbitrary $n$? (For $n = 10$, more than 3,000,000 multiplications are required. If a person were able to multiply two numbers and record the result at a rate of one per second, it would require 126 eight-hour days to find the determinant of a $(10 \times 10)$ matrix using a cofactor expansion.)
7. Let $L = (\ell_{ij})$ be a lower triangular $(5 \times 5)$ matrix such that $\ell_{11}\ell_{22}\ell_{33}\ell_{44}\ell_{55} \neq 0$. By considering the equation $L\mathbf{x} = \boldsymbol{\theta}$, show that $L$ is non-singular. (Hint: Testing $L\mathbf{x} = \boldsymbol{\theta}$ is criterion (1) for non-singularity.) Extend your proof to an arbitrary $(n \times n)$ lower triangular matrix $L$ such that $\ell_{11}\ell_{22} \cdots \ell_{nn} \neq 0$.
8. Let $L$ be as in Problem 7, but this time suppose that $\ell_{44} = 0$ and $\ell_{55} \neq 0$. Show there is a nonzero vector $\mathbf{x}$ such that $L\mathbf{x} = \boldsymbol{\theta}$.

*9. Let $L = (\ell_{ij})$ be an $(n \times n)$ lower triangular matrix. Show that $L$ is singular if and only if $\ell_{11}\ell_{22} \cdots \ell_{nn} = 0$.

10. Let $T$ and $S$ be $(n \times n)$ upper triangular matrices. Use the definition of matrix multiplication to show that the product $ST$ is also upper triangular.

*11. Let $A$ be a lower (upper) triangular non-singular matrix. Show that $A^{-1}$ is also lower (upper) triangular. (Hint: Consider the equation $AA^{-1} = I$, entry by entry.)

12. If $A$ and $B$ are $(n \times n)$ non-singular matrices, show that $AB$ and $BA$ are also non-singular. Furthermore, show that $(AB)^{-1} = B^{-1}A^{-1}$.

*13. a) If $A$ is $(r \times s)$ and $B$ is $(r \times s)$, show that $(A + B)^T = A^T + B^T$.
    b) If $A$ is $(r \times s)$ and $B$ is $(s \times p)$, show that $(AB)^T = B^T A^T$.
    c) Show $(A^T)^T = A$.

## 2.2 DIRECT METHODS

The term *direct method* refers to a numerical procedure which can be executed in a finite number of steps. This is in contrast to *iterative methods* which generate an infinite sequence of approximations which (hopefully) converge to the solution. We will discuss iterative methods in Section 2.4. The two most common types of direct methods for solving the system (2.1) are elimination methods and factorization methods.

### 2.2.1 Gauss Elimination

Gauss elimination is the familiar variable elimination technique whereby the variables are eliminated one at a time to reduce the original system to an equivalent triangular system. The first step in this procedure is to replace the $i$th equation by the equation which is the result of multiplying the first equation by $(-a_{i1}/a_{11})$ and adding it to the original $i$th equation. Proceeding thus for $i = 2, 3, \ldots, n$, we obtain a system of equations, equivalent to (2.1):

$$\begin{aligned} a_{11}x_1 + a_{12}x_2 + \cdots + a_{1n}x_n &= b_1 \\ a_{22}^{(1)}x_2 + \cdots + a_{2n}^{(1)}x_n &= b_2^{(1)} \\ &\vdots \\ a_{n2}^{(1)}x_2 + \cdots + a_{nn}^{(1)}x_n &= b_n^{(1)} \end{aligned} \tag{2.5}$$

(This assumes that $a_{11} \neq 0$. If $a_{11} = 0$ we find an $a_{i1} \neq 0$, interchange the first and $i$th rows and proceed in the same fashion.) We continue in this manner until the system is in an equivalent triangular form:

$$\begin{aligned} a'_{11}x_1 + a'_{12}x_2 + a'_{13}x_3 + \cdots + a'_{1,n-1}x_{n-1} + a'_{1n}x_n &= b'_1 \\ a'_{22}x_2 + a'_{23}x_3 + \cdots + a'_{2,n-1}x_{n-1} + a'_{2n}x_n &= b'_2 \\ a'_{33}x_3 + \cdots + a'_{3,n-1}x_{n-1} + a'_{3n}x_n &= b'_3 \\ &\vdots \\ a'_{n-1,n-1}x_{n-1} + a'_{n-1,n}x_n &= b'_{n-1} \\ a'_{n,n}x_n &= b'_n \end{aligned} \tag{2.6}$$

The solution to this triangular system is now easily found by *backsolving*, that is, by solving the last equation for $x_n$, then the $(n-1)$st equation for $x_{n-1}$, and continuing until each $x_k$; $k = n, n-1, n-2, \ldots 1$; is determined.

**Example 2.6.** As a specific case to demonstrate Gauss elimination, consider the linear system

$$\begin{aligned} 5x_1 + 7x_2 + 6x_3 + 5x_4 &= 23 \\ 7x_1 + 10x_2 + 8x_3 + 7x_4 &= 32 \\ 6x_1 + 8x_2 + 10x_3 + 9x_4 &= 33 \\ 5x_1 + 7x_2 + 9x_3 + 10x_4 &= 31 \end{aligned}$$

Proceeding as indicated by (2.5), multiply the 1st equation by $-7/5$ and add the result to the 2nd equation to eliminate $x_1$ in the new 2nd equation. Proceeding similarly to eliminate $x_1$ from the 3rd and 4th equations, we obtain an equivalent reduced linear system

$$\begin{aligned} 5x_1 + 7x_2 + 6x_3 + 5x_4 &= 23 \\ 0.2x_2 - 0.4x_3 \qquad &= -0.2 \\ -0.4x_2 + 2.8x_3 + 3x_4 &= 5.4 \\ 3x_3 + 5x_4 &= 8 \end{aligned}$$

which corresponds to (2.5). Eliminating $x_2$ from the 3rd equation gives the equivalent linear system

$$\begin{aligned} 5x_1 + 7x_2 + 6x_3 + 5x_4 &= 23 \\ 0.2x_2 - 0.4x_3 \qquad &= -0.2 \\ 2x_3 + 3x_4 &= 5 \\ 3x_3 + 5x_4 &= 8 \end{aligned}$$

Finally, by eliminating $x_3$ from the 4th equation, we obtain a linear system in the triangular form of (2.6)

$$\begin{aligned} 5x_1 + 7x_2 + 6x_3 + 5x_4 &= 23 \\ 0.2x_2 - 0.4x_3 \qquad &= -0.2 \\ 2x_3 + 3x_4 &= 5 \\ 0.5x_4 &= 0.5 \end{aligned}$$

Backsolving the triangular linear system, we find from the 4th equation that $x_4 = 1$. Having $x_4$, the 3rd equation yields $x_3 = 1$; knowing $x_3$ and $x_4$, the 2nd equation says $x_2 = 1$ and, finally, from the 1st equation, we obtain $x_1 = 1$. The coefficient matrix for the original system is

$$A = \begin{bmatrix} 5 & 7 & 6 & 5 \\ 7 & 10 & 8 & 7 \\ 6 & 8 & 10 & 9 \\ 5 & 7 & 9 & 10 \end{bmatrix}.$$

This matrix is due to T. S. Wilson (see Todd, 1962). The matrix has several interesting properties and will serve as one example when we discuss topics such as error analysis and condition numbers.

Let us continue this example of solution by variable elimination. Consider the $(3 \times 4)$ linear system

$$\begin{aligned} 5x_1 + 7x_2 + 6x_3 + 5x_4 &= 23 \\ 7x_1 + 10x_2 + 8x_3 + 7x_4 &= 32 \\ 6x_1 + 8x_2 + 10x_3 + 9x_4 &= 33. \end{aligned}$$

Proceeding exactly as above, we arrive at the equivalent system

$$\begin{aligned} 5x_1 + 7x_2 + 6x_3 + 5x_4 &= 23 \\ 0.2x_2 - 0.4x_3 \qquad &= -0.2 \\ 2x_3 + 3x_4 &= 5. \end{aligned}$$

Here we see that either $x_3$ or $x_4$ can be chosen completely arbitrarily. For example, let $x_4$ be arbitrary, then $x_3 = (1/2)(5 - 3x_4)$. Continuing we obtain $x_2 = 4 - 3x_4$ and $x_1 = 5x_4 - 4$, and we may substitute any value for $x_4$ and have a corresponding solution of the system. In particular, if we let $x_4 = 1$, we obtain $x_3 = x_2 = x_1 = 1$ as above. This example of a $(3 \times 4)$ system shows how Gauss elimination can be used for rectangular systems, as well as for square systems.

In the simple $(4 \times 4)$ example above, no row interchanges were necessary. When Gauss elimination is programmed for use on a digital computer, we will clearly have to include a statement to check whether we might be dividing by zero. Thus, given (2.1) to solve, the first step would be to test $a_{11}$ to determine whether or not $a_{11} = 0$, and then to perform a row interchange if $a_{11} = 0$. Having obtained the first reduced system (2.5), we would again test $a_{22}^{(1)}$ to see if $a_{22}^{(1)} = 0$ and perform a row interchange if necessary. We have included, in Example 2.7, a simple subroutine to show how this rudimentary version of Gauss elimination might be carried out.

**Example 2.7.** Subroutine GAUS (see Fig. 2.1) was written to solve a linear system $A\mathbf{x} = \mathbf{b}$ by Gauss elimination, where $A$ is a square matrix. The subroutine requires: $N$ where $A$ is $(N \times N)$, the entries $A(I, J)$ of $A$ and the entries $B(I)$ of the right-hand side $\mathbf{b}$.

As an example run, the system of Example 2.6 was solved. This program was executed in single precision and, as can be seen, the errors in the computed solution range from 0.000583 to 0.00009. For purposes of comparison, the final triangular form of the matrix was printed out as shown and can be compared with the exact triangular form displayed in Example 2.6.

```
SOLUTION VECTOR:

0.999417E 00
0.100035E 01
0.100015E 01
0.999910E 00

THE TRIANGULARIZED MATRIX:

0.500000E 01   0.700000E 01    0.600000E 01   0.500000E 01
0.000000E 00   0.200003E 00   -0.399998E 00   0.190735E-05
0.000000E 00   0.000000E 00    0.200002E 01   0.300000E 01
0.000000E 00   0.000000E 00    0.000000E 00   0.500038E 00
```

This computer example illustrates the effects of round-off error, and we note that this computer solution would probably not be satisfactory in most practical problems. As we shall see later, the coefficient matrix of Example 2.6 is moderately "ill-conditioned" (see Section 2.3.3). This will mean that even a few small computer-round-off errors will induce relatively large errors in the computed solution.

```
      SUBROUTINE GAUS(A,B,X,N,IERROR)
      DIMENSION A(20,20),B(20),X(20)
C
C   SUBROUTINE GAUS USES GAUSS ELIMINATION (WITHOUT PIVOTING) TO SOLVE
C   THE SYSTEM AX=B.  THE CALLING PROGRAM MUST SUPPLY THE MATRIX A, THE
C   VECTOR B AND AN INTEGER N (WHERE A IS (NXN)).  ARRAYS A AND B ARE
C   DESTROYED IN GAUS, THE SOLUTION IS RETURNED IN X AND A FLAG, IERROR,
C   IS SET TO 1 IF A IS NON-SINGULAR AND IS SET TO 2 IF A IS SINGULAR.
C
      NM1=N-1
      DO 5 I=1,NM1
C
C   SEARCH FOR NON-ZERO PIVOT ELEMENT AND INTERCHANGE ROWS IF NECESSARY.
C   IF NO NON-ZERO PIVOT ELEMENT IS FOUND, SET IERROR=2 AND RETURN.
C
      DO 3 J=I,N
      IF(A(J,I).EQ.0)    GO TO 3
      DO 2 K=I,N
      TEMP=A(I,K)
      A(I,K)=A(J,K)
    2 A(J,K)=TEMP
      TEMP=B(I)
      B(I)=B(J)
      B(J)=TEMP
      GO TO 4
    3 CONTINUE
      GO TO 8
C
C   ELIMINATE THE COEFFICIENTS OF X(I) IN ROWS I+1,...,N
C
    4 IP1=I+1
      DO 5 K=IP1,N
      Q=-A(K,I)/A(I,I)
      A(K,I)=0.
      B(K)=Q*B(I)+B(K)
      DO 5 J=IP1,N
    5 A(K,J)=Q*A(I,J)+A(K,J)
      IF(A(N,N).EQ.0.)    GO TO 8
C
C   BACKSOLVE THE EQUIVALENT TRIANGULARIZED SYSTEM, SET IERROR=1,
C   AND RETURN
C
      X(N)=B(N)/A(N,N)
      NP1=N+1
      DO 7 K=1,NM1
      Q=0.
      NMK=N-K
      DO 6 J=1,K
    6 Q=Q+A(NMK,NP1-J)*X(NP1-J)
    7 X(NMK)=(B(NMK)-Q)/A(NMK,NMK)
      IERROR=1
      RETURN
    8 IERROR=2
      RETURN
      END
```

Figure 2.1

This program was executed again, in double precision, and gave the solution vector correct to as many places as are listed:

```
SOLUTION VECTOR

0.10000000D 01
0.10000000D 01
0.10000000D 01
0.10000000D 01

THE TRIANGULARIZED MATRIX

0.500000D 01    0.700000D 01     0.600000D 01    0.500000D 01
0.000000D 00    0.200000D 00    -0.400000D 00    0.444089D-15
0.000000D 00    0.000000D 00     0.200000D 01    0.300000D 01
0.000000D 00    0.000000D 00     0.000000D 00    0.500000D 00
```

Since linear systems of any appreciable size are normally solved on the computer, round-off errors can cause problems and the technique of Gauss elimination must be examined more carefully. In addition to the question of rounding errors, there are other aspects which must be considered for practical and efficient implementation of numerical procedures on the machine. Three such aspects that can influence the choice of an algorithm are storage requirements, round-off errors, and execution time.

### 2.2.2 Operations Counts

For the system (2.1), storage requirements are clearly linked to the size of the system, $n$. In order to obtain a feeling for the execution time and rounding errors of direct methods, it is customary to make an *operations count*. By this, we mean a calculation of the number of arithmetic operations necessary to carry out the direct method. One rationale for an operations count is that execution time and the error in the computed solution due to round-off are both related to the total number of arithmetic operations. For Gauss elimination, multiplication and addition are (essentially) the only arithmetic operations performed. (We consider "division" as a multiplication and "subtraction" as an addition.) To illustrate how an operations count is made, we will count the multiplications necessary to solve the system (2.1) by Gauss elimination. The reader can quickly verify that the necessary number of additions is about the same as the number of multiplications.

To transform (2.1) to the equivalent system (2.5) requires $(n - 1)(n + 1)$ multiplications. This is seen by noting that to eliminate the variable $x_1$ in the $i$th equation of (2.1), we multiply the 1st equation $a_{11}x_1 + a_{12}x_2 + \cdots + a_{1n}x_n = b_1$ by the scalar $-a_{i1}/a_{11}$ and add the resulting multiple of the 1st equation to the $i$th equation. It takes one multiplication to form the scalar $-a_{i1}/a_{11}$ (we don't count multiplications by $-1$) and an additional $n$ multiplications to form each of the products

$$\frac{-a_{i1}}{a_{11}}a_{12}, \frac{-a_{i1}}{a_{11}}a_{13}, \ldots, \frac{-a_{i1}}{a_{11}}a_{1n}, \qquad \frac{-a_{i1}}{a_{11}}b_1.$$

Note that there is no need to form the product $(-a_{i1}/a_{11})a_{11}$, since we *already* know we are going to have 0 as the scalar multiplying $x_1$ in the new $i$th equation. Thus multiplying the 1st equation by $-a_{i1}/a_{11}$ and adding it to the $i$th equation to produce the new $i$th equation

$$a_{i2}^{(1)}x_2 + a_{i3}^{(1)}x_3 + \cdots + a_{in}^{(1)}x_n = b_i^{(1)}, \tag{2.7}$$

takes $(n+1)$ multiplications (and also, $(n+1)$ additions). We have to perform this variable elimination in rows $2, 3, \ldots, n$ and hence require a total of $(n-1)(n+1)$ multiplications to produce the equivalent reduced system (2.5).

Given the system (2.5), the same analysis shows that $n$ multiplications are needed to eliminate $x_2$ in equations $3, 4, \ldots, n$ of (2.5). Thus the next step of Gauss elimination which results in

$$\begin{aligned} a_{11}^{(2)}x_1 + a_{12}^{(2)}x_2 + a_{13}^{(2)}x_3 + \cdots + a_{1n}^{(2)}x_n &= b_1^{(2)} \\ a_{22}^{(2)}x_2 + a_{23}^{(2)}x_3 + \cdots + a_{2n}^{(2)}x_n &= b_2^{(2)} \\ a_{33}^{(2)}x_3 + \cdots + a_{3n}^{(2)}x_n &= b_3^{(2)} \\ \vdots \qquad\qquad \vdots \quad &\quad \vdots \\ a_{n3}^{(2)}x_3 + \cdots + a_{nn}^{(2)}x_n &= b_n^{(2)} \end{aligned} \tag{2.8}$$

takes $(n-2)n$ multiplications. Continuing until the triangular system (2.6) is obtained, it follows that we have performed a total of $(n-1)(n+1) + (n-2)n + \cdots + 1 \cdot 3$ multiplications. To evaluate this sum, we write it symbolically as

$$\sum_{j=1}^{n-1} j(j+2) = \sum_{j=1}^{n-1} j^2 + 2\sum_{j=1}^{n-1} j. \tag{2.9}$$

In this form, using the well-known formulas for the sum of the first $(n-1)$ integers and the sum of the first $(n-1)$ integers squared, we find that a total of

$$\frac{n(n-1)(2n-1)}{6} + n(n-1) = \frac{n(n-1)(2n+5)}{6} \tag{2.10}$$

multiplications are needed to produce an equivalent triangular system from a square $(n \times n)$ system.

That essentially $n^3/3$ multiplications are needed to triangularize (2.1) turns out to be quite an important observation. For the moment, however, we have not completed the operations count of Gauss elimination. We have yet to count the multiplications necessary to backsolve the triangular system (2.6). We leave as a problem for the reader to show that $n(n+1)/2$ multiplications are required to solve (2.6). This means that the bulk of the computation involved in Gauss elimination is in triangularizing the coefficient matrix.

### 2.2.3 Implementation of Gauss Elimination

There is a modification of Gauss elimination that is relatively easy to program, which should in most cases reduce the effects of round-off error. To illustrate, we first consider the following example from Forsythe (1967) where for simplicity, we consider a three-digit floating-decimal machine. (Recall, we assumed in Chapter 1 that on such a machine the arithmetic operations are done in a six-digit mode and then rounded to three digits. For example, this machine will record $(1 - 10000)$ as $-10000$.)

**Example 2.8.***

$$0.0001x + 1.00y = 1.00$$
$$1.00x + 1.00y = 2.00$$

The true solution rounded to five decimals is $x = 1.00010$ and $y = 0.99990$. If we proceed as above, without rearranging the equations, we obtain $-10000y = -10000$ and so we get $y = 1.00$ and $x = 0.00$, quite different from the true solution. However if we rewrite the system as

$$1.00x + 1.00y = 2.00$$
$$0.0001x + 1.00y = 1.00,$$

then we obtain the answer $x = 1.00$ and $y = 1.00$, a remarkable improvement.

This example demonstrates one source of round-off error, namely that rounding errors can become an important factor when two numbers are added where one is much larger than the other, and the significant digits of the smaller one are essentially lost. A technique to compensate for this is called *pivoting* and is explained below. In ordinary Gauss elimination, if $|a_{11}|$ is relatively small in comparison with some $|a_{i1}|$, then the multiplicative factor $(-a_{i1}/a_{11})$ is a relatively large number in magnitude. Let $\lambda = -a_{i1}/a_{11}$, then the result of multiplying the first equation by $\lambda$ and adding to the $i$th equation is the new $i$th equation

$$(a_{i2} + \lambda a_{12})x_2 + \cdots + (a_{in} + \lambda a_{1n})x_n = b_i + \lambda b_1.$$

If $\lambda a_{1j}$ is large relative to $a_{ij}$, then a significant rounding error can occur. This can be countered by first finding $|a_{I1}| = \max_{1 \le i \le n} |a_{i1}|$, and rewriting the system

* Reprinted with permission from G. E. Forsythe, "Today's computational methods of linear algebra," SIAM Review, Vol. 9, 1967. Copyright © 1967 by Society for Industrial and Applied Mathematics.

so that the $I$th equation becomes the first equation. (Note that this altered system still has the same solution as the original system.) Now the multiplicative constant to eliminate the coefficients of $x_1$ is $(-a_{i1}/a_{I1})$, which is the relatively smallest factor we could obtain in this fashion. Thus the effect of the round-off error is reduced in the remaining equations. This process is repeated for each successive variable elimination until the triangularization is completed. This technique is known as *partial pivoting.*

Further reduction of round-off error can be accomplished in the following manner. As the first step find $|a_{IJ}| = \max_{1 \le i,\, j \le n} |a_{ij}|$. Retain the $I$th equation and eliminate the coefficient of $x_J$ in the $i$th equation, $1 \le i \le n$, $i \ne I$, by replacing the $i$th equation by $(-a_{iJ}/a_{IJ})$ times the $I$th equation plus the original $i$th equation. Repeat this process until the final system can be backsolved in a manner similar to a triangular system. This process, called *total pivoting*, reduces the effect of round-off error over partial pivoting but is more complicated to program, since the order in which the variables are eliminated must be retained.

Besides pivoting, another important consideration is the efficient programming of Gauss elimination. We wish to mention just one aspect of this general organizational problem. Frequently in practice, one is confronted with solving a succession of linear systems which have the same coefficient matrix. That is, solve

$$A\mathbf{x} = \mathbf{b}_i, \qquad i = 1, 2, \ldots, m. \tag{2.11}$$

(For example, this was the case when we were computing $A^{-1}$ in the first section.) If we employ a Gauss elimination routine to solve, say, $A\mathbf{x} = \mathbf{b}_1$, then we first obtain an equivalent triangular system (such as (2.6))

$$A'\mathbf{x} = \mathbf{b}_1' \tag{2.12}$$

and this system is backsolved. If we next employ the same routine to solve $A\mathbf{x} = \mathbf{b}_2$, we obtain, as above,

$$A'\mathbf{x} = \mathbf{b}_2' \tag{2.13}$$

and (2.13) is then backsolved.

A moment's reflection shows that instead of calling the same routine repeatedly to solve the $m$ systems in (2.11) it would be better if we could store $A'$ and just have a procedure to generate $\mathbf{b}_i'$ from $\mathbf{b}_i$. If this could be done, we would have to perform the approximately $n^3/3$ operations necessary to triangularize $A$ only once. This is a programming problem, which can be solved by considering how the triangular system (2.6) is obtained from the original system (2.1). In particular, as the system $A\mathbf{x} = \mathbf{b}$ is transformed to $A'\mathbf{x} = \mathbf{b}'$, the vector $\mathbf{b}$ is transformed to the vector $\mathbf{b}'$ by the same sequence of scalar multiples which transforms $A$ to the triangular matrix $A'$. Thus what is needed is some way to store the sequence of the scalar multiples of the triangularization. Rather than trying to elaborate on this abstractly, we will give an example which makes the procedure clear.

**Example 2.9.** Consider the linear system

$$\begin{aligned} x_1 + \ 2x_2 + \ x_3 &= b_1 \\ 3x_1 + \ 4x_2 \qquad &= b_2. \\ 2x_1 + 10x_2 + 4x_3 &= b_3 \end{aligned} \tag{2.14}$$

We are seeking the scalar multiples to transform $\mathbf{b}$ into $\mathbf{b}'$, so we first determine how $A$ is transformed to $A'$. That is, we focus our attention upon how to triangularize the matrix

$$A = \begin{bmatrix} 1 & 2 & 1 \\ 3 & 4 & 0 \\ 2 & 10 & 4 \end{bmatrix} \tag{2.15}$$

by adding multiples of one row to the next. Multiplying the first row by $-3$ and adding this to the second row, then multiplying the first row by $-2$ and adding this to the third row, we obtain

$$A^{(1)} = \begin{bmatrix} 1 & 2 & 1 \\ 0 & -2 & -3 \\ 0 & 6 & 2 \end{bmatrix}, \qquad M_{21} = -3, \qquad M_{31} = -2, \tag{2.16}$$

where $M_{21}$ and $M_{31}$ are the respective multiples needed to introduce zeros in the first column below the main diagonal. Continuing this simple example, if we multiply the second row of $A^{(1)}$ by 3 and add to the third row, we obtain $A'$, where

$$A' = \begin{bmatrix} 1 & 2 & 1 \\ 0 & -2 & -3 \\ 0 & 0 & -7 \end{bmatrix}, \qquad M_{32} = 3. \tag{2.17}$$

If we had performed Gauss elimination on the system (2.14), the coefficient matrix of the equivalent triangular system would clearly be $A'$ in (2.17).

Moreover, the multiples $M_{21}$, $M_{31}$, and $M_{32}$ tell us how to transform $\mathbf{b}$. To be specific, we first obtain

$$\mathbf{b}^{(1)} = \begin{bmatrix} b_1 \\ b_2 + M_{21}b_1 \\ b_3 + M_{31}b_1 \end{bmatrix} = \begin{bmatrix} b_1^{(1)} \\ b_2^{(1)} \\ b_3^{(1)} \end{bmatrix} \tag{2.18}$$

and finally,

$$\mathbf{b}' = \begin{bmatrix} b_1^{(1)} \\ b_2^{(1)} \\ b_3^{(1)} + M_{32}b_2^{(1)} \end{bmatrix} = \begin{bmatrix} b_1' \\ b_2' \\ b_3' \end{bmatrix}. \tag{2.19}$$

Clearly, in terms of efficient storage, we can overlay the scalar multiples $M_{ij}$ in the zero entries of $A'$ that they introduce during the triangularization.

Thus we proceed as indicated by the arrows:

$$A = \begin{bmatrix} 1 & 2 & 1 \\ 3 & 4 & 0 \\ 2 & 10 & 4 \end{bmatrix} \to \begin{bmatrix} 1 & 2 & 1 \\ -3 & -2 & -3 \\ -2 & 6 & 2 \end{bmatrix} \to \begin{bmatrix} 1 & 2 & 1 \\ -3 & -2 & -3 \\ -2 & 3 & -7 \end{bmatrix} = A'' \tag{2.20}$$

(Note: no additional locations are needed to store $M_{21}$, $M_{31}$, and $M_{32}$.) Moreover, the positions of the multiples in $A''$ describes how they are to operate on vector $\mathbf{b}$ to transform $\mathbf{b}$ to $\mathbf{b}'$. To illustrate, consider

$$\begin{aligned} x_1 + 2x_2 + x_3 &= 3 \\ 3x_1 + 4x_2 \phantom{+ x_3} &= 3 \\ 2x_1 + 10x_2 + 4x_3 &= 10 \end{aligned} \tag{2.21}$$

which we know from above is equivalent to

$$\begin{aligned} x_1 + 2x_2 + x_3 &= b_1' \\ -2x_2 - 3x_3 &= b_2' \\ -7x_3 &= b_3' \end{aligned} \tag{2.22}$$

where we get the coefficients of (2.22) from the upper triangle portion of $A''$. To find $b_1', b_2', b_3'$, we change $\mathbf{b}$ to $\mathbf{b}'$ by the sequence

$$\mathbf{b} = \begin{bmatrix} 3 \\ 3 \\ 10 \end{bmatrix} \to \begin{bmatrix} 3 \\ 3 + (-3)\cdot 3 \\ 10 + (-2)\cdot 3 \end{bmatrix} = \begin{bmatrix} 3 \\ -6 \\ 4 \end{bmatrix} \to \begin{bmatrix} 3 \\ -6 \\ 4 + 3\cdot(-6) \end{bmatrix} = \begin{bmatrix} 3 \\ -6 \\ -14 \end{bmatrix} \tag{2.23}$$

where the numbers below the main diagonal in $A''$ define the sequence. Thus from $A''$, we can read that (2.21) is equivalent to

$$\begin{aligned} x_1 + 2x_2 + x_3 &= 3 \\ -2x_2 - 3x_3 &= -6 \\ -7x_3 &= -14 \end{aligned} \tag{2.24}$$

and backsolving, obtain $x_3 = 2$, $x_2 = 0$, and $x_1 = 1$.

The basic ideas in Example 2.9 are fairly easy to program and the reader is urged to write his own Gauss elimination routine following the pattern of Example 2.9. In that simple example, no row interchanges were necessary, but clearly this possibility must be accounted for. Given an $(n \times n)$ system, an additional column can be adjoined to the coefficient matrix $A$ and this is sufficient to account for all possible row interchanges—we leave the details to the reader. Programming exercises such as these demonstrate that solving linear systems amounts essentially to organizing, modifying, and shifting entries in large arrays of data, a task that the digital computer is well-suited to perform.

One might suppose, if $m$ in (2.11) is very large, that the most efficient way of solving the $m$ linear systems, $A\mathbf{x} = \mathbf{b}_k$, $1 \le k \le m$, is to first calculate $A^{-1}$, and then generate the $k$th solution vector, $\mathbf{y}_k = A^{-1}\mathbf{b}_k$, for each $k$. However, since $A^{-1}$ is an $(n \times n)$ matrix, it follows that $n^2$ multiplications are required to compute $A^{-1}\mathbf{b}_k$. By contrast, it is shown in Problem 6 that a total of $n(n-1)/2$ multiplications are necessary to generate $\mathbf{b}_k'$ from $\mathbf{b}_k$ and a total of $n(n+1)/2$ multiplications are needed to solve $A'\mathbf{x} = \mathbf{b}_k'$. Thus if we know the triangular matrix $A'$ and the $n(n-1)/2$ scalar multiples needed to generate $\mathbf{b}_k'$, we can solve

$A\mathbf{x} = \mathbf{b}_k$ with precisely as many multiplications as it takes to form $A^{-1}\mathbf{b}_k$. Clearly then, the computation of $A^{-1}$ for solving $A\mathbf{x} = \mathbf{b}_i$, $i = 1, 2, \ldots, m$ is not justified, since in order to find $A^{-1}$, we must first solve $n$ systems of equations of the form $A\mathbf{x} = \mathbf{e}_i$, $i = 1, 2, \ldots, n$ (see Section 2.1).

At this point, we would like to offer as a guideline the following: It is seldom necessary, in a practical problem, to compute either the inverse of a matrix, the determinant of a matrix, or the power of a matrix. It is often convenient, in the formulation of a problem, to use inverses, powers and determinants in an abstract mathematical sense (e.g., the solution of $A\mathbf{x} = \mathbf{b}$ is given by $\mathbf{x} = A^{-1}\mathbf{b}$). However, it is rare to encounter a problem where the *solution* requires the computation of inverses, powers or determinants. Thus, for example the solution to $A\mathbf{x} = \mathbf{b}$ is found efficiently by using Gauss elimination and is only *represented mathematically* by $A^{-1}\mathbf{b}$. Exceptions to the above guidelines do occur. For example, in certain problems in statistical analysis, the inverse matrix is of significance and may have to be calculated.

### 2.2.4 Factorization Methods

The idea behind factorization methods is inspired in part by the observation that triangular systems are easy to solve. To exploit this observation, suppose we can factor the $(n \times n)$ matrix $A$ into a product of two matrices $L$ and $U$, so that

$$A = LU, \tag{2.25}$$

where $L$ is an $(n \times n)$ lower triangular matrix and $U$ is an $(n \times n)$ upper triangular matrix.

If $A$ has a factorization like (2.25), then solutions of the system $A\mathbf{x} = \mathbf{b}$ are found by solving $LU\mathbf{x} = \mathbf{b}$. To solve $LU\mathbf{x} = \mathbf{b}$, we proceed in two stages:

1) Solve $L\mathbf{y} = \mathbf{b}$
2) Solve $U\mathbf{x} = \mathbf{y}$.

In this fashion, if $U\mathbf{x} = \mathbf{y}$ then $LU\mathbf{x} = L\mathbf{y} = \mathbf{b}$ and, moreover, both systems $L\mathbf{y} = \mathbf{b}$ and $U\mathbf{x} = \mathbf{y}$ are triangular and hence easy to solve. What is needed, then, is a procedure to generate the factorization (or, as it is often called, the *LU-decomposition*). One such procedure is to consider the equation $A = LU$:

$$\begin{bmatrix} a_{11} & a_{12} & \cdots & a_{1n} \\ a_{21} & a_{22} & \cdots & a_{2n} \\ \vdots & & & \vdots \\ a_{n1} & a_{n2} & \cdots & a_{nn} \end{bmatrix} = \begin{bmatrix} \ell_{11} & 0 & 0 & \cdots & 0 \\ \ell_{21} & \ell_{22} & 0 & \cdots & 0 \\ \vdots & & & & \vdots \\ \ell_{n1} & \ell_{n2} & & \cdots & \ell_{nn} \end{bmatrix} \begin{bmatrix} u_{11} & u_{12} & \cdots & u_{1n} \\ 0 & u_{22} & \cdots & u_{2n} \\ \vdots & & & \vdots \\ 0 & 0 & \cdots & u_{nn} \end{bmatrix} \tag{2.26}$$

Writing (2.26) out at length and equating both sides coordinate-wise gives $n^2$ equations which can then be (possibly) solved to determine $L$ and $U$. We do not intend to do an extensive analysis of the general case represented by (2.26), but we do want to observe that as (2.26) is written, $L$ and $U$ both contain $n(n + 1)/2$ undetermined entries, for a total of $n^2 + n$ undetermined entries. Since we have just $n^2$ equations, it seems feasible that $n$ quantities may be chosen as we wish. In practice, these free parameters are normally specified by either:

$$\text{a) } \ell_{11} = \ell_{22} = \cdots = \ell_{nn} = 1$$

or

$$\text{b) } \ell_{11} = u_{11}, \quad \ell_{22} = u_{22}, \ldots, \ell_{nn} = u_{nn}.$$

The following example should serve to illustrate factorization.

**Example 2.10.** Given

$$A = \begin{bmatrix} 3 & 1 & 2 \\ 6 & 3 & 3 \\ -3 & 2 & -3 \end{bmatrix}$$

find an $LU$-decomposition of the form

$$\begin{bmatrix} 3 & 1 & 2 \\ 6 & 3 & 3 \\ -3 & 2 & -3 \end{bmatrix} = \begin{bmatrix} 1 & 0 & 0 \\ \ell_{21} & 1 & 0 \\ \ell_{31} & \ell_{32} & 1 \end{bmatrix} \begin{bmatrix} u_{11} & u_{12} & u_{13} \\ 0 & u_{22} & u_{23} \\ 0 & 0 & u_{33} \end{bmatrix}. \tag{2.27}$$

The most natural way to set up the equations that must be solved is in a row-by-row fashion:

$$\begin{array}{llll} u_{11} = 3 & u_{12} = 1 & u_{13} = 2 & \\ \ell_{21}u_{11} = 6 & \ell_{21}u_{12} + u_{22} = 3 & \ell_{21}u_{13} + u_{23} = 3 & (2.28) \\ \ell_{31}u_{11} = -3 & \ell_{31}u_{12} + \ell_{32}u_{22} = 2 & \ell_{31}u_{13} + \ell_{32}u_{23} + u_{33} = -3 & \end{array}$$

Solving these equations successively, we obtain

$$\begin{bmatrix} 3 & 1 & 2 \\ 6 & 3 & 3 \\ -3 & 2 & -3 \end{bmatrix} = \begin{bmatrix} 1 & 0 & 0 \\ 2 & 1 & 0 \\ -1 & 3 & 1 \end{bmatrix} \begin{bmatrix} 3 & 1 & 2 \\ 0 & 1 & -1 \\ 0 & 0 & 2 \end{bmatrix}. \tag{2.29}$$

Not all matrices have a factorization. As a simple example, the non-singular matrix

$$A = \begin{bmatrix} 0 & 1 \\ 1 & 0 \end{bmatrix} \tag{2.30}$$

can be shown (Problem 8) to have no $LU$-decomposition, while the singular matrix $I + A$ does have an $LU$-decomposition. It *is* known that if a matrix $A$ is non-singular, then there is some rearrangement of the rows of $A$ such that the rearranged matrix has an $LU$-decomposition. See Forsythe and Moler (1967) *Computer Solution of Linear Algebraic Systems*, for a proof. Programs to implement $LU$-decomposition can be written, utilizing this theorem. Such a program would find a triangular decomposition for any non-singular matrix and also would store the row interchanges that are necessary. The program would be somewhat similar to a Gauss elimination routine that incorporates partial pivoting. A full discussion of this would lead us somewhat afield, so we would prefer to allow the interested reader to pursue this himself.

It is not hard to see from the form of the equations in (2.28), that a factorization procedure will involve exactly as many multiplications as Gauss elimination. Factorization methods do have a feature that makes them valuable for hand computation, namely, that intermediate results do not have to be recorded. Equations (2.28) show that as the procedure progresses, we fill in the entries of $L$ and $U$. By contrast, if we were using a desk calculator to solve a linear system by Gauss elimination, we would have to copy down each of the intermediate steps, such as (2.5) and (2.8).

There are special sorts of problems for which factorization methods are very natural to use. One such problem that occurs in many different applied settings, is that of solving a *tridiagonal* linear system

$$\begin{aligned}
a_{11}x_1 + a_{12}x_2 \qquad\qquad\qquad\qquad &= b_1 \\
a_{21}x_1 + a_{22}x_2 + a_{23}x_3 \qquad\qquad &= b_2 \\
a_{32}x_2 + a_{33}x_3 + a_{34}x_4 \qquad &= b_3 \\
a_{43}x_3 + a_{44}x_4 + a_{45}x_5 &= b_4 \\
\ddots \qquad\qquad &\ \ \vdots \\
a_{n,n-1}x_{n-1} + a_{nn}x_n &= b_n
\end{aligned} \tag{2.31}$$

Such systems are called tridiagonal because the coefficient matrix $A$ for the system (2.31) has three *diagonals*, or more precisely, $a_{ij} = 0$ if $j < i - 1$ or if $j > i + 1$. These systems commonly arise, for example, when numerical methods are used to solve two-point boundary value problems or when cubic spline approximations are used to fit data (see Chapters 7 and 5). In most practical situations, a factorization of the form (2.32) below is possible:

$$LU = \begin{bmatrix} 1 & 0 & 0 & 0 & \cdots & 0 \\ \ell_{21} & 1 & 0 & 0 & \cdots & 0 \\ 0 & \ell_{32} & 1 & 0 & \cdots & 0 \\ 0 & 0 & \ell_{43} & 1 & \cdots & 0 \\ \vdots & & & & & \vdots \\ 0 & 0 & 0 & 0 & \cdots & 1 \end{bmatrix} \begin{bmatrix} u_{11} & a_{12} & 0 & 0 & \cdots & 0 \\ 0 & u_{22} & a_{23} & 0 & \cdots & 0 \\ 0 & 0 & u_{33} & a_{34} & \cdots & 0 \\ 0 & 0 & 0 & u_{44} & \cdots & 0 \\ \vdots & & & & & \vdots \\ 0 & 0 & 0 & 0 & \cdots & u_{nn} \end{bmatrix} \tag{2.32}$$

If $A$ is the coefficient matrix of (2.31), we can express $A$ as $A = LU$, by solving recursively:

$$\begin{aligned}
u_{11} &= a_{11} \\
\ell_{i,i-1} &= a_{i,i-1}/u_{i-1,i-1} \qquad i = 2, 3, \ldots, n \\
u_{ii} &= a_{ii} - \ell_{i,i-1}a_{i-1,i}
\end{aligned} \tag{2.33a}$$

Having the factorization above, we can solve (2.31) by first solving $Ly = \mathbf{b}$ and then $U\mathbf{x} = \mathbf{y}$. The explicit equations are

$$\begin{aligned} y_1 &= b_1 \\ y_i &= b_i - \ell_{i,i-1} y_{i-1}; \qquad i = 2, 3, \ldots, n \\ x_n &= y_n / u_{nn} \\ x_{n-i} &= (y_{n-i} - a_{n-i,n+1-i} x_{n+1-i}) / u_{n-i,n-i}; \qquad i = 1, 2, \ldots, n-1. \end{aligned} \tag{2.33b}$$

```
      SUBROUTINE TRDIAG(A,B,C,X,F,N,IERROR)
      DIMENSION A(19),B(20),C(19),F(20),U(20),Y(20),X(20)
      REAL L(19)
C
C  SUBROUTINE TRDIAG SOLVES THE TRIDIAGONAL LINEAR SYSTEM TX=F BY
C  FACTORIZATION.  THE CALLING PROGRAM MUST SUPPLY ARRAYS
C  A(I),I=2,...,N; B(I),I=1,...,N; C(I),I=1,...,N-1;
C  (WHERE T(I,I-1)=A(I), T(I,I)=B(I), AND T(I,I+1)=C(I)).  THE CALLING
C  PROGRAM MUST ALSO SUPPLY F AND N (WHERE T IS (NXN)).  THE SOLUTION
C  IS RETURNED IN X AND A FLAG, IERROR, IS SET TO 2 IF DIRECT FACTOR-
C  IZATION IS NOT POSSIBLE AND SET TO 1 OTHERWISE.
C
C
C  FACTOR T AS T=LU
C
      U(1)=B(1)
      DO 1 I=2,N
      IF(U(I-1).EQ.0)    GO TO 4
      L(I)=A(I)/U(I-1)
    1 U(I)=B(I)-L(I)*C(I-1)
C
C  SOLVE LUX=F
C
      Y(1)=F(1)
      DO 2 I=2,N
    2 Y(I)=F(I)-L(I)*Y(I-1)
      IF(U(N).EQ.0)    GO TO 4
      X(N)=Y(N)/U(N)
      NM1=N-1
      DO 3 I=1,NM1
    3 X(N-I)=(Y(N-I)-C(N-I)*X(N+1-I))/U(N-I)
      IERROR=1
      RETURN
    4 IERROR=2
      RETURN
      END
```

**Figure 2.2**

The subroutine given in Fig. 2.2 implements the procedure outlined in (2.33a) and (2.33b). Finally, we note (as in Problem 9) that Gauss elimination can be thought of as a factorization method, if the elimination can be carried out without any row interchanges.

## PROBLEMS

1. Modify the Gauss elimination routine of Fig. 2.1, Example 2.7, to include partial pivoting.
2. Use a Gauss elimination routine (such as the program in Example 2.7 or the modification of Problem 1) to solve the $(3 \times 3)$ system of Example 2.2, Section 2.1.
3. Use a Gauss elimination routine to solve the $(3 \times 3)$ system of Example 2.14, Section 2.3.3.
4. Find the inverse of the coefficient matrix for the $(4 \times 4)$ system of Example 2.6, following the procedure developed in Example 2.3 of Section 2.1. An efficient program to find an inverse would use the ideas developed in Example 2.9, where the possibility of row interchanges is accounted for.
5. An $(n \times n)$ Hilbert matrix $H = (h_{ij})$ is a square matrix whose entries are given by $h_{ij} = 1/(i + j - 1)$. Hilbert matrices occur frequently in statistical and data-fitting problems and are considered to be badly behaved or *ill-conditioned* (see Section 2.3.3). Write a program to obtain an $LU$-decomposition for the $(10 \times 10)$ Hilbert matrix, where in Eq. (2.26), $\ell_{11} = u_{11}, \ell_{22} = u_{22}, \ldots, \ell_{10,10} = u_{10,10}$. Having the $LU$-decomposition, find the inverse matrix. It is more efficient to generate the entries $h_{ij}$ of $H$ in your program, rather than to read them in as data. (It is known that *positive definite* matrices ((2.68) Section 2.4.1), such as the Hilbert matrices, have an $LU$-decomposition of this type that can be found without need for row interchanges.)
6. Given the system of equations (2.6), show that $n(n + 1)/2$ multiplications are required to solve for $x_n, x_{n-1}, \ldots, x_2, x_1$.
7. Use Gauss elimination (without pivoting) to show that

$$\begin{aligned}
x_1 + x_2 + \phantom{2}x_3 + \phantom{3}x_4 &= 5\\
x_1 - x_2 + 2x_3 - \phantom{3}x_4 &= 0\\
x_1 + x_2 - \phantom{2}x_3 + 3x_4 &= 7\\
2x_1 \phantom{{}+ x_2} + 3x_3 \phantom{{}+ 3x_4} &= 5\\
3x_1 + x_2 + 2x_3 + 3x_4 &= 12
\end{aligned}$$

   yields $x_3 = x_4 - 1, x_2 = 2 - x_4/2, x_1 = 4 - 3x_4/2$, where $x_4$ can be chosen arbitrarily. *(How could one modify a Gauss elimination computer program to allow for this contingency?)
8. Show that the non-singular matrix $A = \begin{pmatrix} 0 & 1 \\ 1 & 0 \end{pmatrix}$ has no $LU$-decomposition, while the singular matrix $(A + I)$ does.

*9. Given an $(n \times n)$ matrix $A = (a_{ij})$ with $a_{11} \neq 0$, find the $(n \times n)$ lower triangular matrix $T_1$ such that $T_1 A = A_1$, where $A_1$ is the coefficient matrix of (2.5). From this, argue that if there are no row interchanges necessary in the Gauss elimination process there exist non-singular triangular matrices $\{T_k\}_{k=1}^{n-1}$ such that $(T_{n-1}T_{n-2} \cdots T_1)A = A_{n-1}$, where $A_{n-1}$ is the upper triangular coefficient matrix of (2.6). Thus we see by Problems 9, 10, and 11 of Section 2.1, that Gauss elimination without row interchanges generates an $LU$ triangular decomposition of $A$.

*10. In (2.5), if $a_{22}^{(1)} \neq 0$, it is evident that $x_2$ can be eliminated from the first equation by multiplying the second equation by $(-a_{12}/a_{22}^{(1)})$ and adding it to the first equation. Proceeding like this at each step we can eliminate all variables both below *and* above the diagonal. The end result is an equivalent diagonal system, thus eliminating the need to backsolve to obtain the $x_i$'s. This is called the *Gauss-Jordan* elimination process. Count the number of multiplications necessary to solve the linear system by this process and compare it with the operations count of Gauss elimination.

## 2.3 ERROR ANALYSIS AND NORMS

There are many different types of linear systems of equations each having their own special characteristics. Thus we can hardly expect any particular direct method, like Gauss elimination, to be the best possible method to use in all circumstances. Moreover, if we do use a direct method, our computed solution will almost certainly be incorrect due to round-off error. Therefore we need some way to determine the size of the error of any computed solution and also some way to improve this computed solution. In this section, we will briefly develop a little of the theoretical background that is needed both for the analysis of errors and for the analysis of the "iterative" algorithms of Section 2.4 which generate a sequence of approximate solutions to $A\mathbf{x} = \mathbf{b}$.

The subject of error analysis must be approached somewhat carefully, since a particular computed solution (say $\mathbf{x}_c$) to $A\mathbf{x} = \mathbf{b}$ may be considered badly in error or quite acceptable depending on how we intend to use the vector $\mathbf{x}_c$. For example, let $\mathbf{x}_t$ denote the "true" solution of $A\mathbf{x} = \mathbf{b}$ and let $\mathbf{r} = A\mathbf{x}_c - \mathbf{b}$ be the *residual vector*. Then, $\mathbf{r} = A\mathbf{x}_c - \mathbf{b} = A\mathbf{x}_c - A\mathbf{x}_t$ or

$$\mathbf{x}_c - \mathbf{x}_t = A^{-1}\mathbf{r}. \tag{2.34}$$

If $A^{-1}$ has some very large entries, then Eq. (2.34) demonstrates the fact that the residual vector, $\mathbf{r}$, might be small, and yet $\mathbf{x}_c$ might be quite far from $\mathbf{x}_t$. Depending on the context of the problem that gave rise to the equation $A\mathbf{x} = \mathbf{b}$, we might be happy with having $A\mathbf{x}_c - \mathbf{b}$ small or we might need $\mathbf{x}_c$ to be near $\mathbf{x}_t$.

### 2.3.1 Vector Norms

In the discussion above, we were forced in a natural way to use words like "large" and "small" to describe the size of a vector and words like "near" and "far" to describe the proximity of two vectors. Also, in subsequent material we will be developing methods which generate *sequences* of vectors, $\{\mathbf{x}^{(k)}\}_{k=1}^{\infty}$, which hopefully *converge* to some vector $\mathbf{x}$. In these methods we will need to have some idea of how "close" each $\mathbf{x}^{(k)}$ is to $\mathbf{x}$ in order that we may know how large $k$ must be so that $\mathbf{x}^{(k)}$ is an acceptable approximation to $\mathbf{x}$. This will tell us when to terminate the iteration that generates the sequence. Thus we must have some meaningful way to measure the size of a vector or the distance between two vectors. To do this, we extend the concept of *absolute value* or *magnitude* from the real numbers to vectors. For a vector

$$\mathbf{x} = \begin{bmatrix} x_1 \\ x_2 \\ \vdots \\ x_n \end{bmatrix} \tag{2.35}$$

we are already familiar with the Euclidean length $|\mathbf{x}| = \sqrt{(x_1)^2 + (x_2)^2 + \cdots + (x_n)^2}$, as a measure of size. It turns out, as we shall see, that there are other measures of size for a vector that are also *practical* to use in many computational problems. It is this realization that leads us to the definition of a *norm*.

To make our setting precise, let $R^n$ denote the set of all $n$-dimensional vectors with real components

$$R^n = \left\{ \mathbf{x} \middle| \mathbf{x} = \begin{bmatrix} x_1 \\ x_2 \\ \vdots \\ x_n \end{bmatrix} \quad x_1, x_2, \ldots, x_n \text{ real} \right\}. \tag{2.36}$$

A norm on $R^n$ is a real-valued function $\|\cdot\|$ defined on $R^n$, satisfying the three conditions of (2.37) below (where, as before, $\boldsymbol{\theta}$ denotes the zero vector in $R^n$):

$$\begin{aligned} &\text{a)}\ \|\mathbf{x}\| \geq 0, \text{ and } \|\mathbf{x}\| = 0 \text{ if and only if } \mathbf{x} = \boldsymbol{\theta}; \\ &\text{b)}\ \|\alpha\mathbf{x}\| = |\alpha|\ \|\mathbf{x}\|, \text{ for all scalars } \alpha \text{ and vectors } \mathbf{x}; \\ &\text{c)}\ \|\mathbf{x} + \mathbf{y}\| \leq \|\mathbf{x}\| + \|\mathbf{y}\|, \text{ for all vectors } \mathbf{x} \text{ and } \mathbf{y}. \end{aligned} \tag{2.37}$$

As noted above, the quantity $\|\mathbf{x}\|$ is thought of as being a measure of the size of the vector $\mathbf{x}$, and double bars are used to emphasize the distinction between the norm of a vector and the absolute value of a scalar. Three useful examples of norms are the so-called $\ell_p$ norms, $\|\cdot\|_p$, for $R^n$; $p = 1, 2, \infty$; given by

$$\begin{aligned} \|\mathbf{x}\|_1 &= |x_1| + |x_2| + \cdots + |x_n| \\ \|\mathbf{x}\|_2 &= \sqrt{(x_1)^2 + (x_2)^2 + \cdots + (x_n)^2} \\ \|\mathbf{x}\|_\infty &= \max\{|x_1|, |x_2|, \cdots, |x_n|\} \end{aligned} \tag{2.38}$$

**Example 2.11.** By way of illustration for $R^3$,

$$\mathbf{x} = \begin{bmatrix} 1 \\ -2 \\ -2 \end{bmatrix}, \quad \|\mathbf{x}\|_1 = 5, \quad \|\mathbf{x}\|_2 = 3, \quad \|\mathbf{x}\|_\infty = 2. \tag{2.39}$$

To further emphasize the properties of norms, we now show that the definition of the function $\|\mathbf{x}\|_1$ in (2.38) satisfies the three conditions of (2.37). Clearly, for each $\mathbf{x} \in R^n$, $\|\mathbf{x}\|_1 \geq 0$. The rest of (a) is also trivial, for if $\mathbf{x} = \boldsymbol{\theta}$, then

$$\mathbf{x} = \begin{bmatrix} 0 \\ 0 \\ \vdots \\ 0 \end{bmatrix} \quad \text{and} \quad \|\mathbf{x}\|_1 = 0.$$

Conversely, if

$$\mathbf{x} = \begin{bmatrix} x_1 \\ x_2 \\ \vdots \\ x_n \end{bmatrix} \quad \text{and} \quad \|\mathbf{x}\|_1 = 0,$$

then $|x_1| + |x_2| + \cdots + |x_n| = 0$ and hence $x_1 = x_2 = \cdots = x_n = 0$, or $\mathbf{x} = \boldsymbol{\theta}$. For part (b),

$$\alpha\mathbf{x} = \begin{bmatrix} \alpha x_1 \\ \alpha x_2 \\ \vdots \\ \alpha x_n \end{bmatrix}$$

so $\|\alpha\mathbf{x}\|_1 = |\alpha x_1| + |\alpha x_2| + \cdots + |\alpha x_n| = |\alpha| \, \|\mathbf{x}\|_1$. If

$$\mathbf{y} = \begin{bmatrix} y_1 \\ y_2 \\ \vdots \\ y_n \end{bmatrix}, \quad \text{then} \quad \mathbf{x} + \mathbf{y} = \begin{bmatrix} x_1 + y_1 \\ x_2 + y_2 \\ \vdots \\ x_n + y_n \end{bmatrix}$$

and therefore

$$\begin{aligned} \|\mathbf{x} + \mathbf{y}\|_1 &= |x_1 + y_1| + |x_2 + y_2| + \cdots + |x_n + y_n| \\ &\leq (|x_1| + |x_2| + \cdots + |x_n|) + (|y_1| + |y_2| + \cdots + |y_n|) \\ &= \|\mathbf{x}\|_1 + \|\mathbf{y}\|_1. \end{aligned}$$

Similarly (Problem 7) it is easy to show that $\|\cdot\|_\infty$ is a norm for $R^n$. The remaining $\ell_p$ norm, $\|\cdot\|_2$, is not handled quite so easily. In this case, the triangle inequality in condition (c) does not follow immediately from the triangle inequality for absolute values, and (c) is usually demonstrated with an application of the Cauchy-Schwarz inequality (see Section 2.6).

Having the concept of a norm, we can now make precise quantitative statements about size and distance, saying $A\mathbf{x}_c - \mathbf{b}$ is small if $\|A\mathbf{x}_c - \mathbf{b}\|$ is small and saying $\mathbf{x}_c$ is near $\mathbf{x}_t$ if $\|\mathbf{x}_c - \mathbf{x}_t\|$ is small. The definition of a norm also has some flexibility. For example, we might have reason to insist that the first coordinate of the residual vector $\mathbf{r} = A\mathbf{x}_c - \mathbf{b}$ is quite critical and must be small, even at the expense of growth in the other coordinates. In this case we might select a norm like the one below, where $\mathbf{x}$ is as in (2.35)

$$\|\mathbf{x}\| = \max\{10|x_1|, |x_2|, |x_3|, \ldots, |x_n|\}$$

A norm such as this emphasizes the first coordinate. For example, saying that $\|\mathbf{x}\| \leq 10^{-5}$ implies that $|x_i| \leq 10^{-5}$ for $i = 2, 3, \ldots, n$ and $|x_1| \leq 10^{-6}$. Weightings of this sort are quite common in problems that involve fitting curves to data and we shall see some examples when we discuss topics such as least-squares fits.

An idea related to norms which weight different coordinates differently is the concept of *relative error*. This is the simple and practical notion that when the size of the error $\mathbf{x}_c - \mathbf{x}_t$ is measured, we should take into account the size of the components of $\mathbf{x}_t$. That is, if $\mathbf{x}_t$ has components of the order of $10^4$, then an error of 0.01 is probably acceptable, while if $\mathbf{x}_t$ has components which are generally of the order of $10^{-4}$, then an error like 0.01 is disastrous. Thus if $\|\cdot\|$ is a norm for $R^n$, we define $\|\mathbf{x}_c - \mathbf{x}_t\|$ to be the *absolute error* and define the quantity $\|\mathbf{x}_c - \mathbf{x}_t\|/\|\mathbf{x}_t\|$ to be the *relative error*. We will usually be more interested in the relative error than the absolute error.

**Example 2.12.** Consider the two vectors

$$\mathbf{x}_t = \begin{bmatrix} 0.000397 \\ 0.000214 \\ 0.000309 \end{bmatrix} \quad \text{and} \quad \mathbf{x}_c = \begin{bmatrix} 0.000504 \\ 0.000186 \\ 0.000342 \end{bmatrix}.$$

Then the vector $\mathbf{x}_c - \mathbf{x}_t$ is given by

$$\mathbf{x}_c - \mathbf{x}_t = \begin{bmatrix} 0.000107 \\ -0.000028 \\ 0.000033 \end{bmatrix}.$$

One measure of the absolute error is $\|\mathbf{x}_c - \mathbf{x}_t\|_1 = 0.000168$, so $\mathbf{x}_c$ seems a reasonably good approximation to $\mathbf{x}_t$. However, checking the relative error, we see that

$$\frac{\|\mathbf{x}_c - \mathbf{x}_t\|_1}{\|\mathbf{x}_t\|_1} = \frac{0.000168}{0.000920} \approx 0.183$$

which more nearly reflects the true state of affairs, i.e., that our approximation $\mathbf{x}_c$ is in error by nearly 20 percent. Relative errors become particularly important in computer programs where a criterion is needed for terminating an iteration.

### 2.3.2 Matrix Norms

In Eq. (2.34) we see that in order to measure the size of $\mathbf{x}_c - \mathbf{x}_t$, we must also be able to measure the size of the *vector* $(A^{-1}\mathbf{r})$, since $\mathbf{x}_t$ is not known. If we knew $A^{-1}$, then we could simply multiply $A^{-1}$ times $\mathbf{r}$ and measure the size of $(A^{-1}\mathbf{r})$ by one of the vector norms of the previous section. However, since $A^{-1}$ is not known, we must use a different approach, since it would be quite inefficient to accurately compute $A^{-1}$ just to check the accuracy of $\mathbf{x}_c$. We are thus led to consider a way of measuring the "size" or "norms" of matrices as well as vectors. We shall do this in a way such that we are able to estimate the "size" of $A^{-1}$ by knowing the "size" of $A$ and then to estimate the norm of $(A^{-1}\mathbf{r})$. The concept of matrix norms is not limited to this particular problem, i.e., estimating $\|\mathbf{x}_c - \mathbf{x}_t\|$, but is practically indispensable in deriving computational procedures and error estimates for many other problems, as we shall see presently.

By way of notation, let $M_n$ denote the set of all $(n \times n)$ matrices and let $\mathcal{O}$ denote the $(n \times n)$ zero matrix. Then, a *matrix norm* for $M_n$ is a real-valued function $\|\cdot\|$ defined on $M_n$ satisfying for all $(n \times n)$ matrices $A$ and $B$:

$$\begin{aligned} &\text{a) } \|A\| \geq 0 \text{ and } \|A\| = 0 \text{ if and only if } A = \mathcal{O} \\ &\text{b) } \|\alpha A\| = |\alpha|\,\|A\| \text{ for any scalar } \alpha \\ &\text{c) } \|A + B\| \leq \|A\| + \|B\| \\ &\text{d) } \|AB\| \leq \|A\|\,\|B\|. \end{aligned} \tag{2.40}$$

The addition of condition d) should be noted. Thus matrix norms have a triangle inequality for both the operation of addition and multiplication.

Just as there are numerous ways of defining specific vector norms, there are also numerous ways of defining specific matrix norms. We will concentrate on three that are easily computable and are intrinsically related to the three basic vector norms discussed in the previous section. Specifically, if $A = (a_{ij}) \in M_n$, we define

$$\begin{aligned} \|A\|_1 &\equiv \operatorname*{Max}_{1 \leq j \leq n} \left[\sum_{i=1}^{n} |a_{ij}|\right] \text{—(Maximum absolute column sum)} \\ \|A\|_\infty &\equiv \operatorname*{Max}_{1 \leq i \leq n} \left[\sum_{j=1}^{n} |a_{ij}|\right] \text{—(Maximum absolute row sum)} \\ \|A\|_E &\equiv \sqrt{\sum_{i=1}^{n}\sum_{j=1}^{n} (a_{ij})^2}. \end{aligned} \tag{2.41}$$

**Example 2.13.** Let

$$A = \begin{bmatrix} 0 & 0 & 10 & 0 \\ 1 & 1 & 5 & 1 \\ 0 & 1 & 5 & 1 \\ 0 & 0 & 5 & 1 \end{bmatrix},$$

then $\|A\|_1 = \text{Max}\{1, 2, 25, 3\} = 25$, $\|A\|_\infty = \text{Max}\{10, 8, 7, 6\} = 10$, and $\|A\|_E = \sqrt{181} \approx 13.454$.

These three norms are stressed here because they are *compatible* with the $\ell_1$, $\ell_\infty$, and $\ell_2$ vector norms, respectively. This means that given any matrix

$A \in M_n$, then it is true for *all* vectors $\mathbf{x} \in R^n$, that

$$\|A\mathbf{x}\|_1 \le \|A\|_1\|\mathbf{x}\|_1, \quad \|A\mathbf{x}\|_\infty \le \|A\|_\infty\|\mathbf{x}\|_\infty, \quad \text{and} \quad \|A\mathbf{x}\|_2 \le \|A\|_E\|\mathbf{x}\|_2. \qquad \textbf{(2.42)}$$

(The reader should be careful to distinguish between vector norms and matrix norms since they bear the same subscript notation in two of the three cases. This distinction is clear from the usage. For example, $A\mathbf{x}$ is a vector so $\|A\mathbf{x}\|_1$ denotes the use of the $\|\cdot\|_1$ vector norm, whereas $\|A\|_1$ denotes use of the matrix norm. The reader should also note that the pairs of matrix and vector norms are compatible only in the orders given by (2.42) and cannot be mixed. For example, let $A$ be given as in the above example and let $\mathbf{x} = (0, 0, 1, 0)^T$. Then $\|\mathbf{x}\|_1 = 1$, but $\|A\mathbf{x}\|_1 = 25$ and $\|A\|_\infty\|\mathbf{x}\|_1 = 10$. That is, we cannot expect to have the inequality $\|A\mathbf{x}\|_1 \le \|A\|_\infty\|\mathbf{x}\|_\infty$ or $\|A\mathbf{x}\|_1 \le \|A\|_\infty\|\mathbf{x}\|_1$.)

Compatibility is a property that connects vector norms and matrix norms together. For example, in (2.34) we had $(\mathbf{x}_c - \mathbf{x}_t) = A^{-1}\mathbf{r}$. Using the idea of compatible vector and matrix norms, we can estimate $\|\mathbf{x}_c - \mathbf{x}_t\|_1$ in terms of $\|A^{-1}\|_1$ and $\|\mathbf{r}\|_1$:

$$\|\mathbf{x}_c - \mathbf{x}_t\|_1 = \|A^{-1}\mathbf{r}\|_1 \le \|A^{-1}\|_1\|\mathbf{r}\|_1.$$

As we shall see in Section 2.4, compatibility is also crucial to a clear understanding of iterative methods.

Thus far we have not shown that the three matrix norms satisfy the compatibility properties, (2.42), or even that their definitions given by (2.41) satisfy the necessary norm properties given by (2.40). For the sake of brevity we shall supply only the necessary proofs for the $\|\cdot\|_1$ norm, leaving the $\|\cdot\|_\infty$ and $\|\cdot\|_E$ norms to the reader.

Let $A = (a_{ij}) \in M_n$ and write $A$ in terms of its column vectors, $A = [\mathbf{A}_1, \mathbf{A}_2, \ldots, \mathbf{A}_n]$. Let $\mathbf{x} = (x_1, x_2, \ldots, x_n)^T$ be any vector in $R^n$, and recall from Section 2.1 that $A\mathbf{x}$ may be written as $Ax = x_1\mathbf{A}_1 + x_2\mathbf{A}_2 + \cdots + x_n\mathbf{A}_n$. Since $A\mathbf{x}$ is an $(n \times 1)$ vector, we use (2.37) to get

$$\begin{aligned} \|A\mathbf{x}\|_1 &= \|x_1\mathbf{A}_1 + x_2\mathbf{A}_2 + \cdots + x_n\mathbf{A}_n\|_1 \\ &\le \|x_1\mathbf{A}_1\|_1 + \|x_2\mathbf{A}_2\|_1 + \cdots + \|x_n\mathbf{A}_n\|_1 \\ &= |x_1|\,\|\mathbf{A}_1\|_1 + |x_2|\,\|\mathbf{A}_2\|_1 + \cdots + |x_n|\,\|\mathbf{A}_n\|_1 \qquad \textbf{(2.43)} \\ &\le (|x_1| + |x_2| + \cdots + |x_n|)\Big(\max_{1\le j\le n}\|\mathbf{A}_j\|_1\Big) \equiv \|A\|_1\|\mathbf{x}\|_1. \end{aligned}$$

Thus we have shown compatibility for the $\|\cdot\|_1$ norm.

It is trivial to see that parts (a) and (b) of (2.40) hold for $\|\cdot\|_1$. Since the $i$th column of $A + B$ is precisely the $i$th column of $A$ plus the $i$th column of $B$, part (c) of (2.40) is also easily seen to be true. For part (d) of (2.40), we recall from Section 2.1 that the $i$th column of $AB$ equals $A\mathbf{B}_i$, where $\mathbf{B}_i$ is the $i$th column of $B$. By compatibility, $\|A\mathbf{B}_i\|_1 \le \|A\|_1\|\mathbf{B}_i\|_1$. Now choose $i$ such that $\|\mathbf{B}_i\|_1 \ge \|\mathbf{B}_j\|_1$ for

$1 \le j \le n$. Then $\|B\|_1 = \|\mathbf{B}_i\|_1$, and

$$\|AB\|_1 = \max_{1 \le j \le n} \|A\mathbf{B}_j\|_1 \le \max_{1 \le j \le n} \|A\|_1 \|\mathbf{B}_j\|_1 = \|A\|_1 \|\mathbf{B}_i\|_1 = \|A\|_1 \|B\|_1,$$

and thus part (d) of (2.40) is satisfied. We have therefore shown that $\|A\|_1 = \max_{1 \le j \le n} \sum_{i=1}^{n} |a_{ij}|$ is a matrix norm and that it is compatible with the $\|\cdot\|_1$ vector norm.

We conclude this material with an observation on matrix norms, considering the three numbers $K_p$; $p = 1, 2, \infty$; where if $A \in M_n$ is given, then

$$K_p = \inf\{K \in R^1 : \|A\mathbf{x}\|_p \le K\|\mathbf{x}\|_p, \text{ for all } \mathbf{x} \in R^n\} \tag{2.44}$$

(where "inf" denotes *infimum* or *greatest lower bound*). It can be shown that $K_1 = \|A\|_1$ and $K_\infty = \|A\|_\infty$, although it goes beyond our purposes to do so. We introduce these numbers, however, since $K_2 \ne \|A\|_E$, and this explains why we use the subscript "$E$" instead of "2." Thus, although we have $\|A\mathbf{x}\|_2 \le \|A\|_E \|\mathbf{x}\|_2$ (compatibility), there is a matrix norm $K_2 \equiv \|A\|_2$, smaller for most matrices than $\|A\|_E$, such that $\|A\mathbf{x}\|_2 \le \|A\|_2 \|\mathbf{x}\|_2$, for all $\mathbf{x} \in R^n$. $\|A\|_2$ is rather unwieldly in computations and involves some deeper theory to derive, and so we are satisfied to use the easily computable and compatible $\|A\|_E$ in place of $\|A\|_2$ (see Theorem 3.11).

### 2.3.3 Condition Numbers and Error Estimates

In this section, we use the ideas of matrix and vector norms to provide some more information that is useful in helping to determine how good a computed (approximate) solution to the system $A\mathbf{x} = \mathbf{b}$ is. The norms used below can be any pair of compatible matrix and vector norms; for convenience we have omitted the subscripts. The reader can tell from the context whether a particular norm is a matrix norm or a vector norm. The following theorem provides valuable information with respect to the relative error.

**Theorem 2.1.** Suppose $A \in M_n$ is non-singular and $\mathbf{x}_c$ is an approximation to $\mathbf{x}_t$, the exact solution of $A\mathbf{x} = \mathbf{b}$, where $\mathbf{b} \ne \mathbf{\theta}$. Then for any compatible matrix and vector norms

$$\frac{1}{\|A\|\,\|A^{-1}\|} \frac{\|A\mathbf{x}_c - \mathbf{b}\|}{\|\mathbf{b}\|} \le \frac{\|\mathbf{x}_c - \mathbf{x}_t\|}{\|\mathbf{x}_t\|} \le \|A\|\,\|A^{-1}\| \frac{\|A\mathbf{x}_c - \mathbf{b}\|}{\|\mathbf{b}\|}. \tag{2.45}$$

*Proof*: Again by (2.34), $\mathbf{x}_c - \mathbf{x}_t = A^{-1}\mathbf{r}$, where $\mathbf{r} = A\mathbf{x}_c - \mathbf{b}$. Thus by the compatibility conditions, (2.42), $\|\mathbf{x}_c - \mathbf{x}_t\| \le \|A^{-1}\|\,\|\mathbf{r}\| = \|A^{-1}\|\,\|A\mathbf{x}_c - \mathbf{b}\|$. Now, $A\mathbf{x}_t = \mathbf{b}$, so $\|A\|\,\|\mathbf{x}_t\| \ge \|\mathbf{b}\|$, and $\|A\|/\|\mathbf{b}\| \ge 1/\|\mathbf{x}_t\|$. Thus,

$$\frac{\|\mathbf{x}_c - \mathbf{x}_t\|}{\|\mathbf{x}_t\|} \le \|A^{-1}\|\,\|A\mathbf{x}_c - \mathbf{b}\| \frac{\|A\|}{\|\mathbf{b}\|},$$

establishing the right-hand side of (2.45). Now,

$$\|A\mathbf{x}_c - \mathbf{b}\| = \|\mathbf{r}\| = \|A\mathbf{x}_c - A\mathbf{x}_t\| \le \|A\|\,\|\mathbf{x}_c - \mathbf{x}_t\|,$$

and, since $\|A\| > 0$,

$$\|\mathbf{x}_c - \mathbf{x}_t\| \ge \|A\mathbf{x}_c - \mathbf{b}\|/\|A\|.$$

Also, $\mathbf{x}_t = A^{-1}\mathbf{b}$, so $\|\mathbf{x}_t\| \le \|A^{-1}\|\,\|\mathbf{b}\|$, or $1/\|\mathbf{x}_t\| \ge 1/\|A^{-1}\|\,\|\mathbf{b}\|$. Combining these last two inequalities establishes the left-hand side of (2.45). ■

Note the appearance of the term $\|A\|\,\|A^{-1}\|$ in both the upper and lower bounds for the relative error. This term is usually called the *condition number* and its size is a measure of how good a solution we can expect direct methods to generate. It is easy to show (Problem 8) that the condition number satisfies the inequality: $\|A\|\,\|A^{-1}\| \ge 1$. The number $\|A^{-1}\|$ is valuable not only in terms of the condition number, but also in the "primitive" error estimate of (2.34), which we rephrase using compatible norms as

$$\|\mathbf{x}_c - \mathbf{x}_t\| \le \|A^{-1}\|\,\|\mathbf{r}\|. \tag{2.46}$$

Since we can compute $\|\mathbf{r}\|$, any information about $\|A^{-1}\|$ can be utilized, but as mentioned above, $A^{-1}$ is not normally available. A way to get a lower estimate for $\|A^{-1}\|$ is provided by noting that for any $\mathbf{x}$ in $R^n$,

$$\|\mathbf{x}\| = \|A^{-1}A\mathbf{x}\| \le \|A^{-1}\|\,\|A\mathbf{x}\| \tag{2.47}$$

and thus

$$\frac{\|\mathbf{x}\|}{\|A\mathbf{x}\|} \le \|A^{-1}\|. \tag{2.48}$$

**Example 2.14.** Consider the linear system

$$\begin{aligned} 6x_1 + 6x_2 + 3.00001x_3 &= 30.00002 \\ 10x_1 + 8x_2 + 4.00003x_3 &= 42.00006 \\ 6x_1 + 4x_2 + 2.00002x_3 &= 22.00004 \end{aligned}$$

which has the unique solution $x_1 = 1$, $x_2 = 3$, $x_3 = 2$. This example serves to illustrate the notion of an "ill-conditioned" system as well as serving as an example of how we can use matrix norms to analyze the results of a computational solution. (Recall: A system of linear equations is called ill-conditioned if "small" changes in the coefficients produce "large" changes in the solution.) Before solving this system, note that the perturbed system below

$$\begin{aligned} 6x_1 + 6x_2 + 3.00001x_3 &= 30 \\ 10x_1 + 8x_2 + 4.00003x_3 &= 42 \\ 6x_1 + 4x_2 + 2.00002x_3 &= 22 \end{aligned}$$

has a unique solution $x_1 = 1, x_2 = 4, x_3 = 0$, which is substantially different from the solution of the first system, even though the coefficient matrices are the same and the constants on the right-hand side are the same through six significant figures. Therefore we call these two systems ill-conditioned.

The first system also demonstrates that we should be cautious about how we determine whether an estimate to a solution is a good estimate or not. If we try $x_1 = 1$, $x_2 = 4$, and $x_3 = 0$ in the first system, we obtain the residual vector

$$\mathbf{r} = \begin{bmatrix} -2. \times 10^{-5} \\ -6. \times 10^{-5} \\ -4. \times 10^{-5} \end{bmatrix}$$

whereas the actual error in using this estimate is on the order of $10^5$ times as large as the residual vector would indicate. By (2.34) it follows that the inverse of this coefficient matrix has large entries.

To see what happens when Gauss elimination is used to solve the first system of equations, a single precision Gauss elimination routine was employed, and found

$$\begin{bmatrix} x_1 \\ x_2 \\ x_3 \end{bmatrix} = \begin{bmatrix} 0.9999898 \\ 1.907699 \\ 4.184615 \end{bmatrix}, \quad \mathbf{r} = \begin{bmatrix} 0.000019 \\ 0.000076 \\ 0.000049 \end{bmatrix}, \quad \mathbf{x}_t - \mathbf{x}_c = \begin{bmatrix} 0.000010 \\ 1.092301 \\ -2.184615 \end{bmatrix}$$

These results are typical if this system is solved on any digital device which has six to eight place accuracy. A double precision computation on a computer would give almost exactly the correct answer, but going to double precision is obviously not a cure-all, for there are simple examples like the one above for which double-precision arithmetic is not sufficient.

While this example is somewhat contrived (so that the essence of the problem is not hidden in a mass of cumbersome calculations), there are many real-life problems which are ill-conditioned, particularly in areas such as statistical analysis and least-squares fits. Thus it is appropriate that a person who must deal with numerical solutions of linear equations have as many tools at hand as possible in order to test computed results for accuracy. Often a very reliable test is simply a feeling for what the computed results are supposed to represent physically, i.e., do the answers fit the physical problem? In the absence of a physical intuition for what the answer should be, or in terms of a tool for a mathematical analysis of the significance of the computed results, the estimates of Theorem 2.1, (2.46), and (2.48) provide a beginning.

We continue now with Example 2.14 to illustrate how the $\|\cdot\|_1$ and $\|\cdot\|_\infty$ norms can be readily used to analyze errors in computed solutions. Thus $A\mathbf{x} = \mathbf{b}$ has $\mathbf{x}_t$ as its solution vector, where

$$A = \begin{bmatrix} 6 & 6 & 3.00001 \\ 10 & 8 & 4.00003 \\ 6 & 4 & 2.00002 \end{bmatrix}, \quad \mathbf{b} = \begin{bmatrix} 30.00002 \\ 42.00006 \\ 22.00004 \end{bmatrix}, \quad \mathbf{x}_t = \begin{bmatrix} 1 \\ 3 \\ 2 \end{bmatrix}. \tag{2.49}$$

In order not to obscure the point of this example, let us suppose that a computed estimate to the solution, $\mathbf{x}_c$, and hence the residual $\mathbf{r} = A\mathbf{x}_c - \mathbf{b}$, are given by

$$\mathbf{x}_c = \begin{bmatrix} 1 \\ 4 \\ 0 \end{bmatrix} \quad \text{and} \quad \mathbf{r} = \begin{bmatrix} -0.00002 \\ -0.00006 \\ -0.00004 \end{bmatrix} \tag{2.50}$$

For an analysis of $\mathbf{x}_c - \mathbf{x}_t$, let us use the $\ell_\infty$ norm and carry only three significant figures. Thus

$$\frac{\|A\mathbf{x}_c - \mathbf{b}\|_\infty}{\|\mathbf{b}\|_\infty} \approx \frac{6 \times 10^{-5}}{42} \approx 1.43 \times 10^{-6}. \tag{2.51}$$

Clearly, $\|A\|_\infty = 22.00003 \approx 22$, and therefore by Theorem 2.1,

$$\frac{\|\mathbf{x}_c - \mathbf{x}_t\|_\infty}{\|\mathbf{x}_t\|_\infty} \leq (22.)(1.43 \times 10^{-6})\|A^{-1}\|_\infty \leq (3.15 \times 10^{-5})\|A^{-1}\|_\infty. \tag{2.52}$$

Hence, a guess at $\|A^{-1}\|_\infty$ is in order, and we can use (2.48) to obtain lower bounds for $\|A^{-1}\|_\infty$.

Since we are estimating $\|A^{-1}\|_\infty$ from below, we will naturally want as large a lower bound as we can obtain. From the form of (2.48), it appears that we want to vector $\mathbf{x}$ such that $\|A\mathbf{x}\|_\infty$ is as small as possible, where $\mathbf{x} \neq \mathbf{0}$. Reference to (2.49) shows that for $\mathbf{x}'$ given by

$$\mathbf{x}' = \begin{bmatrix} 0 \\ -1 \\ 2 \end{bmatrix}, \qquad A\mathbf{x}' = \begin{bmatrix} 0.00002 \\ 0.00006 \\ 0.00004 \end{bmatrix}. \tag{2.53}$$

Thus a lower bound for $\|A^{-1}\|_\infty$ is provided by

$$\frac{\|\mathbf{x}'\|_\infty}{\|A\mathbf{x}'\|_\infty} = \frac{2}{6. \times 10^{-5}} \approx 3.33 \times 10^4 \leq \|A^{-1}\|_\infty. \tag{2.54}$$

This estimate for $\|A^{-1}\|_\infty$ makes the estimate of (2.52) more meaningful, for now the upper bound for the relative error in (2.52) is at least as large as $(3.15 \times 10^{-5})(3.33 \times 10^4) \approx 1.05$ while, in fact,

$$\frac{\|\mathbf{x}_c - \mathbf{x}_t\|_\infty}{\|\mathbf{x}_t\|_\infty} = \frac{2}{3}.$$

We will make further use of matrix norms in subsequent sections, in such topics as iterative procedures to solve $A\mathbf{x} = \mathbf{b}$, methods for finding eigenvalues of matrices, methods for solving nonlinear systems, methods of optimization, etc.

### 2.3.4 Iterative Improvement

Iterative improvement (sometimes known as the "method of residual correction") is a procedure which is often recommended for refining a computed solution, $\mathbf{x}_c$, of the system $A\mathbf{x} = \mathbf{b}$. The procedure is relatively simple to implement and frequently will yield improved estimates of the solution. In order to describe this method, we again let $\mathbf{x}_t$ denote the true solution of $A\mathbf{x} = \mathbf{b}$ and use $\mathbf{r} = A\mathbf{x}_c - \mathbf{b}$ to denote the residual vector (where $\mathbf{x}_c$ is a computed (approximate) solution to $A\mathbf{x} = \mathbf{b}$). If we let $\mathbf{e} = \mathbf{x}_c - \mathbf{x}_t$, then $A\mathbf{e} = A\mathbf{x}_c - A\mathbf{x}_t = A\mathbf{x}_c - \mathbf{b} = \mathbf{r}$. Thus if we could solve $A\mathbf{e} = \mathbf{r}$, we would be able to find $\mathbf{x}_t$ from the equation $\mathbf{x}_t = \mathbf{x}_c - \mathbf{e}$. The method of iterative improvement is precisely the implementation of this idea. We immediately see the difficulty inherent in such a procedure, since if we cannot solve $A\mathbf{x} = \mathbf{b}$ exactly, we cannot expect to solve $A\mathbf{e} = \mathbf{r}$ exactly.

Iterative improvement, when implemented properly on the computer, attempts to overcome the objection noted above by calculating the residual, $\mathbf{r}$, in double precision. To be precise, the steps in iterative improvement are:

a) Calculate $\mathbf{r} = A\mathbf{x}_c - \mathbf{b}$ in double precision;

b) Solve $A\mathbf{e} = \mathbf{r}$ and let $\mathbf{e}_c$ be the computed solution to this system;

c) Let $\mathbf{x}'_c = \mathbf{x}_c - \mathbf{e}_c$.

We hope that $\mathbf{x}'_c$ is a better approximation to $\mathbf{x}_t$ than was $\mathbf{x}_c$. In order to analyze the method a bit more carefully, let $\mathbf{e}_t$ denote the true solution to $A\mathbf{e} = \mathbf{r}$. First, we can obtain some qualitative information from the steps outlined above by noting that

$$\frac{\|\mathbf{e}_t\|}{\|\mathbf{x}_t\|} = \frac{\|\mathbf{x}_c - \mathbf{x}_t\|}{\|\mathbf{x}_t\|}.$$

Thus, we can hope that given the numbers we have, namely $\|\mathbf{e}_c\|$ and $\|\mathbf{x}_c\|$, that the ratio

$$\frac{\|\mathbf{e}_c\|}{\|\mathbf{x}_c\|}$$

might give an indication of the relative error.

We next note that the relative error of solving $A\mathbf{e} = \mathbf{r}$ is given by

$$\frac{\|\mathbf{e}_c - \mathbf{e}_t\|}{\|\mathbf{e}_t\|}.$$

From (c) we have that $\mathbf{e}_c = \mathbf{x}_c - \mathbf{x}'_c$, so (since $\mathbf{e}_t = \mathbf{x}_c - \mathbf{x}_t$) it follows that

$$\frac{\|\mathbf{e}_c - \mathbf{e}_t\|}{\|\mathbf{e}_t\|} = \frac{\|\mathbf{x}_t - \mathbf{x}'_c\|}{\|\mathbf{x}_t - \mathbf{x}_c\|}.$$

Therefore if we can solve $A\mathbf{e} = \mathbf{r}$ with a relative error less than 1, then $\mathbf{x}'_c$ is a better approximation to $\mathbf{x}_t$ than is $\mathbf{x}_c$.

Finally, we recall from (2.45), that we have an error bound for the relative error of the form

$$\frac{\|\mathbf{x}_c - \mathbf{x}_t\|}{\|\mathbf{x}_t\|} \leq M\|A\mathbf{x}_c - \mathbf{b}\| \tag{2.55a}$$

where $M = \|A\|\,\|A^{-1}\|/\|\mathbf{b}\|$. We can find a similar bound for $\|\mathbf{x}'_c - \mathbf{x}_t\|/\|\mathbf{x}_t\|$ as follows. Let $\mathbf{r}_c = A\mathbf{e}_c - \mathbf{r}$, so that $A^{-1}\mathbf{r}_c = \mathbf{e}_c - \mathbf{e}_t$ and hence, $\|A^{-1}\|\,\|\mathbf{r}_c\| \geq \|\mathbf{e}_c - \mathbf{e}_t\|$. But, as above, $\|\mathbf{e}_c - \mathbf{e}_t\| = \|\mathbf{x}_t - \mathbf{x}'_c\| = \|\mathbf{x}'_c - \mathbf{x}_t\|$, so $\|\mathbf{x}'_c - \mathbf{x}_t\| \leq \|A^{-1}\|\,\|A\mathbf{e}_c - \mathbf{r}\|$. Just as in Theorem 2.1, $\|A\|\,\|\mathbf{x}_t\| \geq \|\mathbf{b}\|$, so we obtain $1/\|\mathbf{x}_t\| < \|A\|/\|\mathbf{b}\|$. Using these two inequalities, we find

$$\frac{\|\mathbf{x}'_c - \mathbf{x}_t\|}{\|\mathbf{x}_t\|} \leq \frac{\|A\|\,\|A^{-1}\|}{\|\mathbf{b}\|}\|A\mathbf{e}_c - \mathbf{r}\|,$$

which is an inequality of the form

$$\frac{\|\mathbf{x}_c' - \mathbf{x}_t\|}{\|\mathbf{x}_t\|} \leq M\|A\mathbf{e}_c - \mathbf{r}\| \tag{2.55b}$$

where M is the same constant that appears in (2.55a). We can calculate both the numbers $\|A\mathbf{x}_c - \mathbf{b}\|$ and $\|A\mathbf{e}_c - \mathbf{r}\|$, and, if $\|A\mathbf{e}_c - \mathbf{r}\| < \|A\mathbf{x}_c - \mathbf{b}\|$, we expect that $\mathbf{x}_c'$ might be a relatively better approximation to $\mathbf{x}_t$ than is $\mathbf{x}_c$. Notice that since $\mathbf{x}_c'$ is an approximation to $\mathbf{x}_t$ just as was $\mathbf{x}_c$, we can apply iterative improvement again, hoping to improve $\mathbf{x}_c'$ by solving $A\mathbf{e} = A\mathbf{x}_c' - \mathbf{b}$. If the corrections, $\mathbf{e}_c^{(i)}$, that we obtain in this fashion do not decrease in size, then we have a strong indication that the matrix $A$ is ill-conditioned. We also notice that the techniques described in Section 2.2.3 are relevant to iterative improvement, since we are solving several systems which have the same coefficient matrix. The subroutine

```
      SUBROUTINE RESCOR(A,B,XC,N,M)
      DOUBLE PRECISION RES(20)
      DIMENSION A(20,20),SA(20,20),B(20),XC(20),SRES(20),E(20)
C
C  SUBROUTINE RESCOR USES ITERATIVE IMPROVEMENT TO REFINE AN APPROXIMATE
C  SOLUTION, XC, TO THE SYSTEM AX=B.  THE CALLING PROGRAM MUST SUPPLY
C  THE MATRIX A, THE APPROXIMATE SOLUTION VECTOR XC, THE VECTOR B, AN
C  INTEGER N (WHERE A IS (NXN)) AND AN INTEGER M=NUMBER OF TIMES
C  XC IS TO BE REFINED.
C
      DO 1 I=1,N
      DO 1 J=1,N
    1 SA(I,J)=A(I,J)
    2 CONTINUE
      DO 3 I=1,N
      DO 3 J=1,N
    3 A(I,J)=SA(I,J)
C
C  COMPUTE RESIDUAL, RES=A*XC-B
C
      DO 4 I=1,N
      RES(I)=-B(I)
      DO 4 J=1,N
    4 RES(I)=RES(I)+A(I,J)*XC(J)
      DO 5 I=1,N
    5 SRES(I)=RES(I)
C
C  SOLVE A*E=RES AND UPDATE XC TO XC=XC-E
C
      CALL GAUS(A,SRES,E,N,IERROR)
      DO 6 I=1,N
    6 XC(I)=XC(I)-E(I)
C
C  TEST TO SEE IF RESIDUAL CORRECTION SHOULD BE
C  APPLIED TO THE UPDATED APPROXIMATION, XC
C
      M=M-1
      IF(M.GT.0)   GO TO 2
      RETURN
      END
```

Figure 2.3

given above uses iterative improvement to refine approximate solutions to the system $A\mathbf{x} = \mathbf{b}$.

To illustrate iterative improvement, subroutine RESCOR was used on the system in Example 2.7. The approximate solution, $\mathbf{x}_c$, was calculated using subroutine GAUS, as in Example 2.7. Displayed below are the results of two applications of iterative improvement. That is, $\mathbf{x}'_c$ is the result of solving $A\mathbf{e} = \mathbf{r}'$, where $\mathbf{r}' = A\mathbf{x}_c - \mathbf{b}$ and $\mathbf{x}''_c$ is the result of solving $A\mathbf{e} = \mathbf{r}''$, where $\mathbf{r}'' = A\mathbf{x}'_c - \mathbf{b}$.

$$\mathbf{x}_c = \begin{bmatrix} 0.999417\text{E}\ \ 00 \\ 0.100035\text{E}\ \ 01 \\ 0.100015\text{E}\ \ 01 \\ 0.999910\text{E}\ \ 00 \end{bmatrix} \quad \mathbf{x}'_c = \begin{bmatrix} 0.100004\text{E}\ \ 01 \\ 0.999977\text{E}\ \ 00 \\ 0.999991\text{E}\ \ 00 \\ 0.100000\text{E}\ \ 01 \end{bmatrix} \quad \mathbf{x}''_c = \begin{bmatrix} 0.100000\text{E}\ \ 01 \\ 0.100000\text{E}\ \ 01 \\ 0.100000\text{E}\ \ 01 \\ 0.100000\text{E}\ \ 01 \end{bmatrix}$$

Thus two steps of the iterative improvement algorithm give the answer correct to as many places as are printed.

## PROBLEMS

1. Let $A$ be the $(4 \times 4)$ coefficient matrix of the system in Example 2.6. Then $A^{-1}$ is given by

$$A^{-1} = \begin{bmatrix} 68 & -41 & -17 & 10 \\ -41 & 25 & 10 & -6 \\ -17 & 10 & 5 & -3 \\ 10 & -6 & -3 & 2 \end{bmatrix}$$

Find the condition number $\|A\|_\infty \|A^{-1}\|_\infty$ of this matrix. Use the inequality (2.45) of Theorem 2.1, and the computer results of Example 2.7, to estimate the relative error $\|\mathbf{x}_c - \mathbf{x}_t\|_\infty / \|\mathbf{x}_t\|_\infty$. What is the true relative error?

2. Let $A_n$ be the $(2 \times 2)$ matrix given by

$$A_n = \begin{bmatrix} 1 & 2 \\ 2 & 4 + 1/n^2 \end{bmatrix}.$$

Find $A_n^{-1}$ and the condition number $\|A_n\|_\infty \|A_n^{-1}\|_\infty$. Let $n = 100$, so that

$$A_n = \begin{bmatrix} 1 & 2 \\ 2 & 4.0001 \end{bmatrix}$$

and let

$$\mathbf{b} = \begin{bmatrix} 1 \\ 2 - 1/n^2 \end{bmatrix} = \begin{bmatrix} 1 \\ 1.9999 \end{bmatrix}.$$

Solve $A_{100}\mathbf{x} = \mathbf{b}$ mathematically and call the answer $\mathbf{x}_t$. Let $\mathbf{x}_c$ be given by

$$\mathbf{x}_c = \begin{bmatrix} 1 \\ 0 \end{bmatrix}.$$

Find $\mathbf{r} = A_{100}\mathbf{x}_c - \mathbf{b}$ and check the error bound $\|\mathbf{x}_c - \mathbf{x}_t\|_\infty \le \|A^{-1}\|_\infty \|\mathbf{r}\|_\infty$. One might expect from this problem and from Example 2.14 that ill-conditioned matrices must have small determinants. Find the determinant of the matrix $A$ in Problem 1 to see that this suspicion is not always valid.

3. For the matrix $A$ of Problem 1, find a $(4 \times 1)$ vector $\mathbf{x}$, $\mathbf{x} \neq \mathbf{0}$, such that $\|A\mathbf{x}\|_\infty = \|A\|_\infty \|\mathbf{x}\|_\infty$. (Hint: Consider only vectors $\mathbf{x}$ such that $\|\mathbf{x}\|_\infty = 1$.) Find a $(4 \times 1)$ vector $\mathbf{y}$, $\mathbf{y} \neq \mathbf{0}$, such that $\|A\mathbf{y}\|_1 = \|A\|_1 \|\mathbf{y}\|_1$.
4. Let $I$ be the $(n \times n)$ identity matrix. What is $\|I\|_E$? Is there any $(n \times 1)$ vector $\mathbf{x}$, $\mathbf{x} \neq \mathbf{0}$, such that $\|I\mathbf{x}\|_2 = \|I\|_E \|\mathbf{x}\|_2$?

*5. Let $\mathbf{x}$ be any vector in $R^n$ as given by (2.36), and let $\{\mathbf{x}^{(j)}\}_{j=1}^{\infty}$ be any sequence of vectors in $R^n$.
   a) Show that $\|\mathbf{x}\|_1 \ge \|\mathbf{x}\|_2 \ge \|\mathbf{x}\|_\infty \ge (1/n)\|\mathbf{x}\|_1$, for all $\mathbf{x} \in R^n$.
   b) From the definition of convergence of a sequence of real numbers, use Part (a) to show that if $\lim_{j\to\infty} \|\mathbf{x}^{(j)}\|_p = 0$ for $p = 1, 2$, or $\infty$; then $\lim_{j\to\infty} \|\mathbf{x}^{(j)}\|_q = 0$ for $q = 1, 2$, or $\infty$; $q \neq p$.
   c) For $p$ given as 1, 2, or $\infty$; show that $\lim_{j\to\infty} \|\mathbf{x}^{(j)} - \mathbf{x}\|_p = 0$ implies that

$$\lim_{j\to\infty} \|A(\mathbf{x}^{(j)} - \mathbf{x})\|_p \equiv \lim_{j\to\infty} \|A\mathbf{x}^{(j)} - A\mathbf{x}\|_p = 0.$$

6. Graph the sets of points,

$$\mathbf{x} = \begin{bmatrix} x_1 \\ x_2 \end{bmatrix} \in R^2,$$

   such that $\|\mathbf{x}\|_1 = 1$, $\|\mathbf{x}\|_2 = 1$, and $\|\mathbf{x}\|_\infty = 1$.
7. If $\|\cdot\|_\infty$ is given by (2.38), show that it satisfies all three properties of (2.37) and thus is a norm for $R^n$.
8. Let $\|\cdot\|_m$ be any matrix norm on $M_n$. Show that $\|A\|_m \|A^{-1}\|_m \ge 1$. (Hint: consider $\|I^2\|_m = \|I\|_m$.)

*9. Estimate $\|H_3^{-1}\|_\infty$ from below via (2.48), (as in Example 2.14, Eq. (2.54)), where $H_3$ is the $(3 \times 3)$ Hilbert matrix (see Problem 5, Section 2.2). You should be able to obtain a lower bound of at least 50.

## 2.4 ITERATIVE METHODS

There are instances when direct methods such as Gauss elimination and LU-decomposition may not be the best methods to use for solving the system of linear equations $A\mathbf{x} = \mathbf{b}$. There is an alternative class of methods that can be used for this problem; namely the *iterative* methods. We will first develop the concept of an iterative method and give some particular examples. Later we will discuss the relative merits of iterative and direct methods.

The methods presented here are called iterative because each method is designed to generate a sequence of vectors, (*iterates*), $\{\mathbf{x}^{(k)}\}_{k=0}^{\infty}$, which converge to the true solution, $\mathbf{x}_t$, of $A\mathbf{x} = \mathbf{b}$. The basic idea of iterative methods can be described as follows:

1) The matrix $A$ is written as the difference of two matrices $N$ and $P$, so that $A = N - P$. This decomposition of $A$ is called a *splitting*.
2) An initial guess $\mathbf{x}^{(0)}$ is made for the solution vector $\mathbf{x}_t$.
3) A sequence $\mathbf{x}^{(1)}, \mathbf{x}^{(2)}, \mathbf{x}^{(3)}, \ldots$, of estimates to $\mathbf{x}_t$ is generated by the formula

$$N\mathbf{x}^{(k+1)} = P\mathbf{x}^{(k)} + \mathbf{b}, \qquad k = 0, 1, 2, \ldots. \tag{2.56}$$

Since the idea of iteration is probably not too familiar, we present an example below before discussing iterative methods further.

**Example 2.15.** Let

$$A = \begin{bmatrix} 4 & 1 & 0 \\ 2 & 5 & 1 \\ -1 & 2 & 4 \end{bmatrix}, \qquad \mathbf{b} = \begin{bmatrix} 1 \\ 0 \\ 3 \end{bmatrix} \text{ and}$$

let $A\mathbf{x} = \mathbf{b}$ be the linear system to be solved. Let the splitting be given by

$$N = \begin{bmatrix} 4 & 0 & 0 \\ 0 & 5 & 0 \\ 0 & 0 & 4 \end{bmatrix} \quad \text{and} \quad P = \begin{bmatrix} 0 & -1 & 0 \\ -2 & 0 & -1 \\ 1 & -2 & 0 \end{bmatrix},$$

so that $A = N - P$. If we denote the vector $\mathbf{x}^{(k)}$ by $\mathbf{x}^{(k)} = \begin{bmatrix} x_k \\ y_k \\ z_k \end{bmatrix}$,

then writing out formula (2.56) we obtain the equations

$$\begin{aligned} 4x_{k+1} &= -y_k + 1 \\ 5y_{k+1} &= -2x_k - z_k \\ 4z_{k+1} &= x_k - 2y_k + 3 \end{aligned}$$

which define $\mathbf{x}^{(k+1)}$ for $k = 0, 1, 2, \ldots$. As an initial guess, let

$$\mathbf{x}^{(0)} = \begin{bmatrix} 1 \\ 1 \\ 1 \end{bmatrix}.$$

Then the first few iterates are

$$\mathbf{x}^{(0)} = \begin{bmatrix} 1 \\ 1 \\ 1 \end{bmatrix}, \qquad \mathbf{x}^{(1)} = \begin{bmatrix} 0 \\ -\frac{3}{5} \\ \frac{1}{2} \end{bmatrix}, \qquad \mathbf{x}^{(2)} = \begin{bmatrix} \frac{2}{5} \\ -\frac{1}{10} \\ \frac{21}{20} \end{bmatrix}.$$

(The true solution in this example is $\mathbf{x}_t = (1/3, -1/3, 1)^T$.)

Let us return now to the general procedure outlined in steps (1) to (3). The first thing to observe is that if $A\mathbf{x}_t = \mathbf{b}$ then $N\mathbf{x}_t = P\mathbf{x}_t + \mathbf{b}$ and vice versa (that is, solving $A\mathbf{x} = \mathbf{b}$ is equivalent to solving $N\mathbf{x} = P\mathbf{x} + \mathbf{b}$). In order to get some idea of what might constitute a good choice $N$ and $P$, we consider formula (2.56). This formula says that if we have $\mathbf{x}^{(k)}$, then we can get the next iterate $\mathbf{x}^{(k+1)}$ provided we can solve the linear system $N\mathbf{x}^{(k+1)} = \mathbf{h}^{(k)}$ where the vector $\mathbf{h}^{(k)}$ is given by $\mathbf{h}^{(k)} = P\mathbf{x}^{(k)} + \mathbf{b}$. Thus it is clear that we must require $N$ to be non-singular in order to be assured that we can implement the iteration. Furthermore, for an iterative procedure to be efficient, $N$ should be chosen so that $N\mathbf{x}^{(k+1)} = \mathbf{h}^{(k)}$ is quite easy to solve. This is the case if, for instance, $N$ is chosen to be a triangular matrix (or a diagonal matrix as in Example 2.15 above).

Lastly, the question of convergence to $\mathbf{x}_t$ must be considered. It is fairly easy to make a start at answering this question. Let $\mathbf{e}^{(k)} = \mathbf{x}^{(k)} - \mathbf{x}_t$ denote the error vector at the $k$th step. As noted above, $N\mathbf{x}_t = P\mathbf{x}_t + \mathbf{b}$, so combining this with (2.56) it follows that $N(\mathbf{x}^{(k+1)} - \mathbf{x}_t) = P(\mathbf{x}^{(k)} - \mathbf{x}_t)$ or $N\mathbf{e}^{(k+1)} = P\mathbf{e}^{(k)}$. Since we have required that $N$ be non-singular, we can multiply by $N^{-1}$ to obtain $\mathbf{e}^{(k+1)} = N^{-1}P\mathbf{e}^{(k)}$, for $k = 0, 1, 2, \ldots$. If we set $M \equiv N^{-1}P$, then a fundamental relationship among the error vectors has been established, namely:

$$\mathbf{e}^{(k+1)} = M\mathbf{e}^{(k)}, \qquad k = 0, 1, 2, \ldots. \tag{2.57}$$

If $M$ is in some sense a "small" matrix, then (2.57) would indicate that the errors are diminishing, or that $\{\mathbf{x}^{(k)}\} \to \mathbf{x}_t$. This is made more precise in Theorem 2.2 below. In this theorem, we use the $\ell_\infty$ vector and matrix norms. It is evident from the proof, however, that we could as well have used any compatible vector and matrix norms.

**Theorem 2.2.** Suppose $A = N - P$ and suppose $\|N^{-1}P\|_\infty \le \lambda < 1$. Then:

a) $A$ is non-singular
b) If $\mathbf{x}_t$ is the solution of $A\mathbf{x} = \mathbf{b}$ and if $\{\mathbf{x}^{(j)}\}$ is given by (2.56) then $\lim_{j\to\infty} \mathbf{x}^{(j)} = \mathbf{x}_t$
c) $\|\mathbf{x}^{(j)} - \mathbf{x}_t\|_\infty \le \lambda^j \|\mathbf{x}^{(0)} - \mathbf{x}_t\|_\infty$ (or $\|\mathbf{e}^{(j)}\|_\infty \le \lambda^j\|\mathbf{e}^{(0)}\|_\infty$)

*Proof*: Before establishing Theorem 2.2, it is worth observing that a theorem of this type has a lot of practical computational value when it is possible to verify the hypotheses. The theorem

1. guarantees a solution to the problem,
2. shows that the numerical method will converge to the solution for any initial guess $\mathbf{x}^{(0)}$,
3. indicates to the user of the numerical method how many steps should be taken to attain a desired accuracy.

Part (a) is easiest to prove by contradiction. If $A$ is singular, then there must be a nonzero vector $\mathbf{y}$ such that $A\mathbf{y} = \mathbf{\theta}$. Therefore $(N - P)\mathbf{y} = \mathbf{\theta}$ or $\mathbf{y} = N^{-1}P\mathbf{y}$. Using the compatibility properties of the $\ell_\infty$ vector and matrix norms it follows that

$$\|\mathbf{y}\|_\infty = \|N^{-1}P\mathbf{y}\|_\infty \le \|N^{-1}P\|_\infty \|\mathbf{y}\|_\infty \le \lambda \|\mathbf{y}\|_\infty. \quad \textbf{(2.58)}$$

Since $\mathbf{y} \ne \mathbf{\theta}$, then $\|\mathbf{y}\|_\infty > 0$ and hence the inequality (2.58) means that $1 \le \lambda$. This is a contradiction of the hypothesis, so it must be that there is no nonzero vector $\mathbf{y}$ such that $A\mathbf{y} = \mathbf{\theta}$. Hence, $A$ is non-singular.

We next establish part (c). If we set $M = N^{-1}P$ and $\mathbf{e}^{(j)} = \mathbf{x}^{(j)} - \mathbf{x}_t$, then a repeated application of (2.57) gives

$$\mathbf{e}^{(j)} = M\mathbf{e}^{(j-1)} = M(M\mathbf{e}^{(j-2)}) = M^2\mathbf{e}^{(j-2)} = \cdots = M^j\mathbf{e}^{(0)}.$$

Since $\|\mathbf{e}^{(j)}\|_\infty \le \|M^j\|_\infty \|\mathbf{e}^{(0)}\|_\infty \le \|M\|_\infty^j \|\mathbf{e}^{(0)}\|_\infty \le \lambda^j \|\mathbf{e}^{(0)}\|_\infty$, part (c) of the theorem is proved. Since $\lim_{j\to\infty} \|\mathbf{x}^{(j)} - \mathbf{x}_t\|_\infty = 0$ and since $\|\mathbf{x}^{(j)} - \mathbf{x}_t\|_\infty$ is the absolute value of the largest component of the vector $\mathbf{x}^{(j)} - \mathbf{x}_t$, it follows that $\lim_{j\to\infty} \mathbf{x}^{(j)} = \mathbf{x}_t$. This proves part (b). (We note from Problem 5, Section 2.3, that $\|\mathbf{x}^{(j)} - \mathbf{x}_t\|_2 \to 0$ and $\|\mathbf{x}^{(j)} - \mathbf{x}_t\|_1 \to 0$ as well). ■

Theorem 2.2 together with the observations made previously allows us to summarize the properties that a splitting should have in order to define a useful iterative method. For $A = N - P$, these properties are:

a) $N$ should be non-singular;
b) the equation $N\mathbf{x} = \mathbf{h}$ should be easy to solve;
c) $\|N^{-1}P\|$ should be less than 1 for some matrix norm.

In this context, condition (a) assures us that the sequence $\{\mathbf{x}^{(j)}\}$ given by formula (2.56) can be generated. Condition (b) assures us that the sequence $\{\mathbf{x}^{(j)}\}$ can be generated efficiently (after all, we do not want to work as hard to perform one step of an iterative method as we would to solve $A\mathbf{x} = \mathbf{b}$ by a direct method). Lastly, condition (c) assures us that the sequence we generate will in fact converge to the solution $\mathbf{x}_t$.

### 2.4.1 Basic Iterative Methods

The first and simplest iterative method described in this section is the Jacobi method. For this method $N$ is taken to be a diagonal matrix, with its main diagonal entries equal to $a_{ii}$. The matrix $P$ is then determined by $P = N - A$. We note that $P$ can be visualized as the sum of a lower triangular matrix and an upper triangular matrix. For the purposes of analyzing the Jacobi method and the ensuing Gauss-Seidel method, it is quite convenient to think of $P$ in this way. To be specific, let $A = (a_{ij})$ be an $(n \times n)$ matrix. Define $L$, $D$, and $U$ to be the lower triangular, diagonal, and upper triangular parts of $A$:

$$L = \begin{bmatrix} 0 & 0 & 0 & \cdots & 0 \\ a_{21} & 0 & 0 & \cdots & 0 \\ a_{31} & a_{32} & 0 & \cdots & 0 \\ \vdots & & & & \vdots \\ a_{n1} & a_{n2} & a_{n3} & \cdots & 0 \end{bmatrix}, \qquad D = \begin{bmatrix} a_{11} & 0 & 0 & \cdots & 0 \\ 0 & a_{22} & 0 & \cdots & 0 \\ 0 & 0 & a_{33} & \cdots & 0 \\ \vdots & & & & \vdots \\ 0 & 0 & 0 & \cdots & a_{nn} \end{bmatrix},$$

$$U = \begin{bmatrix} 0 & a_{12} & a_{13} & \cdots & a_{1n} \\ 0 & 0 & a_{23} & \cdots & a_{2n} \\ 0 & 0 & 0 & \cdots & a_{3n} \\ \vdots & & & & \vdots \\ 0 & 0 & 0 & \cdots & 0 \end{bmatrix}.$$

Thus $A = L + D + U$ and so the Jacobi splitting is given by $N = D$ and $P = -(L + U)$.

The Jacobi method for solving $A\mathbf{x} = \mathbf{b}$ (that is, the splitting defined above) is thus:

$$D\mathbf{x}^{(k+1)} = -(L + U)\mathbf{x}^{(k)} + \mathbf{b}. \tag{2.59}$$

The matrix $M_J = -D^{-1}(L + U)$ is called the *Jacobi matrix*. In actual computation, Eq. (2.59) would have to be written out element-wise. Suppose the vector $\mathbf{x}^{(k)}$ is given by

$$\mathbf{x}^{(k)} = \begin{bmatrix} x_1^{(k)} \\ x_2^{(k)} \\ \vdots \\ x_n^{(k)} \end{bmatrix}, \qquad k = 0, 1, 2, \ldots. \tag{2.60}$$

Then Eq. (2.59) leads to the following iteration for the $i$th component of $\mathbf{x}^{(k)}$:

$$x_i^{(k+1)} = \frac{-1}{a_{ii}} \left[ \sum_{\substack{j=1 \\ j \neq i}}^{n} a_{ij}x_j^{(k)} - b_i \right], \qquad i = 1, 2, \ldots, n. \tag{2.61}$$

This formula shows that the Jacobi iteration is quite easy to program. The only real problem is to determine an efficient test for terminating the iteration. Also, it is obvious from (2.61) or from (2.59), that in order for the Jacobi method to be used, the diagonal elements of $A$ must all be nonzero. In practice, this requirement causes no real difficulty. If $A$ is the coefficient matrix of the system $A\mathbf{x} = \mathbf{b}$ and if $a_{ii} = 0$, then the $i$th equation can be interchanged with another equation which will give a coefficient matrix with a nonzero diagonal entry in the

$i$th row. Thus the situation with the Jacobi method is similar to that of Gauss elimination, where the possibility of zero pivot elements must be guarded against.

Finally, we note that when $D^{-1}$ exists, it is relatively easy (in comparison to a direct method) to carry out each step of the iteration. Thus in those cases where $\|-D^{-1}(L + U)\| < 1$, the Jacobi method provides an alternative to direct methods.

Examination of (2.61) reveals that each component of the vector $\mathbf{x}^{(k+1)}$ is computed entirely from the vector $\mathbf{x}^{(k)}$. Assuming that $x_j^{(k+1)}$ is closer to the true answer than $x_j^{(k)}$, the estimate for $x_i^{(k+1)}$ should be improved by replacing $x_j^{(k)}$ by $x_j^{(k+1)}$ whenever $j < i$. That is, we should use our most recent information as soon as it becomes available. The implementation of this idea leads to the procedure known as the Gauss-Seidel method.

If we use the new information as soon as it is available in (2.61), we obtain the equation (after multiplication by $a_{ii}$):

$$a_{ii}x_i^{(k+1)} = -\sum_{j=1}^{i-1} a_{ij}x_j^{(k+1)} - \sum_{j=i+1}^{n} a_{ij}x_j^{(k)} + b_i, \qquad i = 1, \ldots, n. \tag{2.62}$$

(where we interpret the first sum as zero when $i = 1$). We can write this in matrix form, using $A = L + D + U$ as in the Jacobi method, and obtain

$$D\mathbf{x}^{(k+1)} = -L\mathbf{x}^{(k+1)} - U\mathbf{x}^{(k)} + \mathbf{b}. \tag{2.63}$$

Putting this equation in the standard form Eq. (2.56) for an iterative method, we have

$$(D + L)\mathbf{x}^{(k+1)} = -U\mathbf{x}^{(k)} + \mathbf{b}. \tag{2.64}$$

The matrix $M_G = -(D + L)^{-1}U$ is called the *Gauss-Seidel matrix*. Since the Gauss-Seidel method is a refinement of the Jacobi method, it usually (but not always) converges faster. For deeper results on convergence and comparison of rates of convergence, see the Ostrowski-Reich and Stein-Rosenberg Theorems in Varga (1962). We also wish to note that the choice of the starting vector $\mathbf{x}^{(0)}$ is not particularly critical, and one natural choice is $\mathbf{x}^{(0)} = \boldsymbol{\theta}$. We will have more to say of this in Section 3.4.

**Example 2.16.** As an example to illustrate the sorts of computational results that the Jacobi and Gauss-Seidel methods give, consider the linear system

$$\begin{aligned} 3x_1 + x_2 + x_3 &= 5 \\ 2x_1 + 6x_2 + x_3 &= 9 \\ x_1 + x_2 + 4x_3 &= 6 \end{aligned} \qquad \text{with solution vector } \begin{bmatrix} 1 \\ 1 \\ 1 \end{bmatrix}.$$

With $\mathbf{x}^{(0)} = \boldsymbol{\theta}$, we obtain Tables 2.1 and 2.2.

**Table 2.1** Jacobi iteration.

| $k$ | $x_1^{(k)}$ | $x_2^{(k)}$ | $x_3^{(k)}$ |
|---|---|---|---|
| 1 | 0.166667E 01 | 0.150000E 01 | 0.150000E 01 |
| 2 | 0.666667E 00 | 0.694445E 00 | 0.708333E 00 |
| 3 | 0.119907E 01 | 0.115972E 01 | 0.115972E 01 |
| 4 | 0.893518E 00 | 0.907022E 00 | 0.910301E 00 |
| 5 | 0.106089E 01 | 0.105044E 01 | 0.104986E 01 |
| 6 | 0.966564E 00 | 0.971392E 00 | 0.972166E 00 |
| 7 | 0.101881E 01 | 0.101578E 01 | 0.101551E 01 |
| 8 | 0.989568E 00 | 0.991144E 00 | 0.991350E 00 |
| 9 | 0.100584E 01 | 0.100492E 01 | 0.100482E 01 |
| 10 | 0.996753E 00 | 0.997251E 00 | 0.997312E 00 |
| 11 | 0.100181E 01 | 0.100153E 01 | 0.100150E 01 |
| 12 | 0.998991E 00 | 0.999146E 00 | 0.999165E 00 |
| 13 | 0.100056E 01 | 0.100047E 01 | 0.100047E 01 |
| 14 | 0.999687E 00 | 0.999735E 00 | 0.999741E 00 |
| 15 | 0.100017E 01 | 0.100015E 01 | 0.100014E 01 |
| 16 | 0.999903E 00 | 0.999918E 00 | 0.999919E 00 |
| 17 | 0.100005E 01 | 0.100005E 01 | 0.100004E 01 |
| 18 | 0.999970E 00 | 0.999974E 00 | 0.999975E 00 |
| 19 | 0.100002E 01 | 0.100001E 01 | 0.100001E 01 |
| 20 | 0.999991E 00 | 0.999992E 00 | 0.999992E 00 |

**Table 2.2** Gauss-Seidel iteration.

| $k$ | $x_1^{(k)}$ | $x_2^{(k)}$ | $x_3^{(k)}$ |
|---|---|---|---|
| 1 | 0.166667E 01 | 0.944445E 00 | 0.847222E 00 |
| 2 | 0.106944E 01 | 0.100231E 01 | 0.982060E 00 |
| 3 | 0.100521E 01 | 0.100125E 01 | 0.998385E 00 |
| 4 | 0.100012E 01 | 0.100023E 01 | 0.999913E 00 |
| 5 | 0.999953E 00 | 0.100003E 01 | 0.100000E 01 |
| 6 | 0.999989E 00 | 0.100000E 01 | 0.100000E 01 |
| 7 | 0.999998E 00 | 0.100000E 01 | 0.100000E 01 |
| 8 | 0.100000E 01 | 0.100000E 01 | 0.100000E 01 |

The coefficient matrix of the system is *diagonally dominant*, a condition that is sufficient to guarantee convergence of the Jacobi and Gauss-Seidel iterations (see Theorem 2.3).

As an example in which iteration is not so successful, consider the $(4 \times 4)$ linear system of Example 2.6 (solved by Gauss elimination in Example 2.7). This coefficient matrix is *positive-definite* and hence the Gauss-Seidel iteration will converge (see Theorem 2.4) but, as can be seen, convergence is exceedingly slow. (See Table 2.3.)

Table 2.3

| $k$ | $x_1^{(k)}$ | $x_2^{(k)}$ | $x_3^{(k)}$ | $x_4^{(k)}$ |
|---|---|---|---|---|
| 1 | 0.460000E 01 | −0.199982E−01 | 0.556000E 00 | 0.313602E 00 |
| 2 | 0.364720E 01 | −0.173599E−01 | 0.843327E 00 | 0.529559E 00 |
| 3 | 0.308276E 01 | −0.327911E−02 | 0.976370E 00 | 0.682185E 00 |
| 4 | 0.275076E 01 | 0.158432E−01 | 0.102290E 01 | 0.792917E 00 |
| 5 | 0.255742E 01 | 0.364410E−01 | 0.102277E 01 | 0.875290E 00 |
| 6 | 0.244637E 01 | 0.566254E−01 | 0.999118E 00 | 0.937971E 00 |
| 7 | 0.238381E 01 | 0.754578E−01 | 0.965172E 00 | 0.986620E 00 |
| 8 | 0.234954E 01 | 0.925567E−01 | 0.928278E 00 | 0.102499E 01 |
| 9 | 0.233149E 01 | 0.107840E 00 | 0.892337E 00 | 0.105566E 01 |
| 10 | 0.232256E 01 | 0.121379E 00 | 0.859268E 00 | 0.108042E 01 |
| ⋮ | | | | |
| 50 | 0.220322E 01 | 0.276739E 00 | 0.694661E 00 | 0.117948E 01 |
| 51 | 0.219950E 01 | 0.278992E 00 | 0.695580E 00 | 0.117894E 01 |
| 52 | 0.219578E 01 | 0.281235E 00 | 0.696501E 00 | 0.117840E 01 |
| ⋮ | | | | |
| 100 | 0.203333E 01 | 0.378932E 00 | 0.737633E 00 | 0.115422E 01 |
| 101 | 0.203012E 01 | 0.380858E 00 | 0.738446E 00 | 0.115374E 01 |
| 102 | 0.202693E 01 | 0.382777E 00 | 0.739256E 00 | 0.115326E 01 |
| ⋮ | | | | |
| 196 | 0.176445E 01 | 0.540536E 00 | 0.805902E 00 | 0.111409E 01 |
| 197 | 0.176208E 01 | 0.541962E 00 | 0.806505E 00 | 0.111373E 01 |
| 198 | 0.175972E 01 | 0.543384E 00 | 0.807103E 00 | 0.111338E 01 |
| 199 | 0.175736E 01 | 0.544801E 00 | 0.807701E 00 | 0.111303E 01 |
| 200 | 0.175501E 01 | 0.546213E 00 | 0.808298E 00 | 0.111268E 01 |

The question of how fast an iterative procedure will converge is considered in Section 3.4. Through the theory of the above-mentioned section it can be shown that the Jacobi method will not converge for the above $(4 \times 4)$ system.

The question of convergence for iterative methods is sometimes hard to analyze, for convergence depends on $M = N^{-1}P$ being small in some sense, as is shown in (2.57). For the Jacobi method, it is easy to display the form of the Jacobi matrix $M_J$. Using (2.59) it follows quickly that

$$M_J = -\begin{bmatrix} 0 & \frac{a_{12}}{a_{11}} & \frac{a_{13}}{a_{11}} & \cdots & \frac{a_{1n}}{a_{11}} \\ \frac{a_{21}}{a_{22}} & 0 & \frac{a_{23}}{a_{22}} & \cdots & \frac{a_{2n}}{a_{22}} \\ \vdots & & & & \vdots \\ \frac{a_{n1}}{a_{nn}} & \frac{a_{n2}}{a_{nn}} & \frac{a_{n3}}{a_{nn}} & \cdots & 0 \end{bmatrix}. \tag{2.65}$$

An explicit representation for $M_G$ for a general matrix $A$ is obviously more complicated, and it is of little use to find $M_G$.

While Theorem 2.2 provides a general criterion for convergence, it would be helpful to have some convergence theorems that related specifically to the Jacobi and Gauss-Seidel methods. In particular, the ideal theorem would be one that guaranteed convergence for all matrices in some particular class or for all matrices that satisfy some particular property. Below, we illustrate two theorems of this type, theorems that guarantee convergence for *diagonally dominant* matrices and for *symmetric positive-definite* matrices. Diagonally dominant matrices and symmetric positive-definite matrices, two important types of matrices that occur frequently in practical problems, are defined below.

We say a matrix $A = (a_{ij})$ is diagonally dominant if

$$|a_{ii}| > \sum_{\substack{j=1 \\ j \neq i}}^{n} |a_{ij}|, \text{ for } i = 1, 2, \ldots, n. \tag{2.66}$$

That is, each main diagonal entry is larger in absolute value than the sum of the absolute values of all the off-diagonal entries in that row. An $(n \times n)$ matrix $A = (a_{ij})$ is called *symmetric* if

$$A^T = A. \tag{2.67}$$

Clearly, an equivalent way of stating that $A$ is symmetric is to say that $a_{ij} = a_{ji}$ for $1 \leq i \leq n$, $1 \leq j \leq n$. Finally, an $(n \times n)$ matrix $A = (a_{ij})$ is said to be *symmetric* and *positive-definite* if:

a) $A$ is symmetric;

b) $\mathbf{x}^T A\mathbf{x} > 0$ for all $\mathbf{x} \in R^n$, $\mathbf{x} \neq \mathbf{\theta}$. **(2.68)**

(We note that since $A$ is $(n \times n)$ and $\mathbf{x}$ is $(n \times 1)$, then $A\mathbf{x}$ is $(n \times 1)$. Thus, since $\mathbf{x}$ is $(n \times 1)$, $\mathbf{x}^T$ is $(1 \times n)$ and so we have that $\mathbf{x}^T A\mathbf{x}$ is a scalar quantity. Most readers will recognize $\mathbf{x}^T A\mathbf{x}$ as the familiar "dot product" of the vector $\mathbf{x}$ and the vector $A\mathbf{x}$.)

**Example 2.17.** Consider the simple $(2 \times 2)$ matrix $A$, $A = \begin{pmatrix} 2 & 1 \\ 1 & 3 \end{pmatrix}$. Clearly $A$ is diagonally dominant and also symmetric. To determine whether $A$ is positive-definite, we consider

$$\mathbf{x}^T A\mathbf{x} = [x_1, x_2]\begin{bmatrix} 2 & 1 \\ 1 & 3 \end{bmatrix}\begin{bmatrix} x_1 \\ x_2 \end{bmatrix} = [x_1, x_2]\begin{bmatrix} 2x_1 + x_2 \\ x_1 + 3x_2 \end{bmatrix}.$$

Therefore $\mathbf{x}^T A\mathbf{x} = 2x_1^2 + x_1x_2 + x_2x_1 + 3x_2^2 = (x_1 + x_2)^2 + x_1^2 + 2x_2^2$. Hence, we see that $A$ is positive-definite, since $\mathbf{x}^T A\mathbf{x} > 0$ whenever $\mathbf{x} \neq \mathbf{\theta}$. If we modify $A$ and consider $B = \begin{pmatrix} 2 & 1 \\ 1 & -3 \end{pmatrix}$, then we still have that $B$ is symmetric and diagonally dominant. However, $B$ is not positive-definite, since for $\mathbf{x} = \begin{pmatrix} 0 \\ 1 \end{pmatrix}$, $\mathbf{x}^T B\mathbf{x} = (0, 1)\begin{pmatrix} 1 \\ -3 \end{pmatrix} = -3 < 0$.

Having the concepts of diagonally dominant and positive-definite, we can now state two convergence theorems.

**Theorem 2.3.** If $A$ is diagonally dominant, then $A$ is non-singular and the sequence $\{\mathbf{x}^{(k)}\}$ defined by the Jacobi method (2.59) converges for any initial guess $\mathbf{x}^{(0)}$.

*Proof*: Dividing both sides of inequality (2.66) by $|a_{ii}|$ for $i = 1, 2, \ldots, n$, we obtain

$$1 > \sum_{\substack{j=1 \\ j \neq i}}^{n} \frac{|a_{ij}|}{|a_{ii}|} \text{ for } i = 1, 2, \ldots, n.$$

Referring to (2.65), we see that this means that $\|M_J\|_\infty < 1$. Consequently, by Theorem 2.2, $A$ is non-singular and the Jacobi method will converge. ■

The fact that a diagonally dominant matrix is non-singular will also be important in our study of cubic splines in Chapter 5. The proof of Theorem 2.4 is more difficult and we defer it to Section 3.4.

**Theorem 2.4.** If $A$ is symmetric and positive-definite, then the sequence $\{\mathbf{x}^{(k)}\}$ defined by the Gauss-Seidel method (2.64) converges for any initial guess $\mathbf{x}^{(0)}$.

Both of these theorems are quite useful in various computational settings. For example, some numerical procedures result in large linear systems $A\mathbf{x} = \mathbf{b}$ which must be solved, where $A$ is diagonally dominant. Theorem 2.3 guarantees that the Jacobi method will work on this system. In Section 2.5 we will consider problems involving least-squares fits and the solution of over-determined systems. These problems frequently require the solution of a linear system of the form $A^T A\mathbf{x} = A^T\mathbf{b}$. The matrix $A^T A$ is symmetric and, moreover, in many practical cases $A^T A$ is also positive-definite. Thus the Gauss-Seidel method may be suitable for systems of this type.

### 2.4.2 Implementation of Iterative Methods

There are many other important iterative methods besides Jacobi and Gauss-Seidel. Two other important, but rather more complicated, methods are: successive overrelaxation (SOR) and alternating-direction implicit methods (ADI). A good advanced reference is Varga (1962). Iterative methods are normally used when it can easily be shown in advance that they will converge and that they will require less computation than a direct method. The numerical solution of partial differential equations provides the most common area of application for iterative methods. When finite difference or finite element schemes are used, the end result is to replace the partial differential equation by a large matrix equation $A\mathbf{x} = \mathbf{b}$. The matrix equation is an approximation to the partial differential equation, and the solution of $A\mathbf{x} = \mathbf{b}$ gives approximate values for the function that satisfies the partial differential equation. These matrices have a structure that is known in advance, so it is normally easy to see whether or not an iterative procedure will converge. Moreover, these matrices are often quite large (say $(1000 \times 1000)$

perhaps) and *sparse* (that is, there are relatively few nonzero entries in the matrix). In the following, we indicate how this "sparseness" can be exploited by an iterative method.

Notice that an iterative method written in the form

$$\mathbf{x}^{(k+1)} = M\mathbf{x}^{(k)} + \mathbf{g}, \tag{2.69}$$

where $M$ is $(n \times n)$, requires just $n^2$ multiplications per iteration. This observation forms a part of the reason why iterative methods are valuable, since a direct method like Gauss elimination takes about $(n^3/3)$ multiplications. Thus if we are satisfied with the approximate solution $\mathbf{x}^{(k)}$ where $k < n/3$, we have saved machine time. In the context of partial differential equations, since the equation $A\mathbf{x} = \mathbf{b}$ is itself an approximation, it does not make sense to ask for an extremely good estimate of the true solution $\mathbf{x}_t$. Another point to be made is that if the matrix $A$ is sparse (a typical example is when $A$ is $(1000 \times 1000)$ with five nonzero entries per row), then iterative methods can take advantage of the fact that multiplication by zero is not necessary. For example in (2.61) or (2.62), if only four of the coefficients $a_{ij}/a_{ii}$ are nonzero for each $i$, then only 5 multiplications are required to form $x_i^{(k+1)}$. Thus forming $\mathbf{x}^{(k+1)}$ in the Jacobi method (or the Gauss-Seidel method) would take $5n$ multiplications. The total number of multiplications needed to form $\mathbf{x}^{(j)}$ from $\mathbf{x}^{(0)}$ is then $5nj$. So, for $j \le n^2/15$, less effort is needed to compute $\mathbf{x}^{(j)}$ than is needed to solve the system by Gauss elimination. A final consideration is that of storage, since a matrix with only five nonzero entries per row needs only $5n$ storage locations. If, however, Gauss elimination is used on the same sparse system, it may be that the system becomes less sparse (as a whole) as the elimination proceeds. Thus Gauss elimination might require much more than $5n$ storage locations. We should also remark that normally sparse matrices arising from practical problems have such a regular pattern that programming this information is easy.

**Example 2.18.** Consider a system of linear equations with the coefficient matrix,

$$\begin{bmatrix} 1 & 2 & 3 & 4 & 5 & 6 & 7 & 8 & 9 & 10 \\ 2 & 1 & 0 & 0 & 0 & 0 & 0 & 0 & 0 & 0 \\ 3 & 0 & 1 & 0 & 0 & 0 & 0 & 0 & 0 & 0 \\ 4 & 0 & 0 & 1 & 0 & 0 & 0 & 0 & 0 & 0 \\ 5 & 0 & 0 & 0 & 1 & 0 & 0 & 0 & 0 & 0 \\ 6 & 0 & 0 & 0 & 0 & 1 & 0 & 0 & 0 & 0 \\ 7 & 0 & 0 & 0 & 0 & 0 & 1 & 0 & 0 & 0 \\ 8 & 0 & 0 & 0 & 0 & 0 & 0 & 1 & 0 & 0 \\ 9 & 0 & 0 & 0 & 0 & 0 & 0 & 0 & 1 & 0 \\ 10 & 0 & 0 & 0 & 0 & 0 & 0 & 0 & 0 & 1 \end{bmatrix}. \tag{2.70}$$

This provides a graphic example of a sparse matrix which becomes "dense" as Gauss elimination is used. To illustrate our point, this example is admittedly contrived in that we could interchange the first and last rows and solution by Gauss elimination becomes almost immediate. The point of the example, however, is well taken and remains valid in less contrived situations.

## PROBLEMS

1. Write a computer program to implement the Jacobi method.
2. Write a computer program to implement the Gauss-Seidel method.
3. Test the Gauss-Seidel and Jacobi routines of Problem 1 and Problem 2 on the $(2 \times 2)$ system $A^TA\mathbf{x}^* = A^T\mathbf{b}$ in Example 2.19 of Section 2.5.
4. Let $A = (a_{ij})$ be the $(4 \times 4)$ diagonally dominant matrix given by: $a_{ii} = 10$, $i = 1, 2, 3, 4$, and $a_{ij} = 2$, for $i \neq j$. Use the Gauss-Seidel and Jacobi methods to solve $A\mathbf{x} = \mathbf{b}$, where $\mathbf{b}^T = (16, 24, 16, 8)$. (The exact solution is $\mathbf{x}^T = (1, 2, 1, 0)$.)
5. Let $A$ be an $(m \times n)$ matrix. Using Problem 13, Section 2.1, show that $A^TA$ is an $(n \times n)$ symmetric matrix.
6. If $\mathbf{y}$ is any $(n \times 1)$ vector, $\mathbf{y} \neq \boldsymbol{\theta}$, show that $\mathbf{y}^T\mathbf{y} > 0$.
7. Suppose $A$ is an $(m \times n)$ matrix such that $A\mathbf{y} \neq \boldsymbol{\theta}$ for any $(n \times 1)$ vector $\mathbf{y}$, where $\mathbf{y} \neq \boldsymbol{\theta}$. Using Problems 5 and 6, show that $A^TA$ is a symmetric, positive-definite matrix.
8. Using Problem 7, verify for the $(4 \times 2)$ matrix $A$ in Example 2.19, that $A^TA$ is symmetric and positive-definite. (From Theorem 2.4, the results of this problem have a bearing on Problem 3.)

*9. For the matrix $A^TA$ of Example 2.19, verify that the Jacobi matrix $M_J = -D^{-1}(L + U)$ is given by:

$$M_J = -\begin{bmatrix} 0 & \frac{3}{7} \\ \frac{3}{2} & 0 \end{bmatrix} \quad \text{and} \quad M_J^2 = \begin{bmatrix} \frac{9}{14} & 0 \\ 0 & \frac{9}{14} \end{bmatrix}.$$

Using $M_J^4 = M_J^2M_J^2$, $M_J^6 = M_J^2M_J^2M_J^2$, etc., show that the entries of $M_J^{2n}$ all tend to zero as $n \to \infty$. From this information, give a bound for $\|M_J^i\|_\infty$ and show, using (2.57), that the Jacobi method must converge for any initial guess $\mathbf{x}^{(0)}$. Note that $A^TA$ is not diagonally dominant, so that Theorem 2.3 does not apply. Additionally, $\|M_J\| > 1$ for the three matrix norms of (2.41), so that Theorem 2.2 does not apply either. However, the analysis of this problem shows why the Jacobi method is convergent in Problem 3.

10. As another example of a slowly converging iterative method, use the Gauss-Seidel iteration to solve the $(2 \times 2)$ system $A_{100}\mathbf{x} = \mathbf{b}$ of Problem 2, Section 2.3.

## 2.5 LEAST-SQUARES SOLUTION OF OVER-DETERMINED LINEAR SYSTEMS

We now return our attention to the linear system (2.1) (or Eq. (2.2) in matrix form), where we have $m$ linear equations in $n$ unknowns, and we consider the case where $m > n$. Since we have more equations than unknowns, we do not normally expect a solution, $\mathbf{x}$, of $A\mathbf{x} = \mathbf{b}$ to exist. (As pointed out previously, if such an $\mathbf{x}$ does exist, Gauss elimination can be implemented to find it.) However, in this section we shall assume that $A\mathbf{x} = \mathbf{b}$ does *not* have a solution. The reader may thus feel that this is necessarily an unimportant problem. Quite the opposite is true, however, for we can reformulate the problem to ask for the vector $\mathbf{x}^*$ which somehow "minimizes" the vector expression $(A\mathbf{x}^* - \mathbf{b})$. That is, find $\mathbf{x}^*$ such that $\|A\mathbf{x}^* - \mathbf{b}\|$ is minimized for some vector norm $\|\cdot\|$. In the case where we use the Euclidean norm, $\|\cdot\|_2$, this becomes precisely the important statistical problem of finding the best *least-squares solution* of $A\mathbf{x} = \mathbf{b}$. We shall see later that this is a special case of the famous Fourier series approximation problem, but for now we shall be content to solve it in this one particular context.

Let $A$ be any given $(m \times n)$ matrix and define $Y$ to be the set of vectors $\mathbf{y}$ in $R^m$ such that

$$Y = \{\mathbf{y} \in R^m: \mathbf{y} = A\mathbf{x}, \text{ for some } \mathbf{x} \in R^n\}.$$

Thus $Y$ is in reality the "range of $A$" and is a subset (actually a "subspace") of $R^m$. Since we have assumed that there is no $\mathbf{x} \in R^n$ such that $A\mathbf{x} = \mathbf{b}$, then $\mathbf{b}$ does not belong to $Y$ in this case. (However, the following procedures would be valid even if $\mathbf{b} \in Y$.) Let $\mathbf{x} = (x_1, x_2, \ldots, x_n)^T$ be arbitrary in $R^n$, and write $A = [\mathbf{A}_1, \mathbf{A}_2, \ldots, \mathbf{A}_n]$ where each $\mathbf{A}_k$ is the $k$th $(m \times 1)$ column vector of $A$, $1 \le k \le n$. Recall from Section 2.1 that $A\mathbf{x}$ can be written as

$$A\mathbf{x} = x_1\mathbf{A}_1 + x_2\mathbf{A}_2 + \cdots + x_n\mathbf{A}_n.$$

Thus we see that $Y$ can be equivalently written as

$$Y = \{\mathbf{y} \in R^m: \mathbf{y} = x_1\mathbf{A}_1 + x_2\mathbf{A}_2 + \cdots + x_n\mathbf{A}_n\}.$$

Therefore we can see that our problem of minimizing $\|A\mathbf{x} - \mathbf{b}\|_2$ can be restated as: Find $\mathbf{y}^* \in Y$ such that $\|\mathbf{y}^* - \mathbf{b}\|_2 \le \|\mathbf{y} - \mathbf{b}\|_2$, for all $\mathbf{y} \in Y$, and then find $\mathbf{x}^* \in R^n$ such that $A\mathbf{x}^* = \mathbf{y}^*$. This problem can be formally solved easily by use of the following theorem.

**Theorem 2.5.** Assume that there exists a vector $\mathbf{y}^* \in Y$ such that $(\mathbf{b} - \mathbf{y}^*)^T\mathbf{y} = 0$, for all $\mathbf{y} \in Y$. Then $\|\mathbf{b} - \mathbf{y}^*\|_2 \le \|\mathbf{b} - \mathbf{y}\|_2$, for all $\mathbf{y} \in Y$.

*Proof*: We observe, that if $\mathbf{z}$ is any vector in $R^m$, then $\mathbf{z}^T\mathbf{z} = \|\mathbf{z}\|_2^2$ and $\mathbf{x}^T\mathbf{z} = \mathbf{z}^T\mathbf{x}$. Let $\mathbf{y}$ be any vector in $Y$ and remember in the following that $(\mathbf{b} - \mathbf{y}^*)^T\mathbf{y} = 0$.

$$\begin{aligned}
0 \le \|\mathbf{b} - \mathbf{y}\|_2^2 &= (\mathbf{b} - \mathbf{y})^T(\mathbf{b} - \mathbf{y}) = \mathbf{b}^T\mathbf{b} - 2\mathbf{b}^T\mathbf{y} + \mathbf{y}^T\mathbf{y} \\
&= \mathbf{b}^T\mathbf{b} + 2(\mathbf{b} - \mathbf{y}^*)^T\mathbf{y} - 2\mathbf{b}^T\mathbf{y} + \mathbf{y}^T\mathbf{y} \\
&= \mathbf{b}^T\mathbf{b} + 2\mathbf{b}^T\mathbf{y} - 2\mathbf{y}^{*T}\mathbf{y} - 2\mathbf{b}^T\mathbf{y} + \mathbf{y}^T\mathbf{y} \\
&= \mathbf{b}^T\mathbf{b} - \mathbf{y}^{*T}\mathbf{y}^* + (\mathbf{y}^{*T}\mathbf{y}^* - 2\mathbf{y}^{*T}\mathbf{y} + \mathbf{y}^T\mathbf{y}) \\
&= \|\mathbf{b}\|_2^2 - \|\mathbf{y}^*\|_2^2 + (\mathbf{y}^* - \mathbf{y})^T(\mathbf{y}^* - \mathbf{y}).
\end{aligned}$$

Thus $\|\mathbf{b} - \mathbf{y}\|_2^2 = \|\mathbf{b}\|_2^2 - \|\mathbf{y}^*\|_2^2 + \|\mathbf{y}^* - \mathbf{y}\|_2^2$, and obviously this last expression is minimized if and only if $\mathbf{y} = \mathbf{y}^*$. ■

The geometric rationale underlying this theorem is as follows: Suppose that $Y$ is a plane in $R^3$ (in our particular case it is either a plane or a line through the origin when $m = 3$). If $\mathbf{b}$ is not in this plane, then the closest point on the plane to $\mathbf{b}$ is $\mathbf{y}^*$ where $\mathbf{y}^*$ is such that the projection vector, $(\mathbf{b} - \mathbf{y}^*)$, is perpendicular to every $\mathbf{y}$ in the plane. Since $(\mathbf{b} - \mathbf{y}^*)^T\mathbf{y}$ is the usual "dot product" of $(\mathbf{b} - \mathbf{y}^*)$ and $\mathbf{y}$, $(\mathbf{b} - \mathbf{y}^*)^T\mathbf{y} = 0$ means that $(\mathbf{b} - \mathbf{y}^*)$ and $\mathbf{y}$ are perpendicular. So this theorem is a natural analytical extension of this geometric concept.

Later, when we investigate this problem in a more general setting we will have the necessary mathematical tools to prove that $\mathbf{y}^*$, as given in the above theorem, always exists and is unique. For now we will merely assume that this is always true. Under this assumption, then, there is always at least one set of scalars, $\{x_1^*, x_2^*, \ldots, x_n^*\}$, such that $\mathbf{y}^* = x_1^*\mathbf{A}_1 + x_2^*\mathbf{A}_2 + \cdots + x_n^*\mathbf{A}_n \equiv A\mathbf{x}^*$, where $\mathbf{x}^* \equiv (x_1^*, x_2^*, \ldots, x_n^*)^T$. Since every $\mathbf{y} \in Y$ can be written as $\mathbf{y} = x_1\mathbf{A}_1 + x_2\mathbf{A}_2 + \cdots + x_n\mathbf{A}_n$, then $(\mathbf{b} - \mathbf{y}^*)^T\mathbf{y} = 0$, for all $\mathbf{y} \in Y$ is equivalent to saying: $(\mathbf{b} - \mathbf{y}^*)^T\mathbf{A}_i = \mathbf{A}_i^T(\mathbf{b} - \mathbf{y}^*) = 0$ for $1 \le i \le n$, or $\mathbf{A}_i^T\mathbf{y}^* = \mathbf{A}_i^T\mathbf{b}$, $1 \le i \le n$. This gives a set of $n$ simultaneous equations, which written in vector form becomes, since $\mathbf{y}^* = A\mathbf{x}^*$,

$$\begin{bmatrix} \mathbf{A}_1^T(A\mathbf{x}^*) \\ \mathbf{A}_2^T(A\mathbf{x}^*) \\ \vdots \\ \mathbf{A}_n^T(A\mathbf{x}^*) \end{bmatrix} = \begin{bmatrix} \mathbf{A}_1^T\mathbf{b} \\ \mathbf{A}_2^T\mathbf{b} \\ \vdots \\ \mathbf{A}_n^T\mathbf{b} \end{bmatrix}. \tag{2.71}$$

It is easily verified by the basic rules of matrix multiplication, that if $\mathbf{z}$ is any vector in $R^m$, then since the rows of $A^T$ are the columns of $A$,

$$A^T\mathbf{z} = \begin{bmatrix} \mathbf{A}_1^T\mathbf{z} \\ \mathbf{A}_2^T\mathbf{z} \\ \vdots \\ \mathbf{A}_n^T\mathbf{z} \end{bmatrix}. \tag{2.72}$$

Therefore by (2.72), (2.71) reduces to the simple expression

$$A^TA\mathbf{x}^* = A^T\mathbf{b}. \tag{2.73}$$

This is an $(n \times n)$ linear system commonly known as the *normal equations*, and a *least-squares solution*, $\mathbf{x}^*$, may be obtained by Gauss elimination or Gauss-Seidel. (We note here that if $A\mathbf{x} = \mathbf{b}$ has an exact solution, $\mathbf{x}^*$, that is, $\mathbf{b} \in Y$, then $\mathbf{x}^*$ is still a solution of $A^TA\mathbf{x}^* = A^T\mathbf{b}$.)

A familiar problem that arises in this setting is the following: Assume that a function $f(x)$ is known at the $m$ distinct points $\{x_0, x_1, \ldots, x_{m-1}\}$ and we wish to find an $(n - 1)$st degree polynomial $p(x) = a_0x^{n-1} + a_1x^{n-2} + \cdots + a_{n-2}x + a_{n-1}$ which minimizes the expression $\sum_{i=0}^{m-1} (f(x_i) - p(x_i))^2$, where $m > n$ ($p(x)$ is the $(n - 1)$st degree best discrete least-squares polynomial fit for $f(x)$). The equations, $f(x_i) = p(x_i) = a_0x_i^{n-1} + a_1x_i^{n-2} + \cdots + a_{n-1}$, $0 \le i \le m - 1$, yield the $m \times n$ linear system

$$\begin{bmatrix} 1 & x_0 & \cdots & x_0^{n-1} \\ 1 & x_1 & \cdots & x_1^{n-1} \\ \vdots & & & \\ 1 & x_{m-1} & \cdots & x_{m-1}^{n-1} \end{bmatrix} \begin{bmatrix} a_{n-1} \\ a_{n-2} \\ \vdots \\ a_0 \end{bmatrix} = \begin{bmatrix} f(x_0) \\ f(x_1) \\ \vdots \\ f(x_{m-1}) \end{bmatrix}, \tag{2.74}$$

and the least-squares solution for $\{a_i\}_{i=0}^{n-1}$ is obtained via (2.73).

**Example 2.19.** Suppose we wish to draw the straight line in the plane that comes closest to "fitting" the points: (0, 1), (1, 2.1), (2, 2.9), (3, 3.2). As we indicated above, if we set $f(0) = 1$, $f(1) = 2.1$, $f(2) = 2.9$ and $f(3) = 3.2$, then we want $a_0$ and $a_1$ to minimize $\sum_{i=0}^{3} [f(x_i) - p(x_i)]^2$, where $p(x) = a_0 x + a_1$ and $x_i = i$, $i = 0, 1, 2, 3$. This leads to the over-determined system

$$\begin{aligned} a_1 &= 1. \\ a_0 + a_1 &= 2.1 \\ 2a_0 + a_1 &= 2.9 \\ 3a_0 + a_1 &= 3.2 \end{aligned} \quad \text{or} \quad \begin{bmatrix} 0 & 1 \\ 1 & 1 \\ 2 & 1 \\ 3 & 1 \end{bmatrix} \begin{bmatrix} a_0 \\ a_1 \end{bmatrix} = \begin{bmatrix} 1 \\ 2.1 \\ 2.9 \\ 3.2 \end{bmatrix}.$$

Then

$$A^T A \mathbf{x}^* = A^T \mathbf{b} \quad \text{is} \quad \begin{bmatrix} 14 & 6 \\ 6 & 4 \end{bmatrix} \begin{bmatrix} a_0 \\ a_1 \end{bmatrix} = \begin{bmatrix} 17.5 \\ 9.2 \end{bmatrix}.$$

Thus we find $a_1 = 1.19$ and $a_0 = 0.74$. (See Fig. 2.4.)

We have noted previously that matrices of the form $A^T A$ can be ill-conditioned. For this reason, the particular problem of discrete least-squares curve fitting is usually attacked by a different method which we shall investigate in a later section on polynomial approximations for functions.

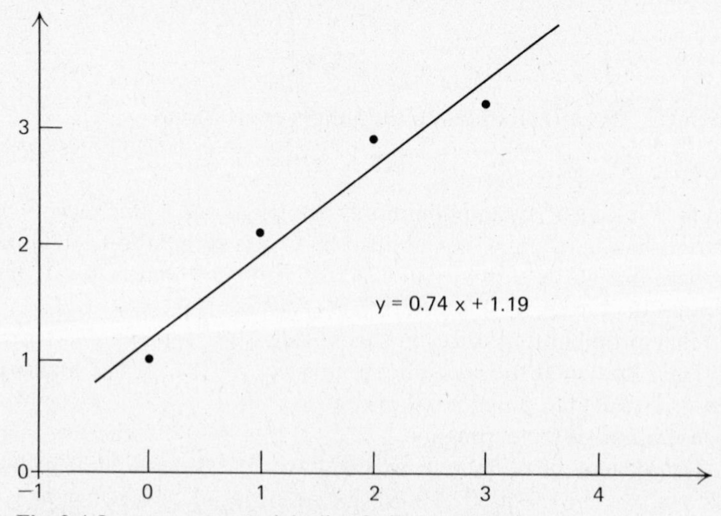


**Fig. 2.4** Least-squares straight-line fit (Example 2.19).

## PROBLEMS

1. Use the methods of this section to find the best least-squares solution of the system

$$\begin{aligned} x_1 + x_2 &= 1 \\ 2x_1 + x_2 &= 0 \\ x_1 - x_2 &= 0 \end{aligned}$$

2. If $A$ denotes the coefficient matrix of the $(3 \times 2)$ system of Problem 1, show that $A^TA$ is symmetric and positive-definite, as in Problem 8, Section 2.4.
3. Consider the $(3 \times 3)$ system

$$\begin{aligned} x_1 + 3x_2 + 7x_3 &= 1 \\ 2x_1 + x_2 - x_3 &= 1 \\ x_1 + 2x_2 + 4x_3 &= 1 \end{aligned}$$

Show the system has no solution (so that coefficient matrix $A$ must be singular). Next, find the best least-squares solution by the techniques of this section. Note that even though $A^TA$ is singular, the equation corresponding to (2.73) is solvable.

4. For the matrix $A$ in Problem 3, find a nonzero vector $\mathbf{x}$ such that $\mathbf{x}^T(A^TA)\mathbf{x} = 0$, and thus conclude that $A^TA$ cannot be positive-definite.

*5. An $(n \times n)$ matrix $B$ is called *positive semi-definite* if $\mathbf{x}^TB\mathbf{x} \geq 0$ for all $(n \times 1)$ vectors $\mathbf{x}$. Show that any matrix $B$ of the form $B = A^TA$ is positive semi-definite.

*6. Write a computer program to find a 4th-degree polynomial approximation to $f(x) = \cos(x)$, for $0 \leq x \leq \pi$, as follows:
   a) Let $x_j = j\pi/20; j = 0, 1, \ldots, 20$.
   b) Set up Eq. (2.74) with $m = 21$ and $n = 5$, to find a best 4th-degree least-squares approximation to $\cos(x)$ on $x_0, x_1, \ldots, x_{20}$.
   c) Having (2.74), form the system (2.73) and solve for $\mathbf{x}^*$ (the Gauss elimination routine of Example 2.7 could be used).
   d) Let $p(x)$ denote the resulting best 4th-degree polynomial approximation. Print a table, listing $p(y_i)$, $f(y_i)$, $|p(y_i) - f(y_i)|/|f(y_i)|$ for $y_i = (0.01)j\pi; j = 0, 1, 2, \ldots, 100; j \neq 50$.

## *2.6 THE CAUCHY-SCHWARZ INEQUALITY.†

Let $\mathbf{x}$ and $\mathbf{y}$ be vectors in $R^n$, with $\mathbf{x} = (x_1, x_2, \ldots, x_n)^T$ and $\mathbf{y} = (y_1, y_2, \ldots, y_n)^T$. We recall the familiar *scalar* or *dot* product, given by

$$\mathbf{x}^T\mathbf{y} = x_1y_1 + x_2y_2 + \cdots + x_ny_n.$$

We next note that the scalar product satisfies the following conditions for all

† This brief section is included for the sake of completeness and may be omitted without loss of continuity. The ambitious reader will find the ideas contained in this section have wide-ranging application.

vectors $\mathbf{x}$, $\mathbf{y}$ and $\mathbf{z}$ in $R^n$ and for all scalars $\alpha$:

$$\begin{aligned} &\text{a) } \mathbf{x}^T\mathbf{y} = \mathbf{y}^T\mathbf{x} \\ &\text{b) } (\alpha\mathbf{x})^T\mathbf{y} = \alpha(\mathbf{x}^T\mathbf{y}) \\ &\text{c) } (\mathbf{x} + \mathbf{y})^T\mathbf{z} = \mathbf{x}^T\mathbf{z} + \mathbf{y}^T\mathbf{z} \\ &\text{d) } \mathbf{x}^T\mathbf{x} \geq 0, \text{ where } \mathbf{x}^T\mathbf{x} = 0 \text{ if and only if } \mathbf{x} = \boldsymbol{\theta}. \end{aligned} \tag{2.75}$$

The idea of a scalar product can be extended to vector spaces other than $R^n$, including even infinite-dimensional vector spaces (see Problem 9). The four properties (a), (b), (c) and (d) serve to define this extension, which is usually called an *inner product* rather than a scalar product. Some of the most fruitful areas in modern analysis, such as Hilbert spaces and Fourier analysis, use inner products as a fundamental tool. Returning to the scalar product $\mathbf{x}^T\mathbf{y}$, we next note (Problem 1) that from formula (2.38),

$$(\mathbf{x}^T\mathbf{x})^{1/2} = \|\mathbf{x}\|_2.$$

Having made this observation, we now are ready to give the main result of this section:

**Theorem 2.6. Cauchy-Schwarz.** Given any $\mathbf{x}$ and $\mathbf{y}$ in $R^n$, $|\mathbf{x}^T\mathbf{y}| \leq \|\mathbf{x}\|_2 \, \|\mathbf{y}\|_2$.

*Proof*: If either $\mathbf{x} = \boldsymbol{\theta}$ or $\mathbf{y} = \boldsymbol{\theta}$, the proof is trivial, so we assume $\mathbf{x} \neq \boldsymbol{\theta}$, $\mathbf{y} \neq \boldsymbol{\theta}$. For any scalar $\alpha$, we have by property (d),

$$0 \leq (\mathbf{x} + \alpha\mathbf{y})^T(\mathbf{x} + \alpha\mathbf{y}). \tag{2.76}$$

Expanding the right-hand side of this inequality, using properties (a), (b) and (c), we obtain

$$0 \leq \mathbf{x}^T\mathbf{x} + 2\alpha\mathbf{x}^T\mathbf{y} + \alpha^2\mathbf{y}^T\mathbf{y}.$$

Choosing $\alpha' = -\,\mathbf{x}^T\mathbf{y}/\mathbf{y}^T\mathbf{y}$, the inequality above reduces to $0 \leq \mathbf{x}^T\mathbf{x} + \alpha'\mathbf{x}^T\mathbf{y}$, which is equivalent to $(\mathbf{x}^T\mathbf{y})(\mathbf{x}^T\mathbf{y}) \leq (\mathbf{x}^T\mathbf{x})(\mathbf{y}^T\mathbf{y})$. Since both sides of the inequality are nonnegative, taking square roots of both sides establishes the theorem. ■

The problems that follow give some indication of the applications of the Cauchy-Schwarz inequality.

## PROBLEMS

1. Show that for $\mathbf{x} \in R^n$, $(\mathbf{x}^T\mathbf{x})^{1/2} = \|\mathbf{x}\|_2$, as in formula (2.38).
2. For $\mathbf{x}$ and $\mathbf{b}$ in $R^n$, show that

$$\|\mathbf{x} + \mathbf{y}\|_2^2 + \|\mathbf{x} - \mathbf{y}\|_2^2 = 2\|\mathbf{x}\|_2^2 + 2\|\mathbf{y}\|_2^2.$$

(This is called the parallelogram law, and a picture of two vectors $\mathbf{x}$ and $\mathbf{y}$ in a plane will show why.)

3. Suppose $\mathbf{x}$ and $\mathbf{y}$ are two vectors in $R^n$ such that $\mathbf{x}^T\mathbf{y} = 0$. Show:

$$\|\mathbf{x} - \mathbf{y}\|_2^2 = \|\mathbf{x}\|_2^2 + \|\mathbf{y}\|_2^2.$$

(Again, a picture of two vectors in a plane will show that this result is essentially an extension of the Pythagorean theorem of plane geometry.)

4. Use the Cauchy-Schwarz inequality to show that

$$\|\mathbf{x} + \mathbf{y}\|_2^2 \leq \|\mathbf{x}\|_2^2 + 2\|\mathbf{x}\|_2 \|\mathbf{y}\|_2 + \|\mathbf{y}\|_2^2.$$

Use this inequality to show that the $\ell_2$ vector norm satisfies Eq. (2.37c). Complete the verification that $\|\cdot\|_2$ is a vector norm by showing that it satisfies (2.37a) and (2.37b).

5. Show that $\|\mathbf{x}\|_2$ and $\|A\|_E$ are compatible vector and matrix norms for $\mathbf{x} \in R^n$ and $A \in M_n$. (Hint: Use the Cauchy-Schwarz inequality on the vector $A\mathbf{x}$, where we note that the $i$th component of $A\mathbf{x}$ is given by the scalar product of the $i$th row of $A$ and $\mathbf{x}$.)

6. Show that equality can hold in the Cauchy-Schwarz inequality if and only if $\mathbf{x} = \beta\mathbf{y}$, for some scalar $\beta$. (Hint: If $\mathbf{x} = \beta\mathbf{y}$, equality follows. By analyzing the proof of Theorem 2.6, show that strict inequality in (2.76) leads to strict inequality in the Cauchy-Schwarz inequality.)

*7. Prove the following stronger form of the Cauchy-Schwarz inequality. For $\mathbf{x} = (x_1, x_2, \ldots, x_n)^T$ and $\mathbf{y} = (y_1, y_2, \ldots, y_n)^T$, show that:

$$\sum_{i=1}^{n} |x_i y_i| \leq \|\mathbf{x}\|_2 \|\mathbf{y}\|_2.$$

To do this, define $\mathbf{z} = (z_1, z_2, \ldots, z_n)^T$, where $z_i = -y_i$ if $x_i y_i < 0$ and $z_i = y_i$ if $x_i y_i \geq 0$. Apply Theorem 2.6 to $\mathbf{x}^T\mathbf{z}$.

*8. Let $\sum_{i=1}^{\infty} \alpha_i$ and $\sum_{i=1}^{\infty} \beta_i$ be two infinite series such that $\sum_{i=1}^{\infty} \alpha_i^2$ and $\sum_{i=1}^{\infty} \beta_i^2$ converge. Using partial sums and Problem 7, show that:

$$\sum_{i=1}^{\infty} |\alpha_i \beta_i| \leq \sqrt{\sum_{i=1}^{\infty} \alpha_i^2}\sqrt{\sum_{i=1}^{\infty} \beta_i^2}.$$

*9. Suppose $f(x)$ and $g(x)$ are continuous functions defined on the finite interval $[a, b]$. Following the proof of Theorem 2.6, show that:

$$\left|\int_a^b f(x)g(x)\,dx\right| \leq \sqrt{\int_a^b [f(x)]^2\,dx}\sqrt{\int_a^b [g(x)]^2\,dx}.$$

You will need to use the fact that if $g(x)$ is not identically zero, then $\int_a^b [g(x)]^2\,dx > 0$. As a loose guide to see how the proof can be constructed, think of $f^Tg = \int_a^b f(x)g(x)\,dx$ where, of course, the symbol $f^Tg$ is used as a hint and does not really make sense mathematically in the context of this chapter. Thinking of Problem 7, prove a stronger version of this inequality, with $\int_a^b |f(x)g(x)|\,dx$ replacing $|\int_a^b f(x)g(x)\,dx|$.

# 3

# THE ALGEBRAIC EIGENVALUE PROBLEM

## 3.1 INTRODUCTION

In this chapter we consider computational procedures for solving the following problem:

> Let $A = (a_{ij})$ be any $(n \times n)$ matrix where each $a_{ij}$ is real. Find all scalars, $\lambda$, (real or complex) and all $(n \times 1)$ nonzero vectors $\mathbf{x}$ (whose components may also be complex) such that the vector equation
>
> $$A\mathbf{x} = \lambda\mathbf{x} \tag{3.1}$$
>
> is satisfied. The scalar $\lambda$ is called an *eigenvalue* of $A$ and the vector $\mathbf{x}$ is called an *eigenvector* corresponding to $\lambda$.

This problem arises in many different physical settings ranging from the social sciences such as economics to the physical sciences such as engineering and physics. Often it occurs in the solution of a system of differential equations modeling some set of physical phenomena, but it can arise in other problems such as structures analysis. As an example which shows how eigenvalues and eigenvectors relate to differential equations, let us consider the simple system of differential equations:

$$\begin{aligned} u'(x) &= u(x) - 2v(x) \\ v'(x) &= u(x) + 4v(x) \end{aligned} \tag{3.2}$$

Thus, in this example, we are seeking functions $u(x)$ and $v(x)$ which satisfy the relationships given in (3.2). It is natural to write this system in matrix form as

$$\mathbf{y}'(x) = A\mathbf{y}(x) \tag{3.3}$$

where

$$\mathbf{y}(x) = \begin{bmatrix} u(x) \\ v(x) \end{bmatrix} \quad \text{and} \quad A = \begin{bmatrix} 1 & -2 \\ 1 & 4 \end{bmatrix}.$$

The equation $\mathbf{y}'(x) = A\mathbf{y}(x)$ is reminiscent of the single variable problem $y' = \alpha y$ which has solutions of the form $y(x) = ce^{\alpha x}$, where $c$ is any nonzero constant. Thus it seems reasonable that a solution of (3.3) might exist in the form

$$\mathbf{y}(x) = e^{\lambda x}\mathbf{y}_0 \tag{3.4}$$

for some scalar $\lambda$ and some nonzero constant vector $\mathbf{y}_0$. Substituting (3.4) into (3.3), we obtain

$$\lambda e^{\lambda x}\mathbf{y}_0 = e^{\lambda x}A\mathbf{y}_0$$

or

$$e^{\lambda x}(A - \lambda I)\mathbf{y}_0 = \boldsymbol{\theta}.$$

Since the function $e^{\lambda x}$ never vanishes, we must have $(A - \lambda I)\mathbf{y}_0 = \boldsymbol{\theta}$ or $A\mathbf{y}_0 = \lambda\mathbf{y}_0$ which is the eigenvalue problem as formulated in (3.1). Thus we can find solutions to (3.2) by finding eigenvalues and eigenvectors of $A$ and then using (3.4).

In the example above, it is easy to verify by direct substitution that $\lambda_1 = 2$ and $\lambda_2 = 3$ are eigenvalues of $A$, with corresponding eigenvectors given by

$$\mathbf{y}_1 = \begin{bmatrix} 2\alpha \\ -\alpha \end{bmatrix}, \quad \mathbf{y}_2 = \begin{bmatrix} \beta \\ -\beta \end{bmatrix}$$

where $\alpha$ and $\beta$ are any nonzero scalars. That is, the identities $A\mathbf{y}_1 = \lambda_1\mathbf{y}_1$ and $A\mathbf{y}_2 = \lambda_2\mathbf{y}_2$ hold no matter what $\alpha$ and $\beta$ are. Hence, the functions $\mathbf{y}_1(x) = \alpha e^{2x}\left[\begin{smallmatrix} 2 \\ -1 \end{smallmatrix}\right]$ and $\mathbf{y}_2(x) = \beta e^{3x}\left[\begin{smallmatrix} 1 \\ -1 \end{smallmatrix}\right]$ are each solutions of (3.3). Furthermore, if $\mathbf{y}_1(x)$ and $\mathbf{y}_2(x)$ are solutions to $\mathbf{y}'(x) = A\mathbf{y}(x)$, then so is $\mathbf{y}(x) = \mathbf{y}_1(x) + \mathbf{y}_2(x)$. Thus we find, for $\mathbf{y}(x)$ given by

$$\mathbf{y}(x) = \begin{bmatrix} 2\alpha e^{2x} + \beta e^{3x} \\ -\alpha e^{2x} - \beta e^{3x} \end{bmatrix}$$

that $\mathbf{y}(x)$ is a solution of (3.3). Equivalently, the choice $u(x) = 2\alpha e^{2x} + \beta e^{3x}$ and $v(x) = -\alpha e^{2x} - \beta e^{3x}$ is a solution to our original problem (3.2). The presence of the arbitrary constants, $\alpha$ and $\beta$, allow us to specify "initial values" for the solution. For example, if we wished to prescribe that $u(0) = 4$ and $v(0) = -3$, we would be led to the system

$$\begin{aligned} 2\alpha + \beta &= 4 \\ -\alpha - \beta &= -3 \end{aligned}$$

which would determine $\alpha = 1$ and $\beta = 2$.

Systems of the sort given in (3.2) occur frequently in problems in engineering and science. For example, suppose we are given a radioactive substance $R$ which decays into a radioactive substance $S$ and suppose that $S$ in turn decays into a stable substance $T$. Next, let us assume that radioactive decay takes place at a rate proportional to the amount of the parent substance present. Thus, if $q_1(t)$ denotes the amount of substance $R$ present at time $t$, then $dq_1/dt = -c_1q_1(t)$. At any time $t$, the rate of change of $q_2(t)$ (where $q_2(t)$ is the amount of $S$ present at time $t$), is equal to the rate of formation of $S$ (from $R$) minus the rate of decay of $S$ (to $T$). Thus, $dq_2/dt = c_1q_1(t) - c_2q_2(t)$. If we suppose at time $t = 0$ that we have $r_0$ grams of $R$, then the following system of equations serves as a mathematical model for the physical system:

$$\frac{dq_1}{dt} = -c_1q_1(t), \qquad q_1(0) = r_0$$

$$\frac{dq_2}{dt} = c_1q_1(t) - c_2q_2(t), \qquad q_2(0) = 0.$$

Returning now to the topic of this chapter, we should first observe that finding eigenvalues and eigenvectors of matrices is a formidable computational problem. The problem is of such practical importance, however, that effective numerical procedures have been developed for its solution. Before presenting some of these, let us recall a few basic mathematical preliminaries that are essential to understanding the problem and to deriving techniques for its solution. As noted above, solving $A\mathbf{x} = \lambda\mathbf{x}$ is equivalent to solving $(A - \lambda I)\mathbf{x} = \mathbf{\theta}$. Obviously $\mathbf{x} = \mathbf{\theta}$ is always a (trivial) solution, and so we ignore it and make the stipulation that *eigenvectors* must be nonzero. We do allow zero eigenvalues, however, and it should be clear to the reader that $A$ is singular if and only if $A$ has a zero eigenvalue.

Considering the eigenvalue problem in the form $(A - \lambda I)\mathbf{x} = \mathbf{\theta}$, $\mathbf{x} \neq \mathbf{\theta}$, we see that $\lambda$ is an eigenvalue of $A$ if and only if the matrix $(A - \lambda I)$ is singular. Therefore the eigenvalues of $A$ are the values of $\lambda$ which satisfy

$$p(\lambda) \equiv \det(A - \lambda I) = 0. \tag{3.5}$$

(Recall the two equivalent criteria for singularity given in Section 1, Chapter 2.) By a cofactor expansion of $\det(A - \lambda I)$ it can be seen that $p(\lambda)$ is an $n$th degree polynomial in $\lambda$ (Problem 4) and, in fact, can be written in the form

$$p(\lambda) = (-1)^n\lambda^n + a_1\lambda^{n-1} + \cdots + a_{n-1}\lambda + a_n.$$

The equation $p(\lambda) = 0$ is called the *characteristic equation* of $A$. Thus the eigenvalues of $A$ are precisely the zeros of $p(\lambda)$ and, since the degree of $p(\lambda)$ equals $n$, there are exactly $n$ eigenvalues, $\{\lambda_i\}_{i=1}^n$, of $A$. (This is a consequence of the well-known "Fundamental Theorem of Algebra" which states that any $n$th degree polynomial has exactly $n$ zeros if we count multiple zeros and allow for the possibility that some of the zeros may be complex, even though the polynomial coefficients are all real. See Section 4.4, for more detail.) Example 3.1 illustrates the most elementary approach for solving the eigenvalue problem. Namely, first find $p(\lambda)$ by expanding the determinant of $(A - \lambda I)$, and then find the zeros, $\{\lambda_i\}_{i=1}^n$, of

$p(\lambda)$ by some polynomial root finding technique (such as those discussed in the next chapter). For each distinct $\lambda_i$, the matrix $(A - \lambda_i I)$ is singular, so there exists at least one nonzero vector $\mathbf{x}_i$ which satisfies $(A - \lambda_i I)\mathbf{x}_i = \mathbf{\theta}$. Eigenvectors $\mathbf{x}_i$ can be found, for example, by Gauss elimination. Though quite illustrative, techniques which follow this procedure are generally computationally ineffective and are seldom used in practice. This is due in part to the problem of *ill-conditioning*, which we shall discuss in detail as we progress, and thus we shall present only one method of this type (see Section 3.3).

**Example 3.1.** Let $A$ be the $(3 \times 3)$ matrix:

$$A = \begin{bmatrix} 1 & 2 & 1 \\ 0 & 1 & 3 \\ 2 & 1 & 1 \end{bmatrix}.$$

Then,

$$\det(A - \lambda I) = \begin{vmatrix} 1-\lambda & 2 & 1 \\ 0 & 1-\lambda & 3 \\ 2 & 1 & 1-\lambda \end{vmatrix}$$

or

$$p(\lambda) = (1-\lambda)\begin{vmatrix} 1-\lambda & 3 \\ 1 & 1-\lambda \end{vmatrix} + 2\begin{vmatrix} 2 & 1 \\ 1-\lambda & 3 \end{vmatrix} = [(1-\lambda)^3 - 3(1-\lambda)] + [12 - 2(1-\lambda)].$$

Thus $p(\lambda) = -\lambda^3 + 3\lambda^2 + 2\lambda + 8 = -(\lambda - 4)(\lambda^2 + \lambda + 2)$. Therefore the eigenvalues of $A$ are $\lambda = 4$, $\lambda = (-1 + i\sqrt{7})/2$ and $\lambda = (-1 - i\sqrt{7})/2$. Solving the equation $(A - 4I)\mathbf{x} = \mathbf{\theta}$ by Gauss elimination yields:

$$\begin{aligned} -3x_1 + 2x_2 + x_3 &= 0 \\ -3x_2 + 3x_3 &= 0; \\ 2x_1 + x_2 - 3x_3 &= 0 \end{aligned} \qquad \begin{aligned} -3x_1 + 2x_2 + x_3 &= 0 \\ -3x_2 + 3x_3 &= 0; \\ 7x_2 - 7x_3 &= 0 \end{aligned} \qquad \begin{aligned} -3x_1 + 2x_2 &= -x_3 \\ -3x_2 &= -3x_3 \\ 0 &= 0 \end{aligned}$$

and so we see that the solution satisfies $x_1 = x_2 = x_3$. Thus any vector of the form

$$\begin{bmatrix} \alpha \\ \alpha \\ \alpha \end{bmatrix}, \qquad \alpha \neq 0,$$

is an eigenvector belonging to $\lambda = 4$. Since the other two eigenvalues are complex and $A$ is real, it follows from (3.1) that their corresponding eigenvectors have complex entries.

In Example 3.1, we saw that there were infinitely many eigenvectors that correspond to the eigenvalue 4. That is, every vector $\mathbf{x}$ of the form $\mathbf{x} = [\alpha, \alpha, \alpha]^T$, where $\alpha$ is a nonzero scalar, satisfies the equation $A\mathbf{x} = 4\mathbf{x}$. This was to be expected, since the matrix $A - 4I$ is singular and hence the equation $(A - 4I)\mathbf{x} = \mathbf{\theta}$ has either no solution or infinitely many solutions. However, all eigenvectors of $A$ corresponding to $\lambda = 4$ are in some sense the same, since each eigenvector $\mathbf{x}$ is of the form $\mathbf{x} = \alpha\mathbf{u}_1$, where $\mathbf{u}_1 = [1, 1, 1]^T$. To consider the important question of how many different eigenvectors a matrix possesses, we must make the following definition: Let $\{\mathbf{x}_k\}_{k=1}^m$ be a collection of $(n \times 1)$ vectors with either real or complex components. Then the set of vectors $\{\mathbf{x}_k\}_{k=1}^m$ is said to be *linearly independent* if for any set of scalars, $\{\alpha_k\}_{k=1}^m$ (real or complex), the vector equation

$$\alpha_1\mathbf{x}_1 + \alpha_2\mathbf{x}_2 + \cdots + \alpha_m\mathbf{x}_m = \boldsymbol{\theta}$$

implies that every $\alpha_k = 0$, $1 \le k \le m$. The set of vectors $\{\mathbf{x}_k\}_{k=1}^m$ is said to be *linearly dependent* if it is not linearly independent. Thus $\{\mathbf{x}_k\}_{k=1}^m$ is linearly dependent if we can find $\beta_1, \beta_2, \ldots, \beta_m$ such that $\beta_1\mathbf{x}_1 + \beta_2\mathbf{x}_2 + \cdots + \beta_m\mathbf{x}_m = \boldsymbol{\theta}$, where not all the scalars $\beta_i$ are zero. Although the concept of linear independence is extremely important in mathematical analysis, it is sufficient for understanding the material of this section to illustrate the concept with the following:

**Example 3.2.** As before, let $\mathbf{e}_j$ denote the $(n \times 1)$ vector with 1 in the $j$th component and zeros elsewhere. By writing the vector equation $\alpha_1\mathbf{e}_1 + \alpha_2\mathbf{e}_2 + \cdots + \alpha_n\mathbf{e}_n = \boldsymbol{\theta}$ componentwise, it is obvious that each $\alpha_j = 0$, $1 \le j \le n$. It is equally obvious that any smaller collection of the $\mathbf{e}_j$'s (say, $\{\mathbf{e}_1, \mathbf{e}_3, \mathbf{e}_5, \ldots, \mathbf{e}_n\}$, for $n$ odd) is also linearly independent.

Now we consider the reverse situation, where $\{\mathbf{x}_k\}_{k=1}^m$ is not linearly independent. Then, by definition, there must exist a set of scalars, $\{\alpha_k\}_{k=1}^m$, where at least one $\alpha_k$ (say $\alpha_1$) is not zero and yet $\alpha_1\mathbf{x}_1 + \alpha_2\mathbf{x}_2 + \cdots + \alpha_m\mathbf{x}_m = \boldsymbol{\theta}$. Thus, we can solve for $\mathbf{x}_1$, finding

$$\mathbf{x}_1 = -\frac{1}{\alpha_1}(\alpha_2\mathbf{x}_2 + \cdots + \alpha_m\mathbf{x}_m).$$

Hence, given any equation involving $\{\mathbf{x}_k\}_{k=1}^m$, the equation can be rewritten using only $\{\mathbf{x}_k\}_{k=2}^m$, and so $\mathbf{x}_1$ is not really needed. In this sense, $\mathbf{x}_1$ is not "different" from the rest and is thus said to be "linearly dependent" on them. As a specific example of linear dependence, consider:

$$\mathbf{x}_1 = \begin{bmatrix} 1 \\ 6 \\ 4 \end{bmatrix}, \quad \mathbf{x}_2 = \begin{bmatrix} 1 \\ 2 \\ 0 \end{bmatrix}, \quad \mathbf{x}_3 = \begin{bmatrix} 1 \\ 14 \\ 12 \end{bmatrix}.$$

Writing the vector equation $\alpha_1\mathbf{x}_1 + \alpha_2\mathbf{x}_2 + \alpha_3\mathbf{x}_3 = \boldsymbol{\theta}$ componentwise, we find that $\alpha_1$, $\alpha_2$ and $\alpha_3$ must satisfy

$$\begin{aligned} \alpha_1 + \alpha_2 + \alpha_3 &= 0 \\ 6\alpha_1 + 2\alpha_2 + 14\alpha_3 &= 0 \\ 4\alpha_1 \qquad\quad + 12\alpha_3 &= 0. \end{aligned}$$

Solving by Gauss elimination, we have that $\alpha_2 = 2\alpha_3$ and $\alpha_1 + \alpha_2 = -\alpha_3$. Thus, for example, a nontrivial solution is provided by choosing $\alpha_3 = 1$, $\alpha_2 = 2$ and $\alpha_1 = -3$. From this we see that $-3\mathbf{x}_1 + 2\mathbf{x}_2 + \mathbf{x}_3 = \boldsymbol{\theta}$ or $\mathbf{x}_1 = \frac{1}{3}(2\mathbf{x}_2 + \mathbf{x}_3)$.

We now return to the concept of eigenvectors of an $(n \times n)$ matrix. Unfortunately, it is not necessarily the case that a given $(n \times n)$ matrix has $n$ different or linearly independent eigenvectors. This is important, since many numerical methods for finding eigenvalues require that $A$ have $n$ linearly independent eigenvectors. As a simple example of a $(2 \times 2)$ matrix which does not have 2 linearly independent eigenvectors, let

$$A = \begin{bmatrix} 1 & 1 \\ 0 & 1 \end{bmatrix}.$$

Then the characteristic equation is $(1 - \lambda)^2 = 0$, and we find every eigenvector must have the form

$$\begin{bmatrix} \alpha \\ 0 \end{bmatrix} = \alpha \begin{bmatrix} 1 \\ 0 \end{bmatrix}.$$

Since $\alpha$ can be any nonzero scalar, it would seem that $A$ has an infinite number of eigenvectors with respect to $\lambda = 1$. Since, however, the only eigenvectors of $A$ are scalar multiples of $\left[\begin{smallmatrix} 1 \\ 0 \end{smallmatrix}\right]$, $A$ has only one linearly independent eigenvector. On the other hand, for

$$B = \begin{bmatrix} 1 & 0 \\ 0 & 1 \end{bmatrix},$$

the characteristic equation is again $(1 - \lambda)^2 = 0$, but two eigenvectors are given by $\alpha\left[\begin{smallmatrix} 1 \\ 0 \end{smallmatrix}\right]$ and $\beta\left[\begin{smallmatrix} 0 \\ 1 \end{smallmatrix}\right]$, where $\alpha$ and $\beta$ are arbitrary. Therefore, $B$ has two linearly independent eigenvectors.

In the above examples both $A$ and $B$ have eigenvalues $\lambda = 1, 1$; i.e., 1 is an eigenvalue of multiplicity two. On the other hand, if $A$ is an $(n \times n)$ matrix with *distinct* eigenvalues, $\{\lambda_j\}_{j=1}^n$, then it can be shown (Problem 9) that any set of eigenvectors, $\{\mathbf{x}_j\}_{j=1}^n$, is linearly independent (we assume, in this context, that $A\mathbf{x}_j = \lambda_j \mathbf{x}_j$, $j = 1, 2, \ldots, n$). Therefore the only instance when the eigenvectors of $A$ can fail to be linearly independent is when at least one eigenvalue has multiplicity greater than one. The matrix $B$, above, illustrates that even with eigenvalues of multiplicity greater than one, the eigenvectors may still be linearly independent. Mathematically, the situation may be summarized by the following (without proof).

**Theorem.** Let $\lambda$ be an eigenvalue of $A$ (so $\det(A - \lambda I) = 0$). Then the number of linearly independent solutions of the equation $(A - \lambda I)\mathbf{x} = \boldsymbol{\theta}$ (thus the number of linearly independent eigenvectors corresponding to $\lambda$) equals $k$, where $k = n - r$ and where $r$ is the number of linearly independent column vectors of the matrix $(A - \lambda I)$. (The number $r$ is known as the *rank* of $(A - \lambda I)$).

## PROBLEMS

1. Find the eigenvalues and eigenvectors for the matrices:

$$A = \begin{bmatrix} -1 & 1 \\ 4 & 2 \end{bmatrix}, \quad B = \begin{bmatrix} 0 & 2 & 1 \\ -1 & 3 & 1 \\ -1 & 1 & 2 \end{bmatrix}.$$

2. For the matrix $A$ of Example 3.1, find an eigenvector corresponding to the eigenvalue $\lambda = (-1 + i\sqrt{7})/2$, using Gauss elimination to solve the equation $(A - \lambda I)\mathbf{x} = \boldsymbol{\theta}$.
3. Show the set of vectors $S_1$ is linearly independent, while the set of vectors $S_2$ is linearly dependent, where:

$$S_1 = \left\{ \begin{bmatrix} 1 \\ 1 \\ 1 \end{bmatrix}, \begin{bmatrix} 1 \\ 0 \\ 1 \end{bmatrix}, \begin{bmatrix} 2 \\ 3 \\ 1 \end{bmatrix} \right\}, \quad S_2 = \left\{ \begin{bmatrix} 1 \\ 3 \\ 2 \end{bmatrix}, \begin{bmatrix} 2 \\ 4 \\ 1 \end{bmatrix}, \begin{bmatrix} 0 \\ 2 \\ 3 \end{bmatrix} \right\}.$$

4. Let $A$ be an $(n \times n)$ matrix. Show the function $p(\lambda)$ defined by $p(\lambda) = \det(A - \lambda I)$ is a polynomial of degree $n$. (This is easily done by mathematical induction, verifying directly that $p(\lambda)$ is a quadratic polynomial when $n = 2$.)
5. Let $A = (a_{ij})$ be a lower triangular $(n \times n)$ matrix. Show the eigenvalues of $A$ are $a_{11}$, $a_{22}, \ldots, a_{nn}$.
6. Using the fact that $\det(B) = \det(B^T)$ for any square matrix $B$, show that the eigenvalues of $A$ and the eigenvalues of $A^T$ are the same, where $A$ is a square matrix. Find a $(2 \times 2)$ matrix $A$ which shows that an eigenvector of $A$ is not necessarily an eigenvector of $A^T$.

*7. A set $\{\mathbf{x}_1, \mathbf{x}_2, \ldots, \mathbf{x}_k\}$ of vectors in $R^n$ is called *orthogonal* if $\mathbf{x}_i^T\mathbf{x}_j = 0$ when $i \neq j$. Show any orthogonal set is a linearly independent set.

8. Find the largest possible set of linearly independent eigenvectors for each of:

$$A_1 = \begin{bmatrix} 1 & 1 & 0 \\ 0 & 1 & 1 \\ 0 & 0 & 1 \end{bmatrix}, \quad A_2 = \begin{bmatrix} 1 & 1 & 0 \\ 0 & 1 & 0 \\ 0 & 0 & 1 \end{bmatrix}, \quad A_3 = \begin{bmatrix} 1 & 0 & 0 \\ 0 & 1 & 0 \\ 0 & 0 & 1 \end{bmatrix}.$$

*9. Let $\lambda_1, \lambda_2, \ldots, \lambda_k$ be $k$ distinct eigenvalues for $A$ and let $\mathbf{u}_1, \mathbf{u}_2, \ldots, \mathbf{u}_k$ be corresponding eigenvectors. Show that $\{\mathbf{u}_1, \mathbf{u}_2, \ldots, \mathbf{u}_k\}$ are linearly independent. (Hint: Suppose $\{\mathbf{u}_1, \mathbf{u}_2, \ldots, \mathbf{u}_r\}$ is the largest linearly independent subset of $\{\mathbf{u}_1, \mathbf{u}_2, \ldots, \mathbf{u}_k\}$, so if $r < k$, then $\{\mathbf{u}_1, \mathbf{u}_2, \ldots, \mathbf{u}_{r+1}\}$ is linearly dependent).

10. Show: If $\lambda$ is a complex eigenvalue of a real $(n \times n)$ matrix $A$, then $\bar{\lambda}$ is also an eigenvalue.

## 3.2 THE POWER METHOD

One of the most widely used procedures to estimate eigenvalues of a matrix $A$ is the power method. Our basic assumptions are that $A$ is a real $(n \times n)$ matrix, its eigenvalues $\lambda_1, \lambda_2, \ldots, \lambda_n$ are numbered so that

$$|\lambda_1| > |\lambda_2| \geq |\lambda_3| \geq \cdots \geq |\lambda_n|, \tag{3.6}$$

and that $A$ has $n$ linearly independent eigenvectors, $\{\mathbf{u}_i\}_{i=1}^n$, where $A\mathbf{u}_i = \lambda_i\mathbf{u}_i$, $1 \leq i \leq n$. When the power method works, it gives us an estimate of the dominant eigenvalue, $\lambda_1$.

We first need the fact that any $(n \times 1)$ vector can be expressed as a linear combination of the eigenvectors $\mathbf{u}_1, \mathbf{u}_2, \ldots, \mathbf{u}_n$. To see this, let $B$ be the $(n \times n)$ matrix defined such that the $k$th column vector of $B$ is $\mathbf{u}_k$, $1 \leq k \leq n$, that is,

$B = [\mathbf{u}_1, \mathbf{u}_2, \ldots, \mathbf{u}_n]$. Let $\mathbf{x} = (x_1, x_2, \ldots, x_n)^T$ be any $(n \times 1)$ vector. Recall from Section 2.1, that $B\mathbf{x} = x_1\mathbf{u}_1 + x_2\mathbf{u}_2 + \cdots + x_n\mathbf{u}_n$. Now if $B\mathbf{x} = \boldsymbol{\theta}$, then $x_1\mathbf{u}_1 + x_2\mathbf{u}_2 + \cdots + x_n\mathbf{u}_n = \boldsymbol{\theta}$. But by the linear independence of $\{\mathbf{u}_i\}_{i=1}^n$, this implies that $x_1 = x_2 = \cdots = x_n = 0$, and thus $B$ is non-singular. (Note that we have essentially just proved the basic result that a square matrix is non-singular if and only if its column vectors are linearly independent. Since $\det(B) = \det(B^T)$, the same result is true for row vectors.) Now let $\mathbf{x}_0$ be an arbitrary $(n \times 1)$ vector, and consider the linear system of equations $B\mathbf{x} = \mathbf{x}_0$. Since $B$ is non-singular this system has a unique solution, $\mathbf{x} = (\alpha_1, \alpha_2, \ldots, \alpha_n)^T$, and so $\mathbf{x}_0$ can be uniquely expressed as

$$\mathbf{x}_0 = \alpha_1\mathbf{u}_1 + \alpha_2\mathbf{u}_2 + \cdots + \alpha_n\mathbf{u}_n. \tag{3.7}$$

(To be able to write *any* $\mathbf{x}_0$ uniquely in the form (3.7) is the reason we require $\{\mathbf{u}_i\}_{i=1}^n$ to be linearly independent. Note that some of the $\alpha_i$'s may be complex if some of the $\mathbf{u}_i$'s have complex entries.)

Next, observe that if $A\mathbf{u}_i = \lambda_i\mathbf{u}_i$ then $A(A\mathbf{u}_i) = \lambda_i A\mathbf{u}_i$ or $A^2\mathbf{u}_i = (\lambda_i)^2\mathbf{u}_i$. Continuing, it follows that $A^3\mathbf{u}_i = (\lambda_i)^3\mathbf{u}_i$, or in general

$$A^k\mathbf{u}_i = (\lambda_i)^k\mathbf{u}_i. \tag{3.8}$$

Thus if we form the sequence

$$\begin{aligned}
\mathbf{x}_1 &= A\mathbf{x}_0 \\
\mathbf{x}_2 &= A\mathbf{x}_1, \quad (\mathbf{x}_2 = A^2\mathbf{x}_0) \\
\mathbf{x}_3 &= A\mathbf{x}_2, \quad (\mathbf{x}_3 = A^3\mathbf{x}_0), \\
&\ \ \vdots \qquad\qquad\quad \vdots \\
\mathbf{x}_k &= A\mathbf{x}_{k-1}, (\mathbf{x}_k = A^k\mathbf{x}_0)
\end{aligned}$$

then by (3.7) and (3.8), we have $\mathbf{x}_k = \alpha_1 A^k\mathbf{u}_1 + \alpha_2 A^k\mathbf{u}_2 + \cdots + \alpha_n A^k\mathbf{u}_n$ or

$$\mathbf{x}_k = \alpha_1(\lambda_1)^k\mathbf{u}_1 + \alpha_2(\lambda_2)^k\mathbf{u}_2 + \cdots + \alpha_n(\lambda_n)^k\mathbf{u}_n. \tag{3.9}$$

Recalling (3.6), it is natural to write $\mathbf{x}_k$ as

$$\mathbf{x}_k = (\lambda_1)^k\left\{\alpha_1\mathbf{u}_1 + \alpha_2\left[\frac{\lambda_2}{\lambda_1}\right]^k\mathbf{u}_2 + \cdots + \alpha_n\left[\frac{\lambda_n}{\lambda_1}\right]^k\mathbf{u}_n\right\}. \tag{3.10}$$

In (3.10), we see for large $k$ that the approximation $\mathbf{x}_k \approx (\lambda_1)^k\alpha_1\mathbf{u}_1$ should be good since $|\lambda_1| > |\lambda_2| \geq \cdots \geq |\lambda_n|$. In order to estimate $\lambda_1$, we replace $k$ by $k + 1$ in (3.10) to obtain $\mathbf{x}_{k+1} \approx (\lambda_1)^{k+1}\alpha_1\mathbf{u}_1$, or $\mathbf{x}_{k+1} \approx \lambda_1\mathbf{x}_k$. This suggests a number of ways to approximate $\lambda_1$. For example, $\mathbf{x}_k^T\mathbf{x}_{k+1} \approx \lambda_1\mathbf{x}_k^T\mathbf{x}_k$ and thus $\lambda_1 \approx \mathbf{x}_k^T\mathbf{x}_{k+1}/\mathbf{x}_k^T\mathbf{x}_k$. Another common procedure is to divide the largest component of $\mathbf{x}_k$ into the corresponding component of $\mathbf{x}_{k+1}$. If the largest component is the $j$th component, then this amounts to $\lambda_1 \approx \mathbf{e}_j^T\mathbf{x}_{k+1}/\mathbf{e}_j^T\mathbf{x}_k$.

We wish to be fairly careful about the description of the power method, for this description has two parts, namely the "practical" part which shows how the method is actually carried out, and the "theoretical" part which shows why the method works, what can go wrong, and how to interpret the output. First of all, powers of $A$ are never actually computed in the implementation of the power method, nor is it necessary to know the $\alpha_i$'s and/or the $\mathbf{u}_i$'s in (3.7). We merely guess a vector $\mathbf{x}_0$ and form the sequence $\{\mathbf{x}_k\}$ from $\mathbf{x}_k = A\mathbf{x}_{k-1}$; $k = 1, 2, 3, \ldots$. The equation $\mathbf{x}_k = A^k\mathbf{x}_0$ is the theoretical basis for the power method, but all we really need is the vector $\mathbf{x}_k$ (note that it is uneconomical to compute $A^k$ and then form $\mathbf{x}_k$ by $\mathbf{x}_k = A^k\mathbf{x}_0$). The estimates for $\lambda_1$ can be found by generating the sequence of scalars $\{\beta_k\}$ from the equation

$$\beta_k = \frac{\mathbf{v}^T\mathbf{x}_{k+1}}{\mathbf{v}^T\mathbf{x}_k} \tag{3.11}$$

where $\mathbf{v}$ is usually $\mathbf{x}_k$ or the $r$th unit-vector $\mathbf{e}_r$ (where the $r$th component of $\mathbf{x}_k$ is the dominant one). When programming the power method, it is a little easier to generate the estimate $\beta_k$ from $\beta_k = \mathbf{x}_k^T\mathbf{x}_{k+1}/\mathbf{x}_k^T\mathbf{x}_k$ than to include a test at each step to determine the maximum component of $\mathbf{x}_k$ (moreover, the program will execute faster). The advantage in finding the maximum component of $\mathbf{x}_k$ is that round-off errors will tend to be reduced. Finally, as $k$ increases, the vectors $\mathbf{x}_k$ are usually growing quite large if $|\lambda_1| > 1$ (or quite small if $|\lambda_1| < 1$), as can be seen from (3.9). As we note below, it is desirable to "scale" or "normalize" $\mathbf{x}_k$ to keep it within computational limits. This can be done by dividing $\mathbf{x}_k$ by $\|\mathbf{x}_k\|$ at each step, where we would use the $\ell_2$ vector norm when $\beta_k = \mathbf{x}_k^T\mathbf{x}_{k+1}/\mathbf{x}_k^T\mathbf{x}_k$ and the $\ell_\infty$ vector norm when $\beta_k = \mathbf{e}_r^T\mathbf{x}_{k+1}/\mathbf{e}_r^T\mathbf{x}_k$. It is easy to show (Problem 8) that this has no effect on the convergence.

In order to study the theoretical convergence of the power method, it is most convenient to consider a variation of the method in which the $\beta_k$ are calculated in a slightly different fashion from the two ways described above. In this variation, we assume $\mathbf{v}$ is a *fixed* vector such that $\mathbf{v}^T\mathbf{u}_1 \neq 0$. We then define $\beta_k$ to be given by $\beta_k = \mathbf{v}^T\mathbf{x}_{k+1}/\mathbf{v}^T\mathbf{x}_k$. If we set $\gamma_j = \mathbf{v}^T\mathbf{u}_j$, Eq. (3.10) shows that

$$\beta_k = \lambda_1 \frac{\alpha_1\gamma_1 + \alpha_2\gamma_2\left[\dfrac{\lambda_2}{\lambda_1}\right]^{k+1} + \cdots + \alpha_n\gamma_n\left[\dfrac{\lambda_n}{\lambda_1}\right]^{k+1}}{\alpha_1\gamma_1 + \alpha_2\gamma_2\left[\dfrac{\lambda_2}{\lambda_1}\right]^{k} + \cdots + \alpha_n\gamma_n\left[\dfrac{\lambda_n}{\lambda_1}\right]^{k}} \tag{3.12}$$

The theoretical result, (3.12), reveals that the sequence $\{\beta_k\}$ converges to $\lambda_1$ at essentially the same rate as $(\lambda_2/\lambda_1)^k$ converges to zero. Equation (3.12) also shows how acceleration procedures like Aitkins $\Delta^2$-process can be used to speed up the convergence of $\{\beta_k\}$ (see Section 4.3.1, Problem 7). We leave as a problem for the reader to show that the sequence $\{\beta_k\}$ generated by $\beta_k = \mathbf{x}_k^T\mathbf{x}_{k+1}/\mathbf{x}_k^T\mathbf{x}_k$ also converges to $\lambda_1$, since the analysis is similar to that in (3.12). When $\beta_k = \mathbf{e}_r^T\mathbf{x}_{k+1}/\mathbf{e}_r^T\mathbf{x}_k$, a rigorous analysis is more difficult since the vector $\mathbf{e}_r$ may change from step to step.

The theoretical analysis in (3.12) also reveals some of the problems that can occur and how to interpret machine output. While we do not wish to examine (3.12) extensively, general observations are easy to make. First, even if our assumption that $A$ has $n$ linearly independent eigenvectors holds and even if (3.6) is valid (neither of which we can check in general), it might be that our initial vector $\mathbf{x}_0$ is chosen so that $\alpha_1 = 0$ in (3.7), or so that $\alpha_1$ is quite small. If $\alpha_1 = 0$ and if $|\lambda_2| > |\lambda_3|$ then mathematically, it follows that $\{\beta_k\} \to \lambda_2$. However, the presence of round-off errors tends to contaminate the computation of $\{\mathbf{x}_k\}$, in that the round-off can begin to introduce a nonzero multiple of $\mathbf{u}_1$ in the $\mathbf{x}_k$'s. This makes it quite likely that some $\mathbf{x}_k$ will not be *only* a linear combination of $\mathbf{u}_2, \mathbf{u}_3, \ldots, \mathbf{u}_n$ but will contain the presence of $\mathbf{u}_1$ as well. Then it is just a matter of time until the dominance of $\lambda_1$ asserts itself. The machine output would manifest this by showing that the sequence $\{\beta_k\}$ is apparently converging to some value for a time ($\lambda_2$, presumably), and then changing directions and converging to another value ($\lambda_1$, this time).

As we mentioned above, it is desirable to scale the vectors $\mathbf{x}_k$, and most programs employing the power method do this. The first step is to select a suitable vector norm to use in the scaling. Usually either the $\ell_\infty$ or $\ell_2$ norms are employed, and we illustrate the procedure with the $\ell_2$ norm. First an initial vector, $\mathbf{x}_0$, is chosen, where $\mathbf{x}_0^T\mathbf{x}_0 = 1$ (that is, $\|\mathbf{x}_0\|_2 = 1$). The power method then proceeds, for $i = 1, 2, \ldots$, by:

a) Let $\mathbf{y}_i = A\mathbf{x}_{i-1}$
b) Set $\beta_i = \mathbf{y}_i^T\mathbf{x}_{i-1}$
c) Let $\eta_i = \sqrt{\mathbf{y}_i^T\mathbf{y}_i}$
d) Set $\mathbf{x}_i = \mathbf{y}_i/\eta_i$ and return to step (a).

Subroutine POWERM given in Fig. 3.1 employs this scaling device. Note that $\beta_i$ is the approximation to $\lambda_1$ and that each $\mathbf{x}_i$ is an approximate eigenvector (recall (3.10)).

**Example 3.3.** In this example, subroutine POWERM was used to estimate the dominant eigenvalue of

$$A = \begin{bmatrix} 1 & -1 & 2 \\ -2 & 0 & 5 \\ 6 & -3 & 6 \end{bmatrix}$$

It is easy to verify that the eigenvalues of $A$ are $\lambda_1 = 5$, $\lambda_2 = 3$ and $\lambda_3 = -1$. Three corresponding eigenvectors are

$$\mathbf{u}_1 = \begin{bmatrix} 5 \\ 16 \\ 18 \end{bmatrix}, \quad \mathbf{u}_2 = \begin{bmatrix} 1 \\ 6 \\ 4 \end{bmatrix}, \quad \text{and} \quad \mathbf{u}_3 = \begin{bmatrix} -1 \\ -2 \\ 0 \end{bmatrix}.$$

```
      SUBROUTINE POWERM(A,X,LAMBDA,TOL,N,M,ITERM)
      REAL LAMBDA
      DIMENSION A(20,20),X(20),Y(20)
C
C   SUBROUTINE POWERM USES THE POWER METHOD WITH SCALING TO ESTIMATE THE
C   DOMINANT EIGENVALUE OF A MATRIX A.  THE CALLING PROGRAM MUST SUPPLY
C   THE MATRIX A, AN INITIAL VECTOR X OF EUCLIDEAN LENGTH 1, A TOLERANCE
C   TOL, AN INTEGER N (WHERE A IS (NXN)) AND AN INTEGER M=MAXIMUM NUMBER
C   OF POWER ITERATIONS DESIRED.  THE SUBROUTINE RETURNS WHEN THE
C   DIFFERENCE OF TWO SUCCESSIVE ESTIMATES IS LESS THAN TOL IN ABSOLUTE
C   VALUE OR WHEN M ITERATIONS HAVE BEEN EXECUTED.  IN THE FIRST CASE, A
C   FLAG, ITERM, IS SET TO 1 AND IN THE SECOND CASE ITERM IS SET TO 2.
C   THE APPROXIMATE EIGENVALUE IS RETURNED AS LAMBDA AND AN APPROXIMATE
C   EIGENVECTOR AS X.  LAMBDA MUST BE DECLARED REAL IN THE CALLING
C   PROGRAM
C
      ITR=1
C
C   CALCULATE THE INITIAL EIGENVALUE APPROXIMATION
C
      DO 1 I=1,N
      Y(I)=0.
      DO 1 J=1,N
    1 Y(I)=Y(I)+A(I,J)*X(J)
      TEMP=0.
      YSCALE=0.
      DO 2 I=1,N
      TEMP=TEMP+Y(I)*X(I)
    2 YSCALE=YSCALE+Y(I)*Y(I)
      YSCALE=SQRT(YSCALE)
      ESTOLD=TEMP
C
C   POWER METHOD ITERATION WITH SCALING
C
    3 ITR=ITR+1
      DO 4 I=1,N
    4 X(I)=Y(I)/YSCALE
      DO 5 I=1,N
      Y(I)=0.
      DO 5 J=1,N
    5 Y(I)=Y(I)+A(I,J)*X(J)
      TEMP=0.
      YSCALE=0.
      DO 6 I=1,N
      TEMP=TEMP+Y(I)*X(I)
    6 YSCALE=YSCALE+Y(I)*Y(I)
      YSCALE=SQRT(YSCALE)
      ESTNEW=TEMP
C
C   TEST FOR TERMINATION OF THE POWER METHOD ITERATION
C
      IF(ABS(ESTNEW-ESTOLD).LE.TOL)    GO TO 7
      IF(ITR.GE.M)    GO TO 8
      ESTOLD=ESTNEW
      GO TO 3
    7 ITERM=1
      LAMBDA=ESTNEW
      RETURN
    8 ITERM=2
      LAMBDA=ESTNEW
      RETURN
      END
```

**Figure 3.1**

For this example, the estimates $\beta_k$ in Table 3.1 were computed from $\beta_k = \mathbf{x}_k^T\mathbf{x}_{k+1}/\mathbf{x}_k^T\mathbf{x}_k$ and initial vector $\mathbf{x}_0 = \gamma(1, 1, 1)^T$, where $\gamma = 1/\sqrt{3}$.

**Table 3.1**

| $k$ | $\beta_k$ | $k$ | $\beta_k$ |
|---|---|---|---|
| 1 | 0.4666666E 01 | 14 | 0.5002243E 01 |
| 2 | 0.7127658E 01 | 15 | 0.5001345E 01 |
| 3 | 0.5770836E 01 | 16 | 0.5000809E 01 |
| 4 | 0.5460129E 01 | 17 | 0.5000484E 01 |
| 5 | 0.5245014E 01 | 18 | 0.5000289E 01 |
| 6 | 0.5142666E 01 | 19 | 0.5000171E 01 |
| 7 | 0.5083027E 01 | 20 | 0.5000103E 01 |
| 8 | 0.5049140E 01 | 21 | 0.5000067E 01 |
| 9 | 0.5029204E 01 | 22 | 0.5000038E 01 |
| 10 | 0.5017425E 01 | 23 | 0.5000021E 01 |
| 11 | 0.5010424E 01 | 24 | 0.5000011E 01 |
| 12 | 0.5006239E 01 | 25 | 0.5000006E 01 |
| 13 | 0.5003741E 01 | 26 | 0.5000005E 01 |

We have listed below some representative values of the approximate eigenvectors, $\mathbf{x}_k$. These vectors $\mathbf{x}_k$ all have Euclidean length 1 and so are converging to $\gamma\mathbf{u}_1$, where $\gamma = 1/\|\mathbf{u}_1\|_2$. A quick calculation shows that $\gamma\mathbf{u}_1 = [0.2032789, 0.6504925, 0.7318041]^T$.

| $k$ | $x_k^T$ | | |
|---|---|---|---|
| 6 | 0.2058211E 00 | 0.6420183E 00 | 0.7385463E 00 |
| 12 | 0.2033920E 00 | 0.6501137E 00 | 0.7321092E 00 |
| 18 | 0.2032841E 00 | 0.6504748E 00 | 0.7318184E 00 |
| 24 | 0.2032790E 00 | 0.6504918E 00 | 0.7318048E 00 |

To illustrate the point made previously about how round-off error may begin to introduce a nonzero multiple of $\mathbf{u}_1$, the program was run again with an initial vector $\mathbf{x}_0 = \mathbf{y}_0/\|\mathbf{y}_0\|_2$, where $\mathbf{y}_0 = \alpha_2\mathbf{u}_2 + \alpha_3\mathbf{u}_3$ and where $\alpha_2 = 1/3$ and $\alpha_3 = -2/3$. In this case, the iterates begin converging to $\lambda_2 = 3$ and then change direction and begin converging to $\lambda_1 = 5$. (See Table 3.2.)

Finally, (3.12) reveals that if $\lambda_1 = \lambda_2$, and $|\lambda_1| > |\lambda_3| \geq \cdots \geq |\lambda_n|$, then the $\beta_i$ still converge to $\lambda_1$ (that is, multiple eigenvalues do not affect the power method). If $\lambda_1 = -\lambda_2$ and $|\lambda_1| > |\lambda_3| \geq \cdots \geq |\lambda_n|$, then the $\beta_i$ would exhibit a periodic behavior while $\mathbf{v}^T\mathbf{x}_{i+2}/\mathbf{v}^T\mathbf{x}_i$ would provide an estimate for $(\lambda_1)^2$. Dominant complex eigenvalues cause the most problems, for if $\lambda_1$ is a complex root of $p(\lambda) = 0$ then so is the complex conjugate, $\bar{\lambda}_1$. One further circumstance that must be accounted for is that $|\lambda_1| = |\lambda_2| = \cdots = |\lambda_k|$, $k \leq n$. Adjustments for all of these

**Table 3.2**

| $k$ | $\beta_k$ | $k$ | $\beta_k$ | $k$ | $\beta_k$ |
|---|---|---|---|---|---|
| 1 | 0.1527997E 01 | 17 | 0.3002388E 01 | 33 | 0.4635723E 01 |
| 2 | 0.3680351E 01 | 18 | 0.3003982E 01 | 34 | 0.4765195E 01 |
| 3 | 0.2787491E 01 | 19 | 0.3006630E 01 | 35 | 0.4852588E 01 |
| 4 | 0.3073232E 01 | 20 | 0.3011026E 01 | 36 | 0.4909041E 01 |
| 5 | 0.2975838E 01 | 21 | 0.3018322E 01 | 37 | 0.4944480E 01 |
| 6 | 0.3008090E 01 | 22 | 0.3030365E 01 | 38 | 0.4966341E 01 |
| 7 | 0.2997321E 01 | 23 | 0.3050138E 01 | 39 | 0.4979676E 01 |
| 8 | 0.3000922E 01 | 24 | 0.3082280E 01 | 40 | 0.4987756E 01 |
| 9 | 0.2999737E 01 | 25 | 0.3133698E 01 | 41 | 0.4992640E 01 |
| 10 | 0.3000160E 01 | 26 | 0.3213868E 01 | 42 | 0.4995575E 01 |
| 11 | 0.3000079E 01 | 27 | 0.3333886E 01 | 43 | 0.4997344E 01 |
| 12 | 0.3000196E 01 | 28 | 0.3502944E 01 | 44 | 0.4998403E 01 |
| 13 | 0.3000305E 01 | 29 | 0.3721320E 01 | 45 | 0.4999041E 01 |
| 14 | 0.3000515E 01 | 30 | 0.3973443E 01 | 46 | 0.4999423E 01 |
| 15 | 0.3000856E 01 | 31 | 0.4229045E 01 | 47 | 0.4999656E 01 |
| 16 | 0.3001437E 01 | 32 | 0.4456168E 01 | 48 | 0.4999791E 01 |

situations can be made, but we shall not go further into this here. (See Faddeev and Faddeeva 1963.)

We conclude this section with one particular variation of the power method called the *Rayleigh quotient method.* For this method we assume that the matrix $A$ is symmetric, that is $A^T = A$, and that $|\lambda_1| > |\lambda_2| \geq \cdots \geq |\lambda_n|$. We will show in Section 3.4, that when $A$ is symmetric it has $n$ linearly independent real eigenvectors, $\{\mathbf{u}_i\}_{i=1}^n$, and furthermore the eigenvectors can be chosen such that

$$\mathbf{u}_i^T \mathbf{u}_j = \begin{Bmatrix} 1, & i = j \\ 0, & i \neq j \end{Bmatrix}, \qquad 1 \leq i, j \leq n.$$

(Later we shall see that a set of vectors with this property is called *orthonormal.*)

Now, making use of this property in the power method and expressing $\mathbf{x}_0$ as in (3.7) with $\mathbf{x}_k = A\mathbf{x}_{k-1}$, $k \geq 1$, we can easily verify from (3.9) that $\mathbf{x}_k^T A \mathbf{x}_k = \mathbf{x}_k^T \mathbf{x}_{k+1} = \sum_{j=1}^n \alpha_j^2 \lambda_j^{2k+1}$ and $\mathbf{x}_k^T \mathbf{x}_k = \sum_{j=1}^n \alpha_j^2 \lambda_j^{2k}$ (Problem 5). Now we form the sequence of *Rayleigh quotients*, $\{\mu_k\}$, where

$$\mu_k = \frac{\mathbf{x}_k^T A \mathbf{x}_k}{\mathbf{x}_k^T \mathbf{x}_k}.$$

Then, precisely as in (3.12), we see that the sequence $\{\mu_k\}$ converges to $\lambda_1$ at essentially the same rate as $(\lambda_2/\lambda_1)^{2k}$ converges to zero. Thus the Rayleigh quotient method for symmetric matrices will converge to $\lambda_1$ approximately twice as fast as the power method in the general case. When $A$ is not symmetric, the Rayleigh quotients, $\mu_k$, will still converge to $\lambda_1$ (although establishing convergence is a bit tedious). In fact, subroutine POWERM (given in Fig. 3.1) employs Rayleigh quotients. However, convergence for nonsymmetric matrices is no faster than $(\lambda_2/\lambda_1)^k$ when Rayleigh quotients are used.

As additional motivation for the Rayleigh quotient method, let $\alpha$ be any scalar, let $\mathbf{x}$ be any nonzero $(n \times 1)$ vector and define $\mathbf{r}(\alpha) = A\mathbf{x} - \alpha\mathbf{x}$ where $A$ is symmetric. Thus, for any $\alpha$, $\mathbf{r}(\alpha)$ is an $(n \times 1)$ vector and furthermore $\mathbf{r}(\alpha) = \boldsymbol{\theta}$ if and only if $\alpha$ is an eigenvalue of $A$ and $\mathbf{x}$ is a corresponding eigenvector. Thus the size of $\mathbf{r}(\alpha)$ is somewhat a measure of how well $\alpha$ approximates an eigenvalue of $A$. We next show for a given vector $\mathbf{x}$, that $\|\mathbf{r}(\alpha)\|_2$ is minimized if we choose $\alpha = \mathbf{x}^T A\mathbf{x}/\mathbf{x}^T\mathbf{x}$. To see this, let $(\mathbf{r}(\alpha))^T(\mathbf{r}(\alpha)) = \|\mathbf{r}(\alpha)\|_2^2 \equiv g(\alpha) \geq 0$. Now

$$g(\alpha) = (A\mathbf{x} - \alpha\mathbf{x})^T(A\mathbf{x} - \alpha\mathbf{x}) = \mathbf{x}^T A^2\mathbf{x} - 2\alpha\mathbf{x}^T A\mathbf{x} + \alpha^2\mathbf{x}^T\mathbf{x},$$

since $A$ is symmetric. Then $g'(\alpha) = -2\mathbf{x}^T A\mathbf{x} + 2\alpha\mathbf{x}^T\mathbf{x}$, so $g'(\tilde{\alpha}) = 0$ when $\tilde{\alpha} = (\mathbf{x}^T A\mathbf{x})/(\mathbf{x}^T\mathbf{x})$. Also, $g''(\tilde{\alpha}) = 2\mathbf{x}^T\mathbf{x} > 0$, and thus $g(\alpha)$ (and $\|\mathbf{r}(\alpha)\|_2$) is *minimized* for $\tilde{\alpha} = (\mathbf{x}^T A\mathbf{x})/(\mathbf{x}^T\mathbf{x})$, the *Rayleigh quotient*. We will return to this observation in Section 3.2.3.

## PROBLEMS

1. Program the power method, using (3.11) with $\mathbf{v} = \mathbf{e}_r$ to generate an estimate to the dominant eigenvalue. Thus, at each step, $\beta_k = x_{k+1}^{(r)}/x_k^{(r)}$ where $x_k^{(r)}$ is the dominant component of $\mathbf{x}_k$. Also, at each step, "normalize" $\mathbf{x}_k$ by dividing each component of $\mathbf{x}_k$ by $x_k^{(r)}$ before forming $\mathbf{x}_{k+1}$. Thus, $\|\mathbf{x}_k\|_\infty = 1$ for $k = 0, 1, 2, \ldots,$ and $\beta_k = x_{k+1}^{(r)}$. In addition, at each step, print the vector $\mathbf{x}_k$.
   a) Test your program on the matrix $A$ of Example 3.3 with $\mathbf{x}_0 = (1, 1, 1)^T$, and note there is some improvement in convergence.
   b) Note that the normalized vectors $\mathbf{x}_k$ are converging to an eigenvector that corresponds to $\lambda_1 = 5$, as would be expected from (3.10).
   c) Defining $\varepsilon_i = \lambda_1 - \beta_i$, note that $\varepsilon_{i+1} \approx \delta\varepsilon_i$, where $\delta = \lambda_2/\lambda_1 = 3/5$. This would be expected from (3.12).
2. Write a program that calls subroutine POWERM and test your program on the symmetric matrix $A$ given by
$$A = \begin{bmatrix} 1 & 2 & 1 \\ 2 & 3 & 0 \\ 1 & 0 & 1 \end{bmatrix}.$$
   (Recall the POWERM uses Rayleigh quotients.)
   Use the power method program developed in Problem 1 on the matrix $A$, and note the improvement in convergence provided by the Rayleigh quotient method.
3. The power method generates a sequence of vectors $\{\mathbf{x}_i\}$ where (theoretically) $\mathbf{x}_k = A^k\mathbf{x}_0$. Suppose $A$ is an $(n \times n)$ matrix and suppose that $A^{i+1}$ is formed by multiplying $A$ times $A^i$. How many multiplications are necessary to find $\mathbf{x}_k$ if we use the formula $\mathbf{x}_k = A^k\mathbf{x}_0$? How many multiplications are necessary to find $\mathbf{x}_k$ if we generate $\mathbf{x}_k$ from $\mathbf{x}_{i+1} = A\mathbf{x}_i$, $i = 0, 1, \ldots, k - 1$?

4. a) In the notation of (3.9), let $\mathbf{v}$ be any *fixed* vector such that $\mathbf{v}^T\mathbf{u}_1 \neq 0$. Let $\{\beta_k\}$ be formed as in (3.11) and show that $\lim_{k\to\infty} \beta_k = \lambda_1$. (Assume $\alpha_1 \neq 0$ in (3.7).)

   b) In the following three cases, suppose that $|\lambda_2| > |\lambda_3|$, $\mathbf{v}^T\mathbf{u}_2 \neq 0$ and $\alpha_2 \neq 0$. Establish the *theoretical limits*:

   (1) If $\mathbf{v}^T\mathbf{u}_1 = 0$, then $\lim_{k\to\infty} \beta_k = \lambda_2$

   (2) If $\lambda_1 = \lambda_2$, $\alpha_1 \neq 0$, $\mathbf{v}^T\mathbf{u}_1 \neq 0$, then $\lim_{k\to\infty} \beta_k = \lambda_1$

   (3) If $\lambda_1 = -\lambda_2$, $\alpha_1 \neq 0$, $\mathbf{v}^T\mathbf{u}_1 \neq 0$, then $\lim_{k\to\infty} (\mathbf{v}^T\mathbf{x}_{k+2}/\mathbf{v}^T\mathbf{x}_k) = (\lambda_1)^2$.

5. Let $A$ be an $(n \times n)$ symmetric matrix, with eigenvectors $\mathbf{u}_i$ chosen to be orthonormal. Using the notation of (3.9), show that

$$\mathbf{x}_k^T\mathbf{x}_k = \sum_{j=1}^{n} \alpha_j^2\lambda_j^{2k} \quad \text{and} \quad \mathbf{x}_k^T\mathbf{x}_{k+1} = \sum_{j=1}^{n} \alpha_j^2\lambda_j^{2k+1}.$$

*6. Let $A$ be a symmetric matrix and suppose $\lambda_i$ and $\lambda_j$ are two eigenvalues of $A$ with corresponding eigenvectors $\mathbf{u}_i$ and $\mathbf{u}_j$. Assume that $\lambda_i \neq \lambda_j$, and show $\mathbf{u}_i^T\mathbf{u}_j = 0$. (To be precise, $\mathbf{u}_i$ and $\mathbf{u}_j$ are called *orthogonal*. Hint: Consider $\mathbf{u}_i^T A\mathbf{u}_j = \lambda_j\mathbf{u}_i^T\mathbf{u}_j$.)

*7. In (3.12), assume that $|\lambda_1| > |\lambda_2| \geq \cdots \geq |\lambda_n|$ and that $\alpha_1\gamma_1 \neq 0$. Let the estimates $\beta_i$ defined by (3.11) be generated with a fixed vector $\mathbf{v}$.

   a) Show: $\lim_{i\to\infty} (\beta_i - \lambda_1)(\lambda_1/\lambda_2)^i = C$, where $C$ is some constant. (This means, for $i$ sufficiently large, that $\beta_i \approx \lambda_1 + C(\lambda_2/\lambda_1)^i$.)

   b) Using the notation of Problem 1(c), show from part (a) of this problem, that

$$\lim_{i\to\infty} \frac{\varepsilon_{i+1}}{\varepsilon_i} = \frac{\lambda_2}{\lambda_1}.$$

   This substantiates mathematically the empirical evidence suggested by Problem 1(c). Also note that this result shows that Aitkens $\Delta^2$-process can be applied to accelerate convergence. (Again see Section 4.3.1.)

8. Show that scaling does not affect convergence of the power method. To do this, suppose that $\mathbf{x}_0$ is given by (3.7), where $\|\mathbf{x}_0\| = 1$. If scaling is used, the power method takes the form: $\mathbf{y}_i = A\mathbf{x}_{i-1}$, $\beta_i = \mathbf{v}^T\mathbf{y}_i/\mathbf{v}^T\mathbf{x}_{i-1}$, $\mathbf{x}_i = \mathbf{y}_i/\|\mathbf{y}_i\|$, for $i = 1, 2, \ldots$. (Hint: Show that $\mathbf{v}^T\mathbf{y}_i/\mathbf{v}^T\mathbf{x}_{i-1} = \mathbf{v}^T\mathbf{w}_i/\mathbf{v}^T\mathbf{w}_{i-1}$, where $\mathbf{w}_0 = \mathbf{x}_0$ and $\mathbf{w}_i = A\mathbf{w}_{i-1}$, $i = 1, 2, \ldots$.)

*9. Show that the Rayleigh quotient method will converge for matrices that are not symmetric. Assume, as usual, that $|\lambda_1| > |\lambda_2| \geq \cdots \geq |\lambda_n|$ and $\mathbf{x}_0^T\mathbf{u}_1 \neq 0$.

10. An important problem that arises when solving partial differential equations is the general eigenvalue problem $A\mathbf{x} = \lambda B\mathbf{x}$, where $A$ and $B$ are $(n \times n)$. If $B$ is non-singular, show how the power method (or the inverse power method, see Section 3.2.2) can be used for this problem. What can be done if $B$ is singular, but $A$ is non-singular?

### 3.2.1 Deflation

The power method (if it works) finds the dominant eigenvalue, and this is frequently all that is required in many practical problems. However, should we want several more of the eigenvalues ($\lambda_2, \lambda_3, \ldots, \lambda_p$; in the notation of (3.6)), a process called *deflation* can be used. The basic idea of deflation is to find $\lambda_1$ and then transform the original matrix $A$ into another matrix $B$, where $B$ has eigenvalues $\lambda_2, \lambda_3, \ldots, \lambda_n, 0$. The power method can then be used on $B$ and should generate an estimate to $\lambda_2$ if $|\lambda_2| > |\lambda_3|$ (that is, if $\lambda_2$ is the dominant eigenvalue of $B$).

There are a number of deflation procedures, and we present just one here. Suppose $A$ is $(n \times n)$ and let us denote the first row of $A$ by $\mathbf{a}_1$, so that $\mathbf{a}_1$ is the $(1 \times n)$ vector:

$$\mathbf{a}_1 = (a_{11}, a_{12}, \ldots, a_{1n}).$$

Also suppose $A\mathbf{u}_j = \lambda_j \mathbf{u}_j$ for $j = 1, 2, \ldots, n$, where:

a) the set $\{\mathbf{u}_j\}_{j=1}^n$ is linearly independent;
b) no $\mathbf{u}_j$ is the zero vector;
c) the eigenvector $\mathbf{u}_1$ is

$$\mathbf{u}_1 = \begin{bmatrix} 1 \\ \alpha_2 \\ \alpha_3 \\ \vdots \\ \alpha_n \end{bmatrix}. \tag{3.13}$$

Recall that any scalar multiple of an eigenvector is again an eigenvector. Thus, if the first component of an eigenvector $\mathbf{u}_1$ belonging to $\lambda_1$ is nonzero, we can always assume that $\mathbf{u}_1$ has been normalized to make the first component 1. The case where the first entry is zero can be easily handled by premultiplication and post-multiplication of $A$ by a *permutation matrix* (Problem 3, Section 3.3.2). If we know $\lambda_1$ and $\mathbf{u}_1$, where $\mathbf{u}_1$ is normalized as in Eq. (3.13), we next form $A_1$, where $A_1 = A - \mathbf{u}_1\mathbf{a}_1$. (Note that $\mathbf{u}_1$ is $(n \times 1)$ and $\mathbf{a}_1$ is $(1 \times n)$, so $\mathbf{u}_1\mathbf{a}_1$ is an $(n \times n)$ matrix). It follows immediately that the first row of $A_1$ consists entirely of zeros, so let $A_1$ be given by

$$A_1 = \begin{bmatrix} 0 & 0 & \cdots & 0 \\ b_{21} & b_{22} & & b_{2n} \\ \vdots & & & \\ b_{n1} & b_{n2} & \cdots & b_{nn} \end{bmatrix}. \tag{3.14}$$

We next claim that $A_1$ has eigenvalues $\lambda_2, \lambda_3, \ldots, \lambda_n, 0$. Clearly, 0 is an eigenvalue since $A_1$ is singular. To verify the other eigenvalues, we proceed by noting (as above) that any eigenvector $\mathbf{u}_k$ of $A$ may be assumed to have either form (a) or form (b) below:

$$\text{(a)} \quad \mathbf{u}_k = \begin{bmatrix} 1 \\ \beta_2 \\ \beta_3 \\ \vdots \\ \beta_n \end{bmatrix} \qquad \text{(b)} \quad \mathbf{u}_k = \begin{bmatrix} 0 \\ \gamma_2 \\ \gamma_3 \\ \vdots \\ \gamma_n \end{bmatrix} \tag{3.15}$$

If $\mathbf{u}_k$ has form (3.15a), then $A_1(\mathbf{u}_1 - \mathbf{u}_k) = A(\mathbf{u}_1 - \mathbf{u}_k) - \mathbf{u}_1\mathbf{a}_1(\mathbf{u}_1 - \mathbf{u}_k)$. Now $A(\mathbf{u}_1 - \mathbf{u}_k) = \lambda_1\mathbf{u}_1 - \lambda_k\mathbf{u}_k$ and we note that while $\mathbf{u}_1\mathbf{a}_1$ is an $(n \times n)$ matrix, the product $\mathbf{a}_1(\mathbf{u}_1 - \mathbf{u}_k)$ is a scalar. By the associative property, we can write the $(n \times 1)$ vector $\mathbf{u}_1\mathbf{a}_1(\mathbf{u}_1 - \mathbf{u}_k)$ either as $(\mathbf{u}_1\mathbf{a}_1)(\mathbf{u}_1 - \mathbf{u}_k)$ or as $\mathbf{u}_1(\mathbf{a}_1(\mathbf{u}_1 - \mathbf{u}_k))$. Taking the latter course, we note that $\mathbf{a}_1\mathbf{u}_j$ is the first component of the vector $A\mathbf{u}_j$, $1 \le j \le n$, and hence $\mathbf{a}_1(\mathbf{u}_1 - \mathbf{u}_k) = \lambda_1 - \lambda_k$ by (3.15a). Putting all this together, $A_1(\mathbf{u}_1 - \mathbf{u}_k) = \lambda_1\mathbf{u}_1 - \lambda_k\mathbf{u}_k - (\lambda_1 - \lambda_k)\mathbf{u}_1 = \lambda_k(\mathbf{u}_1 - \mathbf{u}_k)$. Thus $\lambda_k$ is an eigenvalue of $A_1$ with corresponding eigenvector $(\mathbf{u}_1 - \mathbf{u}_k)$, and the first component of $\mathbf{u}_1 - \mathbf{u}_k$ is zero by (3.15a).

In the case $\mathbf{u}_k$ has the form (3.15b), we see immediately that $A_1\mathbf{u}_k = \lambda_k\mathbf{u}_k$, since in this case $\mathbf{a}_1\mathbf{u}_k = 0$ and so $\mathbf{u}_1\mathbf{a}_1\mathbf{u}_k = \mathbf{u}_1(\mathbf{a}_1\mathbf{u}_k) = \boldsymbol{\theta}$. The next thing to note is that in either case (3.15a) or (3.15b), the eigenvectors belonging to the eigenvalues $\lambda_2, \ldots, \lambda_n$ of $A_1$ all have a zero in the first component. This means the first column of $A_1$ is irrelevant as far as the Eq. (3.1) for $\lambda_2, \ldots, \lambda_n$ are concerned. In particular, let us define $B$ from (3.14) by

$$B = \begin{bmatrix} b_{22} & \cdots & b_{2n} \\ \vdots & & \vdots \\ b_{n2} & \cdots & b_{nn} \end{bmatrix} \tag{3.16}$$

so that $B$ is the lower $(n - 1) \times (n - 1)$ block of $A_1$. Then $B$ has eigenvalues $\lambda_2, \ldots, \lambda_n$ and the entire process of the power method can be repeated on $B$, including the above deflation scheme.

**Example 3.4.** We use the matrix of Example 3.1 to illustrate the process. One eigenvalue of $A$ is $\lambda = 4$, with corresponding eigenvector of the form

$$\begin{bmatrix} \alpha \\ \alpha \\ \alpha \end{bmatrix}, \qquad \alpha \neq 0.$$

Thus we set

$$\mathbf{u}_1 = \begin{bmatrix} 1 \\ 1 \\ 1 \end{bmatrix}$$

and observe that $A\mathbf{u}_1 = 4\mathbf{u}_1$, where

$$A = \begin{bmatrix} 1 & 2 & 1 \\ 0 & 1 & 3 \\ 2 & 1 & 1 \end{bmatrix}.$$

Then

$$A_1 = A - \mathbf{u}_1\mathbf{a}_1 = \begin{bmatrix} 1 & 2 & 1 \\ 0 & 1 & 3 \\ 2 & 1 & 1 \end{bmatrix} - \begin{bmatrix} 1 \\ 1 \\ 1 \end{bmatrix}(1 \quad 2 \quad 1) = \begin{bmatrix} 0 & 0 & 0 \\ -1 & -1 & 2 \\ 1 & -1 & 0 \end{bmatrix},$$

and the deflated matrix $B$ is

$$B = \begin{bmatrix} -1 & 2 \\ -1 & 0 \end{bmatrix}.$$

Note that the characteristic equation for $B$ is $\lambda^2 + \lambda + 2 = 0$ and thus $B$ has all the eigenvalues of $A$, except for $\lambda = 4$ which was "divided out."

The process of deflation is subject to substantial numerical difficulties in practice. In most cases, we do not know $\lambda_1$ precisely (perhaps we have a very good estimate of $\lambda_1$ from the power method, but we still do not know it exactly). Consequently, we do not have $u_1$ precisely, and so the deflated matrix $B$ will not have precisely the eigenvalues $\lambda_2, \ldots, \lambda_n$. If we in turn deflate $B$, we introduce additional errors. We cannot be sure that the matrix that results after several deflations has eigenvalues that are very near those of the original matrix $A$. Because of all of the contingencies that may arise, (eigenvalues equal in magnitude, multiple, or complex), the power method and deflation necessitate a very sophisticated program for finding *all* eigenvalues. Thus the method is not generally recommended for this purpose, although it can be quite useful in determining the first few dominant eigenvalues.

### 3.2.2 The Inverse Power Method

The inverse power method gives us a way to approximate eigenvalues other than the dominant one. The price we pay is that the inverse power method requires more computation. The method is based on the observation that if $A\mathbf{u}_i = \lambda_i\mathbf{u}_i$, $i = 1, 2, \ldots, n$; then for a scalar $\alpha$, not equal to an eigenvalue, the non-singular matrix $A - \alpha I$ has eigenvalues $(\lambda_i - \alpha)$ (since if $A\mathbf{u}_i = \lambda_i\mathbf{u}_i$, then $(A - \alpha I)\mathbf{u}_i = (\lambda_i - \alpha)\mathbf{u}_i$). Furthermore, the matrix $(A - \alpha I)^{-1}$ has eigenvalues $(\lambda_i - \alpha)^{-1}$ for $i = 1, 2, \ldots, n$ (for if $(A - \alpha I)\mathbf{u}_i = (\lambda_i - \alpha)\mathbf{u}_i$, then $(\lambda_i - \alpha)^{-1}\mathbf{u}_i = (A - \alpha I)^{-1}\mathbf{u}_i$). If we were to apply the power method to the matrix $(A - \alpha I)^{-1}$, we would converge to the eigenvalue of $(A - \alpha I)^{-1}$ that is largest in magnitude. Since the eigenvalues of $(A - \alpha I)^{-1}$ are $1/(\lambda_1 - \alpha), 1/(\lambda_2 - \alpha), \ldots, 1/(\lambda_n - \alpha)$, it follows that we would obtain estimates to $1/(\lambda_j - \alpha)$, where $|\lambda_j - \alpha| \leq |\lambda_i - \alpha|$, $i = 1, 2, \ldots, n$.

To be a bit more precise, suppose that $\alpha$ is any scalar and that we form the sequence of vectors

$$\mathbf{y}_{k+1} = (A - \alpha I)^{-1}\mathbf{y}_k, \qquad k = 0, 1, \ldots. \tag{3.17}$$

Then, under assumptions similar to those of the last section, the sequence $\{\beta_k\}$ defined by

$$\beta_k = \frac{\mathbf{y}_{k+1}^T\mathbf{y}_k}{\mathbf{y}_k^T\mathbf{y}_k}, \qquad k = 0, 1, \ldots. \tag{3.18}$$

converges to $1/(\lambda_j - \alpha)$, where $\lambda_j$ is the eigenvalue of $A$ nearest to $\alpha$. In particular, the numbers $(1/\beta_k) + \alpha$ converge to the eigenvalue of $A$ that is nearest $\alpha$. As an important special case, we note that the choice $\alpha = 0$ would lead to estimates

```
      SUBROUTINE POWINV(A,X,ALPHA,LAMBDA,TOL,N,M,ITERM)
      REAL LAMBDA
      DIMENSION A(20,20),SA(20,20),X(20),SX(20),Y(20)
C
C   SUBROUTINE POWINV USES THE INVERSE POWER METHOD WITH SCALING TO
C   ESTIMATE EIGENVALUES OF A MATRIX A.  THE CALLING PROGRAM MUST SUPPLY
C   THE MATRIX A, AN INITIAL VECTOR X OF EUCLIDEAN LENGTH 1, A SCALAR
C   ALPHA, A TOLERANCE TOL, AN INTEGER N (WHERE A IS (NXN)) AND AN
C   INTEGER M=MAXIMUM NUMBER OF ITERATIONS DESIRED.  THE SUBROUTINE
C   RETURNS WHEN THE DIFFERENCE OF 2 SUCCESSIVE ESTIMATES IS LESS THAN
C   TOL IN ABSOLUTE VALUE OR WHEN M ITERATIONS HAVE BEEN EXECUTED.  IN
C   THE FIRST CASE, A FLAG, ITERM, IS SET TO 1 AND IN THE SECOND CASE
C   ITERM IS SET TO 2.  THE APPROXIMATE EIGENVALUE IS RETURNED AS LAMBDA
C   AND AN APPROXIMATE EIGENVECTOR AS X.  LAMBDA MUST BE DECLARED REAL
C   IN THE CALLING PROGRAM
C
      ITR=1
      DO 1 I=1,N
      DO 1 J=1,N
    1 SA(I,J)=A(I,J)
C
C   CALCULATE THE INITIAL EIGENVALUE APPROXIMATION
C
      DO 2 I=1,N
      SX(I)=X(I)
    2 A(I,I)=A(I,I)-ALPHA
      CALL GAUS(A,X,Y,N,IERROR)
      IF(IERROR.EQ.2)    GO TO 10
      TEMP=0.
      YSCALE=0.
      DO 3 I=1,N
      TEMP=TEMP+Y(I)*SX(I)
    3 YSCALE=YSCALE+Y(I)*Y(I)
      YSCALE=SQRT(YSCALE)
      ESTOLD=1./TEMP+ALPHA
C
C   INVERSE POWER METHOD ITERATION WITH SCALING
C
    4 ITR=ITR+1
      DO 5 I=1,N
      DO 5 J=1,N
    5 A(I,J)=SA(I,J)
      DO 6 I=1,N
      X(I)=Y(I)/YSCALE
      SX(I)=X(I)
    6 A(I,I)=A(I,I)-ALPHA
      CALL GAUS(A,X,Y,N,IERROR)
      IF(IERROR.EQ.2)    GO TO 10
      TEMP=0.
      YSCALE=0.
      DO 7 I=1,N
      TEMP=TEMP+Y(I)*SX(I)
    7 YSCALE=YSCALE+Y(I)*Y(I)
      YSCALE=SQRT(YSCALE)
      ESTNEW=1./TEMP+ALPHA
C
C   TEST FOR TERMINATION OF THE INVERSE POWER METHOD ITERATION
C
```

Figure 3.2

```
      IF(ABS(ESTNEW-ESTOLD).LE.TOL)    GO TO 8
      IF(ITR.GE.M)    GO TO 9
      ESTOLD=ESTNEW
      GO TO 4
    8 ITERM=1
      LAMBDA=ESTNEW
      RETURN
    9 ITERM=2
      LAMBDA=ESTNEW
      RETURN
C
C  SET ITERM=3 IF THE GAUSS ELIMINATION STEP FAILS
C
   10 ITERM=3
      RETURN
      END
```

**Figure 3.2** (*continued*)

of the eigenvalue of $A$ that is smallest in magnitude. We also observe that the analysis of the last section shows that convergence will be quite rapid when $\alpha$ is close to an eigenvalue.

In order to implement the inverse power method efficiently, we observe that the vectors $\mathbf{y}_{k+1}$ in (3.17) should be computed from the equation

$$(A - \alpha I)\mathbf{y}_{k+1} = \mathbf{y}_k, \qquad k = 0, 1, \ldots. \tag{3.19}$$

Gauss elimination would be a suitable procedure to solve (3.19) and the remarks of Section 2.2.3, are applicable since we are solving a series of problems with the same coefficient matrix, as in (2.11). We also note that, while approximately $n^3/3$ multiplications are needed for Gauss elimination and while finding each $\mathbf{y}_k$ requires $n^2$ multiplications (as opposed to $n^2$ multiplications to get each vector $\mathbf{x}_k$ in the power method), convergence will be quite rapid given that $\alpha$ is a good estimate of an eigenvalue $\lambda_j$. Another complication that may occur to the reader is that if $\alpha$ *is* a good approximation to an eigenvalue then the matrix $A - \alpha I$ is nearly singular and thus may be ill-conditioned. This turns out to be not as much of a problem as we might suspect. This is in part due to the fact that we are looking for a constant of proportionality, not an accurate solution of $(A - \alpha I)\mathbf{y}_k = \mathbf{y}_{k-1}$ (see, for example, Fox 1965).

The ideas of Sections 3.2, 3.2.1 and 3.2.2 can be used together in a number of combinations. For example, the power method can be used to estimate the dominant eigenvalue $\lambda_1$ of $A$, and $A$ can be deflated and the dominant eigenvalue, $\lambda_2$, of the deflated matrix found. Then $\lambda_2$ can be removed by another deflation and $\lambda_3$ estimated by the power method, etc. The inverse power method can then be used on the *original* matrix $A$ with each of these estimates substituted for $\alpha$ in (3.19). Also, we can set $\alpha = 0$ in (3.19) and find the smallest eigenvalue in magnitude, deflate $A$, find the next smallest, etc. These estimates can also then be refined by the inverse power method. The inverse power method is programmed in Fig. 3.2, as subroutine POWINV.

**Example 3.5.** This example illustrates the use of the inverse power method, as applied to the matrix $A$ of Example 3.3. Setting $\alpha = 0$ in (3.17) so as to find the eigenvalue of smallest magnitude, and choosing $\mathbf{y}_0 = \gamma[1, 1, 1]^T$ where $\gamma = 1/\sqrt{3}$, we obtained the results in Table 3.3 from subroutine POWINV (where $\mu_k$ is the eigenvalue estimate).

**Table 3.3**

| $k$ | $\mu_k$ | $k$ | $\mu_k$ | $k$ | $\mu_k$ |
|---|---|---|---|---|---|
| 1 | −0.1184212E 01 | 5 | −0.9719747E 00 | 9 | −0.9996110E 00 |
| 2 | −0.1527612E 01 | 6 | −0.1010056E 01 | 10 | −0.1000133E 01 |
| 3 | −0.8223332E 00 | 7 | −0.9965851E 00 | 11 | −0.9999580E 00 |
| 4 | −0.1082315E 01 | 8 | −0.1001162E 01 | 12 | −0.1000016E 01 |

Since we have the dominant eigenvalue of $A$ from Example 3.3 ($\lambda_1 = 5$) and the smallest eigenvalue of $A$ above ($\lambda_3 = -1$), we set $\alpha = (\lambda_1 + \lambda_3)/2 = 2$ and use subroutine POWINV, finding after 11 iterations that $\lambda_2 \approx 2.999995$. To demonstrate the rapid convergence that results when $\alpha$ is a good estimate of an eigenvalue, we let $\alpha = 5.245$ (from Example 3.3, we have that this value of $\alpha$ approximates that obtained by the power method after five iterations). The inverse power method gave the results in Table 3.4. These results show a striking improvement over those in Example 3.3.

**Table 3.4**

| $k$ | $\mu_k$ |
|---|---|
| 1 | 0.5122054E 01 |
| 2 | 0.5018278E 01 |
| 3 | 0.5002143E 01 |
| 4 | 0.5000240E 01 |
| 5 | 0.5000027E 01 |
| 6 | 0.5000003E 01 |

### 3.2.3 The Rayleigh Quotient Iteration

In Section 3.2.2, we saw that the inverse power iteration, defined by

$$(A - \alpha I)\mathbf{y}_{k+1} = \mathbf{y}_k, \qquad k = 0, 1, \ldots$$

can be used to estimate the eigenvalue of $A$ that is nearest $\alpha$. We also observed that convergence is rapid when $\alpha$ is close to an eigenvalue of $A$. This suggests that we might use a few steps of the power method to provide an estimate $\alpha$ to the dominant eigenvalue, $\lambda_1$, of $A$ and then use inverse power iteration to converge rapidly to $\lambda_1$.

When $A$ is a symmetric matrix, this idea can be exploited even further. Recall (from the concluding remarks of Section 3.2) that given a vector $\mathbf{y}$, the "best" estimate of $\lambda_1$ is found from the Rayleigh quotient, $\mathbf{y}^T A\mathbf{y}/\mathbf{y}^T\mathbf{y}$. Thus we can choose an initial vector $\mathbf{y}_0$, form $\alpha_0 = \mathbf{y}_0^T A \mathbf{y}_0 / \mathbf{y}_0^T \mathbf{y}_0$, and then find $\mathbf{y}_1$ from $(A - \alpha_0 I)\mathbf{y}_1 = \mathbf{y}_0$. We would next form $\alpha_1 = \mathbf{y}_1^T A \mathbf{y}_1 / \mathbf{y}_1^T \mathbf{y}_1$ and find $\mathbf{y}_2$ from $(A - \alpha_1 I)\mathbf{y}_2 = \mathbf{y}_1$, etc. This procedure is called the *Rayleigh quotient iteration* and is usually carried out as follows:

Choose an initial vector $\mathbf{y}_0$ such that $\mathbf{y}_0^T \mathbf{y}_0 = 1$, then for $i = 0, 1, 2, \ldots$:

a) form the scalar $\alpha_i = \mathbf{y}_i^T A \mathbf{y}_i$;

b) solve the equation $(A - \alpha_i I)\mathbf{w}_{i+1} = \mathbf{y}_i$;

c) set $\eta_{i+1} = \mathbf{w}_{i+1}^T \mathbf{w}_{i+1}$;

d) let $\mathbf{y}_{i+1} = \mathbf{w}_{i+1}/\eta_{i+1}$ and return to step (a).

While a rigorous analysis of the Rayleigh quotient iteration is beyond the scope of this text, we can make some remarks of an intuitive nature. In the normal course of events, the vectors $\mathbf{y}_i$ converge to $\mathbf{u}_1$, where $\mathbf{u}_1$ is an eigenvector corresponding to the dominant eigenvalue $\lambda_1$ of $A$. To see this, suppose that $\mathbf{y}_i$ is a good approximation to $\mathbf{u}_1$ and let

$$\mathbf{y}_i = \gamma_1 \mathbf{u}_1 + \gamma_2 \mathbf{u}_2 + \cdots + \gamma_n \mathbf{u}_n$$

where $\gamma_2, \gamma_3, \ldots, \gamma_n$ are small with respect to $\gamma_1$. If $\alpha_i$ is a good approximation to $\lambda_1$, then we see that

$$\begin{aligned} \mathbf{w}_{i+1} &= \gamma_1 \frac{1}{\lambda_1 - \alpha_i} \mathbf{u}_1 + \gamma_2 \frac{1}{\lambda_2 - \alpha_i} \mathbf{u}_2 + \cdots + \gamma_n \frac{1}{\lambda_n - \alpha_i} \mathbf{u}_n \\ &= \mu_1 \mathbf{u}_1 + \mu_2 \mathbf{u}_2 + \cdots + \mu_n \mathbf{u}_n \end{aligned} \tag{3.20}$$

Clearly, if $\gamma_2, \gamma_3, \ldots, \gamma_n$ are small with respect to $\gamma_1$, then $\mu_2, \mu_3, \ldots, \mu_n$ are yet smaller with respect to $\mu_1$. Thus $\mathbf{w}_{i+1}$ (and hence $\mathbf{y}_{i+1}$) is a better approximation to $\mathbf{u}_1$. It can be shown that when $\{\mathbf{y}_i\} \to \mathbf{u}_1$, that convergence is "cubic" (that is, there exists a constant $C$ such that $\|\mathbf{y}_{i+1} - \mathbf{u}_1\| \le C\|\mathbf{y}_i - \mathbf{u}_1\|^3$, where the vector norm is the $\ell_2$ norm).

## PROBLEMS

1. Deflate the matrix $A$ of Example 3.3, finding a $(2 \times 2)$ matrix $A_1$ with eigenvalues 3 and $-1$.

*2. As we noted in the discussion of the Rayleigh quotient method in Section 3.2.1, an $(n \times n)$ symmetric matrix $A$ has a set of $n$ orthonormal eigenvectors $\{\mathbf{u}_1, \mathbf{u}_2, \ldots, \mathbf{u}_n\}$. This fact forms a basis for a deflation technique that applies to *symmetric* matrices. In particular, if $A\mathbf{u}_i = \lambda_i \mathbf{u}_i$ for $i = 1, 2, \ldots, n$; show that the matrix $A_1 = A - \lambda_1 \mathbf{u}_1 \mathbf{u}_1^T$ has eigenvalues $\lambda_2, \lambda_3, \ldots, \lambda_n, 0$.

3. Write a program that utilizes subroutine POWINV. Test your program on the matrix $A$ of Example 3.3, using $\alpha = 4.5$. You should note a dramatic improvement in the rate of convergence, compared to that of Example 3.3. Estimate the rate of convergence, as in (3.10), and check this estimate with your computed results.

4. Use the program of Problem 3 to find the smallest eigenvalue $\lambda_3$ of the matrix $A$ in Problem 2, Section 3.2.1. Having the dominant eigenvalue $\lambda_1$ of $A$ and a corresponding eigenvector $\mathbf{u}_1$ such that $\mathbf{u}_1^T\mathbf{u}_1 = 1$, deflate this symmetric matrix using the ideas developed in Problem 2 above. Using the power method on the deflated matrix, find $\lambda_2$. Refine your estimate of $\lambda_2$, using the inverse power method on the original matrix $A$.
5. Program the Rayleigh quotient iteration and test your program on the matrix $A$ in Problem 2, Section 3.2.1. Contrast your results with those of the first part of Problem 4 above.

## 3.3 SIMILARITY TRANSFORMATIONS AND THE CHARACTERISTIC EQUATION

Before developing some other methods for solving the algebraic eigenvalue problem, we need some elementary theoretical tools. The first is the idea of a *similarity transformation*. Let $A$ be $(n \times n)$ and let $S$ be any non-singular $(n \times n)$ matrix. We say that $B = S^{-1}AS$ is *similar* to $A$. We call the process of transforming $A$ into $B$ a *similarity transformation*. The most significant fact about similar matrices is that they have the same eigenvalues. To see this, note that if $B\mathbf{x} = \lambda\mathbf{x}$, where $\mathbf{x}$ is a nonzero vector, then $S^{-1}AS\mathbf{x} = \lambda\mathbf{x}$ or $A(S\mathbf{x}) = \lambda(S\mathbf{x})$. Since $S$ is non-singular and $\mathbf{x} \neq \mathbf{0}$, then the vector $S\mathbf{x}$ is nonzero, and hence $\lambda$ is an eigenvalue of $A$. Conversely, suppose $A\mathbf{y} = \lambda\mathbf{y}$ where $\mathbf{y}$ is a nonzero vector (so that $\mathbf{x} = S^{-1}\mathbf{y}$ is nonzero). But then, $A(S\mathbf{x}) = \lambda(S\mathbf{x})$ or $S^{-1}AS\mathbf{x} = \lambda\mathbf{x}$ and we have that $\lambda$ is an eigenvalue of $B$ also.

Similarity transformations are very important in both theoretical and applied problems and they provide the foundation for many numerical procedures for determining eigenvalues. One example of the theoretical value of similarity transformations is that they can be used to provide a very nice proof of the famous Cayley-Hamilton theorem stated below. (See Fröberg 1969.)

**Theorem 3.1.** Let the $(n \times n)$ matrix $A$ have characteristic equation $\lambda^n + a_1\lambda^{n-1} + \cdots + a_{n-1}\lambda + a_n = 0$. Then the matrix defined by $A^n + a_1A^{n-1} + \cdots + a_{n-1}A + a_nI$ is the zero matrix.

This theorem is usually stated as "A matrix satisfies its own characteristic equation." The theorem is a good example of a theoretical result that can be used in a practical way. To see how, we need only note that if we know the characteristic polynomial, $p(\lambda)$, for $A$, then we can use procedures introduced in Chapter 4 to find the roots of $p(\lambda) = 0$, and thereby obtain the eigenvalues of $A$. The only difficulty with this idea is developing an efficient and accurate way to find $p(\lambda)$. From the definition, $p(\lambda) = \det(A - \lambda I)$, but clearly the evaluation of a determinant is not practical. For example, it would require more than $10^{18}$ multiplications to evaluate the determinant of a $(20 \times 20)$ matrix using a cofactor expansion.

The Cayley-Hamilton theorem provides a more practical procedure. Let $\mathbf{y}_0$ be any $(n \times 1)$ vector. Then, although we do not know the coefficients $a_1, \ldots, a_n$, we do know by Theorem 3.1 that $(A^n + a_1A^{n-1} + \cdots + a_{n-1}A + a_nI)\mathbf{y}_0 = \mathbf{0}$, or

$$A^n\mathbf{y}_0 + a_1A^{n-1}\mathbf{y}_0 + \cdots + a_{n-1}A\mathbf{y}_0 + a_nI\mathbf{y}_0 = \mathbf{0} \tag{3.21}$$

Now, if we write out the known vectors, $A^{n-i}\mathbf{y}_0$, in Eq. (3.21), we have a system of $n$ linear equations in the $n$ unknowns $a_1, a_2, \ldots, a_n$

$$a_1 A^{n-1}\mathbf{y}_0 + \cdots + a_{n-1}A\mathbf{y}_0 + a_n I\mathbf{y}_0 = -A^n\mathbf{y}_0 \tag{3.22}$$

which we can solve to find the coefficients of $p(\lambda)$. Again, we note the obvious: There is no need to form the powers $A^2, A^3, \ldots, A^{n-1}, A^n$; since all we want is $A^2\mathbf{y}_0 = A(A\mathbf{y}_0)$, $A^3\mathbf{y}_0 = A(A(A\mathbf{y}_0))$, etc. We can illustrate this method (called *Krylov's method*) by reference again to the matrix of Example 3.1.

**Example 3.6.** Choose

$$\mathbf{y}_0 = \begin{bmatrix} 1 \\ 0 \\ 0 \end{bmatrix},$$

then form

$$\mathbf{y}_1 = A\mathbf{y}^0, \qquad \mathbf{y}_2 = A\mathbf{y}_1 = A^2\mathbf{y}_0, \qquad \text{and} \qquad \mathbf{y}_3 = A\mathbf{y}_2 = A^3\mathbf{y}_0.$$

Hence, as

$$A = \begin{bmatrix} 1 & 2 & 1 \\ 0 & 1 & 3 \\ 2 & 1 & 1 \end{bmatrix},$$

we have

$$\mathbf{y}_0 = \begin{bmatrix} 1 \\ 0 \\ 0 \end{bmatrix}, \quad \mathbf{y}_1 = \begin{bmatrix} 1 \\ 0 \\ 2 \end{bmatrix}, \quad \mathbf{y}_2 = \begin{bmatrix} 3 \\ 6 \\ 4 \end{bmatrix}, \quad \mathbf{y}_3 = \begin{bmatrix} 19 \\ 18 \\ 16 \end{bmatrix}.$$

Using (3.22), we obtain the system

$$\begin{aligned} 3a_1 + \ a_2 + a_3 &= -19 \\ 6a_1 \qquad\qquad &= -18 \\ 4a_1 + 2a_2 \qquad &= -16 \end{aligned} \tag{3.23}$$

Thus $a_1 = -3$, $a_2 = -2$, and $a_3 = -8$ and hence $p(\lambda) = \lambda^3 - 3\lambda^2 - 2\lambda - 8$. (The difference in signs between $p(\lambda)$ and $\det(A - \lambda I)$ comes from the form in which the Cayley-Hamilton theorem is usually stated. This clearly makes no difference to us, since the roots of $p(\lambda) = 0$ and $-p(\lambda) = 0$ are the same.) While this method is conceptually quite simple and easy to implement, we must be cautious in its use. The reason for caution, is that the coefficients $a_1, a_2, \ldots, a_n$ cannot, in general, be calculated without some error due to rounding. (See Example 3.9, for an instance of a polynomial where *small* perturbations in the values of the coefficients can cause *large* perturbations in the values of its roots.)

There are many other methods for estimating eigenvalues which are based more directly on the idea of a similarity transformation. These procedures quite often proceed by transforming a matrix $A$ into another (similar) matrix whose characteristic polynomial can be more readily evaluated, although the polynomial itself is not usually explicitly computed. For these procedures (and for other purposes as well), it is desirable to have some feeling for where the eigenvalues of a matrix are located in the complex plane. This is the topic of the next section.

### 3.3.1 Localization of Eigenvalues

In numerical analysis, theorems which describe the location of zeros of polynomials are called localization theorems. Clearly, if we are searching for the roots of a polynomial equation, it is quite important to have an idea of where in the complex plane to begin this search. In this section, we state and prove a number of these theorems and give an indication of how to apply them in specific instances. In all these theorems, we assume $A$ is a real, $(n \times n)$ matrix.

**Theorem 3.2.** If $A$ is symmetric, then all the eigenvalues of $A$ are real.

*Proof*: Suppose $A\mathbf{x} = \lambda\mathbf{x}$ where $\mathbf{x} \neq \boldsymbol{\theta}$. Let $\overline{\mathbf{x}}$ denote the vector whose entries are the complex conjugates of those of $\mathbf{x}$. It is easy to show that $(\overline{A\mathbf{x}}) = A\overline{\mathbf{x}}$ (since $A$ is real) and that $(\overline{\lambda\mathbf{x}}) = \overline{\lambda}\overline{\mathbf{x}}$. Therefore, if we have

$$\mathbf{x} = \begin{bmatrix} x_1 \\ x_2 \\ \vdots \\ x_n \end{bmatrix},$$

then

$$\overline{\mathbf{x}} = \begin{bmatrix} \overline{x}_1 \\ \overline{x}_2 \\ \vdots \\ \overline{x}_n \end{bmatrix},$$

and

$$\overline{\mathbf{x}}^T\mathbf{x} = \mathbf{x}^T\overline{\mathbf{x}} = x_1\overline{x}_1 + x_2\overline{x}_2 + \cdots + x_n\overline{x}_n. \tag{3.24}$$

Recall, if $\beta = r + is$, then $\beta\overline{\beta} = (r + is)(r - is) = r^2 + s^2$. Thus, since $\mathbf{x} \neq \boldsymbol{\theta}$, $\mathbf{x}^T\overline{\mathbf{x}}$ is a real number and, moreover, $\mathbf{x}^T\overline{\mathbf{x}} > 0$. Putting all this information together: From $A\overline{\mathbf{x}} = \overline{\lambda}\overline{\mathbf{x}}$ we obtain $\mathbf{x}^T A\overline{\mathbf{x}} = \overline{\lambda}\mathbf{x}^T\overline{\mathbf{x}}$, and from $A\mathbf{x} = \lambda\mathbf{x}$ we obtain $\overline{\mathbf{x}}^T A\mathbf{x} = \lambda\overline{\mathbf{x}}^T\mathbf{x}$, where $\overline{\mathbf{x}}^T\mathbf{x} = \mathbf{x}^T\overline{\mathbf{x}} > 0$. Now, $\mathbf{x}^T A\overline{\mathbf{x}}$ is a $(1 \times 1)$ matrix and hence its transpose is itself, so $(\mathbf{x}^T A\overline{\mathbf{x}})^T = \mathbf{x}^T A\overline{\mathbf{x}}$. Recalling that the transpose of a product is the product of the transposes in reverse order, $(\mathbf{x}^T A\overline{\mathbf{x}})^T = \overline{\mathbf{x}}^T A^T\mathbf{x}$, so $\overline{\mathbf{x}}^T A^T\mathbf{x} = \mathbf{x}^T A\overline{\mathbf{x}}$. Since $A$ is symmetric, then $\overline{\mathbf{x}}^T A\mathbf{x} = \mathbf{x}^T A\overline{\mathbf{x}}$. Therefore $\lambda\overline{\mathbf{x}}^T\mathbf{x} = \overline{\lambda}\mathbf{x}^T\overline{\mathbf{x}}$ or $\lambda = \overline{\lambda}$ which shows that $\lambda$ is real. ■

This is a significant result since it shows that the search for eigenvalues of a symmetric matrix can be restricted to the real axis (as opposed to the entire complex plane). The following two theorems yield bounds on the magnitude of the eigenvalues of an arbitrary $(n \times n)$ matrix.

**Theorem 3.3.** Let $\lambda$ be an eigenvalue of $A$, then $|\lambda| \leq \|A\|_1$, $|\lambda| \leq \|A\|_E$, and $|\lambda| \leq \|A\|_\infty$.

*Proof*: The proof is trivial and we do it only for the first inequality, $|\lambda| \leq \|A\|_1$. Note that the proof is valid for any compatible pair of vector and matrix norms.

Suppose $A\mathbf{x} = \lambda\mathbf{x}$, where $\mathbf{x} \neq \mathbf{\theta}$. Then $|\lambda| \, \|\mathbf{x}\|_1 = \|\lambda\mathbf{x}\|_1 = \|A\mathbf{x}\|_1 \leq \|A\|_1\|\mathbf{x}\|_1$ and since $\|\mathbf{x}\|_1 > 0$, the theorem follows. ■

The applications of this theorem are obvious. Given a matrix $A$, we can compute easily a number $\|A\|$, which gives us the radius of a circular disk in the complex plane (centered at the origin) in which all the eigenvalues of $A$ must lie. Returning to Example 3.1, we find $\|A\|_1 = 5$, $\|A\|_\infty = 4$ and $\|A\|_E = \sqrt{22}$. Our best bound from the above theorem is then that all the eigenvalues are in a disk of radius 4, centered at the origin. For this particular matrix, it happens that one of these three norms provides the best possible disk, but this is not usually the case.

**Theorem 3.4. Gerschgorin Circle Theorem.** For $A = (a_{ij})$, define the absolute off-diagonal row and column sums by

$$r_k = \sum_{\substack{j=1 \\ j \neq k}}^{n} |a_{kj}| \qquad \text{and} \qquad c_k = \sum_{\substack{j=1 \\ j \neq k}}^{n} |a_{jk}|, \qquad \text{respectively.}$$

For $k = 1, 2, \ldots, n$, the sets $R_k = \{z\colon |z - a_{kk}| \leq r_k\}$ and $C_k = \{z\colon |z - a_{kk}| \leq c_k\}$ are circular disks in the complex plane, centered at $a_{kk}$, of radius $r_k$ and $c_k$, respectively. If $\lambda$ is any eigenvalue of $A$, then $\lambda \in R_k$ for some $k$, $1 \leq k \leq n$, and $\lambda \in C_m$ for some $m$, $1 \leq m \leq n$.

*Proof*: Suppose $A\mathbf{x} = \lambda\mathbf{x}$ where $\mathbf{x} \neq \mathbf{\theta}$, and

$$\mathbf{x} = (x_1, x_2, \ldots, x_n)^T \tag{3.25}$$

Choose $i$ such that $|x_i| \geq |x_j|$, $j = 1, 2, \ldots, n$, ($\|\mathbf{x}\|_\infty = |x_i|$). Then, the $i$th row of the equation $A\mathbf{x} = \lambda\mathbf{x}$ is

$$a_{i1}x_1 + a_{i2}x_2 + \cdots + a_{ii}x_i + \cdots + a_{in}x_n = \lambda x_i \tag{3.26}$$

and therefore

$$(a_{ii} - \lambda)x_i = -\sum_{\substack{j=1 \\ j \neq k}}^{n} a_{ij}x_j. \tag{3.27}$$

Dividing by $x_i$, taking absolute values, and using the triangle inequality on the right, we obtain the fact that $|a_{ii} - \lambda| \leq r_i$. To see that $\lambda$ is also in a disk $C_m$ for some $m$, we note that $\det(A) = \det(A^T)$ and thus $A$ and $A^T$ have the same eigenvalues (but not necessarily the same eigenvectors). It can further be shown that if exactly $p$ of these disks intersect then there are exactly $p$ eigenvalues in the union of the $p$ disks. ■

**Example 3.7.** Consider the $(3 \times 3)$ matrix

$$A = \begin{bmatrix} 1 & 2 & -1 \\ 2 & 7 & 0 \\ -1 & 0 & -5 \end{bmatrix}.$$

As $A$ is symmetric, all the eigenvalues of $A$ are real. By Gerschgorin's theorem, any eigenvalues must be in one of the three intervals $[-2, 4]$, $[5, 9]$ and $[-6, -4]$. See Problem 4 for a procedure to sharpen this estimate.

## PROBLEMS

1. Use Krylov's method, as in Example 3.6, to determine the characteristic equation $p(\lambda) = 0$ for the matrix of Example 3.3. Use $\mathbf{y}_0 = (1, 1, 1)^T$.
2. An example of a permutation matrix $P$ is given by the $(n \times n)$ matrix $P$ where for a particular choice $r$ and $s$:

$$p_{rs} = p_{sr} = 1; \qquad p_{rk} = 0, \quad k \neq s; \qquad p_{sk} = 0, \quad k \neq r$$

and for $i \neq r, s$; $p_{ii} = 1$ and $p_{ij} = 0$ for $j \neq i$. That is, $P$ looks like the $(n \times n)$ identity matrix except for the $r$th and $s$th rows. Show:

a) $P$ is symmetric and $PP = I$;

b) the matrix $AP$ is the same as $A$, except with the $r$th and $s$th columns interchanged;

c) the matrix $PA$ is the same as $A$, except with the $r$th and $s$th rows interchanged.

*3. Referring to the deflation technique of Section 3.2.2, as demonstrated in Example 3.4, suppose that $A\mathbf{u}_1 = \lambda_1\mathbf{u}_1$ and that the first component of $\mathbf{u}_1$ is zero. Since $\mathbf{u}_1$ is an eigenvector, $\mathbf{u}_1$ has at least one nonzero component, say the second component. Let $P$ be a permutation matrix, as in Problem 2, with $r = 1$ and $s = 2$. Let $\mathbf{x}_1 = P\mathbf{u}_1$ and show:

a) the first component of the vector $\mathbf{x}_1$ is nonzero;

b) the matrix $PAP$ is similar to $A$;

c) $\mathbf{x}_1$ is an eigenvector of $PAP$ corresponding to $\lambda_1$.

4. In Example 3.7, show that the Gerschgorin circle theorem applied to $D^{-1}AD$ gives sharper bounds than those obtained in the example, where $D$ is a diagonal matrix, $d_{11} = 1$, $d_{22} = 1/7$, and $d_{33} = -45$.
5. Construct a $(4 \times 4)$ permutation matrix $P$ to interchange the first and third rows of a $(4 \times 4)$ matrix $A$. Thus, the matrix $PA$ is the same as $A$, except with the third and first rows interchanged. Verify by direct multiplication that $PP = I$ and that forming the product $AP$ has the effect of interchanging the first and third columns of $A$.

*6. Recall from the discussion of the Rayleigh quotient method, that an $(n \times n)$ symmetric matrix $A$ has a set of $n$ orthonormal eigenvectors $\mathbf{u}_1, \mathbf{u}_2, \ldots, \mathbf{u}_n$ (see Section 3.4 for a proof of this important fact). Also recall that each eigenvector $\mathbf{u}_i$ is real (that is, each $\mathbf{u}_i$ has real components), and any $(n \times 1)$ vector $\mathbf{x}$ can be written as $\mathbf{x} = \alpha_1\mathbf{u}_1 + \alpha_2\mathbf{u}_2 + \cdots + \alpha_n\mathbf{u}_n$. From Theorem 3.2, every eigenvalue of $A$ is real, but more can be said. Show: if $\mathbf{x} \neq \mathbf{0}$, then

$$\lambda_n \leq \frac{\mathbf{x}^T A\mathbf{x}}{\mathbf{x}^T\mathbf{x}} \leq \lambda_1$$

where we have ordered the eigenvalues of $A$ so that $\lambda_n \leq \lambda_{n-1} \leq \cdots \leq \lambda_1$.

7. Consider the matrix $A$ given by

$$A = \begin{bmatrix} 5 & 4 & 1 & 1 \\ 4 & 5 & 1 & 1 \\ 1 & 1 & 4 & 2 \\ 1 & 1 & 2 & 4 \end{bmatrix}.$$

This matrix has eigenvalues $\lambda_4 = 1$, $\lambda_3 = 2$, $\lambda_2 = 5$ and $\lambda_1 = 10$. Use Theorem 3.4 to find an interval $[a, b]$ that contains all the eigenvalues of $A$. Use Problem 6 to obtain an upper bound for $\lambda_4$ and a lower bound for $\lambda_1$, trying various vectors $\mathbf{x}$.

### 3.3.2 Transformation Methods

The reader by now can begin to see the truth in our original statement that the determination of *all* of the eigenvalues of an arbitrary $(n \times n)$ real matrix is quite a formidable task. For finding certain isolated eigenvalues, the power method or the inverse power method are quite effective. But the methods we have discussed up to now may suffer greatly from round-off when we are trying to find all of the eigenvalues, and much deeper mathematical theory is necessary to attack this problem efficiently. This section is a preliminary which provides a background for the further study of modern methods for determining the eigenvalues of an arbitrary matrix. For the details of these methods, the interested reader may consult a number of advanced texts. The eigenvalue problem for a symmetric matrix, however, is considerably simpler and more stable with respect to round-off. The reader may wish to cover this present section more lightly than the rest and be more diligent again in Section 3.3.3 on symmetric matrices. The reader should be sure, however, not to neglect Example 3.9 of this section because of its significant illustration of the effects of round-off error.

In this section, we see how to transform $A$ by similarity transformations into a matrix whose characteristic polynomial, $p(\lambda)$, can be readily evaluated for any fixed scalar $\alpha$. (For example, if $\alpha = 2$, we want an efficient way to find the value, $p(2)$, *without* needing to know the coefficients of $p(\lambda)$.) The form we want to consider is the Hessenberg form. A matrix $B$ is called a *Hessenberg matrix* if $B$ has the form

$$B = \begin{bmatrix} b_{11} & b_{12} & 0 & 0 & 0 & \cdots & 0 \\ b_{21} & b_{22} & b_{23} & 0 & 0 & \cdots & 0 \\ b_{31} & b_{32} & b_{33} & b_{34} & 0 & \cdots & 0 \\ \vdots & & & & & & \\ b_{n1} & b_{n2} & b_{n3} & & & \cdots & b_{nn} \end{bmatrix}. \tag{3.28}$$

Thus $b_{ij} = 0$ when $j \geq i + 2$ and $B$ would be a lower triangular matrix except for the presence of the *super diagonal.* We will see shortly how to evaluate the characteristic polynomial of a Hessenberg matrix for any given scalar $\alpha$, but first we wish to see how a general matrix $A$ can be put into Hessenberg form.

Let $A$ be $(n \times n)$ and suppose $a_{12} \neq 0$. We construct a matrix $S_1$ as follows:

$$S_1 = \begin{bmatrix} 1 & 0 & 0 & 0 & \cdots & 0 \\ 0 & 1 & -\dfrac{a_{13}}{a_{12}} & -\dfrac{a_{14}}{a_{12}} & \cdots & -\dfrac{a_{1n}}{a_{12}} \\ 0 & 0 & 1 & 0 & \cdots & 0 \\ \vdots & & & & & \\ 0 & 0 & 0 & 0 & \cdots & 1 \end{bmatrix}. \tag{3.29}$$

That is, $S_1$ is the $(n \times n)$ identity matrix, except for the 2nd row, where $s_{21} = 0$, $s_{22} = 1$, and $s_{2j} = -a_{1j}/a_{12}$ for $j = 3, 4, \ldots, n$. If we form the product $AS_1$, the product matrix has zeros in the (1, 3), (1, 4), ..., (1, $n$) positions, since post-multiplying $A$ by $S_1$ has the effect of multiplying the 2nd column of $A$ by scalars $-a_{1j}/a_{12}$ and adding it to the $j$th column, for $j = 3, 4, \ldots, n$. Thus the first row of $AS_1$ looks like the first row of the Hessenberg matrix (3.28), and the procedure resembles Gauss elimination.

The next step is to verify (Problem 1) that $S_1^{-1}$ is given by

$$S_1^{-1} = \begin{bmatrix} 1 & 0 & 0 & 0 & \cdots & 0 \\ 0 & 1 & \dfrac{a_{13}}{a_{12}} & \dfrac{a_{14}}{a_{12}} & \cdots & \dfrac{a_{1n}}{a_{12}} \\ 0 & 0 & 1 & 0 & \cdots & 0 \\ \vdots & & & & & \vdots \\ 0 & 0 & 0 & 0 & \cdots & 1 \end{bmatrix}. \tag{3.30}$$

Therefore, $S_1^{-1}$ is given by the formula $S_1 + S_1^{-1} = 2I$. Now, the matrix $B_1 = S_1^{-1}AS_1$ is similar to $A$ and, moreover, premultiplication of $AS_1$ by $S_1^{-1}$ does not disturb the zeros in the first row of $AS_1$. As a particular example, to illustrate the idea behind the method, let

$$A = \begin{bmatrix} 2 & 1 & 2 & 4 & -3 \\ 1 & 2 & 5 & 3 & 1 \\ 6 & 0 & -3 & 4 & 2 \\ 1 & 1 & 3 & 2 & 0 \\ 0 & -1 & 1 & 2 & 1 \end{bmatrix}, \qquad S_1 = \begin{bmatrix} 1 & 0 & 0 & 0 & 0 \\ 0 & 1 & -2 & -4 & 3 \\ 0 & 0 & 1 & 0 & 0 \\ 0 & 0 & 0 & 1 & 0 \\ 0 & 0 & 0 & 0 & 1 \end{bmatrix}.$$

then

$$AS_1 = \begin{bmatrix} 2 & 1 & 0 & 0 & 0 \\ 1 & 2 & 1 & -5 & 7 \\ 6 & 0 & -3 & 4 & 2 \\ 1 & 1 & 1 & -2 & 3 \\ 0 & -1 & 3 & 6 & -2 \end{bmatrix} \quad \text{and} \quad S_1^{-1}AS_1 = \begin{bmatrix} 2 & 1 & 0 & 0 & 0 \\ 17 & 9 & -10 & -23 & 29 \\ 6 & 0 & -3 & 4 & 2 \\ 1 & 1 & 1 & -2 & 3 \\ 0 & -1 & 3 & 6 & -2 \end{bmatrix}$$

The next step in the reduction of a general matrix to Hessenberg form is to obtain zeros in the (2, 4), (2, 5), . . . , (2, $n$) positions by another similarity transformation. So, if $S_1^{-1}AS_1$ is given by

$$S_1^{-1}AS_1 = \begin{bmatrix} a'_{11} & a'_{12} & 0 & 0 & 0 & \cdots & 0 \\ a'_{21} & a'_{22} & a'_{23} & a'_{24} & a'_{25} & \cdots & a'_{2n} \\ \vdots & & & & & & \vdots \\ a'_{n1} & a'_{n2} & & & & \cdots & a'_{nn} \end{bmatrix} \tag{3.31}$$

where we assume that $a'_{23} \neq 0$, we define $S_2$ by

$$S_2 = \begin{bmatrix} 1 & 0 & 0 & 0 & \cdots & 0 \\ 0 & 1 & 0 & 0 & \cdots & 0 \\ 0 & 0 & 1 & -\dfrac{a'_{24}}{a'_{23}} & \cdots & -\dfrac{a'_{2n}}{a'_{23}} \\ \vdots & & & & & \vdots \\ 0 & 0 & 0 & 0 & \cdots & 1 \end{bmatrix} \tag{3.32}$$

and $S_2^{-1}$ by $S_2^{-1} + S_2 = 2I$. (If $a'_{23} = 0$, but $a'_{2j} \neq 0$ for some $j > 3$, we can modify this procedure, as suggested below, to interchange columns). Clearly, by construction, if $B_1 = S_1^{-1}AS_1$ and if $B_2 = S_2^{-1}B_1S_2$, then the first two rows of $B_2$ look like (3.28), that is, $B_2$ has zeros in the (1, 3), (1, 4), . . . , (1, $n$); (2, 4), (2, 5), . . . , (2, $n$) positions. Moreover, since $B_1$ is similar to $A$ and $B_2$ is similar to $B_1$, then $B_2$ and $A$ have the same eigenvalues. Continuing, we arrive at a matrix $B$ of the form (3.28) which is similar to $A$ (Problem 4), by way of the sequence:

$$B_1 = S_1^{-1}AS_1, \qquad B_2 = S_2^{-1}B_1S_2, \ldots, B = S_{n-2}^{-1}B_{n-3}S_{n-2}.$$

In implementing this procedure, we can always interchange columns to obtain the maximum pivot element (as long as we do not introduce nonzero terms where they are not wanted, see Problem 3 of the last section). In this manner, just as in the case of pivoting in Gauss elimination, we can hope to reduce the effects of round-off error. If at the $i$th stage of the process, we find that $a_{ij} = 0$ for $j = i + 1$, $i + 2, \ldots, n$, then the $i$th row is already like the $i$th row of (3.28) and we immediately proceed to the $(i + 1)$st step.

Having found a Hessenberg matrix $B$ which is similar to $A$ it remains to find the eigenvalues of $B$. The usual procedure is to use some root-finding method (as in Chapter 4) to estimate the roots of $p(\lambda) = 0$, where now $p(\lambda)$ is the characteristic polynomial of $B$. To use most root-finding methods, all that is needed is a way to evaluate $p(\lambda)$ for various values of $\lambda$, that is, we do *not* need to know the coefficients of the polynomial to use a root-finding method. Therefore we now consider how to evaluate the characteristic polynomial of a Hessenberg matrix for some value, say $\lambda = \alpha$. Let $\alpha$ be a fixed scalar, then by (3.28) we see that the equation $(B - \alpha I)\mathbf{x} = \mathbf{0}$ has the form

$$\begin{aligned}
(b_{11} - \alpha)x_1 + b_{12}x_2 \qquad\qquad\qquad\qquad\qquad &= 0\\
b_{21}x_1 \quad + (b_{22} - \alpha)x_2 + b_{23}x_3 \qquad\qquad &= 0\\
\vdots \qquad\qquad\qquad\qquad\qquad\qquad\qquad &= \vdots\\
b_{n1}x_1 \quad + b_{n2}x_2 \quad + b_{n3}x_3 + \cdots + (b_{nn} - \alpha)x_n &= 0
\end{aligned} \tag{3.33}$$

For the present, we will assume that $b_{12}, b_{23}, \ldots, b_{n-1,n}$ are all nonzero. This may seem as if we are restricting the problem, but we will show later that if $b_{j,j+1} = 0$ for some $j$, $1 \le j \le n - 1$, then the problem actually "uncouples" (see Example 3.8) and simplifies into two smaller problems, each of which can be attacked in the fashion explained below. Since we have a specific value for $\alpha$, if we set $x_1 = 1$, then the first equation determines $x_2$ uniquely. Having $\alpha$, $x_1$ and $x_2$, the second equation determines $x_3$ uniquely. Finally, the $(n - 1)$st equation determines $x_n$, so by using the first $(n - 1)$ equations, we can determine a unique vector $x$ such that multiplication by $B - \alpha I$ yields:

$$(B - \alpha I)\mathbf{x} = \begin{bmatrix} 0 \\ 0 \\ \vdots \\ 0 \\ f(\alpha) \end{bmatrix} \tag{3.34}$$

where $f(\alpha) = b_{n1} + b_{n2}x_2 + \cdots + b_{n,n-1}x_{n-1} + (b_{nn} - \alpha)x_n$. We now present a theorem which serves as a basis for determining the eigenvalues of $B$.

**Theorem 3.5.** Let the Hessenberg matrix $B$ (with $b_{j,j+1} \neq 0$, $1 \le j \le n - 1$), the scalar $\alpha$, the vector $\mathbf{x}$, and the function $f(\alpha)$ be given as above. Then $\alpha$ is an eigenvalue of $B$ if and only if $f(\alpha) = 0$.

*Proof*: a) If $f(\alpha) = 0$ then clearly $\alpha$ is an eigenvalue of $B$ by virtue of formula (3.34).

b) Let $\alpha$ be an eigenvalue of $B$ with corresponding eigenvector, $\mathbf{x} = (\gamma_1, \gamma_2, \ldots, \gamma_n)^T$. If $\gamma_1 \neq 0$, then $\tilde{\mathbf{x}} = (1, \gamma_2', \gamma_3', \ldots, \gamma_n')$ where $\gamma_i' = \gamma_i/\gamma_1$ is also an eigenvector corresponding to $\alpha$, and thus its components must satisfy (3.33). Also, since $b_{j,j+1} \neq 0$, $1 \le j \le n - 1$, the above transition from (3.33) to (3.34) is unique. Thus the vector $\mathbf{x}$ found in (3.34) satisfies $x_1 = 1$, $x_2 = \gamma_2', \ldots, x_n = \gamma_n'$; and by (3.34),

$$\mathbf{0} = (B - \alpha I)\mathbf{x} = (0, 0, \ldots, 0, f(\alpha))^T.$$

Therefore in this case $(\gamma_1 \neq 0)$, $f(\alpha) = 0$. We next show that $\gamma_1 = 0$ cannot happen, for, if $\gamma_1 = 0$, we still have $(B - \alpha I)\mathbf{x} = \mathbf{0}$. Now in the first equation of (3.33), we let $x_1 = \gamma_1 = 0$ instead of $x_1 = 1$ as before. Since $b_{12} \neq 0$, we get $x_2 = 0$.

Using $x_1 = x_2 = 0$ in the second equation, we get $x_3 = 0$, since $b_{23} \neq 0$. Continuing in this fashion, and using the uniqueness of the transition from (3.33) to (3.34), we get $x_j = \gamma_j = 0$ for $1 \leq j \leq n$. But this says that the eigenvector $\mathbf{x}$ is $\mathbf{0}$, which is impossible. Thus $\gamma_1 = 0$ cannot happen, and this completes the proof. ■

From the proof of the above theorem we see that all of the eigenvectors of a Hessenberg matrix $B$, with $b_{j,j+1} \neq 0$, have nonzero first components. It can be further shown, in this case, that $p(\alpha) = (-1)^{n+1} b_{12} b_{23} \cdots b_{n-1,n} f(\alpha)$ for all $\alpha$ (Problem 3). Thus, even though we do not know the coefficients of $p(x)$, we can use the above process to evaluate $p(\alpha)$ for any $\alpha$. (This process can be modified to generate $p'(\alpha)$ as well). The ability to do this is the basis for several numerical procedures for approximating eigenvalues. Before going further, however, we give the following example to illustrate the fact that the process is actually simplified if one of the superdiagonal terms, $b_{j,j+1}$, is zero.

**Example 3.8.** This example serves to illustrate how the problem of finding the eigenvalues of a Hessenberg matrix is broken down into two subproblems when a superdiagonal term is zero. Let

$$B = \begin{bmatrix} 1 & 1 & 0 & 0 & 0 \\ 1 & 2 & 0 & 0 & 0 \\ -1 & 5 & 3 & 4 & 0 \\ 2 & 6 & 1 & 2 & 2 \\ 7 & 1 & 3 & 1 & 4 \end{bmatrix} \qquad \text{(the term } b_{23} = 0\text{)}$$

We can write $B$ in a partitioned form:

$$B = \left[\begin{array}{c|c} B_{11} & B_{12} \\ \hline B_{21} & B_{22} \end{array}\right], \qquad B_{11} = \begin{bmatrix} 1 & 1 \\ 1 & 2 \end{bmatrix}, \qquad \text{and } B_{22} = \begin{bmatrix} 3 & 4 & 0 \\ 1 & 2 & 2 \\ 3 & 1 & 4 \end{bmatrix}.$$

If $\lambda$ is an eigenvalue of $B_{22}$ with corresponding eigenvector $\mathbf{x} = (x_1, x_2, x_3)^T$, then, since $B_{12}$ is a zero matrix, $B\mathbf{y} = \lambda\mathbf{y}$, where $\mathbf{y} = (0, 0, x_1, x_2, x_3)^T$. Since $\mathbf{x}$ is a nonzero vector $\mathbf{y}$ is also. In a similar manner we can show that $\lambda$ being an eigenvalue of $B_{11}^T$ implies that $\lambda$ is an eigenvalue of $B^T$ and thus of $B$ (see Problem 5). Notice also, that $B_{11}$ and $B_{22}$ are both Hessenberg matrices and hence the problem of finding the eigenvalues of the Hessenberg matrix $B$ is reduced to finding eigenvalues of two smaller Hessenberg matrices.

As we have mentioned, the ability to evaluate the characteristic polynomial of a Hessenberg matrix suggests that polynomial root-finding methods can be applied. However, there are sophisticated ways of finding the eigenvalues of a Hessenberg matrix, for example, the QR Algorithm of Francis (see Ralston 1965). As we have mentioned, all of the methods presented thus far suffer from contamination by round-off if we are searching for *all* of the eigenvalues. To demonstrate the disastrous effect round-off can have, we give here a well-known example of polynomial ill-conditioning due to Wilkinson (1963).*

**Example 3.9.** Let $p(\lambda) = \lambda^n + a_1\lambda^{n-1} + \cdots + a_{n-1}\lambda + a_n$ where $p(\lambda)$ has zeros $\lambda_1, \lambda_2, \ldots, \lambda_n$. We say that $p(\lambda)$ is ill-conditioned if *small* changes in the coefficients $a_i$ lead to *large* changes in the zeros $\lambda_i$ of $p(\lambda)$. As such an example, let $p(\lambda) = (\lambda - 1)(\lambda - 2)(\lambda - 3)\cdots(\lambda - 20) = \lambda^{20} + a_1\lambda^{19} + \cdots + a_{19}\lambda + a_{20}$. Let $p_\varepsilon(\lambda) = \lambda^{20} + (a_1 - 2^{-23})\lambda^{19} + \cdots + a_{19}\lambda + a_{20} \equiv p(\lambda) - 2^{-23}\lambda^{19}$. The zeros of $p_\varepsilon(\lambda)$ are:

| | |
|---|---|
| 1.00000 | 8.91725 |
| 2.00000 | $10.09527 \pm 0.64350i$ |
| 3.00000 | $11.79363 \pm 1.65233i$ |
| 4.00000 | $13.99236 \pm 2.51883i$ |
| 5.00000 | $16.93074 \pm 2.81262i$ |
| 6.00000 | $19.50244 \pm 1.94033i$ |
| 6.99970 | 20.84691 |
| 8.00727 | |

This is a rather remarkable shift in the zeros of $p(\lambda)$ due to a small perturbation of $a_1$. Moreover, $p(\lambda)$ does not seem to present what we would think of as a pathological case.

### 3.3.3 Householder's Method

Most of the eigenvalue methods presented previously have been for arbitrary $(n \times n)$ real matrices. If we restrict ourselves to symmetric matrices, however, we are able to derive a much more efficient numerical procedure. This method, due to Householder and Bauer (1959), is less sensitive to round-off error, since it does not require any deflations or the calculation of the coefficients of the characteristic polynomial.

Given a symmetric matrix, $A$, this method generates a *tridiagonal* matrix $T$ which is similar to $A$ through a series of similarity transformations. Since a tridiagonal matrix is a special case of a Hessenberg matrix, we could find the eigenvalues of $T$ using the method above. However, we do not take this course because we can derive, in this case, a more efficient way to evaluate the characteristic polynomial, $\det(T - \lambda I)$, for $\lambda = \alpha$, $\alpha$ an arbitrary constant. Householder's technique is to reduce an entire row and column to zeros (except for the tridiagonal elements) at each step. To see how this is done, let $\mathbf{v}$ be any vector such that $\mathbf{v}^T\mathbf{v} = 1$. Then it is easily shown that the matrix $P = I - 2\mathbf{v}\mathbf{v}^T$ is symmetric and, moreover, that $PP = I$ (that is, $P = P^{-1}$). We consider a sequence of matrices

$$A_k = P_k A_{k-1} P_k, \qquad 2 \le k \le n - 1, \qquad A_1 \equiv A. \tag{3.35}$$

* From J. H. Wilkinson, *Rounding Errors in Algebraic Processes*, 1963. Reprinted by permission of Prentice-Hall, Inc.

At the $(k-1)$st step, assume that the matrix $A_{k-1} = (\alpha_{ij})$ as shown in Fig. 3.3 has zero nontridiagonal elements in its first $(k-2)$ rows and columns:

$$A_{k-1} = \begin{bmatrix} \alpha_{11} & \alpha_{12} & 0 & & & & \cdots & 0 \\ \alpha_{21} & \alpha_{22} & \alpha_{23} & 0 & & & \cdots & 0 \\ 0 & \alpha_{32} & \alpha_{33} & \alpha_{34} & 0 & & \cdots & 0 \\ & 0 & \ddots & \ddots & \ddots & \ddots & & \vdots \\ & & & \ddots & \alpha_{k-2,k-2} & \alpha_{k-2,k-1} & 0 \cdots & 0 \\ & & & & \alpha_{k-1,k-2} & \alpha_{k-1,k-1} & \cdots & \alpha_{k-1,n} \\ & & & & 0 & & \ddots & \\ \vdots & \vdots & & & \vdots & \vdots & & \vdots \\ 0 & 0 & \cdots & & 0 & \alpha_{n,k-1} & \cdots & \alpha_{nn} \end{bmatrix}$$

**Figure 3.3**

Now let $\mathbf{v}_k^T = (0, \ldots, 0, v_k^{(k)}, v_k^{(k+1)}, \ldots, v_k^{(n)})$, and define $P_k = I - 2\mathbf{v}_k\mathbf{v}_k^T$. The upper left $(k-1) \times (k-1)$ submatrix of $P_k$ is the $(k-1) \times (k-1)$ identity and the lower right $(n-k+1) \times (n-k+1)$ submatrix of $P_k$ has the form:

$$\begin{bmatrix} (1 - 2(v_k^{(k)})^2) & \cdots & -2v_k^{(n)}v_k^{(k)} \\ \vdots & & \vdots \\ -2v_k^{(n)}v_k^{(k)} & \cdots & (1 - 2(v_k^{(n)})^2) \end{bmatrix},$$

and all other elements are zero. By simple matrix multiplication we may easily verify that $A_k = P_k A_{k-1} P_k$ has zero nontridiagonal elements in its first $(k-2)$ rows and columns. We now wish to choose the nonzero components of $\mathbf{v}_k$ so that the off-tridiagonal elements in the $(k-1)$st row (and thus column) of $A_k$ are also zero. We leave as Problem 8 to verify that this is true if

$$(v_k^{(k)})^2 = 1/2\,[1 \pm (\alpha_{k-1,k}/\sqrt{S})]$$

and

$$v_k^{(j)} = \pm \alpha_{k-1,j}/(2v_k^{(k)}\sqrt{S}), \qquad k+1 \le j \le n, \tag{3.36}$$

where $S = \sum_{j=k}^{n} \alpha_{k-1,j}^2$. Since $v_k^{(k)}$ is used as a divisor, we can minimize round-off errors by choosing the sign in $(v_k^{(k)})^2$ to maximize its magnitude and then using the same sign in $v_k^{(j)}$. Continuing in this fashion, we finally obtain $T = A_{n-1}$ which is tridiagonal and similar to $A$.

The implementation of Householder's procedure to tridiagonalize $A$ does not actually require the formation of the products $P_k A_{k-1} P_k$ in Eq. (3.35)—that is—it is inefficient to compute $P_k$, then form the matrix product $P_k A_{k-1}$, and then form another matrix product $(P_k A_{k-1})P_k$. This should not surprise us since the matrix $P_k$ is completely determined by a single vector $\mathbf{v}_k$. It can, in fact, be shown that $A_k$ can be generated from $A_{k-1}$ by approximately $2(n-k)^2$ multiplications by a clever reorganization of the computations involved in (3.35) and (3.36). A description of this more efficient procedure, however, we feel would unnecessarily complicate the presentation of this method of tridiagonalization, and we omit it and refer the interested reader to Isaacson and Keller (1966), pp. 166–167. The power of Householder's method is due to a result of Givens (1954) which provides a scheme for finding the roots of the characteristic equation of the symmetric tridiagonal matrix which we have just generated from $A$ by a finite number of similarity transformations, via (3.35) and (3.36).

Now let us consider the determination of the eigenvalues $\{\beta_i\}_{i=1}^n$ of a real symmetric tridiagonal matrix $B \equiv (b_{ij})$. Set $\alpha_i = b_{ii}$ for $1 \le i \le n$, and $\gamma_i = b_{i,i+1} = b_{i+1,i}$ for $1 \le i \le n-1$, and let $p(\lambda) \equiv \det|B - \lambda I|$. Note that if any $\gamma_i = 0$, then this problem reduces to the eigenvalue problem for two distinct smaller real symmetric tridiagonal matrices. Thus we assume that $\gamma_i \neq 0$ for $1 \le i \le n-1$ and consider the following sequence of *Sturm polynomials*:

$$\begin{aligned} p_0(\lambda) &\equiv 1 \\ p_1(\lambda) &\equiv (\lambda - \alpha_1) \\ p_2(\lambda) &\equiv (\lambda - \alpha_2)(\lambda - \alpha_1) - \gamma_1^2 \\ &\vdots \\ p_i(\lambda) &\equiv (\lambda - \alpha_i)p_{i-1}(\lambda) - \gamma_{i-1}^2 p_{i-2}(\lambda) \qquad \text{for } i = 3, 4, \ldots, n. \end{aligned} \tag{3.37}$$

By Problem 9, we see that $p_n(\lambda) = (-1)^n p(\lambda)$, and thus for any scalar $\alpha$ we can set $\lambda = \alpha$ in (3.37) and find the value $p(\alpha)$ *without* having to know the coefficients of $p(\lambda)$. Thus we can use the following theorem (due to Givens) as a basis for an algorithm to approximate the $\beta_i$'s. (We omit the proof of this theorem since it is long and somewhat complicated but the theorem is actually quite easy to understand and use (see Example 3.10)).

**Theorem 3.6.** Let $B$ and $p_i(\lambda)$, $0 \le i \le n$, be given as above. Then the zeros of each $p_i(\lambda)$, $2 \le i \le n$, are real and distinct and are separated by the zeros of $p_{i-1}(\lambda)$. Furthermore, if $p_n(\alpha) \neq 0$ for any real $\alpha$, the number of eigenvalues greater than $\alpha$ equals the number of sign variations in the sequence $\{p_n(\alpha), p_{n-1}(\alpha), \ldots, p_1(\alpha), 1\}$.

**Example 3.10.** Let a symmetric tridiagonal matrix $T$ and its Sturm sequence be given by:

$$T = \begin{bmatrix} 3 & 2 & 0 & 0 \\ 2 & 1 & 1 & 0 \\ 0 & 1 & 1 & -2 \\ 0 & 0 & -2 & 2 \end{bmatrix}, \qquad \begin{aligned} p_0(\lambda) &= 1 \\ p_1(\lambda) &= \lambda - 3 \\ p_2(\lambda) &= (\lambda - 1)p_1(\lambda) - 4p_0(\lambda) \\ p_3(\lambda) &= (\lambda - 1)p_2(\lambda) - 1p_1(\lambda) \\ p_4(\lambda) &= (\lambda - 2)p_3(\lambda) - 4p_2(\lambda). \end{aligned}$$

Note that, while we can explicitly carry out these multiplications and find the polynomial, $p_4(\lambda)$, we need not do so in order to evaluate $p_4(\lambda)$. Using Theorem 3.3 we find that all the eigenvalues of $T$ are in $[-5, 5]$. The Gerschgorin circle theorem provides a sharper estimate, showing that each eigenvalue of $T$ is in one of the intervals $[1, 5]$, $[-2, 4]$, $[0, 4]$ and hence we restrict our search to $[-2, 5]$.

We now consider the sequence $\{p_4(\alpha), p_3(\alpha), p_2(\alpha), p_1(\alpha), p_0(\alpha)\}$ for various numbers $\alpha$, where $d(\alpha)$ denotes the number of sign variations. For $\alpha = -2$, the sequence is $\{96, -28, 11, -5, 1\}$; $d(-2) = 4$. For $\alpha = 5$, the sequence is $\{36, 14, 4, 2, 1\}$; $d(5) = 0$. For $\alpha = 0$, the sequence is $\{-4, 4, -1, -3, 1\}$; $d(0) = 3$. Thus there are three eigenvalues in $(0, 5)$ and one eigenvalue in $(-2, 0)$.

For $\alpha = 1$, the sequence is $\{14, 2, -4, -2, 1\}$; $d(1) = 2$. In this case, there is one eigenvalue in $(0, 1)$, two eigenvalues in $(1, 5)$, and one eigenvalue in $(-2, 0)$. We have proceeded in a somewhat unsystematic fashion with this example in order to demonstrate how the eigenvalues can be isolated.

To find the eigenvalue in $(0, 1)$, we proceed more systematically and check at the midpoint of $(0, 1)$ and find that $d(1/2) = 2$. Thus $d(0) = 3$, $d(1/2) = 2$, and $d(1) = 2$, showing that the eigenvalue in $(0, 1)$ is in fact in $(0, 1/2)$. We continue this process by checking the midpoints of the intervals we find which contain the eigenvalue. After $n$ steps we will have generated an interval, $I_n$, containing the eigenvalue, whose length equals the length of the original interval ($(0, 1)$ in this instance) divided by $2^n$. Thus we have approximated the eigenvalue in $(0, 1)$ to within $1/2^n$. This more systematic approach is actually the bisection method, which is discussed in further detail in Chapter 4.

Householder's method for symmetric matrices was actually preceded by two older methods, the Jacobi method and the Givens method. The Jacobi method generates an *infinite* sequence of similarity transformations which converge in the limit to a diagonal matrix. Givens recognized that the matrix could be transformed into tridiagonal form with a *finite* number of steps, and that the eigenvalues of the resultant similar tridiagonal matrix could be obtained via the Sturm process described above. Householder refined the Givens method to make the tridiagonalization more efficient, and the result is the method presented in this section. The Jacobi method is still used but is not usually competitive with Householder's method. Its main advantage over Householder's method is that it converges to a diagonal matrix with the eigenvalues on the diagonal and thus the Sturm process need not be used. This convergence can be quite slow, however, and the method is not presented here. (See Section 3.4 and Fox (1965) for more details of the Jacobi method.)

The example below illustrates the output and results of a typical computer program using Householder's method to find the eigenvalues of a symmetric matrix (in this case, a $(5 \times 5)$ matrix). The output consists of the original matrix $A$, the tridiagonalized form of $A$ and an interval containing the eigenvalues (obtained by applying Theorem 3.4 to the tridiagonalized form). Next, the eigenvalues are isolated and then refined, using the inverse power method.

**Example 3.11.** (This example is copied from the print-out of a computer program for Householder's method.)

Matrix $A$:

| | | | | |
|---|---|---|---|---|
| 1.000000 | 2.000000 | 1.000000 | 2.000000 | −0.000000 |
| 2.000000 | 1.000000 | 2.000000 | −1.000000 | 3.000000 |
| 1.000000 | 2.000000 | −0.000000 | 3.000000 | 1.000000 |
| 2.000000 | −1.000000 | 3.000000 | 1.000000 | −0.000000 |
| −0.000000 | 3.000000 | 1.000000 | −0.000000 | 4.000000 |

Tridiagonalized matrix is:

| | | | | |
|---|---|---|---|---|
| 1.000000 | −3.000000 | 0.000000 | −0.000000 | 0.000000 |
| −3.000000 | 2.222222 | 3.614101 | 0.000000 | 0.000000 |
| 0.000000 | 3.614101 | 0.508402 | −3.064175 | −0.000000 |
| −0.000000 | −0.000000 | −3.064175 | 2.310238 | −2.194545 |
| 0.000000 | −0.000000 | 0.000000 | −2.194545 | 0.959138 |

The Eigenvalues are between −6.170875 and 8.836324.

Search of interval using Sturm sequences gives:

Eigenvalue is between −4.229939 and −4.079864
Eigenvalue is between −1.078428 and −0.928356
Eigenvalue is between 0.722436 and 0.872508
Eigenvalue is between 4.174092 and 4.324164
Eigenvalue is between 6.875388 and 7.025460

Further refinement of these intervals gives:

Eigenvalue is −4.09049970 to within 0.00000050
Eigenvalue is −0.98757371 to within 0.00000050
Eigenvalue is 0.87207458 to within 0.00000050
Eigenvalue is 4.23464279 to within 0.00000050
Eigenvalue is 6.97135555 to within 0.00000050

## PROBLEMS

1. Verify Eq. (3.30) and that $S_1 + S_1^{-1} = 2I$.

*2. The eigenvalues of $A$ are $5 \pm 12i$, $8 \pm 6i$, $\pm 9i$, where

$$A = \begin{bmatrix} -31 & 96 & 84 & -96 & -240 & 384 \\ -30 & 74 & 45 & -60 & -150 & 240 \\ -12 & 6 & -25 & 24 & 60 & -96 \\ -12 & 18 & -6 & 43 & 85 & -145 \\ -36 & 54 & -18 & 66 & 102 & -174 \\ -24 & 36 & -12 & 50 & 80 & -137 \end{bmatrix}$$

Develop a computer program that can find these eigenvalues. (This problem entails extra reference work.)

3. Show that $p(\alpha) = (-1)^{n+1} b_{12} b_{23} \cdots b_{n-1,n} f(\alpha)$, where $f(\alpha)$ and $p(\alpha)$ are defined in Theorem 3.5.
4. Let $B_k = S_k^{-1} B_{k-1} S_k$ for $k = 1, 2, \ldots, n-2$; where $B_0 = A$. Show that $B_{n-2}$ is similar to $A$.
5. In Example 3.8 show that if $\lambda$ is an eigenvalue of $B_{11}^T$, then $\lambda$ is also an eigenvalue of $B^T$.
6. Use the Sturm procedure to approximate the eigenvalues of

$$A = \begin{bmatrix} \sqrt{3} & \sqrt{2} & 0 & 0 \\ \sqrt{2} & -\sqrt{3} & 1 & 0 \\ 0 & 1 & \sqrt{3} & \sqrt{2} \\ 0 & 0 & \sqrt{2} & -\sqrt{3} \end{bmatrix}.$$

7. Reduce $A$ to tridiagonal form, using the Householder procedure

$$A = \begin{bmatrix} 1 & 2 & 1 & 2 \\ 2 & 1 & 2 & -1 \\ 1 & 2 & 0 & 3 \\ 2 & -1 & 3 & 1 \end{bmatrix}.$$

8. Verify that $\mathbf{v}_k$ as defined in Eq. (3.36) makes the off-tridiagonal terms of the $(k-1)$st row of $A_k$ all zero.
9. Verify that $p_n(\lambda)$ as defined in Eq. (3.37) is equal to $(-1)^n p(\lambda)$ where $p(\lambda)$ is the characteristic polynomial of $B$.

## *3.4 SCHUR'S THEOREM AND RELATED TOPICS

In this section, we discuss a few of the more advanced aspects of matrix theory, paying particular attention to those which have practical applications. Most of the results of this section are presented with only a sketch of their proofs. The interested reader should attempt to work through these proofs, since they lend a great deal of insight into how numerical methods are derived and how well they work. This section is not required for any of the following chapters and may be omitted should the reader desire.

The focus of this section will be Schur's theorem. Schur's theorem is a theoretical result which is of great value in the analysis of numerical methods. As a preliminary, we establish the following characterization of $(n \times n)$ matrices that have a *complete* set of eigenvectors (that is, a set of $n$ linearly independent eigenvectors). Recall, that this was a matter of concern in developing procedures to find eigenvalues.

**Theorem 3.7.** Let $A$ be an $(n \times n)$ matrix. Then $A$ has a set of $n$ linearly independent eigenvectors if and only if $A$ is similar to a diagonal matrix.

*Proof*: Suppose $A\mathbf{u}_i = \lambda_i\mathbf{u}_i$, $i = 1, 2, \ldots, n$ where $\mathbf{u}_1, \mathbf{u}_2, \ldots, \mathbf{u}_n$ are linearly independent. Form the $(n \times n)$ matrix $C = [\mathbf{u}_1, \mathbf{u}_2, \ldots, \mathbf{u}_n]$ and note (Problem 1) that for

$$D = \begin{bmatrix} \lambda_1 & 0 & 0 & \cdots & 0 \\ 0 & \lambda_2 & 0 & \cdots & 0 \\ 0 & 0 & \lambda_3 & \cdots & 0 \\ \vdots & & & & \\ 0 & 0 & 0 & \cdots & \lambda_n \end{bmatrix}$$

we have $AC = CD$. Since the columns of $C$ are linearly independent, $C$ is nonsingular and hence the inverse matrix, $C^{-1}$, exists. Therefore $C^{-1}AC = D$, where $D$ is diagonal. The converse is left to the reader (Problem 2). ■

This theorem, while characterizing matrices with a complete set of eigenvectors, is of little immediate practical use since a knowledge of the eigenvectors is necessary in order to construct the similarity transformation. The theorem does, however, suggest a numerical procedure called Jacobi's method. As mentioned above, Householder's method of the previous section is actually a refinement of Jacobi's method and is usually more computationally efficient. We therefore only present a sketch of the basic ideas underlying Jacobi's method here, in order that the reader may gain a more complete insight into Householder's method. The method proceeds by an infinite sequence of similarity transformations which produce matrices which converge to a diagonal matrix $D$ and is applicable only to symmetric matrices.

To be more precise, let $A$ be symmetric. Suppose $a_{pq}$ is the largest off-diagonal element of $A$ in absolute value, so $|a_{pq}| \geq |a_{ij}|$, $1 \leq i, j \leq n$, $p \neq q$, $i \neq j$. There is a matrix $U$ such that:

a) $U^TU = I$

b) If $U^TAU = A_1$, then the $(p,q)$th entry of $A_1$ is zero (Problem 3).

If $r_i$ denotes the sum of the squares of the off-diagonal elements of $A_i$; $i = 0, 1, 2, \ldots,$ (where $A_0 = A$), then it can be shown that $r_0 > r_1$ and that $A_1$ is also symmetric. This leads to the sequence

$$A_{k+1} = U_{k+1}^T A_k U_{k+1} \tag{3.38}$$

where $A_0 = A$, $r_{k+1} < r_k$, $U_{k+1}^T U_{k+1} = I$, and thus, inductively, $A_{k+1}$ is symmetric. It can further be shown that $\lim_{k\to\infty} r_k = 0$, so that the matrices $A_k$ tend to a diagonal matrix $D$. Since $A_k$ is similar to $A$, the diagonal entries of $A_k$ are tending to the eigenvalues of $A$.

The proof of the convergence of Jacobi's method would provide a proof that every symmetric matrix is similar to a diagonal matrix, and hence that symmetric matrices have a complete set of eigenvectors. We will see, however, that this important result is also a simple consequence of Schur's theorem. To establish Schur's theorem, we need a few preliminaries which are necessary because real matrices may have complex eigenvalues (and hence eigenvectors with complex components). Given an $(m \times n)$ matrix $B = (b_{ij})$, we define the matrix $\overline{B}$ to be the $(m \times n)$ matrix whose $ij$th entry is $\overline{b}_{ij}$, where the bar denotes the complex conjugate. We next set $B^* = (\overline{B})^T$. Finally, if $U$ is an $(n \times n)$ matrix such that $U^*U = I$, we say $U$ is *unitary* (if $U$ is a real matrix, it is customary to call $U$ *orthogonal*).

**Theorem 3.8. Schur.** If $A$ is any $(n \times n)$ matrix, then there is a unitary matrix $U$ such that $U^*AU = T$, where $T$ is upper triangular.

*Proof*: We use induction on $n$. Let $\mathbf{u}$ be an eigenvector of $A$ such that $A\mathbf{u} = \lambda\mathbf{u}$ and such that $\mathbf{u}^*\mathbf{u} = 1$. By Problem 4, we can obtain a set of vectors $\{\mathbf{u}, \mathbf{w}_2, \mathbf{w}_3, \ldots, \mathbf{w}_n\}$ such that

a) $\mathbf{u}^*\mathbf{w}_i = 0, \qquad i = 2, 3, \ldots, n$

b) $\mathbf{w}_i^*\mathbf{w}_j = \delta_{ij}, \qquad 2 \le i, j \le n$

(such a set is called *orthonormal*). Let $S$ be the matrix whose first column is $\mathbf{u}$ and whose $k$th column is $\mathbf{w}_k$, $2 \le k \le n$. Then $S$ is unitary (Problem 5). Now the first column of $S^*AS$ equals $(S^*A)\mathbf{u} = S^*(A\mathbf{u}) = \lambda S^*\mathbf{u}$. But since $S^*S = I$ and $\mathbf{u}$ is the first column of $S$, then $S^*\mathbf{u} = \mathbf{e}_1$, where $\mathbf{e}_1^T = (1, 0, 0, \ldots, 0)$. Thus

$$S^*AS = \left[\begin{array}{c|c} \lambda & \mathbf{w} \\ \hline 0 & A_1 \end{array}\right] \tag{3.39}$$

where $\mathbf{w}$ is a $1 \times (n-1)$ vector and $A_1$ is an $(n-1) \times (n-1)$ matrix. By the induction hypothesis there exists an $(n-1) \times (n-1)$ unitary matrix $W$ such that $W^*A_1W$ is upper triangular. Now let

$$R = \left[\begin{array}{c|c} 1 & 0 \\ \hline 0 & W \end{array}\right] \tag{3.40}$$

and let $U = SR$. Then $U^*AU = R^*S^*ASR$ (Problem 5), and

$$\begin{aligned} U^*AU &= \left[\begin{array}{c|c} 1 & 0 \\ \hline 0 & W^* \end{array}\right] \times \left[\begin{array}{c|c} \lambda & \mathbf{w} \\ \hline 0 & A_1 \end{array}\right] \times \left[\begin{array}{c|c} 1 & 0 \\ \hline 0 & W \end{array}\right] \\ &= \left[\begin{array}{c|c} \lambda & \mathbf{w} \\ \hline 0 & W^*A_1 \end{array}\right] \times \left[\begin{array}{c|c} 1 & 0 \\ \hline 0 & W \end{array}\right] = \left[\begin{array}{c|c} \lambda & \mathbf{z} \\ \hline 0 & W^*A_1W \end{array}\right], \end{aligned} \tag{3.41}$$

with $\mathbf{z} = \mathbf{w}W$. Thus, $U^*AU$ is upper triangular, and the induction is complete. ■

We note here that if $A$ is a real matrix with all real eigenvalues then $U$ is real and orthogonal. One important theorem involving the iterative solution of linear systems, whose proof utilizes Schur's Theorem, is given below. We say an $(n \times n)$ matrix $A$ is *convergent* if $\lim_{m \to \infty} A^m = \mathcal{O}$ (where $\mathcal{O}$ denotes the zero matrix). Finally, if $A$ has eigenvalues $\lambda_1, \lambda_2, \ldots, \lambda_n$ with $|\lambda_1| \ge |\lambda_2| \ge \cdots \ge |\lambda_n|$, then we define the *spectral radius of* $A$, $\rho(A)$ by $\rho(A) = |\lambda_1|$.

**Theorem 3.9.** An $(n \times n)$ matrix $A$ is convergent if and only if $\rho(A) < 1$.

*Proof*: If $A$ is convergent, then clearly $\lim_{m\to\infty} \|A^m\|_1 = 0$. By Theorem 3.3, and since $\rho(A^m) = \rho(A)^m$ (Problem 6),

$$\lim_{m\to\infty} \|A^m\|_1 \geq \lim_{m\to\infty} \rho(A)^m = \lim_{m\to\infty} \rho(A^m) = 0.$$

Thus it must be that $\rho(A) < 1$.

Suppose now that $\rho(A) < 1$. By Schur's Theorem, there is a unitary matrix $U$ such that $U^*AU = T$, where $T$ is triangular. Observing that $T^m = U^*A^mU$ (and hence $UT^mU^* = A^m$, as in Problem 7), it follows that if $T$ is convergent, then so is $A$. Now, an $(n \times n)$ matrix $B$ has $n$ linearly independent eigenvectors if and only if $B$ is similar to a diagonal matrix. Thus, if $T$ has $n$ linearly independent eigenvectors, then there is a matrix $C$ such that $C^{-1}TC = D$, where $D$ is diagonal. As before, if $D$ is convergent then so is $T$, and hence $A$. Since $D$ is diagonal, $D$ is convergent if and only if the diagonal entries of $D$ are all less than 1 in magnitude. Therefore (since $\rho(D) = \rho(T) = \rho(A) < 1$) if $T$ has $n$ linearly independent eigenvectors, then $A$ is convergent.

To handle the case where $T$ does not have $n$ linearly independent eigenvectors, we construct a matrix $T_1$ that "majorizes" $T$. For $T = (t_{ij})$, choose $T_1$ to be an $(n \times n)$ triangular matrix of the form:

$$T_1 = \begin{bmatrix} \alpha_1 & \beta & \beta & \beta & \cdots & \beta \\ 0 & \alpha_2 & \beta & \beta & \cdots & \beta \\ 0 & 0 & \alpha_3 & \beta & \cdots & \beta \\ \vdots & & & & & \\ 0 & 0 & 0 & & \cdots & \alpha_n \end{bmatrix} \tag{3.42}$$

where $1 > \alpha_i \geq |t_{ii}|$ and $\beta \geq |t_{ij}|$, $1 \leq i, j \leq n$. Clearly we can make choices for $T_1$ such that the eigenvalues $\alpha_1, \alpha_2, \ldots, \alpha_n$ of $T_1$ are distinct. From Problem 9, Section 3.1, we see that $T_1$ is similar to a diagonal matrix, and thus, arguing as we did above, it follows that $T_1$ is convergent since $\rho(T_1) < 1$. If the $ij$th entry of $T_1^m$ is denoted by $w_{ij}^{(m)}$, it is an easy exercise to verify that $w_{ij}^{(m)} \geq |t_{ij}^{(m)}|$, where $t_{ij}^{(m)}$ is the $ij$th entry of $T^m$. To summarize, if $\rho(A) < 1$ and if $U^*AU = T$, then $T$ is majorized by a convergent matrix $T_1$ and, hence, $A$ is convergent. ■

As an application of Theorem 3.9 we return briefly to the iterative methods of Section 2.4. Thus, if

$$\mathbf{x} = M\mathbf{x} + \mathbf{g} \tag{3.43}$$

is an equivalent fixed-point form of the equation $A\mathbf{x} = \mathbf{b}$ with solution $\mathbf{x}_t$, then the errors $\mathbf{e}^{(k)} = \mathbf{x}^{(k)} - \mathbf{x}_t$ are given by

$$\mathbf{e}_k^{(k)} = M^k\mathbf{e}_0, \qquad k = 1, 2, \ldots. \tag{3.44}$$

(See formula (2.57)). Thus, it follows that the sequence $\{\mathbf{x}^{(k)}\}$ converges to $\mathbf{x}_t$ for every initial guess $\mathbf{x}^{(0)}$ if and only if $M$ is convergent (Problem 8). This observation is what provides the basis for the proof of Theorem 2.4.

We next extend the idea of symmetric matrices to matrices with complex entries, and say that $A$ is *Hermitian* if $A^* = A$. (For example, a real symmetric matrix is a Hermitian matrix). We can use Schur's theorem to prove:

**Theorem 3.10.** Let $A$ be an $(n \times n)$ Hermitian matrix with eigenvalues $\lambda_1, \lambda_2, \ldots, \lambda_n$. Then we can choose eigenvectors $\mathbf{u}_1, \mathbf{u}_2, \ldots, \mathbf{u}_n$ for $A$ such that:

a) $A\mathbf{u}_i = \lambda_i\mathbf{u}_i$; $i = 1, 2, \ldots, n$; and $\{\mathbf{u}_i\}_{i=1}^n$ are linearly independent;

b) $\mathbf{u}_i^*\mathbf{u}_j = \delta_{ij}$, $1 \le i, j \le n$ (the eigenvectors can be chosen to be orthonormal).

The proof of this theorem is left to the reader (Problem 9). The reader will recall that this theorem is needed to derive the improved version of the power method, the Rayleigh quotient method, which is applicable to Hermitian matrices. Note also that, if $A$ is real and symmetric, then conclusion (b) becomes $\mathbf{u}_i^T\mathbf{u}_j = \delta_{ij}$.

Another consequence of Theorem 3.10 is to find the matrix norm which is the natural extension of the vector norm $\|\mathbf{x}\|_2$. Recall from Section 2.3, that for any real matrix $A$, we would like to find the nonnegative real number $g(A) \equiv L_2$ such that $\|A\mathbf{x}\|_2 \le L_2\|\mathbf{x}\|_2$ for all $x$ in $R^n$ and where $L_2 < L$ for any number $L$ which also satisfies $\|A\mathbf{x}\|_2 \le L\|\mathbf{x}\|_2$ for all $\mathbf{x}$ in $R^n$. (That is, $g(A) \equiv L_2 = \min\{L: \|A\mathbf{x}\|_2 \le L\|\mathbf{x}\|_2$, for all $\mathbf{x}$ in $R^n\}$.) It can be shown that $g(A)$ satisfies all the properties of a matrix norm, and thus we write $g(A) \equiv \|A\|_2$. (Note: From the above definition of $g(A)$, it follows that $g(A) \equiv \|A\|_2 \le \|A\|_E$, for all $A$ in $M_n$, and that $\|A\|_2$ is compatible with the $\|\cdot\|_2$ vector norm.) It can easily be shown (Problem 10) that the matrix $A^TA$ is Hermitian and its eigenvalues $\{v_i\}_{i=1}^n$ are real and nonnegative, that is, $A^TA$ is *positive semidefinite*. Let $v_1 \ge v_2 \ge \cdots \ge v_n \ge 0$. By Theorem 3.10 the eigenvectors $\mathbf{u}_i$ of $A^TA$, can be chosen to be orthonormal, $\mathbf{u}_i^T\mathbf{u}_j = \delta_{ij}$, $1 \le i, j \le n$. Then any vector $\mathbf{x} \ne \mathbf{0}$ can be written as

$$\mathbf{x} = c_1\mathbf{u}_1 + c_2\mathbf{u}_2 + \cdots + c_n\mathbf{u}_n.$$

Thus we can verify that

$$\left(\frac{\|A\mathbf{x}\|_2}{\|\mathbf{x}\|_2}\right)^2 \equiv \frac{\mathbf{x}^TA^TA\mathbf{x}}{\mathbf{x}^T\mathbf{x}} = \frac{\sum_{i=1}^n |c_i|^2 v_i}{\sum_{i=1}^n |c_i|^2}, \tag{3.45}$$

which implies

$$0 \le v_n \le \left(\frac{\|A\mathbf{x}\|_2}{\|\mathbf{x}\|_2}\right)^2 \le v_1, \tag{3.46}$$

or

$$\|A\mathbf{x}\|_2 \le v_1^{1/2}\|\mathbf{x}\|_2, \tag{3.47}$$

for all vectors $\mathbf{x}$. Thus $v_1^{1/2} \equiv \rho(A^TA)^{1/2} \ge L_2 = \|A\|_2$. Now substitute $\mathbf{u}_1$ for $\mathbf{x}$ in (3.45) and (3.47) becomes $\|A\mathbf{u}_1\|_2 = v_1^{1/2}\|\mathbf{u}_1\|_2$. This implies that $\rho(A^TA)^{1/2} = L_2$, and we have proved the following theorem.

**Theorem 3.11.** $\|A\|_2 \equiv \min\{L: \|A\mathbf{x}\|_2 \le L\|\mathbf{x}\|_2, \text{ for all } \mathbf{x} \text{ in } R^n\} = \rho(A^TA)^{1/2}$.

Note that if $A$ is real and symmetric, then $A^TA = A^2$ and so $\|A\|_2 = \rho(A^TA)^{1/2} = \rho(A^2)^{1/2} = \rho(A)$ (see Problem 6).

## PROBLEMS

1. In the proof of Theorem 3.7, establish that $AC = CD$.
2. Complete the converse of Theorem 3.7, showing that if $A$ is similar to a diagonal matrix, then $A$ has $n$ linearly independent eigenvectors. (Hint: Consider the column vectors of the matrix of the similarity transformation.)
3. Let $A = (a_{ij})$ be symmetric and suppose the $(p, q)$th entry of $A$, $a_{pq}$, is nonzero (where also, $a_{pq}$ is not on the diagonal). Define $U$ by:

   A) $u_{pp} = u_{qq} = \cos(\theta)$, $u_{pq} = -u_{qp} = \sin(\theta)$;

   B) $u_{ij} = \delta_{ij}$, for all other elements.

   a) Show $U^TU = I$ and $U^TAU$ is symmetric;

   b) Choose $\theta$, $-\pi/4 \le \theta \le \pi/4$, such that $\tan(2\theta) = 2a_{pq}/(a_{qq} - a_{pp})$. Show the $(p, q)$th entry of $U^TAU$ is 0. (Note: for $\alpha = \tan(2\theta)$, $\cos^2(\theta) = 1/2 + 1/(2\sqrt{\alpha^2 - 1})$.)
4. This problem describes the Gram-Schmidt process of constructing an orthonormal set of $k$ vectors from a given set of $k$ linearly independent vectors. Let $\{\mathbf{t}_1, \mathbf{t}_2, \ldots, \mathbf{t}_k\}$ be a set of $k$ linearly independent, $n$-dimensional column vectors (which may have complex components). To derive a set $\{\mathbf{w}_1, \mathbf{w}_2, \ldots, \mathbf{w}_k)$ of orthonormal vectors from $\{\mathbf{t}_1, \mathbf{t}_2, \ldots, \mathbf{t}_k\}$, we proceed in a step-by-step fashion:

   A) Let $\mathbf{v}_1 = \mathbf{t}_1$ and $\beta_1 = 1/\sqrt{\mathbf{v}_1^*\mathbf{v}_1}$. Next, set $\mathbf{w}_1 = \beta_1\mathbf{v}_1$.

   B) For $2 \le j \le k$, let $\mathbf{v}_j = \mathbf{t}_j + \alpha_{j-1}\mathbf{w}_{j-1} + \cdots + \alpha_1\mathbf{w}_1$ where $\alpha_i = -\mathbf{w}_i^*\mathbf{t}_j$ and $\beta_j = 1/\sqrt{\mathbf{v}_j^*\mathbf{v}_j}$. Then set $\mathbf{w}_j = \beta_j\mathbf{v}_j$ for each $j$.

   Show that:

   a) each step of this process is always defined (that is, we can always form $\beta_i$, $i = 1, 2, \ldots, k$);

   b) each $\mathbf{t}_r$ is expressible (uniquely) as a linear combination $\mathbf{t}_r = \gamma_1\mathbf{w}_1 + \gamma_2\mathbf{w}_2 + \cdots + \gamma_r\mathbf{w}_r$, $r \le k$;

   c) for $1 \le i, j \le k$, $\mathbf{w}_i^*\mathbf{w}_j = \delta_{ij}$;

   d) to complete the proof of Theorem 3.8, if $\mathbf{u}$ is any eigenvector of $A$ such that $\mathbf{u}^*\mathbf{u} = 1$, then we can find $\mathbf{w}_2, \ldots, \mathbf{w}_n$ such that $\{\mathbf{u}, \mathbf{w}_2, \mathbf{w}_3, \ldots, \mathbf{w}_n\}$ is an orthonormal set. To do this, we need only find $\mathbf{t}_2, \mathbf{t}_3, \ldots, \mathbf{t}_n$ so that $\{\mathbf{u}, \mathbf{t}_2, \ldots, \mathbf{t}_n\}$ is linearly independent. Prove that this can always be done. (Hint: Consider the vectors $\{\mathbf{e}_1, \mathbf{e}_2, \ldots, \mathbf{e}_n\}$).
5. In the proof of Theorem 3.8, show:

   a) The matrix $S$ is unitary;

   b) If $B$ is $(m \times n)$ and $C$ is $(n \times p)$, then $(BC)^* = C^*B^*$ (only outline the proof of this).
6. If $A$ is $(n \times n)$, show $\rho(A^k) = [\rho(A)]^k$.
7. Show by induction, that if $C^{-1}AC = B$ then $C^{-1}A^rC = B^r$.
8. In (3.44), show that $\{\mathbf{e}^{(k)}\} \to \mathbf{0}$ for every initial guess $\mathbf{x}^{(0)}$ if and only if $\rho(M) < 1$.

9. Show that an $(n \times n)$ Hermitian matrix $A$ has an orthonormal set of $n$ eigenvectors. (Hint: Use Schur's theorem to show that $A$ is "unitarily" similar to a diagonal matrix.)
10. Verify that the following steps in Theorem 3.11 are valid.
    a) Show, in analogy to Theorem 3.2, that a Hermitian matrix has real eigenvalues.
    b) Show the eigenvalues of $A^*A$ are real and nonnegative.
    c) Verify formula (3.45) by direct computation.
    d) Show that (3.45) implies (3.46).
11. Consider $A\mathbf{x} = \mathbf{b}$, where

$$A = \begin{bmatrix} 2 & -1 \\ -1 & 2 \end{bmatrix}, \quad \mathbf{b} = \begin{bmatrix} 1 \\ 1 \end{bmatrix}, \quad \mathbf{x}_t = \begin{bmatrix} 1 \\ 1 \end{bmatrix}.$$

    a) Show the Jacobi iteration method will converge for any initial guess, $\mathbf{x}^{(0)}$.
    b) Show the Gauss-Seidel method will converge for any initial guess, $\mathbf{x}^{(0)}$.
    c) Show $0 < \rho(M_G) < \rho(M_J)$. What does this say about the relative rates of convergence of these two methods?
    d) Carry out a few iterations of these methods and compare the estimates to the true solution.
12. Let $U$ be an $(n \times n)$ orthogonal matrix. Show:
    a) if $\mathbf{x} \in R^n$, then $\|U\mathbf{x}\|_2 = \|\mathbf{x}\|_2$;
    b) if $\lambda$ is an eigenvalue of $U$, then $|\lambda| = 1$.
13. Use Theorem 3.9 and Problem 8 to show why the Jacobi method cannot converge for the $(4 \times 4)$ matrix $A$ in Example 2.16. (Hint: Show $\det(M_J) > 1$. What does this say about $\rho(M_J)$?)
14. Let $A = (a_{ij})$ be a symmetric, positive definite $(n \times n)$ matrix (recall formula (2.68)). Show that $a_{ii} > 0$ for $i = 1, 2, \ldots, n$. (Hint: Consider $\mathbf{e}_i^T A \mathbf{e}_i$). Show that the eigenvalues of $A$ are all positive.
15. Let $D = (d_{ij})$ be an $(n \times n)$ diagonal matrix, ($d_{ij} = 0$ for $i \neq j$), such that $d_{ii} > 0$ for $i = 1, 2, \ldots, n$. Show that $D$ is positive definite and symmetric.

*16. This problem outlines the proof of Theorem 2.4, showing that the Gauss-Seidel iteration converges for any matrix $A$ that is symmetric and positive definite. As in Section 2.4.1, let $A = L + D + U$ and note that $A = L + D + L^T$ since $A$ is symmetric. The first step in the proof is to show that the matrix $P = A - M_G^T A M_G$ is symmetric and positive definite, where $M_G = -(D + L)^{-1}L^T$:
    a) show $P$ is symmetric;
    b) show $M_G = I - Q$, $Q = (D + L)^{-1}A$ (Hint: $A = L + D + L^T$);
    c) show $P = Q^T(AQ^{-1} + (Q^T)^{-1}A - A)Q$ (Hint: Use Part (b)); Now show $P = Q^T D Q$, using the above and the definition of $Q$;
    d) using Problems 14 and 15 above, show that $P$ is positive-definite symmetric. To finish the proof, suppose $\lambda$ is an eigenvalue of $M_G$ and $\mathbf{u}$ is a corresponding eigenvector. Show that $\mathbf{u}^T P \mathbf{u} > 0$ implies that $|\lambda| < 1$. This shows that $\rho(M_G) < 1$ and completes the proof by virtue of Theorem 3.9.

# 4

# SOLUTION OF NONLINEAR EQUATIONS

## 4.1 INTRODUCTION

One of the oldest and most basic problems encountered in practice is that of finding the roots of an equation, i.e., given $y = f(x)$, find all values, $s$, such that $f(s) = 0$. In this chapter we shall study several techniques for the numerical approximation of these values of $s$ and will demonstrate how one of these techniques (Newton's method) can be easily extended to the problem of simultaneously solving a system of $n$ equations in $n$ unknowns, for $n > 1$. Before delving into this particular problem, however, we pause to give the following general illustration.

In previous chapters we considered some relatively simple physical examples which only involved the illustration of the particular numerical method we were studying at the moment. Most real-world examples are more complicated than this and involve different numerical procedures to solve the different problems that arise at various stages of the solution process of the problem as a whole. The following example provides an illustration of a realistic problem of this type. We have given this example in some detail since there are problems contained in it that require procedures for numerically solving ordinary differential equations (Chapter 7), numerically evaluating integrals (Chapter 6), least-squares data fitting (Chapter 5), as well as numerically solving nonlinear equations (the topic of this chapter). We shall return to this example as we discuss these procedures in subsequent sections.

**Example 4.0.** In this example, we consider the ballistic reentry of a body into the earth's atmosphere (for instance, the return of a spacecraft or the recovery of a package from an orbiting satellite). In a ballistic trajectory, the only forces acting on the body are gravity and drag (that is, lift can be assumed to be zero). The equations of motion governing a ballistic reentry trajectory can be expressed by the equation (4.0a)

$$\ddot{\mathbf{R}} = \frac{-\bar{q}C_D A}{M}\frac{\dot{\mathbf{R}}}{|\dot{\mathbf{R}}|} + \mathbf{G}(\mathbf{R}); \qquad \mathbf{R}(0) = \mathbf{R}_0, \quad \dot{\mathbf{R}}(0) = \mathbf{V}_0 \tag{4.0a}$$

where the three-dimensional position vector $\mathbf{R} = \mathbf{R}(t)$ gives the $x$, $y$, and $z$ coordinates of the body at any time $t$ (in an earth-centered coordinate frame). In this equation, $\mathbf{G}(\mathbf{R})$ denotes the gravitational force and the scalar quantity $\bar{q}C_DA/M$ represents the drag forces exerted on the vehicle due to the atmosphere. The term $\bar{q}$ is the *dynamic pressure*, which is defined by $\bar{q} = \rho(H)|\dot{\mathbf{R}}|^2/2$, where $\rho(H)$ is the atmospheric density at an altitude $H$ above the earth's surface and $|\dot{\mathbf{R}}|$ is the magnitude of the velocity vector, $\dot{\mathbf{R}}(t)$. The drag coefficient, $C_D$, is normally an experimentally determined function, usually depending on velocity, altitude and the orientation of the body along the path of the trajectory. The evaluation of $C_D$ in the equations of motion (4.0a) typically requires table look-ups and/or the evaluation of a series of empirical formulas. The term $A$ represents a reference cross-section area of the body. Drag forces can be increased by orienting the vehicle so that a larger frontal area is exposed. This serves as a drag-brake, slowing the vehicle down (or increasing the velocity, if a smaller frontal area is exposed). The term $M$ in (4.0a) is the mass of the body. Finally, to avoid undue complication in the form of (4.0a), we have assumed a nonrotating earth so that the trajectory will lie in the plane determined by the initial position and velocity vectors, $\mathbf{R}_0$ and $\mathbf{V}_0$.

During the initial phases of engineering design for a vehicle (or for its guidance system, etc.), simplifying approximations are often made in order to obtain a rough idea of the performance characteristics of a particular design. For example, if we assume that $C_DA/M$ is constant, that gravitational forces are negligble with respect to drag and that atmospheric density is given by the formula $\rho(H) = \rho_0 e^{-\lambda H}$, then we obtain a reentry model which approximates (4.0a), namely

$$\ddot{\mathbf{R}} = -Ke^{-\lambda H}|\dot{\mathbf{R}}|\dot{\mathbf{R}}; \qquad \mathbf{R}(0) = \mathbf{R}_0, \quad \dot{\mathbf{R}}(0) = \mathbf{V}_0 \tag{4.0b}$$

where the constant $K$ is given by $K = \rho_0 C_D A/2M$. This approximate model, which is usually called the Allen and Eggers reentry model, can be solved in closed form by the elementary calculus technique of separation of variables (see Thomas 1972 and Problem 5, Section 7.1). This closed form solution gives an expression for velocity as a function of altitude:

$$\dot{\mathbf{R}}(H) = \mathbf{V}_0 e^{Q[e^{-\lambda H} - e^{-\lambda H_0}]}$$

where $H_0$ denotes the initial altitude at $t = 0$ and $Q = -(C_D A\rho_0/2\lambda M \sin(\alpha))$, with $\alpha$ being the angle the velocity vector makes with the local horizontal (see Figure 4.0)).

The Allen and Eggers reentry model in Eq. (4.0b) gives a trajectory that is a straight-line path, since the body is initially traveling along the path defined by the vector $\mathbf{V}_0$ and the only force acting on the body is drag (which is also along the vector $\dot{\mathbf{R}}$, in the opposite direction). The inclusion of a force term involving gravity would, of course, cause the trajectory to deviate from a straight-line path, but including a gravity term in (4.0b) would yield a differential equation which could not be solved in closed form. (In particular, Eq. (4.0a) cannot be solved except by numerical means (such as those in Chapter 7) and the reason for the approximate model, (4.0b), is to get an indication of the solution of (4.0a) for design purposes.)

Some sort of control system is often required for vehicles during reentry, in order to preserve the structural integrity of the vehicle, to keep heating within a tolerable range, etc. Such control systems could, for example, involve the deployment of parachutes and/or reorientation of the body along the path of the trajectory. A device called an accelerometer (which can be visualized, roughly, as a mass attached to a spring) is frequently used as a part of a control system. The accelerometer can be used to measure the magnitude of the drag forces and, on the basis of the output of the accelerometer, a control maneuver can be initiated.

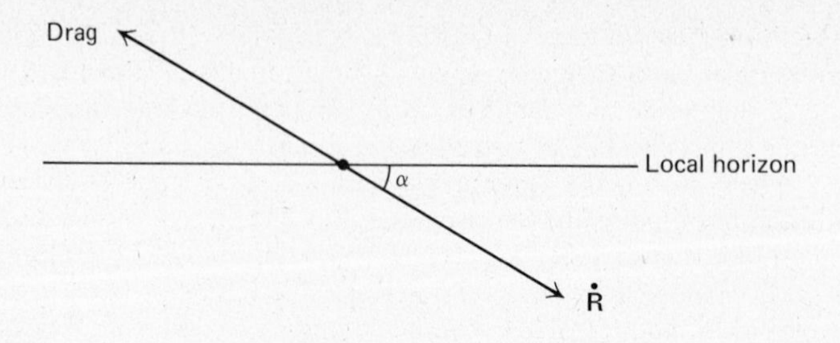


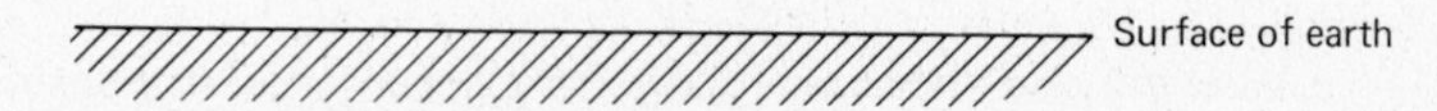


**Fig. 4.0** Atmospheric reentry, neglecting gravity.

In the reentry model (4.0b), the magnitude of the drag is proportional to

$$\bar{q} = \frac{|\dot{\mathbf{R}}(H)|^2 \rho_0 e^{-\lambda H}}{2},$$

so we can use the closed form expression for $\dot{\mathbf{R}}(H)$ to approximate the output of an accelerometer. In particular, to determine design parameters (or timer settings, etc.), we can estimate the altitude $H$ at which the magnitude of drag is some specified value, say $\gamma$. This would then amount to solving an equation of the form

$$\frac{\gamma}{K} = |\mathbf{V}_0|^2 e^{2Q[e^{-\lambda H} - e^{-\lambda H_0}]} e^{-\lambda H}$$

for $H$, where $\gamma/K$, $|\mathbf{V}_0|$, $H_0$, $\lambda$ and $Q$ are constants. We cannot normally expect to be able to solve such an equation by elementary means, and so we must use some numerical technique to approximate the solution. The construction of such root-finding procedures is the topic of this chapter.

As we mentioned earlier, many of the problems that arise in practice require numerical approximations of one sort or another at various stages of the solution process. For instance, in this example an exponential atmosphere, $\rho(H) = \rho_0 e^{-\lambda H}$, was assumed in order to derive (4.0b). Since the standard atmospheric density models are not exponential, the constants $\rho_0$ and $\lambda$ have to be selected so that $\rho_0 e^{-\lambda H}$ provides a good approximation to the standard atmospheric density model in the altitude range under consideration. Choosing $\rho_0$ and $\lambda$ is then a problem in data-fitting (see Example 5.13). As another instance, (4.0a) must be solved to provide a benchmark against which to compare the approximate model (4.0b). Solving (4.0a) requires numerical techniques for the solution of ordinary differential equations which in turn require us to be able to evaluate the right-hand side of (4.0a) at various points $\mathbf{R}(t)$, $\dot{\mathbf{R}}(t)$ along the trajectory. This evaluation in turn will require interpolation in tables, developing a good approximate model for gravity, etc., (see Section 7.2.2).

## 4.2 BRACKETING METHODS

As is so often true in numerical analysis, the simplest methods are the ones which converge most slowly. However, because of their inherent simplicity, one is often willing to use them even at the sacrifice of some speed of convergence.

One of the oldest and most geometrically intuitive root-finding methods is the bisection method. In this method, we assume $f(x)$ is a continuous function on the closed interval $[a, b]$ such that $f(a)f(b) < 0$. By the intermediate value theorem, there must exist at least one value $s \in [a, b]$ such that $f(s) = 0$. In the bisection method, we merely evaluate $f(x)$ at the midpoint $x = (a + b)/2$. If $f((a + b)/2) \neq 0$, then either $f(a)f((a + b)/2) < 0$ or $f((a + b)/2)f(b) < 0$. Without loss of generality let us assume that $f(a)f((a + b)/2) < 0$, and let $a = x_0$ and $(a + b)/2 = x_1$. Again by the intermediate value theorem, there must exist at least one value $s \in [x_0, x_1] \equiv I_1$, such that $f(s) = 0$. We next repeat this very same process on $I_1$, and find an interval, $I_2$, containing at least one root of $f(x) = 0$. We note here that the length of $I_1$ is $(b - a)/2$ and the length of $I_2$ is $(b - a)/4$. After this process is repeated $n$ times we will obtain an interval $I_n$, of length $(b - a)/2^n$, which contains at least one root, $s$, of $f(x) = 0$. If $x$ is any point of $I_n$, then $|x - s| \leq (b - a)/2^n$. Because every $I_n$ must contain at least one such $s$, we say that each step "brackets" the root, and the error of the approximation at the $n$th step is bounded by $(b - a)/2^n$. We continue, repeating this process until $n$ is sufficiently large so that we are satisfied with any approximation $x^*$ for $s$, which satisfies $|x^* - s| \leq (b - a)/2^n$. Thus once the initial $a$ and $b$ are found, the method is guaranteed to converge for any continuous function. The error bound for the $n$th step, $(b - a)/2^n$, however, is quite large compared to the errors of more sophisticated methods.

We next describe one further "bracketing" method, which usually converges faster than the bisection method. This technique is known as *Regula Falsi* or the method of false position, and its derivation is quite geometric, as shown in Fig. 4.1. Once again we must find an interval $[a, b]$ such that $f(a)f(b) < 0$. Next we find

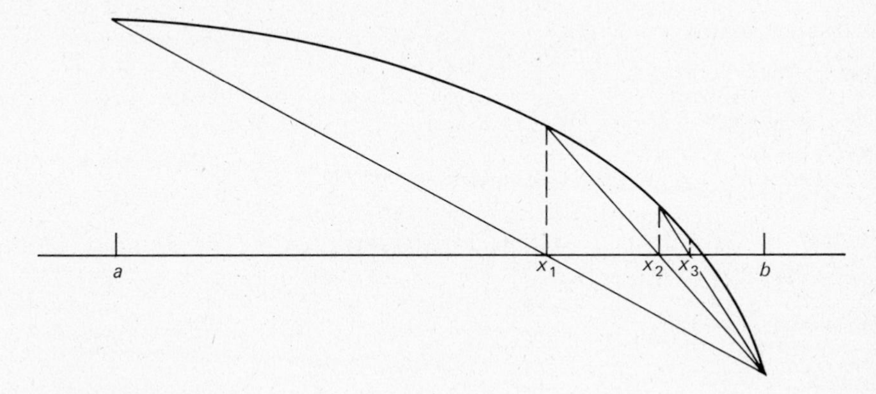


**Fig. 4.1** *Regula Falsi.*

the point, $x_1$, where the secant line between the points $(a, f(a))$ and $(b, f(b))$ intersects the $x$ axis. The equation of this secant line is

$$\frac{y - f(b)}{x - b} = \frac{f(a) - f(b)}{a - b}.$$

The secant line crosses the $x$-axis when $x = x_1$, where

$$x_1 = \frac{af(b) - bf(a)}{f(b) - f(a)}$$

and we take $x_1$ as our first approximation to the root $s$. As in the bisection method, if $f(x_1) \neq 0$, then either $f(a)f(x_1) < 0$ or $f(x_1)f(b) < 0$ (say, $f(a)f(x_1) < 0$). Then again there must be at least one root of $f(x) = 0$ in the interval $I_1 = [x_0, x_1]$, (in this case, $x_0 = a$). This process is repeated again and again generating intervals, $I_n$, each of which contains at least one root of $f(x) = 0$, until we are satisfied with the approximation for $s$. It can happen that the length of each $I_n$ is greater than $(b - a)/2$. In this case, as opposed to the bisection method, the lengths of $I_n$ do not converge to zero as $n \to \infty$. This is what happens for the particular function illustrated in Fig. 4.1, and yet it is clear that for functions of this form, *Regula Falsi* usually converges much faster than bisection (see Problem 2, Section 4.3.5). Furthermore, it can be shown that *Regula Falsi* will converge for any continuous

```
      SUBROUTINE BISECT(A,B,ROOT,TOL)
      REAL MIDPT
C
C  THIS SUBROUTINE USES THE BISECTION METHOD TO FIND A ROOT OF F(X)=0.
C  THE CALLING PROGRAM MUST SUPPLY NUMBERS A AND B SUCH THAT F(A)*F(B)<0
C  AND A STOPPING CRITERIA, TOL.  THE SUBROUTINE RETURNS AN
C  APPROXIMATION TO A ZERO OF F(X) (NAMED ROOT) WHEN AN INTERVAL OF
C  LENGTH TOL OR LESS IS FOUND IN WHICH F(X) CHANGES SIGN.  THE
C  SUBROUTINE ALSO REQUIRES A FUNCTION SUBPROGRAM, F(X)
C
      FA=F(A)
C
C  THE BISECTION ITERATION
C
    1 MIDPT=(A+B)/2.
      FM=F(MIDPT)
      IF(FA*FM.LE.0.)   GO TO 2
      A=MIDPT
      FA=FM
      GO TO 3
    2 B=MIDPT
C
C  TEST TO SEE IF EXIT CRITERION IS SATISFIED
C
    3 IF((B-A).LE.TOL)   GO TO 4
      GO TO 1
    4 ROOT=(A+B)/2.
      RETURN
      END
```

**Fig. 4.2** Subroutine BISECT.

function for which $f(a)f(b) < 0$, even though the lengths of the $I_n$ may not go to zero. The *Regula Falsi* method is based on the premise that if $|f(a)| > |f(b)|$, then we expect $|s - b| < |s - a|$, that is, $s$ is closer to $b$ than $a$. Since in each successive step, $x_k$ is computed using knowledge of the size of $|f(x)|$ at the endpoints of $I_{k-1}$, we would expect the *Regula Falsi* method to usually converge faster than the bisection method. It can be shown (see Section 3.4), that if $f''(x)$ is continuous then $|x_k - s| \leq \lambda^{k-1}\varepsilon$, where $0 < \lambda < 1$ and $\varepsilon$ is a fixed constant.

We have listed two subroutines which use the methods described in this section to find zeros of $f(x)$. Subroutine BISECT (Fig. 4.2) employs the bisection procedure while subroutine REGFAL (Fig. 4.3) uses *Regula Falsi.* Note that REGFAL

```
      SUBROUTINE REGFAL(A,B,ROOT,TOL,M,ITERM)
C
C     THIS SUBROUTINE USES REGULA FALSI TO FIND A ROOT OF F(X)=0.  THE
C     CALLING PROGRAM MUST SUPPLY NUMBERS A AND B SUCH THAT F(A)*F(B)<O
C     AND TWO STOPPING CRITERIA, TOL AND M.  THE SUBROUTINE RETURNS AN
C     APPROXIMATION TO A ZERO OF F(X) (NAMED ROOT) WHEN THE ABSOLUTE VALUE
C     OF F(ROOT) IS LESS THAN TOL OR WHEN M ITERATIONS HAVE BEEN EXECUTED.
C     IN THE FIRST CASE, A FLAG, ITERM, IS SET TO 1 AND IN THE SECOND CASE
C     ITERM IS SET TO 2.  THE SUBROUTINE ALSO REQUIRES A FUNCTION
C     SUBPROGRAM, F(X).
C
      ITR=1
      FA=F(A)
      FB=F(B)
C
C     REGULA FALSI ITERATION
C
    1 XN=(A*FB-B*FA)/(FB-FA)
      FN=F(XN)
C
C     TEST TO SEE IF MAXIMUM NUMBER OF ITERATIONS HAS BEEN EXECUTED
C
      IF(ITR.GE.M)   GO TO 4
C
C     TEST TO SEE IF A ROOT HAS BEEN FOUND
C
      IF(ABS(FN).LE.TOL)   GO TO 5
      IF(FA*FN.LE.0.)   GO TO 2
      A=XN
      FA=FN
      GO TO 3
    2 B=XN
      FB=FN
C
C     INCREMENT THE ITERATION COUNTER
C
    3 ITR=ITR+1
      GO TO 1
    4 ITERM=2
      ROOT=XN
      RETURN
    5 ITERM=1
      ROOT=XN
      RETURN
      END
```

**Fig. 4.3** Subroutine REGFAL.

requires two stopping criteria, where REGFAL terminates when $|f(x_n)| \leq \text{TOL}$ or when $n \geq M$. An upper bound for the number of iterations is normally required for most root-finding programs since we cannot usually be certain that we will be able to find a number $x_n$ such that $|f(x_n)| < \varepsilon$ for a prescribed $\varepsilon$ (nor can we be certain that we can find two successive iterates, $x_n$ and $x_{n+1}$, such that $|x_n - x_{n+1}| < \delta$, for a prescribed $\delta$).

**Example 4.1.** Let $f(x) = \cos(x) - x$ and note that $f(0) > 0 > f(1)$, so that $f(x) = 0$ has at least one root in $[0, 1]$. Subroutines BISECT and REGFAL were employed to estimate a root of $f(x) = 0$, with $a = 0$ and $b = 1$ in both cases. The results (Table 4.1) show the superiority of *Regular Falsi* for this particular function. The exact answer to eight places is $s = 0.73908513$. The fact that both methods give a final answer that is incorrect in the 7th place is due to round-off error, since we are near the limits of single precision for the machine used.

**Table 4.1**

| *Bisection Method* | | *Regula Falsi* | |
|---|---|---|---|
| $x_i$ | $f(x_i)$ | $x_i$ | $f(x_i)$ |
| 0.5000000E 00 | 0.3775825E 00 | 0.6850733E 00 | 0.8929932E−01 |
| 0.7500000E 00 | −0.1831114E−01 | 0.7362989E 00 | 0.4660129E−02 |
| 0.6250000E 00 | 0.1859630E 00 | 0.7389453E 00 | 0.2340078E−03 |
| 0.6875000E 00 | 0.8533489E−01 | 0.7390780E 00 | 0.1186132E−04 |
| 0.7187500E 00 | 0.3387933E−01 | 0.7390846E 00 | 0.8344650E−06 |
| 0.7343750E 00 | 0.7874667E−02 | 0.7390849E 00 | 0.2384185E−06 |
| 0.7421875E 00 | −0.5195736E−02 | 0.7390850E 00 | 0.1788139E−06 |
| 0.7382812E 00 | 0.1345098E−02 | | |
| 0.7402343E 00 | −0.1923918E−02 | | |
| 0.7392578E 00 | −0.2890229E−03 | | |
| 0.7387695E 00 | 0.5281567E−03 | | |
| 0.7390136E 00 | 0.1195669E−03 | | |
| 0.7391357E 00 | −0.8475780E−04 | | |
| 0.7390747E 00 | 0.1740455E−04 | | |
| 0.7391052E 00 | −0.3367662E−04 | | |
| 0.7390899E 00 | −0.8106231E−05 | | |
| 0.7390823E 00 | 0.4649162E−05 | | |
| 0.7390861E 00 | −0.1728534E−05 | | |
| 0.7390842E 00 | 0.1430511E−05 | | |
| 0.7390851E 00 | −0.1192092E−06 | | |
| 0.7390847E 00 | 0.6556510E−06 | | |
| 0.7390847E 00 | 0.6556510E−06 | | |

## PROBLEMS

1. If we wish to compute the cube root of 5, we could proceed by finding the zeros of $f(x) = x^3 - 5$. Let $a = 0$ and $b = 3$ and calculate an approximation for $\sqrt[3]{5}$ using four steps of the bisection and *Regula Falsi* methods. Which method is converging faster? ($\sqrt[3]{5} \simeq 1.70998$)
2. One difficulty in implementing a bracketing method is finding points $a$ and $b$ such that $f(a)f(b) < 0$. As an illustration, consider the polynomial $f(x) = x^3 - 4.78x^2 + 6.3931x - 2.615718$, which has 3 real roots in $[0, 3]$. Write a computer program that:
   a) finds three subintervals of $[0, 3]$, each of which contains a root of $f(x) = 0$;
   b) uses *Regula Falsi* to obtain each root, terminating the iteration when $|x_{i+1} - x_i| \leq 10^{-3}$.
3. Considering Fig. 4.1, it appears that all the secant lines will pass through the point $(b, f(b))$. This suggests a modification to speed up *Regula Falsi*, namely: After $x_1$ is found, use a line joining $(b, f(b)/2)$ and $(x_1, f(x_1))$ to find $x_2$. If $f(x_2)f(x_1) > 0$, find $x_3$ from the line joining $(b, f(b)/4)$ and $(x_2, f(x_2))$. If $f(x_2)f(x_1) < 0$, then $x_2$ and $x_1$ bracket the root and the procedure can start anew. Program this modification of subroutine REGFAL and test your program with the function $f(x)$ in Example 4.1.
4. For the function $f(x)$ of Example 4.1, show that $f(x) = 0$ has just one root in $[0, 1]$ by considering $f'(x)$. Next, since $f(x) = \cos(x) - x$, consider the equation $x = \cos(x)$ and show that $f(x) = 0$ can have but one root for $-\infty < x < \infty$. (Hint: A solution of $x - \cos(x)$ occurs only when the graphs $y = x$ and $y = \cos(x)$ intersect.)

## 4.3 FIXED-POINT METHODS

Two methods which normally converge more rapidly than bisection and *Regula Falsi*, are Newton's method and the secant method. Although the Newton and secant methods have a simple geometric interpretation for "nice" functions (see, for instance, Figures 4.4 and 4.8), they do not have the root bracketing property, and do not guarantee convergence for all continuous functions, as do the bisection and *Regula Falsi* methods. When they do converge, however, they are generally much faster. In order to better understand these convergence properties, we shall derive these methods via a concept known as the *fixed-point problem* and illustrate them geometrically after they are derived. Another important reason for taking this approach is that it can easily be extended to solving systems of equations in several variables. Furthermore, we shall see that other problems such as the iterative methods of Chapter 2 and the predictor-corrector methods of Chapter 7 are special cases of fixed-point iterations and can be analyzed by the procedures of the next section.

To illustrate a specific fixed-point problem, we consider the geometric rationale for Newton's method, as given in most calculus texts. This geometric interpretation is quite simple (see Fig. 4.4). Given an initial estimate $x_0$ to a root $s$ of $f(x) = 0$,

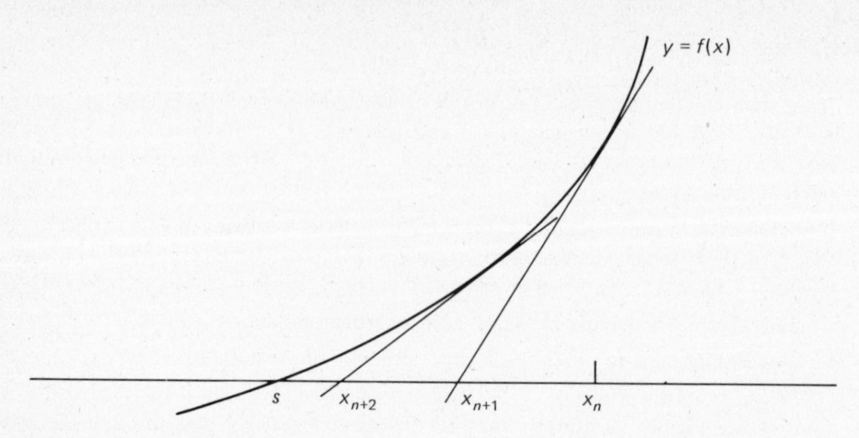


**Fig. 4.4** Newton's method.

we first construct the tangent line to the curve $y = f(x)$ through the point $(x_0, f(x_0))$ and find that the equation of the tangent line is $y = f'(x_0)(x - x_0) + f(x_0)$. We then find the point $x_1$ where the tangent line intersects the $x$-axis, and $x_1$ is taken as the next approximation to $s$. The process is then repeated, with $x_1$ playing the role of $x_0$ and a new approximation to $s$, $x_2$, is found. Since $x_1 = x_0 - f(x_0)/f'(x_0)$, $x_2 = x_1 - f(x_1)/f'(x_1)$, etc., we are generating the sequence

$$x_{n+1} = x_n - \frac{f(x_n)}{f'(x_n)}, \qquad n = 0, 1, \ldots.$$

Thus Newton's method is a special case of an iteration of the form $x_{n+1} = g(x_n)$ (where for the case of Newton's method, $g(x) = x - f(x)/f'(x)$). The analysis of this general iteration, $x_{n+1} = g(x_n)$, is the topic of the next few sections.

### 4.3.1 The Fixed-Point Problem

Throughout this analysis let us be careful not to lose sight of our prime objective: Given a function $f(x)$, where $a \le x \le b$, find values $s$ such that $f(s) = 0$. Given such a function, $f(x)$, we now construct an auxiliary function, $g(x)$, such that $s = g(s)$ whenever $f(s) = 0$. The construction of $g(x)$ is not unique. For example, if $f(x) = x^3 - 13x + 18$, then possible choices for $g(x)$ might be: (a) $g(x) = (x^3 + 18)/13$ (b) $g(x) = (13x - 18)^{1/3}$, (c) $g(x) = (13x - 18)/x^2$ and (d) $g(x) = x^3 - 12x + 18$, just to list a few. In each of these cases, if $s = g(s)$, then $f(s) = 0$.

The problem of finding $s$ such that $s = g(s)$ is known as the *fixed-point problem*, and $s$ is said to be a *fixed-point* of $g(x)$. Thus if we develop an efficient procedure for finding a fixed-point for $g(x)$, $a \le x \le b$, then we automatically have an efficient procedure for finding a zero of $f(x)$, $a \le x \le b$. The fixed-point problem turns out to be quite simple, both theoretically and geometrically. It is immediate

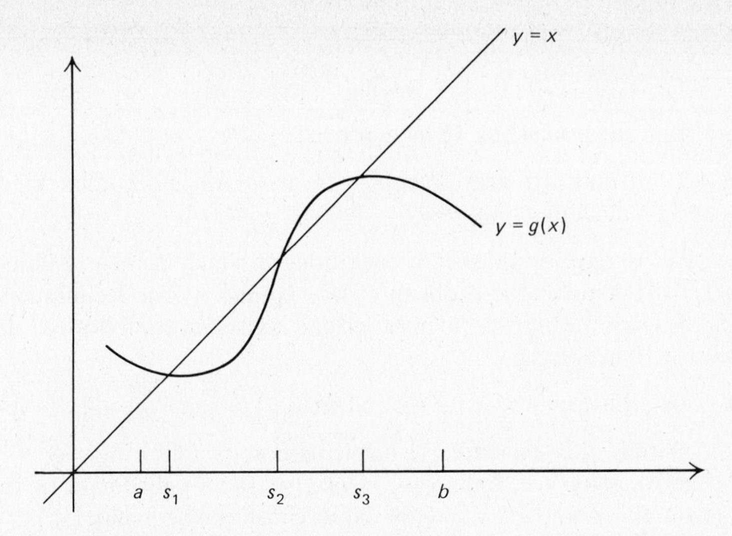


**Fig. 4.5** $s_i = g(s_i)$; $i = 1, 2, 3$.

from Fig. 4.5, that $g(x)$ has a fixed-point on the interval $[a, b]$ whenever the graph of $g(x)$ intersects the line $y = x$.

It is also obvious that on any given interval, $I \equiv [a, b]$, $g(x)$ may have many fixed-points, or none at all. Thus, in order to insure that $g(x)$ has a fixed point in $I$, certain restrictions must be placed on $g(x)$. First of all, let us assume for each $x \in I$, that $g(x) \in I$. This is plausible since, for $s \in I$, $s$ cannot equal $g(s)$ if no $g(x)$ belongs to $I$. (Hereafter this condition will be written as $g(I) \subseteq I$.) Even with this restriction, $g(x)$ may still not have a fixed-point in $I$. For example, if $g(x)$ is discontinuous, part of its graph may lie above $y = x$ and part may be below. However, if we also require $g(x)$ to be continuous, we can prove that $g(x)$ must have a fixed-point in $I$. To see this, suppose that $g(I) \subseteq I$ and $g(x)$ is continuous. Observe that $g(I) \subseteq I$ means $a \leq g(a) \leq b$ and $a \leq g(b) \leq b$. If either $a = g(a)$ or $b = g(b)$, then that endpoint is a fixed point. Let us assume that this is not the case, so that $(g(a) - a) > 0$ and $(g(b) - b) < 0$. For $F(x) \equiv g(x) - x$, $F(x)$ is continuous and $F(a) > 0$ and $F(b) < 0$. Thus, by the intermediate value theorem, there exists at least one $s \in I$ such that $F(s) \equiv g(s) - s = 0$. So we have established:

**Theorem 4.1.** If $g(I) \subseteq I$ and $g(x)$ is continuous, then $g(x)$ has at least one fixed point in $I$.

In order to insure that $g(x)$ has a unique fixed point in $I$, we must not allow $g(x)$ to vary too rapidly. For this we make the additional assumption that $g'(x)$ exists on $I$ and that $|g'(x)| \leq L < 1$, for all $x \in I$. (Note that this condition implies that $g(x)$ is continuous on $I$.) Now, let us assume that $s_1 \in I$, $s_2 \in I$, $s_1 \neq s_2$, and $s_1 = g(s_1)$ and $s_2 = g(s_2)$. Then, by the mean-value theorem, with $\xi$ between $s_1$

100606
YORK COLLEGE PENNSYLVANIA

and $s_2$,

$$|s_2 - s_1| = |g(s_2) - g(s_1)| = |g'(\xi)(s_2 - s_1)| \le L|s_2 - s_1| < |s_2 - s_1|,$$

which is a contradiction. Thus we have proved:

**Theorem 4.2.** If $g(I) \subseteq I$ and $|g'(x)| \le L < 1$, for all $x \in I$, then there exists exactly one $s \in I$ such that $g(s) = s$.

Now that we have established a condition for which $g(x)$ has a unique fixed point in $I$, there remains the problem of how to find it. The technique we shall employ is known as the fixed-point iteration and is given by the following algorithm and is illustrated in Fig. 4.6.

Let $x_0$ be arbitrary in $I = [a, b]$, and let $x_{n+1} = g(x_n)$ for all $n \ge 0$.

Geometrically, this sequence can be pictured in the following way: Given any $x_n$ of the above sequence, then $x_{n+1} = g(x_n)$ is the $y$-coordinate of the point $(x_n, g(x_n))$ on the graph of $y = g(x)$. Now consider the point $(x_{n+1}, x_{n+1}) = (x_{n+1}, g(x_n))$ which lies on the graph of $y = x$. The vertical projection from this point to the $x$-axis yields the point $(x_{n+1}, 0)$, so we know the position of $x_{n+1}$ on the $x$-axis and are ready to repeat the process. This is illustrated in Fig. 4.6.

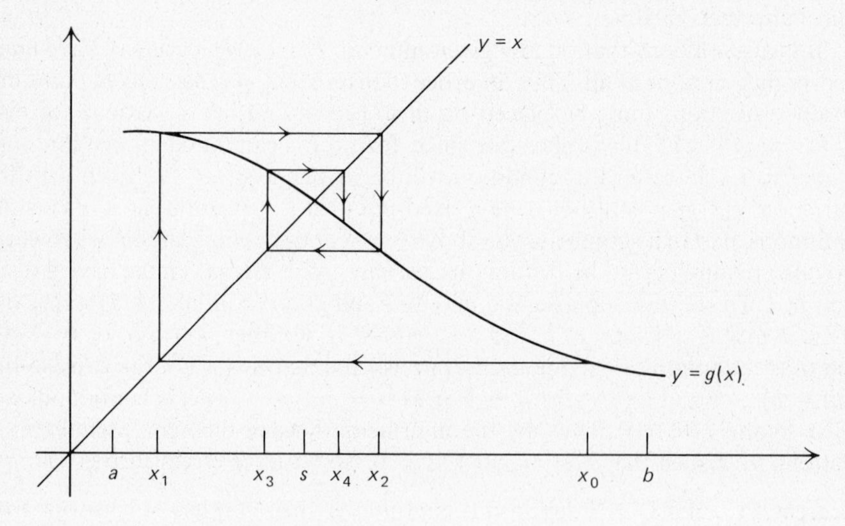


**Fig. 4.6** The fixed-point iteration.

Note that if for any $n$, $x_n = s$, then $x_{n+1} = g(x_n) = g(s) = s = x_n$. Similarly, $x_m = s$, for all $m \ge n$, and the sequence stays "fixed" at $s$. We shall now show that under the conditions of Theorem 4.2, the fixed-point iteration converges, and we shall give a bound on the error after $n$ steps.

**Theorem 4.3.** Let $g(I) \subseteq I \equiv [a, b]$ and $|g'(x)| \leq L < 1$, for all $x \in I$. For $x_0 \in I$, the sequence $x_n = g(x_{n-1})$, $n = 1, 2, \ldots$ converges to the fixed-point $s$, and the $n$th error, $e_n \equiv x_n - s$, satisfies

$$|e_n| \leq \frac{L^n}{(1 - L)} |x_1 - x_0|.$$

*Proof*: By Theorem 4.2, we know there is precisely one fixed-point $s$ in $I$. Given any $n$, there exists a mean-value point $\xi_n$ between $x_{n-1}$ and $s$ such that

$$|x_n - s| = |g(x_{n-1}) - g(s)| = |g'(\xi_n)|\,|x_{n-1} - s| \leq L|x_{n-1} - s|.$$

Successive repetition of this inequality yields $|x_n - s| \leq L^n|x_0 - s|$. Since $0 \leq L < 1$,

$$\lim_{n \to \infty} L^n = 0, \text{ and so } \lim_{n \to \infty} x_n = s.$$

Thus the method is convergent. To establish the error bound, consider for any $n$,

$$\begin{aligned} |x_{n+1} - x_n| &= |g(x_n) - g(x_{n-1})| \\ &= |g'(\eta_{n+1})|\,|x_n - x_{n-1}| \leq L|x_n - x_{n-1}| \\ &= L|g(x_{n-1}) - g(x_{n-2})| \leq L^2|x_{n-1} - x_{n-2}| \leq \cdots \leq L^n|x_1 - x_0| \end{aligned}$$

Now, let $n$ be *fixed* and let $m$ be any integer where $m > n$. Then

$$\begin{aligned} |x_m - x_n| &\leq |x_m - x_{m-1}| + |x_{m-1} - x_{m-2}| + \cdots + |x_{n+1} - x_n| \\ &\leq (L^{m-1} + L^{m-2} + \cdots + L^n)|x_1 - x_0| \leq \left(L^n \sum_{j=0}^{\infty} L^j\right)|x_1 - x_0| \\ &= \frac{L^n}{(1 - L)}|x_1 - x_0|. \end{aligned}$$

Letting $m \to \infty$ so that $x_m \to s$ establishes the inequality for $|e_n|$. ∎

Theorem 4.3 bears a striking resemblance to Theorem 2.2 in Chapter 2 and the proofs are essentially the same. One important feature of Theorem 4.3 is that it provides an error estimate at each step and, hence, can be used as a test for terminating the iteration. Lacking an estimate, such as the one provided by this theorem, we are left only with the possibly unattractive alternative of testing $|x_{n+1} - x_n|$ to determine whether or not to end the iteration.

Theorem 4.3 is called a "nonlocal" convergence theorem because it specifies a fixed, *known* interval, $I = [a, b]$, and displays convergence for any $x_0 \in I$. Often it is not possible to specify such an interval ahead of time, but we might still hope that the fixed-point iteration would converge if we could manage to make our initial guess, $x_0$, "sufficiently close" to the fixed point $s$. Any theorem which says: "If the initial guess is 'very close' to the solution, then the method will converge," is called a "local" convergence theorem, because it does not specify beforehand precisely how close $x_0$ must be to $s$. The following is an example of a local convergence theorem:

**Theorem 4.4.** Let $g'(x)$ be continuous in some open interval containing $s$, where $s$ is a fixed point of $g(x)$. If $|g'(s)| < 1$, there exists an $\varepsilon > 0$ such that the fixed-point iteration is convergent whenever $|x_0 - s| < \varepsilon$.

*Proof*: Since $g'(x)$ is continuous in an open interval containing $s$ and $|g'(s)| < 1$, then for any constant $K$ satisfying $|g'(s)| < K < 1$, there exists an $\varepsilon > 0$ such that if $x \in [s - \varepsilon, s + \varepsilon] \equiv I_\varepsilon$, then $|g'(x)| \leq K$. By the mean-value theorem, given any $x \in I_\varepsilon$, there exists an $\eta$ between $x$ and $s$ such that $|g(x) - s| = |g(x) - g(s)| = |g'(\eta)|\,|x - s| \leq K\varepsilon < \varepsilon$, and thus $g(I_\varepsilon) \subseteq I_\varepsilon$. Therefore by using $I_\varepsilon$ in Theorem 4.3, our result is proved. ■

(Note that this theorem does not say what the value of $\varepsilon$ is; it merely assures us that there is an $\varepsilon$-neighborhood of $s$ where the fixed-point iteration will converge.) Problem 5 gives a contrasting result, showing that if $|g'(s)| > 1$, then there is a neighborhood of $s$ in which no initial guess (except $x_0 = s$) will work.

## PROBLEMS

1. Let $I = [1, 3]$ and consider the four choices, (i), (ii), (iii), and (iv) for $g(x)$.

   i. $g(x) = \dfrac{x^3 + 18}{13}$

   ii. $g(x) = (13x - 18)^{1/3}$

   iii. $g(x) = \dfrac{(13x - 18)}{x^2}$

   iv. $g(x) = x^3 - 12x + 18$

   Which of these satisfy the conditions of Theorem 4.3?

   b) For these same functions find, if possible, an $\varepsilon$-neighborhood of $s = 2$ wherein the fixed-point iteration will converge.

2. Given that $g(I) \subseteq I \equiv [a, b]$ and $0 \leq g'(x) \leq L < 1$ for all $x \in I$, show that all of the fixed-point iterates remain on the same side of $s$ as $x_0$.

3. Transform the problem of solving $f(x) = x^2 - c\ (c > 0)$ into the fixed-point problem $x = (x^2 + c)/2x$ and find an interval for which Theorem 4.3 guarantees convergence. (To what value will the iteration converge?)

4. Evaluate: $s = \sqrt[3]{6 + \sqrt[3]{6 + \sqrt[3]{6 + \cdots}}}$. Hint: Let $x_0 = 0$ and consider $g(x) = \sqrt[3]{6 + x}$.

5. Suppose $g(s) = s$, $g(x)$ is continuously differentiable in an interval containing $s$ and $|g'(s)| > 1$. Show there is $\varepsilon > 0$ such that if $0 < |x_0 - s| < \varepsilon$ then $|x_0 - s| < |x_1 - s|$ (thus, no matter how close $x_0$ is to $s$, the next iterate is farther away).

6. Figure 4.6 illustrates the case where $-1 < g'(s) < 0$. Draw similar graphs illustrating the cases:

$$0 \le g'(s) < 1, \qquad 1 < g'(s), \qquad g'(s) < -1.$$

7. Let $g(x) = x^2$. From a graph (as in Problem 6), deduce for what values $x_0$, $-\infty < x_0 < \infty$, the iteration $x_{i+1} = g(x_i)$ will converge to a fixed-point of $g(x)$ and for what values $x_0$ will the iteration diverge.
8. For the equation $f(x) = 0$ with $f(x) = \cos(x) - x$ (as in Example 4.1), a natural associated fixed-point problem is $x = \cos(x)$. As in Problem 7, graph $g(x) = \cos(x)$ and convince yourself from the picture that the sequence $x_{i+1} = g(x_i)$ will converge for any $x_0 \in [0, 1]$. Program this iteration and test your program with $x_0 = 1$ and $x_0 = 0.1$. Use Theorem 4.3 to prove that the sequence converges for any $x_0 \in [0, 1]$ and print out both the theoretical error bounds and actual errors at each step of the iteration.

### 4.3.2 Rate of Convergence of the Fixed-Point Algorithm

In this section, let $s$ be a fixed-point of $g(x)$ where $g(x)$ satisfies the hypotheses of Theorem 4.4 in an interval $I$. Let $x_0 \in I$ and, for each $k$, let $e_k = x_k - s$. Further, let us suppose that the $(k + 1)$st derivative of $g(x)$ is continuous on $I$ (hereafter written: $g \in C^{k+1}(I)$). To determine how the errors decrease at each step of the iteration, we expand $g(x)$ in a Taylor's series about $x = s$, obtaining

$$e_{n+1} = g(x_n) - g(s) = g'(s)e_n + \frac{g''(s)}{2!} e_n^2 + \cdots + \frac{g^{(k)}(s)}{k!} e_n^k + E_{k,n} \qquad \textbf{(4.1)}$$

where

$$E_{k,n} = g^{(k+1)}(\alpha_n)e_n^{k+1}/(k+1)!$$

(and where the mean-value point $\alpha_n$ is between $x_n$ and $s$). Suppose first that $g'(x) \neq 0$ for all $x \in I$. From (4.1) with $k = 0$, we have

$$e_{n+1} = g'(\alpha_n)e_n \qquad \text{or} \qquad \frac{e_{n+1}}{e_n} = g'(\alpha_n).$$

By Theorem 4.4, we have $\lim_{n\to\infty} x_n = s$, so $\lim_{n\to\infty} \alpha_n = s$ as well. Using the continuity of $g'(x)$, we find

$$\lim_{n\to\infty} \frac{e_{n+1}}{e_n} = g'(s). \qquad \textbf{(4.1a)}$$

Since we have assumed that $g'(s) \neq 0$, this means (for sufficiently large $n$) that $e_{n+1} \approx g'(s)e_n$. Such a rate of convergence is called *linear* or *first-order* convergence. By contrast, if $g'(s) = 0$ and $g''(x) \neq 0$ for all $x$ in $I$, then (from (4.1) with $k = 1$) we obtain the stronger result

$$\lim_{n\to\infty} \frac{e_{n+1}}{e_n^2} = \frac{g''(s)}{2!}. \qquad \textbf{(4.1b)}$$

This case, where the $(n+1)$st error is approximately proportional to the square of the $n$th error, is called *quadratic* or *second-order* convergence.

Similarly, if $g'(s) = g''(s) = \cdots = g^{(k)}(s) = 0$ and $g^{(k+1)}(x)$ does not vanish on $I$, then we have "$(k+1)$st order" convergence

$$\lim_{n\to\infty} \frac{e_{n+1}}{e_n^{k+1}} = \frac{g^{(k+1)}(s)}{(k+1)!}. \tag{4.1c}$$

Thus, the more derivatives of $g(x)$ which vanish at $x = s$, the faster the rate of convergence of the fixed-point iteration. In the analysis above, we implicitly assumed that $e_n \neq 0$ for $n = 0, 1, 2, \ldots$. This assumption is always a valid one if $x_0 \neq s$ and if (as we required) $g^{(i)}(x) \neq 0$ for $x \in I$, when $g'(s) = \cdots = g^{(i-1)}(s) = 0$. For example, consider the linear case, where $g'(s) \neq 0$. Let $m$ be the first positive integer such that $e_m = 0$. Then

$$0 = e_m = g(x_{m-1}) - g(s) = g'(\alpha)(x_{m-1} - s) = g'(\alpha)e_{m-1}$$

where $\alpha$ is between $x_{m-1}$ and $s$. As $g'(x) \neq 0$ for all $x$ in $I$, then $e_{m-1} = 0$, which contradicts our choice of $m$. Thus if $e_0 \neq 0$, then $e_n \neq 0$ for $n = 1, 2, \ldots$.

We shall be particularly interested in quadratic convergence, since this leads to the derivation of Newton's method. For now, however, we pause to consider the slowest case, $g'(x) \neq 0$ for all $x \in I$, and we shall develop a way to speed up the convergence of these sequences. The technique we shall employ is called the *Aitken's $\Delta^2$-method.*

We shall emphasize that Aitken's method can be used on sequences other than those generated by fixed-point iteration and that we can employ the $\Delta^2$-method in different procedures in other chapters. (For example, we note by Problem 7 that the power-method sequence (see Chapter 3) can be accelerated by Aitken's method by virtue of Theorem 4.5.) Thus we present Aitken's method in a context completely divorced from the fixed-point iteration and in its full generality. This method is given in the following algorithm.

Let $\{t_n\}_{n=1}^{\infty}$ be any sequence that converges to $t^*$. Form a new sequence $\{t'_n\}_{n=1}^{\infty}$ by the formula

$$t'_n = t_n - \frac{(\Delta t_n)^2}{\Delta^2 t_n}, \tag{4.2}$$

where $\Delta t_n = t_{n+1} - t_n$ and $\Delta^2 t_n = t_{n+2} - 2t_{n+1} + t_n$. The symbols $\Delta$ and $\Delta^2$ are *forward differences* (see Problem 2 of this section), but we will defer the theory of differences until later. Equation (4.2) is an "extrapolation" procedure which is suggested by considering sequences $\{x_n\}$ which are converging in a "fairly regular" fashion, such as the sequences generated by a linearly converging fixed-point iteration, where

$$\lim_{n\to\infty} \frac{x_{n+1} - s}{x_n - s} = g'(s)$$

(that is, $x_{n+1} - s \approx g'(s)(x_n - s)$, when $n$ is large). If we knew, for example, that $x_{n+1} - s = B(x_n - s)$ and $x_{n+2} - s = B(x_{n+1} - s)$, then we could use these two equations (in the unknowns $B$ and $s$) to solve for $s$, yielding $s = x_n - (x_{n+1} - x_n)^2/(x_{n+2} - 2x_{n+1} + x_n)$, which is exactly analogous to Eq. (4.2) (see Problem 4). In practice, we use the fact that whenever we have regularity of convergence we can take three successive terms of a sequence and "extrapolate" to the limit. The manner in which Aitken's $\Delta^2$-method speeds convergence is given by the following theorem, which also explains more precisely what we mean by "regular convergence."

**Theorem 4.5.** Let $\{t_n\}_{n=1}^{\infty}$ and $\{t'_n\}_{n=1}^{\infty}$ be as in Eq. (4.2), where $\lim_{n\to\infty} t_n = t^*$. Further assume for all $n$, that $\varepsilon_n = t_n - t^*$ satisfies $\varepsilon_{n+1} = (B + \beta_n)\varepsilon_n$, where $\varepsilon_n \neq 0$, $|B| < 1$ and $\lim_{n\to\infty} \beta_n = 0$. Then, for $n$ sufficiently large, $t'_n$ is well defined and the new sequence converges to $t^*$ faster than the old sequence in the sense that

$$\lim_{n\to\infty} \frac{t'_n - t^*}{t_n - t^*} = 0.$$

*Proof*: Observe that $\varepsilon_{n+2} = (B + \beta_{n+1})\varepsilon_{n+1} = (B + \beta_{n+1})(B + \beta_n)\varepsilon_n$. Therefore

$$\begin{aligned}\Delta^2 t_n &= \varepsilon_{n+2} - 2\varepsilon_{n+1} + \varepsilon_n \\ &= (B + \beta_{n+1})(B + \beta_n)\varepsilon_n - 2(B + \beta_n)\varepsilon_n + \varepsilon_n \\ &= [(B - 1)^2 + \beta'_n]\varepsilon_n,\end{aligned}$$

where

$$\beta'_n = B(\beta_n + \beta_{n+1}) - 2\beta_n + \beta_n\beta_{n+1}.$$

Moreover, $\lim_{n\to\infty} \beta'_n = 0$ since $\beta_n \to 0$ as $n \to \infty$. Thus, for $n$ sufficiently large, $\Delta^2 t_n \neq 0$, and $t'_n$ is well defined. Now, setting

$$\Delta\varepsilon_n = \varepsilon_{n+1} - \varepsilon_n$$

and

$$\Delta^2\varepsilon_n = \varepsilon_{n+2} - 2\varepsilon_{n+1} + \varepsilon_n,$$

we have

$$\begin{aligned}t'_n - t^* &= t_n - t^* - \frac{(\Delta t_n)^2}{\Delta^2 t_n} = t_n - t^* - \frac{(\Delta\varepsilon_n)^2}{\Delta^2\varepsilon_n} \\ &= t_n - t^* - \frac{(B - 1 + \beta_n)^2\varepsilon_n^2}{[(B - 1)^2 + \beta'_n]\varepsilon_n}.\end{aligned}$$

Therefore,

$$\lim_{n\to\infty} \frac{t'_n - t^*}{t_n - t^*} = \lim_{n\to\infty} \left\{1 - \frac{(B - 1)^2 + 2\beta_n(B - 1) + \beta_n^2}{(B - 1)^2 + \beta'_n}\right\} = 0. \quad \blacksquare$$

**Corollary.** Let $g(I) \subseteq I = [a, b]$ and $g'(x)$ be continuous on $I$, where $0 < |g'(x)| \leq L < 1$. Then Theorem 4.5 may be applied to the fixed-point sequence, $\{x_n\}_{n=0}^{\infty}$, to speed its convergence.

*Proof*: In Theorem 4.5, identify $\{t_n\}_{n=0}^{\infty}$ with $\{x_n\}_{n=0}^{\infty}$ and $t^*$ with $s$, the fixed point of $g(x)$. As noted previously, if $x_0 \neq s$, then $\varepsilon_n = t_n - t^* = x_n - s = e_n \neq 0$, and the theorem applies. ■

## PROBLEMS

1. In formula (4.1) assume that $x_0 \neq s$, $g'(s) = 0$ and that $g''(x)$ is continuous and nonzero on $I$. Show that $e_n \neq 0$, for all $n \geq 1$.

*2. Given a sequence $\{t_n\}_{n=1}^{\infty}$, define $\Delta t_n = t_{n+1} - t_n$, for all $n$, and $\Delta^k t_n = \Delta(\Delta^{k-1} t_n)$, for $k > 1$. Show by induction that

$$\Delta^k t_n = \sum_{j=0}^{k} \binom{k}{j} (-1)^{k-j} t_{n+j},$$

where the binomial coefficient

$$\binom{k}{j} = \frac{k!}{(k-j)!j!}.$$

Also show, that if $\varepsilon_n = t_n - s$, for all $n$, then $\Delta^k \varepsilon_n = \Delta^k t_n$ for $k \geq 1$.

3. Let $\{t_n\}_{n=1}^{\infty} = \{(1/2)^n\}_{n=1}^{\infty}$ and $\{r_n\}_{n=1}^{\infty} = \{1/n^2\}_{n=1}^{\infty}$. Determine whether Aitken's $\Delta^2$-method can be applied to either of these sequences.
4. Suppose $\{x_n\}$ is a sequence and suppose that $x_{i+1} - s = B(x_i - s)$ and $x_{i+2} - s = B(x_{i+1} - s)$ for some $i$. Under the assumption that $B \neq 1$, solve these two equations for $s$, thus giving an intuitive derivation for Aitken's $\Delta^2$-method.
5. Let $g(x) = x^2 - 2x + 2$. What are the fixed points of $g(x)$? For each fixed point $s$, determine whether there is an $\varepsilon > 0$ such that if $|x_0 - s| < \varepsilon$ then the sequence $x_{i+1} = g(x_i)$ is convergent to $s$. What is the order of convergence of the iteration at those fixed points for which convergence occurs?
6. If some care is exercised, Aitken's $\Delta^2$-method can be used to accelerate the convergence of infinite series and power series. For example, a power series for $e^x$ is given by

$$e^x = 1 + x + x^2/2! + \cdots + x^n/n! + \cdots = \sum_{k=0}^{\infty} \frac{x^k}{k!}.$$

For a *fixed* $x$, let $s_n(x) = 1 + x + x^2/2! + \cdots + x^n/n!$; so that $\{s_n(x)\}_{n=0}^{\infty}$ is a sequence converging to $e^x$. We observe that $\varepsilon_n = e^x - s_n(x) = e^{\alpha} x^{n+1}/(n+1)!$, where $\alpha$ depends on $n$ and is between 0 and $x$. Show that

$$\lim_{n \to \infty} \frac{\varepsilon_{n+1}}{\varepsilon_n} = 0$$

and thus conclude that Theorem 4.5 can be applied. For $x = 2$, use the ideas developed above the approximate $e^2$, extrapolating from $s_8(2)$, $s_9(2)$, and $s_{10}(2)$ using (4.2). How does $s_{10}(2)$ compare with the extrapolated estimate?

7. Consider the sequence $\{\beta_k\}$ generated by the power method as in formula (3.12). Use (3.12) to show that

$$\lim_{n\to\infty} \frac{\beta_{k+1} - \lambda_1}{\beta_k - \lambda_1} = r, \qquad |r| < 1.$$

Thus the sequence $\{\beta_k\}$ satisfies the conditions of Theorem 4.5, and can be accelerated by Aitken's method.

*8. (Steffensen's method). Consider the fixed-point problem, $x = g(x)$, with an initial guess $x_0 = t_0$, and solution $s = g(s)$. Let $x_1 = g(x_0)$ and $x_2 = g(x_1) = g(g(x_0))$. Now apply the Aitken formula (4.2) to obtain

$$t_1 = x_0 - \frac{(\Delta x_0)^2}{\Delta^2 x_0} = x_0 - \frac{(x_1 - x_0)^2}{x_2 - 2x_1 + x_0}.$$

Now repeat the process, that is, let $x_0' = t_1$, $x_1' = g(t_1)$, $x_2' = g(x_1') = g(g(t_1))$, and $t_2 = t_1 - (g(t_1) - t_1)^2/[g(g(t_1)) - 2g(t_1) + t_1]$. Thus we see we are generating a sequence $\{t_k\}$, where each $t_k$ is obtained by three steps of a fixed-point iteration and then an Aitken's method correction. We formalize this procedure by defining $R(x) = g(g(x)) - 2g(x) + x$ and let $G(x) = x - (g(x) - x)^2/R(x)$, if $R(x) \neq 0$, and let $G(x) = x$, if $R(x) = 0$. Thus, with $t_0$ given, the sequence $\{t_k\}$ is generated by the fixed-point formula: $t_{k+1} = G(t_k)$, $k \geq 0$.

a) Let $g(x) = \sqrt[3]{6 + x}$ and $x_0 = t_0 = 3$. Write a program which generates *both* the fixed-point sequence $x_{k+1} = g(x_k)$, and the Steffensen sequence $t_{k+1} = G(t_k)$. Compare the rates of convergence. (Obviously $s = 2$ satisfies $s = g(s)$.)

*b) Assume that an arbitrary $g(x)$ satisfies $g'(s) \neq 1$ and $g'''(x)$ is continuous. Prove that $G'(s) = 0$, and thus Steffensen's method is quadratically convergent by (4.1) and (4.1b).

### 4.3.3 Newton's Method

We are now ready to apply the above fixed-point analysis to our principal problem of finding the zeros of a given function, $f(x)$. One natural choice of the fixed-point function would be $g(x) = x + cf(x)$, where $c$ is a nonzero constant. Then $f(s) = 0$ if and only if $s = g(s)$. From Section 4.3.2, we see that the fixed-point iteration is accelerated by making as many derivatives of $g(x)$ at $x = s$ equal to zero as possible. Now, $g'(s) = 1 + cf'(s)$, thus $g'(s) \neq 0$ unless $c = -(1/f'(s))$. Unfortunately, since we do not know the value of $s$, it is impossible to make this choice for $c$, *a priori*.

Thus we are led to make a different choice for $g(x)$. This time we let $g(x) = x + h(x)f(x)$, and try to select $h(x)$ such that $g'(s) = 0$. Now, $g'(s) = 1 + h'(s)f(s) + h(s)f'(s) = 1 + h(s)f'(s)$. Thus, for $g'(s) = 0$, we must select $h(x)$ such that $h(s) = -(1/f'(s))$. Immediately, we see that $h(x) \equiv -(1/f'(x))$ has this property, and furthermore we need not know the value of $s$ to make this choice. Therefore, we select $g(x) \equiv x - f(x)/f'(x)$, and the fixed-point algorithm yields the following iteration, known as Newton's method. Given the function, $f(x)$, let $x_0$ be an initial guess for $s$, where $f(s) = 0$. Then let

$$x_{n+1} = x_n - f(x_n)/f'(x_n), \qquad n = 0, 1, \ldots \tag{4.3}$$

The above analysis along with Theorem 4.4 yields the following local convergence theorem for Newton's method. (The proof is left to the reader.)

**Theorem 4.6.** Let $f''(x)$ be continuous and $f'(x) \neq 0$ in some open interval containing $s$, where $f(s) = 0$. Then there exists an $\varepsilon > 0$ such that Newton's method is quadratically convergent whenever $|x_0 - s| < \varepsilon$.

Subroutine NEWTON (Fig. 4.7) employs Newton's method to find an approximate root of the equation $f(x) = 0$. One difficulty that is encountered in programming a root-finding procedure that does not possess the bracketing property of the bisection method or *Regula Falsi*, is that of selecting an appropriate test for terminating the iteration. When it is not practical to use a guaranteed error bound, such as that displayed in Theorem 4.3, we are left with the alternatives of prescribing either a tolerance $\varepsilon > 0$ or a tolerance $\delta > 0$, and stopping the iteration when $|x_{n+1} - x_n| < \delta$ or when $|f(x_n)| < \varepsilon$. Subroutine NEWTON uses both of these criteria and an upper bound on the number of iterations to be executed, as well.

```
      SUBROUTINE NEWTON(XO,XTERM,FTERM,N,ITERM)
C
C  THIS SUBROUTINE USES NEWTON'S METHOD TO FIND A ROOT OF F(X)=0.  THE
C  CALLING PROGRAM MUST SUPPLY AN INITIAL GUESS, XO, AND 3 TERMINATION
C  CRITERIA, XTERM, FTERM AND N.  THE SUBROUTINE RETURNS AN APPROXIMATE
C  ROOT IN XO WHEN ONE OF THE TERMINATION REQUIREMENTS IS MET.  A FLAG,
C  ITERM, IS SET TO 1 WHEN THE ABSOLUTE VALUE OF F(XO) IS LESS THAN
C  FTERM; ITERM IS SET TO 2 WHEN 2 SUCESSIVE ITERATES DIFFER BY LESS
C  THAN XTERM AND ITERM IS SET TO 3 WHEN THE MAXIMUM NUMBER OF
C  ITERATIONS, N, IS REACHED.  TWO FUNCTION SUBPROGRAMS NAMED F(X)
C  AND FPRIM(X) MUST BE SUPPLIED TO CALCULATE F(X) AND THE DERIVATIVE
C  OF F(X).
C
      DO 1 I=1,N
      FO=F(XO)
      IF(ABS(FO).LT.FTERM)   GO TO 2
      FPO=FPRIM(XO)
      CORREC=FO/FPO
      IF(ABS(CORREC).LT.XTERM)    GO TO 3
    1 XO=XO-CORREC
      ITERM=3
      RETURN
    2 ITERM=1
      RETURN
    3 ITERM=2
      XO=XO-CORREC
      RETURN
      END
```

**Fig. 4.7** Subroutine NEWTON.

**Example 4.2.** Let $f(x)$ be the simple function of Example 4.1, $f(x) = \cos(x) - x$. Starting with $x_0 = 0$, the following sequence of iterates was generated by subroutine NEWTON. (See Table 4.2.) For this example, convergence is quite rapid.

**Table 4.2** Newton's method.

| $x_i$ | $f(x_i)$ |
|---|---|
| 0.0000000E 00 | 0.1000000E 01 |
| 0.1000000E 01 | −0.4596977E 00 |
| 0.7503638E 00 | −0.1892304E−01 |
| 0.7391128E 00 | −0.4643201E−04 |
| 0.7390850E 00 | 0.5960464E−07 |

We note, that while the geometric derivation illustrated in Figure 4.4 is quite simple, it says nothing about quadratic convergence, sufficient conditions on $f(x)$ to insure convergence, and the question of convergence when $f'(s) = 0$. Recall that we say the root $s$ has *multiplicity* $p$ if $f(s) = f'(s) = \cdots = f^{(p-1)}(s) = 0$, but $f^{(p)}(s) \neq 0$. We shall consider the case when $s$ has multiplicity 2, that is, $s$ is a *double root* of $f(x)$, and consider higher multiplicities in the problems.

Let us assume that $f(s) = f'(s) = 0$, and $f^{(4)}(x)$ is continuous. Again let $g(x) = x - f(x)/f'(x)$, so $g'(x) = 1 - (f'(x)^2 - f(x)f''(x))/f'(x)^2$, or $g'(x) = f(x)f''(x)/f'(x)^2$. We easily verify by l'Hôpital's rule that $g'(s) = 1/2$. Since $|g'(s)| < 1$, and the rest of the hypotheses of Theorem 4.4 are satisfied, we see that Newton's method still converges locally to $s$. The convergence, however, is not quadratic. Instead we get only linear convergence, where $\lim_{n\to\infty} (e_{n+1}/e_n) = g'(s) = 1/2$.

In the above analysis, however, if we choose $g(x) = x - 2f(x)/f'(x)$, then $g'(s) = 0$. Thus, if we know *a priori* that the multiplicity of $s$ is 2, then the sequence

$$x_{n+1} = x_n - 2f(x_n)/f'(x_n)$$

will converge to $s$ quadratically for $x_0$ sufficiently close to $s$.

In general, if $s$ has multiplicity $p$, then we can show (Problem 3) that if $g(x) = x - f(x)/f'(x)$ then $g'(s) = 1 - 1/p$, and if $g(x) = x - pf(x)/f'(x)$ then $g'(s) = 0$. Thus, for $x_0$ sufficiently close to $s$, the sequence

$$x_{n+1} = x_n - pf(x_n)/f'(x_n) \tag{4.4}$$

is quadratically convergent to $s$. The formula (4.4) is of little practical use, however, since we rarely know the multiplicity of a root in advance.

**Example 4.3.** In this example, we illustrate the behavior of Newton's method near a point $s$ where $f(s) = f'(s) = 0$. For $f(x) = x^3 + x^2 - 5x + 3$, we have $f(x) = (x - 1)^2(x + 3)$ so that $f(1) = f'(1) = 0$ and $f(-3) = 0$. Newton's method was run with an initial guess of $x_0 = 4$. Convergence to $s = 1$ is seen to be quite slow. As the iterates near 1, they exhibit an oscillatory behavior that is typical when the limits of machine accuracy are approached. The same program was run with an initial guess of $x_0 = -6$, with rapid convergence to $s = -3$. (See Table 4.3.)

**Table 4.3**

| $x_i$ | $f(x_i)$ | $x_i$ | $f(x_i)$ |
|---|---|---|---|
| 0.4000000E 01 | 0.6300000E 02 | −0.6000000E 01 | −0.1470000E 03 |
| 0.2764706E 01 | 0.1795237E 02 | −0.4384615E 01 | −0.4014566E 02 |
| 0.1999479E 01 | 0.4994273E 01 | −0.3470246E 01 | −0.9396978E 01 |
| 0.1545152E 01 | 0.1350779E 01 | −0.3081737E 01 | −0.1361792E 01 |
| 0.1287998E 01 | 0.3556594E 00 | −0.3003147E 01 | −0.5043125E−01 |
| 0.1148676E 01 | 0.9170532E−01 | −0.3000004E 01 | −0.7629394E−04 |
| 0.1075647E 01 | 0.2332305E−01 | −0.3000000E 01 | 0.0000000E 00 |
| 0.1038170E 01 | 0.5883216E−02 | | |
| 0.1019176E 01 | 0.1478195E−02 | | |
| 0.1009609E 01 | 0.3700256E−03 | | |
| 0.1004812E 01 | 0.9250640E−04 | | |
| 0.1002412E 01 | 0.2384185E−04 | | |
| 0.1001178E 01 | 0.5722045E−05 | | |
| 0.1000572E 01 | 0.1907348E−05 | | |
| 0.1000155E 01 | 0.9536743E−06 | | |
| 0.9993886E 00 | 0.1907348E−05 | | |
| 0.9997784E 00 | 0.9536743E−06 | | |
| 0.1000315E 01 | 0.9536743E−06 | | |
| 0.9999380E 00 | 0.0000000E 00 | | |

The same program, this time run in double precision with $x_0 = 4$, yields results (Table 4.4) that are in agreement with geometric intuition (See Problem 7 of this section).

Newton's method is very useful in problems where $f'(x)$ is easily evaluated. (Such is the case where $f(x)$ is a polynomial, as we shall see in Section 4.4). We shall illustrate here a case of particular importance. Suppose we are given a constant $c > 0$ and wish to find the real, positive $k$th root of $c$. We then let $f(x) = x^k - c$ and the Newton iteration becomes:

$$x_{n+1} = x_n - \frac{x_n^k - c}{kx_n^{k-1}} = \frac{(k-1)x_n + c/x_n^{k-1}}{k} \tag{4.5}$$

**Example 4.4.** In order to approximate $\sqrt{2}$, let $f(x) = x^2 - 2$ and $x_0 = 1$. Formula (4.5) reduces in this case to: $x_{n+1} = (x_n + 2/x_n)/2$. Then, to eight significant digits, $x_1 = 1.5000000$, $x_2 = 1.4166667$, $x_3 = 1.4142156$, and $x_4 = 1.4142135$. In fact, $\sqrt{2} = 1.4142136$ (to eight places).

For practical computational purposes, some modifications of Newton's method are desirable. These modifications are based on the observation that while

Table 4.4

| $x_i$ | $f(x_i)$ |
|---|---|
| 0.40000000D 01 | 0.63000000D 02 |
| 0.27647059D 01 | 0.17952371D 02 |
| 0.19994794D 01 | 0.49942757D 01 |
| 0.15451534D 01 | 0.13507843D 01 |
| 0.12879985D 01 | 0.35566004D 00 |
| 0.11486779D 01 | 0.91707018D−01 |
| 0.10756476D 01 | 0.23323111D−01 |
| 0.10381716D 01 | 0.58838948D−02 |
| 0.10191756D 01 | 0.14778606D−02 |
| 0.10096106D 01 | 0.37034232D−03 |
| 0.10048111D 01 | 0.92696274D−04 |
| 0.10024070D 01 | 0.23187972D−04 |
| 0.10012038D 01 | 0.57987349D−05 |
| 0.10006020D 01 | 0.14499018D−05 |
| 0.10003010D 01 | 0.36250271D−06 |
| 0.10001505D 01 | 0.90629086D−07 |
| 0.10000753D 01 | 0.22657698D−07 |
| 0.10000376D 01 | 0.56644780D−08 |
| 0.10000188D 01 | 0.14161259D−08 |
| 0.10000094D 01 | 0.35403258D−09 |

Newton's method converges rapidly for a starting value near a simple root of $f(x) = 0$, the method may diverge rapidly (or exhibit other erratic behavior) in the presence of a zero of $f'(x)$ or for a starting value somewhat removed from the root. In order to account for this in a general root-finding program, we might include a test of whether $|f(x_{n+1})| < |f(x_n)|$ at each step (such a test would determine whether it is profitable to continue the iteration using Newton's method). A somewhat more comprehensive program might include a combination of bisection and Newton's method. For example, suppose $f(a)f(b) < 0$ and $m = (a + b)/2$. If $f(m)f(b) < 0$, then there is a zero of $f(x)$ in $(m, b)$. If Newton's method starting with $x_0 = b$ (or $x_0 = a$) does not produce an estimate $x_1$ in $(m, b)$, then $x_1$ can be rejected and several steps of the bisection method can be executed before Newton's method is tried again. Similar modifications may also be made to other "fast" procedures, such as the secant method, which do not possess the root-bracketing property of bisection.

## PROBLEMS

1. Use Theorem 4.4 and formula (4.1) to prove Theorem 4.6, with the additional assumption that $f'''(x)$ is continuous.
2. Prove that the tangent line to the graph of $y = f(x)$ at the point $(x_n, f(x_n))$ intersects the $x$-axis when $x = x_n - f(x_n)/f'(x_n)$.

*3. Assume that $f(s) = f'(s) = \cdots = f^{(p-1)}(s) = 0$, $f^{(p)}(x)$ is continuous, and $f^{(p)}(s) \neq 0$. Let $g(x) = x - f(x)/f'(x)$, and show that $g'(s) = 1 - 1/p$. (Hint: Let $h = x - s$ and expand both $f(x)$ and $f'(x)$ in a Taylor's expansion around $s$. Then let $h \to 0$.)

4. A method which is quadratically convergent is said to "essentially" double the number of correct decimal places at each iteration. Give verification for this statement.

5. Approximate the real root of $x^3 - 2x^2 + 2x - 1 = 0$ by Newton's method with $x_0 = 0$ and $x_0 = 10$.

6. To illustrate the necessity of $x_0$ being "sufficiently close" to $s$ for the convergence of the Newton iteration, we define

$$f(x) = \begin{cases} \sin x, & -\pi/2 \le x \le \pi/2 \\ 1, & x \ge \pi/2 \\ -1, & x \le -\pi/2 \end{cases}$$

Then $f'(x) = \begin{cases} \cos x, & -\pi/2 \le x \le \pi/2 \\ 0, & \text{otherwise.} \end{cases}$

Let $t^*$, $0 < t^* < \pi/2$, satisfy $\tan t^* = 2t^*$. (Since $\tan \pi/4 = 1 < \pi/2 = 2(\pi/4)$ and $\infty = \tan \pi/2 > \pi = 2(\pi/2)$, we know that such a $t^*$ exists and $\pi/4 < t^* < \pi/2$.)

a) Assume that $x_0 = t^*$, then evaluate the rest of the Newton method iterates. Are they converging to $s = 0$? Are they diverging?

b) What happens to the Newton method iterates if $t^* < x_0 < \pi/2$?

c) What happens to the Newton method iterates if $0 < x_0 < t^*$?

*7. Suppose $f(x) = (x - r_1)(x - r_2) \cdots (x - r_n)$ where $r_1 < r_2 < \cdots < r_n$. That is, $f(x)$ is an $n$th degree polynomial with $n$ distinct real roots and with leading coefficient equal to 1. By a geometric argument, convince yourself that if $x_0 > r_n$ then the sequence $\{x_i\}$ generated by Newton's method satisfies:

$$r_n < \cdots < x_{i+1} < x_i < \cdots < x_0 \qquad \text{for } i = 1, 2, \ldots.$$

Prove this mathematically, using Rolle's Theorem to observe that $f'(x) > 0$ for $x \ge r_n$.

8. As an extreme case of a function for which Newton's method is slowly convergent, consider $f(x) = (x - \alpha)^n$, for $n$ some positive integer and $\alpha$ some real number. Show that Newton's method generates the sequence

$$x_{i+1} = (1 - 1/n)x_i + \alpha/n$$

and then show that $x_{i+1} - \alpha = (1 - 1/n)(x_i - \alpha)$. This is a special case of Problem 3 above.

### 4.3.4 The Secant Method

In general Newton's method converges much faster than the bracketing methods, but it has serious disadvantages such as how close $x_0$ must be to $s$ before convergence and the need to evaluate $f'(x_n)$, for each $n$. This derivative evaluation can be quite a cumbersome task, and yet it must be done accurately to avoid excessive round-off, especially when $|f'(x_n)|$ is small. The secant method is, in a way, a compromise between Newton's method and the bracketing methods. Its rate of convergence is stronger than linear (called *superlinear*), but not quadratic. This, plus the fact that it does not require derivative evaluations, makes it a very attractive practical method.

The secant method iteration comes directly from the Newton iteration by simply replacing $f'(x_n)$ by the difference quotient $(f(x_n) - f(x_{n-1}))/(x_n - x_{n-1})$. Note that when $x_n$ and $x_{n-1}$ are "close," then this difference quotient is an approximation to $f'(x_n)$. The secant method is represented by the following algorithm.

Given $f(x)$ such that $f(s) = 0$, let $x_{-1}$ and $x_0$ be initial guesses for $s$. Then, for $n \geq 0$,

$$x_{n+1} = x_n - \frac{f(x_n)(x_n - x_{n-1})}{f(x_n) - f(x_{n-1})} = \frac{f(x_n)x_{n-1} - f(x_{n-1})x_n}{f(x_n) - f(x_{n-1})}. \tag{4.6}$$

This algorithm should remind the reader of the *Regula Falsi* method. In fact if $x_{-1} = a$ and $x_0 = b$ bracket $s$, then $x_1$ is generated by exactly the same formula that generates the first iterate of *Regula Falsi*. (See Problem 1.) However, $x_2$ will not (in general) be the second iterate of *Regula Falsi*, since $x_0$ and $x_1$ will not necessarily bracket the root. This is illustrated in Fig. 4.8, which has the same object function as Fig. 4.1. We further note that since $x_{n+1}$ depends explicitly on $x_{n-1}$ and $x_n$, the secant method is not a fixed-point iteration, although it was directly derived from the Newton iteration which is. The secant method does have the property, similar to the fixed-point iteration, that, if $x_{n-1} \neq s$ but $x_n = s$, then $x_{n+1} = s$. For the secant iteration to be well defined we must assume that $f(x_n) - f(x_{n-1}) \neq 0$.

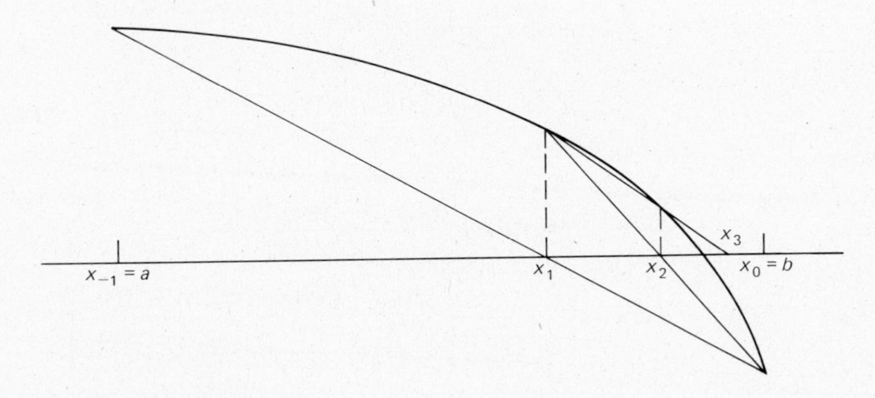


**Fig. 4.8** Secant method.

**Example 4.5.** The secant method was programmed in single precision for $f(x) = \cos(x) - x$, with $x_0 = 0$ and $x_1 = 1$. Note from equation (4.6) that there are two mathematically equivalent ways to generate the sequence $\{x_n\}$:

**Form 1**

$$x_{n+1} = x_n - f(x_n)\frac{x_n - x_{n-1}}{f(x_n) - f(x_{n-1})};$$

**Form 2**

$$x_{n+1} = \frac{f(x_n)x_{n-1} - f(x_{n-1})x_n}{f(x_n) - f(x_{n-1})}.$$

Most numerical analysts feel that Form 1 is preferable to Form 2 and is less subject to rounding error. This particular example was computed using Form 1. (See Table 4.5.)

**Table 4.5**

| $x_i$ | $f(x_i)$ |
|---|---|
| 0.0000000E 00 | 0.1000000E 01 |
| 0.1000000E 01 | −0.4596977E 00 |
| 0.6850733E 00 | 0.8929920E−01 |
| 0.7362989E 00 | 0.4660129E−02 |
| 0.7391192E 00 | −0.5722045E−04 |
| 0.7390850E 00 | 0.5960464E−07 |

To analyze the rate of convergence for the secant method, let us suppose that $f''(x)$ is continuous, $s \neq x_{n-1} \neq x_n$, and define $e_n = s - x_n$ for all $n$. Then, after algebraic reduction,

$$e_{n+1} = s - x_{n+1} = \frac{f(x_n)e_{n-1} - f(x_{n-1})e_n}{f(x_n) - f(x_{n-1})}$$

$$= \frac{f(x_n)e_{n-1} - f(x_{n-1})e_n}{x_{n-1} - x_n} \cdot \frac{x_{n-1} - x_n}{f(x_n) - f(x_{n-1})}. \qquad (4.7)$$

Now,

$$\frac{f(x_n)e_{n-1} - f(x_{n-1})e_n}{x_{n-1} - x_n} = e_n e_{n-1} \frac{f(x_n)/e_n - f(x_{n-1})/e_{n-1}}{x_{n-1} - x_n}$$

$$= e_n e_{n-1} \left( \frac{\dfrac{f(x_n) - f(s)}{x_n - s} - \dfrac{f(x_{n-1}) - f(s)}{x_{n-1} - s}}{x_n - x_{n-1}} \right)$$

$$\equiv e_n e_{n-1} f[x_n, s, x_{n-1}].$$

Let $G(x) = (f(x) - f(s))/(x - s)$, then by the mean value theorem, $f[x_n, s, x_{n-1}] = (G(x_n) - G(x_{n-1}))/(x_n - x_{n-1}) = G'(\zeta_n)$. Now,

$$G'(x) = \frac{f'(x)(x - s) + f(s) - f(x)}{(x - s)^2}$$

and

$$f(s) = f(x) + f'(x)(s - x) + \frac{f''(\eta)}{2}(s - x)^2$$

by Taylor's Theorem and the continuity of $f''(x)$. Putting these two equations together, we get

$$G'(x) = \frac{f''(\eta)}{2}, \quad \text{or} \quad f[x_n, s, x_{n-1}] = G'(\zeta_n) \equiv \frac{f''(\eta_n)}{2}. \tag{4.8}$$

Thus, from Eqs. (4.7) and (4.8) and the mean-value theorem,

$$\begin{aligned} e_{n+1} &= e_n e_{n-1} f[x_n, s, x_{n-1}](x_{n-1} - x_n)/(f(x_n) - f(x_{n-1})) \\ &= e_n e_{n-1}\left(\frac{f''(\eta_n)}{2}\right)\left(\frac{-1}{f'(\xi_n)}\right), \end{aligned} \tag{4.9}$$

where both $\eta_n$ and $\xi_n$ lie in the smallest interval containing $x_{n-1}$, $x_n$, and $s$. With the aid of (4.9) we are able to prove the following local convergence theorem for the secant method.

**Theorem 4.7.** Let $f(s) = 0$, $f'(s) \neq 0$, and let $f''(x)$ be continuous in a neighborhood of $s$. Then there exists an $\varepsilon > 0$ such that if $x_{-1}, x_0 \in I_\varepsilon \equiv [s - \varepsilon, s + \varepsilon]$, then the secant method converges to $s$.

*Proof*: In this proof, we will let $M_\alpha$ denote an upper bound for $|f''(x)/f'(x')|$, where $x$ and $x'$ are any points in $[s - \alpha, s + \alpha]$. Since $f''(x)$ and $f'(x)$ are continuous at $s$ and since $f'(s) \neq 0$, we see for $\beta > 0$ but sufficiently small, that the inequality $|f''(x)/f'(x')| \leq M_\beta$ is valid for $x$ and $x'$ in $[s - \beta, s + \beta]$. Choose $\varepsilon > 0$ such that $\varepsilon M_\beta \equiv K < 1$, where $\varepsilon < \beta$ so that $M_\varepsilon \leq M_\beta$. Let $|x_{-1} - s| \leq \varepsilon$ and $|x_0 - s| \leq \varepsilon$, then $|e_1| \leq |e_{-1}e_0|M_\beta \leq \varepsilon^2 M_\beta = \varepsilon K < \varepsilon$. Also, $|e_2| \leq |e_0 e_1|M_\beta = |e_1|(|e_0|M_\beta) < (\varepsilon K)(\varepsilon M_\beta) < \varepsilon K^2$.

Now, with the induction hypotheses that $e_i < \varepsilon K^i$ and $e_{i-1} < \varepsilon K^{i-1} < \varepsilon$, we can easily see that $|e_{i+1}| < \varepsilon K^{i+1}$ and that $x_{i+1} \in [s - \varepsilon, s + \varepsilon]$, for all $i$. Therefore $\lim_{i\to\infty} e_i = 0$, and we have convergence on $I_\varepsilon$. (The reader can easily see that the inequality on $e_{i+1}$ can be sharpened (see below), but this was not necessary for the proof of simple convergence.) ■

It is beyond the scope of this text to rigorously establish the exact rate of convergence of the secant method (for a thorough coverage, see Ostrowski (1966)). We can, however, present a convincing intuitive argument. Let $M_\beta$ be given as in the above proof, and let $d_n \equiv M_\beta e_n$. Then

$$d_{n+1} = M_\beta e_{n+1} \leq e_{n-1} e_n M_\beta^2 = d_n d_{n-1}.$$

Let $d \equiv \max\{d_{-1}, d_0\}$, then

$$d_1 \le d_{-1}d_0 \le d^2, \qquad d_2 \le d_0 d_1 \le d^3, \qquad d_3 \le d_1 d_2 \le d^5,$$

and clearly by induction we see that $d_n \le d^{\alpha_n}$, where $\alpha_{n+1} = \alpha_n + \alpha_{n-1}$, $\alpha_0 = 1$, $\alpha_1 = 2$. It is left for the reader (as Problem 3), to show that these integers, $\alpha_n$, are given by

$$\alpha_n = \frac{3+\sqrt{5}}{2\sqrt{5}}\left(\frac{1+\sqrt{5}}{2}\right)^n - \frac{3-\sqrt{5}}{2\sqrt{5}}\left(\frac{1-\sqrt{5}}{2}\right)^n$$

which is the famous *Fibonacci sequence* with the so-called *golden ratio*,

$$r^* = \left(\frac{1+\sqrt{5}}{2}\right).$$

The rate of convergence, as shown in Ostrowski, is super-linear and, under the conditions of Theorem 4.7, $\lim_{n\to\infty} (e_{n+1}/e_n^{r^*})$ exists and is usually not zero, so the order of convergence is $r^*$. (Note: $1 < r^* < 2$.)

### 4.3.5 Newton's Method in Two Variables

In this section, we will derive Newton's method for solving a system of nonlinear equations, concentrating on the system of two equations in two unknowns:

$$\begin{aligned} f(x, y) &= 0 \\ g(x, y) &= 0 \end{aligned} \tag{4.10}$$

In (4.10), $f(x, y)$ and $g(x, y)$ are real-valued functions of two variables. We will consider this problem in more detail in Section 4.5, being content in this section to present an intuitive derivation. As we shall see in Section 4.4.2, Newton's method for the special case of the system, (4.10), can be used to find zeros of polynomials where the zeros may be either real or complex.

When solving (4.10), we are looking for a simultaneous solution, that is, numbers $s$ and $t$ such that:

$$\begin{aligned} f(s, t) &= 0 \\ g(s, t) &= 0 \end{aligned}$$

In order to derive Newton's method, let us suppose that we have approximations $x_0$ and $y_0$ to $s$ and $t$, respectively. Expanding $f(x, y)$ and $g(x, y)$ in a Taylor's expansion (in two variables) about the point $(x_0, y_0)$, we find

$$\begin{aligned} f(x, y) &= f(x_0, y_0) + f_x(x_0, y_0)(x - x_0) + f_y(x_0, y_0)(y - y_0) + R_f(x, y) \\ g(x, y) &= g(x_0, y_0) + g_x(x_0, y_0)(x - x_0) + g_y(x_0, y_0)(y - y_0) + R_g(x, y) \end{aligned} \tag{4.11a}$$

where $f_x(x_0, y_0)$ denotes the partial derivative of $f(x, y)$ with respect to $x$ evaluated at the point $(x_0, y_0)$, and where $R_f(x, y)$ is given by

$$R_f(x, y) = \frac{[f_{xx}(\alpha, \beta)(x - x_0)^2 + 2f_{xy}(\alpha, \beta)(x - x_0)(y - y_0) + f_{yy}(\alpha, \beta)(y - y_0)^2]}{2}$$

with the mean-value point, $(\alpha, \beta)$, being somewhere on the line segment joining the points $(x, y)$ and $(x_0, y_0)$. If we substitute $(s, t)$ for $(x, y)$ in (4.11a), we find (since $(s, t)$ is a solution of (4.10)):

$$\begin{aligned} 0 &= f(x_0, y_0) + f_x(x_0, y_0)(s - x_0) + f_y(x_0, y_0)(t - y_0) + R_f(s, t) \\ 0 &= g(x_0, y_0) + g_x(x_0, y_0)(s - x_0) + g_y(x_0, y_0)(t - y_0) + R_g(s, t) \end{aligned} \tag{4.11b}$$

If we now suppose that the point $(x_0, y_0)$ is close to $(s, t)$, then the factors $(s - x_0)^2$, $(s - x_0)(t - y_0)$ and $(t - y_0)^2$, which appear in $R_f(s, t)$ and $R_g(s, t)$, are small with respect to the first-order terms, $(s - x_0)$ and $(t - y_0)$. Thus, we consider the *linear* system of equations, (4.11c), obtained by neglecting $R_f(s, t)$ and $R_g(s, t)$:

$$\begin{aligned} 0 &= f(x_0, y_0) + f_x(x_0, y_0)(s - x_0) + f_y(x_0, y_0)(t - y_0) \\ 0 &= g(x_0, y_0) + g_x(x_0, y_0)(s - x_0) + g_y(x_0, y_0)(t - y_0) \end{aligned} \tag{4.11c}$$

If we solve (4.11c) for $s$ and $t$, we should obtain an approximation to the solution of (4.10) that is better than $(x_0, y_0)$. Note, however, that a solution of (4.11c) will not be precisely the pair $(s, t)$, since Eq. (4.11c) is not the same as Eq. (4.11b). If we let $(x_1, y_1)$ denote the solution of (4.11c), then

$$\begin{aligned} f_x(x_0, y_0)(x_1 - x_0) + f_y(x_0, y_0)(y_1 - y_0) &= -f(x_0, y_0) \\ g_x(x_0, y_0)(x_1 - x_0) + g_y(x_0, y_0)(y_1 - y_0) &= -g(x_0, y_0) \end{aligned} \tag{4.11d}$$

It is convenient to write (4.11d) in matrix form as $J_0(\mathbf{x}_1 - \mathbf{x}_0) = -\mathbf{F}(\mathbf{x}_0)$, where

$$\mathbf{x}_0 = \begin{bmatrix} x_0 \\ y_0 \end{bmatrix}, \quad \mathbf{x}_1 = \begin{bmatrix} x_1 \\ y_1 \end{bmatrix}, \quad \mathbf{F}(\mathbf{x}_0) = \begin{bmatrix} f(x_0, y_0) \\ g(x_0, y_0) \end{bmatrix},$$

$$J_0 = \begin{bmatrix} f_x(x_0, y_0) & f_y(x_0, y_0) \\ g_x(x_0, y_0) & g_y(x_0, y_0) \end{bmatrix}.$$

If $\mathbf{c}_0 = \left[\begin{smallmatrix} u_0 \\ v_0 \end{smallmatrix}\right]$ is the solution of $J_0\mathbf{x} = -\mathbf{F}(\mathbf{x}_0)$, then $\mathbf{x}_1 = \mathbf{x}_0 + \mathbf{c}_0$ is an updated (or corrected) estimate to $(s, t)$.

We note that (4.11c) is normally written in an equivalent form (obtained by multiplying by $J_0^{-1}$): $\mathbf{x}_1 = \mathbf{x}_0 - J_0^{-1}\mathbf{F}(\mathbf{x}_0)$. Having the new estimate, $\mathbf{x}_1$, to the solution, we repeat this correcting process, which leads to the iteration

$$\mathbf{x}_{i+1} = \mathbf{x}_i - J_i^{-1}\mathbf{F}(\mathbf{x}_i), \qquad i = 0, 1, \ldots. \tag{4.11e}$$

The algorithm described by (4.11e) is Newton's method in several variables. The matrix, $J_i$, of partial derivatives evaluated at $\mathbf{x}_i$, is called the *Jacobian* matrix and plays the role of a derivative (see Section 4.5). Thus the form of Newton's method displayed in (4.11e) is similar to Newton's method for a single function of one variable. This algorithm and variations of it are the basis for Bairstow's method (Section 4.4.2) and also can be effectively used in optimization problems (Chapter 8) and collocation methods for differential equations (Chapter 7) as well as in the problem of finding solutions to systems of nonlinear equations.

## PROBLEMS

1. If $x_{-1} = a$ and $x_0 = b$, where $f(a)f(b) < 0$, verify that the first iteration of the secant method is the same as for *Regula Falsi.*
2. Note that Eq. (4.9) can be rewritten as

$$s - \alpha = -(s - a)(s - b)\frac{f''(\eta)}{2f'(\xi)}, \tag{4.9a}$$

where $\alpha$ is the point where the line segment through $(a, f(a))$ and $(b, f(b))$ crosses the $x$-axis. Thus (4.9) can be used for an analysis of *Regula Falsi.* Suppose $x_0$ and $x_1$ are such that $f(x_0) > 0 > f(x_1)$ and suppose $f''(x)f'(x) > 0$ for $x_0 \le x \le x_1$ (see Fig. 4.1). If $x_2, x_3, \ldots, x_n, \ldots$ are the iterates of *Regula Falsi*, and if $f(s) = 0$, then:

a) show $s > x_i$, for $i = 2, 3, \ldots$;

b) let $s - x_i = e_i$ and show (in the notation of Theorem 4.7) that $e_{i+1} \le M_\beta e_i e_0$;

c) for $\varepsilon = \text{Max}\{|e_0|, |e_1|\}$, show $e_{i+1} \le \lambda^i \varepsilon$, where $\lambda = \varepsilon M_\beta$ (see Section 2).

3. Consider the sequence, $\{\alpha_n\}_{n=0}^{\infty}$, such that $\alpha_{n+1} = \alpha_n + \alpha_{n-1}$, $\alpha_0 = 1$, $\alpha_1 = 2$. Find the only two values of $\gamma$ ($\gamma_1$ and $\gamma_2$) that satisfy $\gamma^{n+1} = \gamma^n + \gamma^{n-1}$ for *all* integers $n \ge 0$. Verify that $c_1\gamma_1^n + c_2\gamma_2^n \equiv \alpha_n$ satisfies the equation $\alpha_{n+1} = \alpha_n + \alpha_{n-1}$ for any two values of $c_1$ and $c_2$. Verify that, for $\alpha_0 = 1$ and $\alpha_1 = 2$, $c_1 = (3 + \sqrt{5})/2\sqrt{5}$ and $c_2 = -(3 - \sqrt{5})/2\sqrt{5}$.
4. Let $f(x) = x^2 - 2$ and $[a, b] \equiv [1, 3]$. Compute the first four iterates of both the secant method and *Regula Falsi*, noting how they differ.
5. Program the secant method and test your program on the function $f(x) = \cos(x) - x$.
6. Program Newton's method for two variables and test your program on the system

$$x^2 + y^2 - 9 = 0$$
$$x + y - 1 = 0$$

(This system has two solutions, where a solution represents the intersection of the line and the circle described by the equations.)

7. Let $p(x) = x^3 - x^2 - x - 2$ and let $u$ and $v$ be any real numbers.

a) Verify that $p(x) = (x^2 - ux - v)(x + u - 1) + f_1(u, v)(x - u) + f_2(u, v)$, where

$$f_1(u, v) = u^2 - u + v - 1 \quad \text{and} \quad f_2(u, v) = u^3 - u^2 + 2uv - u - v - 2.$$

Obviously, $p(x) = (x^2 - u^*x - v^*)(x + u^* - 1)$ if and only if $u^*$ and $v^*$ are solutions of $f_1(u^*, v^*) = 0 = f_2(u^*, v^*)$ (see Section 4.4.2). Thus, if we know $u^*$ and $v^*$, we can find two zeros of $p(x)$ by merely applying the quadratic formula to find the two zeros of $x^2 - u^*x - v^*$. Note that even if these two zeros are complex, we need not use any complex arithmetic in finding them.

b) Let $\mathbf{x}_0 \equiv (u_0, v_0) \equiv (0, 0)$, and apply Newton's method for four iterations to find approximations $u_4$ and $v_4$ for $u^*$ and $v^*$, respectively. Apply the quadratic formula to $x^2 - u_4x - v_4 = 0$ to obtain approximations for two zeros of $p(x)$. (The true answers are $(u^*, v^*) = (-1, -1)$ and the resulting zeros are $\frac{1}{2}(-1 \pm \sqrt{3}i)$. This problem presents the basic idea of Bairstow's method which is presented in Section 4.4.2.

## 4.4 ZEROS OF POLYNOMIALS

In Section 4.3, it was noted that a disadvantage of Newton's method was that the derivative of the object function, $f(x)$, had to be calculated at each iterate. For many functions, the calculation of $f'(x)$ is a formidable task. This is not true, however, if the function is an $n$th degree polynomial,

$$p(x) = a_0x^n + a_1x^{n-1} + \cdots + a_{n-1}x + a_n, \qquad a_0 \neq 0. \tag{4.12}$$

Finding the zeros of an $n$th degree polynomial is one of the oldest and most studied problems in mathematics. It is also one of the more important problems in applied mathematics and has extensive practical applications. For $n = 1$, the only zero of $p(x)$ is given by $r = -a_1/a_0$. If $n = 2$, it is well known that the zeros are given by the quadratic formula,

$$r = \frac{-a_1 \pm \sqrt{a_1^2 - 4a_0a_2}}{2a_0}. \tag{4.13}$$

There are similar, but more cumbersome, formulas for $n = 3$ and $n = 4$. But what makes the problem so difficult in general is the fact that for $n \geq 5$, there is *no* algebraic formula such as (4.13) which gives the zeros of $p(x)$ in terms of the coefficients. Thus we are forced to develop numerical procedures to approximate the zeros of $p(x)$. There are numerous such numerical procedures in the literature, but because of space and time considerations, we shall concentrate on just two—the Newton and the Bairstow methods. (There is no reason, however, why the other methods we have discussed, such as the secant method, cannot be used for polynomials.) We shall also limit ourselves to the case where all of the coefficients $a_i$ in (4.12) are real.

Before proceeding further, we shall develop some basic and well-known theory about polynomials that is necessary for understanding and utilizing numerical root-finding techniques. The most basic result, which we recall from Chapter 1, is the following: the *Fundamental Theorem of Algebra.*

**Theorem 4.8.** Given $p(x)$ as in (4.12), with $n \geq 1$, there exists at least one value $r$ (possibly complex), such that $p(r) = 0$.

Given an $r_1$ such that $p(r_1) = 0$, it is then easily shown (Problem 1), that $p(x)$ can be written as $p(x) = (x - r_1)q_1(x)$, where $q_1(x)$ is an $(n - 1)$st degree polynomial. Applying Theorem 4.8 to $q(x)$, we obtain a value $r_2$ (not necessarily distinct from $r_1$) such that $q_1(r_2) = 0$. Thus there exists an $(n - 2)$nd degree polynomial $q_2(x)$, such that $q_1(x) = (x - r_2)q_2(x)$, so $p(x) = (x - r_1)(x - r_2)q_2(x)$. Repeating this process, we finally arrive at $n$ values; $r_1, r_2, \ldots, r_n$ such that $p(r_i) = 0$, $1 \leq i \leq n$. Furthermore it can be shown that this set of roots is unique, and so $p(x)$ can be written as:

$$p(x) = a_0(x - r_1)(x - r_2) \cdots (x - r_n). \tag{4.14}$$

This argument also shows that an $n$th degree polynomial can have no more than $n$ zeros. As noted above, the $r_j$'s need not all be distinct. If any $r_j$ appears in (4.14) exactly $m$ times, $r_j$ is said to have "multiplicity $m$." The reader may easily verify that if $r_j$ is a root of multiplicity $m$, then $p(r_j) = p'(r_j) = \cdots = p^{(m-1)}(r_j) = 0$ and $p^{(m)}(r_j) \neq 0$. Note that this agrees with our earlier definition of "multiplicity $m$" for a zero of an arbitrary function.

The example, $p(x) = x^2 + 1 = (x + i)(x - i)$, demonstrates that even though the coefficients $a_i$ are real, the zeros $\{r_j\}_{j=1}^n$ may be complex. It is well known that complex zeros of a polynomial with real coefficients occur in conjugate pairs. To be more precise, if $r = a + ib$, $b \neq 0$, is a complex number, then $\bar{r}$ (called the *conjugate* of $r$) is defined by $\bar{r} = a - ib$. If $z_1$ and $z_2$ are two complex numbers, then an elementary result is that $\overline{z_1 + z_2} = \overline{z_1} + \overline{z_2}$ and $\overline{z_1 z_2} = \overline{z_1}\,\overline{z_2}$. Using induction, it is easy to show that $\overline{z_1 + z_2 + \cdots + z_n} = \overline{z_1} + \overline{z_2} + \cdots + \overline{z_n}$ and $\overline{(r^k)} = (\bar{r})^k$. Finally, for any real number $a$, we have $\bar{a} = a$. Combining these properties for a polynomial $p(x)$ with real coefficients, we find that $\overline{p(r)} = p(\bar{r})$ for any complex number $r$. Thus if $p(r) = 0$, then

$$p(r) = 0 = \bar{0} = \overline{p(r)} = p(\bar{r}), \tag{4.15}$$

which shows that if $r$ is a zero of $p(x)$, then so is $\bar{r}$. If $r$ is a complex zero of $p(x)$ of multiplicity $m$, then $\bar{r}$ is also a zero of $p(x)$ of multiplicity $m$. This follows since $p^{(i)}(x)$ is a polynomial with real coefficients for $i = 1, 2, \ldots, m - 1$, and therefore, if $p^{(i)}(r) = 0$, then $p^{(i)}(\bar{r}) = 0$. Note that this implies that if $n$ is odd, $p(x)$ must have at least one real zero. The last fundamental result we shall need here is the well-known *division algorithm*:

**Theorem 4.9.** Let $P(x)$ and $Q(x)$ be polynomials of degree $n$ and $m$, respectively, where $1 \leq m \leq n$. Then there exists a unique polynomial $S(x)$ of degree $n - m$ and a unique polynomial $R(x)$ of degree $m - 1$ or less such that

$$P(x) = Q(x)S(x) + R(x). \tag{4.16}$$

In this formula, a polynomial of degree 0 is a constant. Thus if $Q(x)$ is a linear polynomial, the result of dividing $P(x)$ by $Q(x)$ is a quotient $S(x)$ and a remainder $R(x)$, where $R(x)$ is a constant.

## PROBLEMS

1. Let $p(x)$ be given by (4.12) and let $r$ be a value such that $p(r) = 0$. Show there exists a polynomial $q(x)$ of degree $n - 1$ such that $p(x) = (x - r)q(x)$. (Hint: Consider (4.16) with $Q(x) = (x - r)$.)
2. In (4.12), let $n = 4$ and $a_0 = 2$, and assume that $r_1 = 2 - 3i$ and $r_2 = -2 - 3i$ are zeros of $p(x)$. Find $a_4$, $a_3$, $a_2$, and $a_1$.
3. What are the relationships between the coefficients of (4.12) and the derivatives of $p$ evaluated at zero. (Consider a Maclaurin expansion for $p(x)$.)

4. Using (4.16), show that if $P(r) = P'(r) = 0$, then $P(x) = (x - r)^2 Q(x)$. Thus from the remarks following (4.14), $r$ is a zero of multiplicity 2 of $P(x)$ if and only if $P(r) = P'(r) = 0$, $P''(r) \neq 0$. It should be clear that this extends to zeros of multiplicity $m > 2$.

### 4.4.1 Efficient Evaluation of a Polynomial and Its Derivatives

In order to use Newton's method on a polynomial $p(x)$, as given in Eq. (4.12), we must be able to evaluate $p(\alpha)$ and $p'(\alpha)$ for any $\alpha$. Direct substitution of $\alpha$ into $p(x)$ and $p'(x)$, however, is far from being the most efficient way to meet this objective. Furthermore, as we have noted before, if the initial iterate, $x_0$, of Newton's method is not "close" to a zero, the Newton iteration may not converge. (This is obviously the case if all of the zeros of $p(x)$ are complex and $x_0$ is real.) Thus we need some means of "localizing" a zero, $r$, of $p(x)$ in order to choose $x_0$ "close" to $r$. Many of these "localization techniques" (see Section 4.4.3) require that we make a change of variable, $t = x - \alpha$ and write $p(x)$ as

$$p(x) = \beta_0(x - \alpha)^n + \beta_1(x - \alpha)^{n-1} + \cdots + \beta_{n-1}(x - \alpha) + \beta_n \qquad \textbf{(4.17)}$$

instead of in its original form, (4.12). By a Taylor's expansion, we can write

$$p(x) = p(\alpha) + p'(\alpha)(x - \alpha) + \frac{p''(\alpha)}{2!}(x - \alpha)^2 + \cdots + \frac{p^{(n)}(\alpha)}{n!}(x - \alpha)^n. \qquad \textbf{(4.18)}$$

Equating (4.17) and (4.18), we see that

$$\beta_j = p^{(n-j)}(\alpha)/(n - j)!, \qquad \text{for } 0 \leq j \leq n. \qquad \textbf{(4.19)}$$

This shows that it will be useful to have an efficient way of evaluating not only $p(\alpha)$ and $p'(\alpha)$, but $p^{(j)}(\alpha)$, $2 \leq j \leq n$, as well.

The technique we shall employ is known as *nested multiplication* or *synthetic division* and is based on the division algorithm (4.16). We first note there is an alternative way of writing $p(x)$. For example, with $n = 4$, we can write (4.12) as:

$$p(x) = x(x(x(a_0 x + a_1) + a_2) + a_3) + a_4. \qquad \textbf{(4.20)}$$

We see that $p(x)$ in (4.20) can be evaluated at $x = \alpha$ by only four multiplications and four additions, whereas (Problem 2) direct substitution of $\alpha$ into (4.12) for $n = 4$ requires at least seven multiplications and four additions (assuming that no $a_i = 0$). We can easily see that Eq. (4.20) can be extended for any value of $n$, and this extension is given by the following algorithm.

Let $p(x)$ be given by (4.12) and let $\alpha$ be any constant. Let $b_0 = a_0$, and generate $\{b_j\}_{j=1}^n$ by

$$b_j = \alpha b_{j-1} + a_j, \qquad 1 \leq j \leq n,$$

then $p(\alpha) = b_n$.

This algorithm is known as synthetic division and is exactly analogous to (4.20), where $b_n = p(\alpha)$. A less intuitive but more rigorous development of synthetic division is given in Problem 3. Note that the synthetic division algorithm requires only $n$ multiplications to form $b_n = p(\alpha)$. This is an efficient means of evaluating $p(\alpha)$ and, moreover, the iteration scheme can also be used to evaluate *all* derivatives, $p^{(m)}(\alpha)$, $0 \le m \le n$.

To see this, we consider the division algorithm (4.16) and write $p(x)$ in the form

$$p(x) = (x - \alpha)q_{n-1}(x) + r_0(x). \qquad \textbf{(4.21)}$$

In (4.12), $a_0 \ne 0$, so $q_{n-1}(x)$ has degree $n - 1$ and its leading coefficient is $a_0$. Further, $r_0(x)$ is the constant $p(\alpha)$ which can be seen by setting $x = \alpha$ in (4.21). Letting $q_{n-1}(x) = \gamma_0 x^{n-1} + \gamma_1 x^{n-2} + \cdots + \gamma_{n-1}$, substituting into Eq. (4.21) and equating like powers of $x$, we easily see (Problem 3) that

$$q_{n-1}(x) = b_0 x^{n-1} + b_1 x^{n-2} + \cdots + b_{n-1}$$

where $\{b_j\}_{j=0}^{n-1}$ are given by the synthetic division algorithm. Thus this algorithm generates the coefficients of $q_{n-1}(x)$, as well as $p(\alpha)$. By the same reasoning as above, the algorithm can be repeated and will yield $q_{n-1}(\alpha)$ and the coefficients of $q_{n-2}(x)$ (of degree $n - 2$), where

$$q_{n-1}(x) = (x - \alpha)q_{n-2}(x) + r_1, \qquad r_1 = q_{n-1}(\alpha). \qquad \textbf{(4.22)}$$

Substituting (4.22) into (4.21), we obtain

$$p(x) = (x - \alpha)^2 q_{n-2}(x) + r_1(x - \alpha) + r_0. \qquad \textbf{(4.23)}$$

Differentiating (4.23), we see that $p'(\alpha) = q_{n-1}(\alpha) = r_1$. Obviously this procedure may be continued, using synthetic division on each successive $q_k(x)$, $k = n - 2, n - 3, \ldots, 0$ to yield

$$p(x) = r_n(x - \alpha)^n + r_{n-1}(x - \alpha)^{n-1} + \cdots + r_1(x - \alpha) + r_0. \qquad \textbf{(4.24)}$$

Equating (4.24) with the unique Taylor's expansion (4.18), we finally obtain our desired result,

$$r_m = p^{(m)}(\alpha)/m!, \qquad 0 \le m \le n.$$

**Example 4.6.** Let $p(x) = x^6 + 5x^5 + 4x^4 + 3x^3 + 2x^2 + x + 1$ and let $\alpha = 2$. We illustrate synthetic division in Table 4.6, finding $p^{(j)}(\alpha)/j!$ for $j = 0, 1, \ldots, 6$. The entries in any row are the coefficients for $q_{6-j}(x)$, with the last entry in the row being $p^{(j)}(\alpha)/j!$. Thus, $p'(2) = 765$, $p''(2)/2! = 756$, etc.

**Table 4.6**

| $p(x)$ | 1 | 5 | 4 | 3 | 2 | 1 | 1 |
|---|---|---|---|---|---|---|---|
| $q_5(x)$ | 1 | 7 | 18 | 39 | 80 | 161 | 323 |
| $q_4(x)$ | 1 | 9 | 36 | 111 | 302 | 765 | |
| $q_3(x)$ | 1 | 11 | 58 | 227 | 756 | | |
| $q_2(x)$ | 1 | 13 | 84 | 395 | | | |
| $q_1(x)$ | 1 | 15 | 114 | | | | |
| $q_0(x)$ | 1 | 17 | | | | | |
| | 1 | | | | | | |

In this illustration, we see, for example, that

$$p(x) = (x - 2)(x^5 + 7x^4 + 18x^3 + 39x^2 + 80x + 161) + 323$$

and

$$p(x) = (x - 2)^6 + 17(x - 2)^5 + 114(x - 2)^4 + 395(x - 2)^3 + 756(x - 2)^2 + 765(x - 2) + 323.$$

It is obvious how Newton's method should utilize synthetic division. In each iteration, let $x_m = \alpha$ and generate $p(x_m) \equiv r_0$ and $p'(x_m) \equiv r_1$ in exact analogy to the procedure in Example 4.6. Then proceed by setting $x_{m+1} = x_m - p(x_m)/p'(x_m)$. There are many other methods (such as the secant method) which involve evaluation of $p(x)$ or its derivatives at specific points and synthetic division can be utilized by these methods as well. Once again we emphasize that the initial guess of Newton's method, $x_0$, should be "close" to $r$, a zero of $p(x)$, in order to insure convergence. Section 4.4.3 will consider the problem of the approximate location of the zeros of $p(x)$, and will be of immense aid in making a good choice for $x_0$.

If our goal is to find *all* the zeros of an $n$th degree polynomial, $p(x)$, then it seems that a reasonable way to do this is first to find one zero, say $r_1$, of $p(x)$. We can next say that $p(x) = (x - r_1)p_1(x)$, and, hence, any zero of the $(n - 1)$st degree polynomial $p_1(x)$ is a zero of $p(x)$. We now search for a zero of $p_1(x)$, say $r_2$, and write $p_1(x) = (x - r_2)p_2(x)$ and then search for a zero of $p_2(x)$, etc. This process of finding a zero and then dividing it out is called *deflation.* Polynomial deflation must be used judiciously, for large errors can result due to our inability to find roots exactly. Subroutine POLRT, listed in Fig. 4.9, uses Newton's method and deflation to find all the real roots of a polynomial $p(x)$. Synthetic division is used for evaluation and to find the coefficients of the deflated polynomials. An initial guess of $x_0 = 0$ is used for Newton's method at each stage of the deflation. Note that subroutine POLRT cannot determine complex roots, but modifications are easily made which enable the subroutine to find complex roots (see Problem 1).

**Example 4.7.** Subroutine POLRT was used to find the zeros of $p(x) = x^5 + x^4 - 9x^3 - x^2 + 20x - 12$, where $p(x) = (x - 2)(x + 2)(x + 3)(x - 1)^2$. A tolerance TOL = 0.00001 was used in this example. The program found the following estimates to the roots, listed in the order found:

0.9998991E 01
0.2000000E 01
0.1000101E 01
−0.2000031E 01
−0.2999979E 01

To illustrate the effect of not having these roots precisely, we listed the coefficients of the deflated polynomial of degree 4, found after the first root was divided out. These coefficients

```
      SUBROUTINE POLRT(A,ROOT,TOL,N,MTOL,NROOT)
      DIMENSION A(21),B(21),C(21),ROOT(20)
C
C  SUBROUTINE POLRT USES NEWTON'S METHOD AND DEFLATION TO FIND THE
C  REAL ROOTS OF A POLYNOMIAL.  THE CALLING PROGRAM MUST SUPPLY THE
C  DEGREE OF THE POLYNOMIAL, N, THE COEFFICIENTS A(I) (WHERE A(I) IS
C  THE COEFFICIENT OF X**(N+1-I)), A TOLERANCE TOL, AND AN INTEGER
C  MTOL.  THE SUBROUTINE RETURNS AN INTEGER NROOT=NUMBER OF REAL ROOTS
C  FOUND AND THE ROOTS (IN AN ARRAY ROOT).  AT EACH STAGE OF THE
C  DEFLATION, NEWTON'S METHOD TERMINATES WHEN MTOL ITERATIONS HAVE
C  BEEN EXECUTED OR WHEN TWO SUCCESSIVE ITERATES ARE LESS THAN TOL
C  IN ABSOLUTE VALUE.  N MUST BE GREATER THAN 2.
C
      NROOT=0
      NP1=N+1
    1 ITR=0
      X=0.
C
C  SYNTHETIC DIVISION ALGORITHM TO EVALUATE POLYNOMIAL AND ITS
C  DERIVITIVE
C
    2 B(1)=A(1)
      C(1)=B(1)
      DO 3 I=2,N
      B(I)=X*B(I-1)+A(I)
    3 C(I)=X*C(I-1)+B(I)
      B(NP1)=X*B(N)+A(NP1)
C
C  NEWTON'S METHOD UPDATE TO OLD ESTIMATE OF ROOT
C
      XCORR=B(NP1)/C(N)
      X=X-XCORR
      IF(ABS(XCORR).LT.TOL)   GO TO 4
      ITR=ITR+1
      IF(ITR.LT.MTOL)   GO TO 2
      RETURN
    4 NROOT=NROOT+1
      ROOT(NROOT)=X
C
C  SET UP COEFFICIENTS OF DEFLATED POLYNOMIAL
C
      NP1=N
      N=N-1
      DO 5 I=1,NP1
    5 A(I)=B(I)
C
C  USE QUADRATIC FORMULA WHEN DEFLATED POLYNOMIAL HAS DEGREE 2
C
      IF(N.GT.2)   GO TO 1
      DISCRM=A(2)*A(2)-4.*A(1)*A(3)
      IF(DISCRM.GE.0.)   GO TO 6
      RETURN
    6 ROOT(NROOT+1)=(-A(2)+SQRT(DISCRM))/2.
      ROOT(NROOT+2)=(-A(2)-SQRT(DISCRM))/2.
      NROOT=NROOT+2
      RETURN
      END
```

**Fig. 4.9** Subroutine POLRT.

were:

| | |
|---|---|
| 0.1000000E 01 | (coefficient of $x^4$) |
| 0.1999899E 01 | (coefficient of $x^3$) |
| −0.7000303E 01 | (coefficient of $x^2$) |
| −0.7999597E 01 | (coefficient of $x$) |
| 0.1200121E 02 | (constant term) |

These coefficients indicate that the deflated polynomial does not have precisely the remaining zeros of $p(x)$ as its zeros. For example, the constant term, 12.00121, is the product of the roots of the deflated polynomial, whereas the product of 2, $-2$, $-3$, and 1 is 12.

To analyze the effects of not knowing $r_1$ exactly, as in Example 4.7, suppose the Newton iterates $\{x_n\}_{n=0}^{\infty}$ are converging to a zero, $r_1$, of $p(x)$. No matter how large $n$ may be, we can expect that $x_n \neq r_1$. Thus we must settle for some $x_n \equiv r'_1$ as an approximation for $r_1$. (We will say more about when to terminate an iteration in Section 4.4.3) By Theorem 4.9, there is a polynomial $p_1(x)$ of degree $n - 1$ such that $p(x) = (x - r'_1)p_1(x) + p(r'_1)$. Even if we had no round-off error in computing the coefficients of $p_1(x)$, $p(r'_1)$ is probably not zero. Thus the zeros of $p_1(x)$ are not, in general, exactly the same as the remaining zeros, $\{r_j\}_{j=2}^{n}$, of $p(x)$. As a matter of fact, it can happen that even though $r'_1$ is extremely close to $r_1$, the zeros of $p_1(x)$ can be quite different from $\{r_j\}_{j=2}^{n}$. If this phenomenon occurs, $p(x)$ is said, as before, to be ill-conditioned (once again we refer to Wilkinson's example of ill-conditioning given in Section 3.3.2). Therefore the most practical strategy to find $r_2$ is to use the Newton iteration on $p_1(x)$ until it seems to be converging to some approximate zero, $r''_2$, of $p_1(x)$. Then we let $x_0 = r''_2$, and use Newton's iteration on the *original* polynomial $p(x)$ to generate a corrected approximation, $r'_2$, for the actual zero, $r_2$. This correction process should be repeated for all of the other approximates as well. This correction strategy is useful in any root-finding method which finds roots one by one.

When deflation is employed, the error incurred by accepting $r'_1$ as an approximation to $r_1$ (which is due both to rounding errors and to truncation of the Newton iteration) not only influences the approximation $r'_2 \approx r_2$, but also influences the approximation, $r'_3 \approx r_3$, and all successive approximations. Furthermore, the error incurred in generating $r'_2$ also influences $r'_3 \approx r_3$ and all successive approximations. We can also see that the smaller roots (in magnitude) are more sensitive to error than the larger ones. For example, let $r_1$ and $r_2$ be two roots of $p(x)$ with $|r_1| < |r_2|$, and assume that $r'_1$ and $r'_2$ can be found such that $|r_1 - r'_1| = \varepsilon$ and $|r_2 - r'_2| = \varepsilon$. Then the relative errors (which are of more practical importance than the absolute errors) satisfy

$$\frac{|r_1 - r'_1|}{|r_1|} > \frac{|r_2 - r'_2|}{|r_2|}$$

Thus we see that the accumulation of error with respect to the deflation process has less effect on our approximations if we are able to approximate the smaller roots first. Once again this requires some *a priori* knowledge of the approximate location of the roots, which is the subject of Section 4.4.3. However, an intuitive approach to finding the smallest zero first would be to select the initial guess quite close to 0.

**Example 4.8.** To illustrate some of the effects of deflation, the polynomial $p(x) = x^4 - 4x^3 + 6x^2 - 4x + 1$ was run using subroutine POLRT with TOL = 0.00001. (In this case, $p(x) = (x - 1)^4$.) After 19 iterations, Newton's method provided $r_1' = 1.021412$ as one root. The program deflated $p(x)$ using $r_1'$ as an assumed root, and after 23 iterations provided $r_2'' = 0.9776155$ as a second root. The next deflated polynomial was $x^2 - 2.000972x + 1.001454$, which is seen to have no real zeros (in fact, the zeros of this quadratic are approximately $1.000486 \pm 0.021949i$). Newton's method applied to this quadratic gave a sequence of iterates which exhibited a characteristic oscillatory behavior. The first 20 iterates are shown in Table 4.7.

**Table 4.7**

| $x_i$ | $f(x_i)$ |
|---|---|
| 0.0000000E 00 | 0.1001454E 01 |
| 0.5004840E 00 | 0.2504845E 00 |
| 0.7509677E 00 | 0.6274182E−01 |
| 0.8766937E 00 | 0.1580685E−01 |
| 0.9405380E 00 | 0.4076481E−02 |
| 0.9745383E 00 | 0.1156807E−02 |
| 0.9968296E 00 | 0.4968643E−03 |
| 0.1064779E 01 | 0.4615963E−02 |
| 0.1028881E 01 | 0.1289368E−02 |
| 0.1006177E 01 | 0.5149841E−03 |
| 0.9609319E 00 | 0.2047658E−02 |
| 0.9868165E 00 | 0.6704330E−03 |
| 0.1011340E 01 | 0.6008148E−03 |
| 0.9836637E 00 | 0.7667542E−03 |
| 0.1006454E 01 | 0.5187988E−03 |
| 0.9629857E 00 | 0.1888931E−02 |
| 0.9881714E 00 | 0.6341934E−03 |
| 0.1013921E 01 | 0.6637573E−03 |
| 0.9892181E 00 | 0.6103516E−03 |
| 0.1016302E 01 | 0.7333755E−03 |

## PROBLEMS

1. A number of modifications of subroutine POLRT are possible and desirable. Either modify this program, or write your own root-finding program that incorporates two obvious improvements:

a) Use the root-refinement idea, taking each zero of a deflated polynomial and use it as an initial guess for Newton's method applied to the *original* polynomial $p(x)$. This will refine both the root and the deflated polynomial.

b) If after a number of iterations, Newton's method does not appear to be converging, it is possible that $p(x)$ has a complex zero. Include a provision to start the iteration with a complex initial guess, (say $x_0 = i$), when this happens. The program will need a capability to perform complex arithmetic if a complex initial guess is used.

Test your program with $p(x) = (x - 1)^4$ and with $p(x) = x^6 - 2x^5 + 5x^4 - 6x^3 + 2x^2 + 8x - 8$ (which has zeros $1 + i, 1 - i, 1, -1, 2i, -2i$).

2. Let $p(x)$ be given by (4.12) and assume that none of its coefficients are zero. Show that evaluation of $p(\alpha)$ by direct substitution requires at least $2n - 1$ multiplications.
3. Establish that the number, $b_n$, generated by the synthetic division algorithm satisfies $b_n = p(\alpha)$. (Hint: In (4.16), let $p(x) = P(x)$ and $Q(x) = (x - \alpha)$. Let the coefficients of $Q(x)$ be $b_0, b_1, \ldots, b_{n-1}$ and equate like powers on both sides of (4.16).)
4. Start with Eq. (4.23) and show that $r_2 = p''(\alpha)/2$, where $r_2$ is obtained in the same manner as $r_0$ and $r_1$.
5. Let $p(x) = x^6 + 5x^5 + 4x^4 + 3x^3 + 2x^2 + x + 1$. Utilize synthetic division to find $\{\gamma_j\}_{j=0}^{6}$, and write $p(x)$ in the form:

$$p(x) = \gamma_0(x + 2)^6 + \gamma_1(x + 2)^5 + \gamma_2(x + 2)^4 + \gamma_3(x + 2)^3 + \gamma_4(x + 2)^2 + \gamma_5(x + 2) + \gamma_6.$$

What does this value of $\gamma_6$ along with the last entry in the first row of Table 4.6 tell you?

### 4.4.2 Bairstow's Method

The basic purpose of Bairstow's method is to find a quadratic factor of a polynomial, $p(x)$. Let $p(x)$ be given by (4.12) and let $u$ and $v$ be any two real numbers. Then $p(x)$ can be written in the form:

$$p(x) = (x^2 - ux - v)q(x) + b_{n-1}(x - u) + b_n \tag{4.25}$$

$$q(x) \equiv b_0x^{n-2} + b_1x^{n-3} + \cdots + b_{n-3}x + b_{n-2} \tag{4.26}$$

We first note that $b_0 = a_0$ and the degree of $q(x)$ is $n - 2$. We also emphasize that each $b_k$ is actually a function of $u$ and $v$ which we could find explicitly, should we desire (see Problem 1). Obviously, for $n$ very large, the explicit calculation of each $b_k$ in terms of $u$ and $v$ could become quite cumbersome. Fortunately, however, we can derive an algorithm which calculates each $b_k$ quite efficiently.

Let $p(x)$ be given by (4.12), let $u$ and $v$ be arbitrary real numbers, and let $b_{-2} = b_{-1} = 0$. Then generate $\{b_k\}_{k=0}^{n}$ by

$$b_k = a_k + ub_{k-1} + vb_{k-2}, \qquad 0 \leq k \leq n. \tag{4.27}$$

We leave to the reader (Problem 2) to verify that the $\{b_k\}_{k=0}^{n}$ given by (4.27) satisfy (4.26) and (4.25). (This is merely a problem of comparing like powers of $x$

in (4.25)). Before we lose sight of our principal problem of finding zeros of $p(x)$, we state the following:

**Theorem 4.10.** Let $u$, $v$, and $p(x)$ be given as above, and let $r(x) = x^2 - ux - v = (x - s_1)(x - s_2)$. Then $s_1$ and $s_2$ are zeros of $p(x)$ if and only if $p(x) = r(x)q(x)$. (Therefore we see that finding a quadratic factor $r(x)$ of $p(x)$ is equivalent to finding $u$ and $v$ so that $b_{n-1} = b_n = 0$.)

The proof of Theorem 4.10 is quite straightforward and is also left to the reader (Problem 3). (We caution the reader to be careful to analyze the special case where $s_1 = s_2$, as well as the case where $s_1 \neq s_2$.) We emphasize the fact that $b_{n-1} \equiv b_{n-1}(u, v)$ and $b_n \equiv b_n(u, v)$, that is, they are functions of $u$ and $v$. Because of Theorem 4.10 and Eq. (4.25), our problem reduces to finding $u^*$ and $v^*$ such that $b_{n-1}(u^*, v^*) = b_n(u^*, v^*) = 0$ and then using the quadratic formula to find the two roots, $t^*$ and $s^*$ such that $r^*(x) \equiv x^2 - u^*x - v^* = 0$ for $x = t^*$ and $x = s^*$.

In general, $b_{n-1}(u, v)$ and $b_n(u, v)$ will be nonlinear functions in $u$ and $v$, as, for example, they are in Problem 1. We are then led to use Newton's method for two equations in two unknowns as given by (4.11e). We note the apparent difficulty here, however, that (4.11e) requires the evaluation of the functions $\partial b_{n-1}/\partial u$, $\partial b_{n-1}/\partial v$, $\partial b_n/\partial u$, and $\partial b_n/\partial v$ at each step of the iteration. This actually presents no real problem, however, as these quantities can be evaluated by differentiating formula (4.27). (Recall that *every* $b_k$ in (4.27) is actually a function of $u$ and $v$.)

Now $b_0 = a_0$, so $\partial b_0/\partial u = \partial b_0/\partial v = 0$. Also, since $b_1 = a_1 + ub_0$, we see that $\partial b_1/\partial u = b_0$ and $\partial b_1/\partial v = 0$. Now we define $c_k \equiv \partial b_{k+1}/\partial u$, $0 \leq k \leq n - 1$, and for simplicity of notation let $c_{-2} = c_{-1} = 0$. From above, $c_0 = \partial b_1/\partial u = b_0$, and, by (4.27),

$$c_1 = \frac{\partial b_2}{\partial u} = b_1 + u\frac{\partial b_1}{\partial u} + v\frac{\partial b_0}{\partial u} = b_1 + uc_0$$

$$c_2 = \frac{\partial b_3}{\partial u} = b_2 + u\frac{\partial b_2}{\partial u} + v\frac{\partial b_1}{\partial u} = b_2 + uc_1 + vc_0.$$

Continuing in this fashion we see that for $0 \leq k \leq n - 1$,

$$c_k = \frac{\partial b_{k+1}}{\partial u} = b_k + u\frac{\partial b_k}{\partial u} + v\frac{\partial b_{k-1}}{\partial u} = b_k + uc_{k-1} + vc_{k-2}. \qquad \textbf{(4.28)}$$

Thus from (4.28), with $k = n - 2$ and $n - 3$,

$$c_{n-2} = \frac{\partial b_{n-1}(u, v)}{\partial u} \quad \text{and} \quad c_{n-1} = \frac{\partial b_n(u, v)}{\partial u}. \qquad \textbf{(4.29)}$$

Furthermore, the iteration (4.28) is precisely the same as (4.27), replacing $\{a_k\}_{k=0}^{n-1}$ by $\{b_k\}_{k=0}^{n-1}$ and $\{b_k\}_{k=-2}^{n-1}$ by $\{c_k\}_{k=-2}^{n-1}$.

Since (4.27) can be used to generate $\partial b_{n-1}/\partial u$ and $\partial b_n/\partial u$ for any $u$ and $v$, it is not surprising that this algorithm can also be used to generate $\partial b_{n-1}/\partial v$ and $\partial b_n/\partial v$ for any $u$ and $v$. This can be seen once again by differentiating (4.27), this time with respect to $v$. First we define $d_k \equiv \partial b_{k+2}/\partial v$, $0 \le k \le n-2$. We have already observed that

$$d_{-2} = \frac{\partial b_0}{\partial v} = d_{-1} = \frac{\partial b_1}{\partial v} = 0.$$

Now,

$$d_0 = \frac{\partial b_2}{\partial v} = u\frac{\partial b_1}{\partial v} + b_0 + v\frac{\partial b_0}{\partial v} = b_0$$

$$d_1 = \frac{\partial b_3}{\partial v} = u\frac{\partial b_2}{\partial v} + v\frac{\partial b_1}{\partial v} + b_1 = ud_0 + b_1$$

$$d_2 = \frac{\partial b_4}{\partial v} = u\frac{\partial b_3}{\partial v} + v\frac{\partial b_2}{\partial v} + b_2 = ud_1 + vd_0 + b_2.$$

Therefore, for $0 \le k \le n-2$, we obtain

$$d_k = \frac{\partial b_{k+2}}{\partial v} = u\frac{\partial b_{k+1}}{\partial v} + v\frac{\partial b_k}{\partial v} + b_k = ud_{k-1} + vd_{k-2} + b_k. \quad \textbf{(4.30)}$$

So, for $k = n-3$ and $n-2$ in (4.30),

$$d_{n-3} = \frac{\partial b_{n-1}(u, v)}{\partial v} \quad \text{and} \quad d_{n-2} = \frac{\partial b_n(u, v)}{\partial v} \quad \textbf{(4.31)}$$

where again (4.27) generates (4.30) and (4.31). Furthermore, by comparing (4.30) with (4.31) (Problem 4) we see that $d_k = c_k$ for $-2 \le k \le n-2$. Thus, (4.27) need only be used once and yields:

$$c_{n-3} = \frac{\partial b_{n-1}(u, v)}{\partial v}, \quad c_{n-2} = \frac{\partial b_{n-1}(u, v)}{\partial u} = \frac{\partial b_n(u, v)}{\partial v}, \quad c_{n-1} = \frac{\partial b_n(u, v)}{\partial u}. \quad \textbf{(4.32)}$$

Given the partials in (4.32), it is easy to write down Newton's method in two variables for the system

$$b_{n-1}(u, v) = 0$$
$$b_n(u, v) = 0$$

Using (4.11e), we see that Bairstow's method generates the sequences $\{u_i\}$ and $\{v_i\}$, where

$$u_{k+1} = u_k - [b_{n-1}c_{n-2} - b_n c_{n-3}]/\lambda$$
$$v_{k+1} = v_k - [b_n c_{n-2} - b_{n-1}c_{n-1}]/\lambda$$

and where $\lambda = c_{n-2}c_{n-2} - c_{n-1}c_{n-3}$ (that is, $\lambda$ is the determinant of the Jacobian matrix in (4.11e)). In the above iteration, of course, the terms $b_{n-1}$, $b_n$, $c_{n-3}$, $c_{n-2}$, and $c_{n-1}$ are all obtained using $u = u_k$ and $v = v_k$.

**Example 4.9.** Let $p(x) = x^3 - 6x^2 + 11x - 6 = (x-1)(x-2)(x-3)$. (Note: $(x-1)(x-2) = x^2 - 3x + 2$, $(x-1)(x-3) = x^2 - 4x + 3$, and $(x-2)(x-3) = x^2 - 5x + 6$.) Bairstow's method was programmed in single precision with the following results:

| Initial guess | $(u, v)$ Convergents | Roots | Number of iterations |
|---|---|---|---|
| (8, 3) | (4, −3) | (1, 3) | 12 |
| (10, 5) | (5, −6) | (2, 3) | 12 |
| (50, 50) | (3, −2) | (1, 2) | 9 |

We note here, how sensitive Bairstow's method is, with the three different initial guesses leading to the three different possible convergents. We ran this program with numerous initial guesses with magnitudes as large as 1000, and the program did not fail to converge for this simple polynomial for any iteration. Given a particular initial guess, $(u_0, v_0)$, however, there is no readily discernible pattern as to which of the three $(u, v)$ convergents the iteration finds. For example, we considered the three possible line segments connecting the points $(u, v) = (4, -3), (5, -6)$, and $(3, -2)$. (See Fig. 4.10.) Each segment was divided into ten equal parts and each of the twenty-seven partition points was used as an initial point, $(u_0, v_0)$, in the iteration. The points to which these iterations converged are indicated on the graph below. One somewhat suprising result of this numerical experiment is that all of the initial

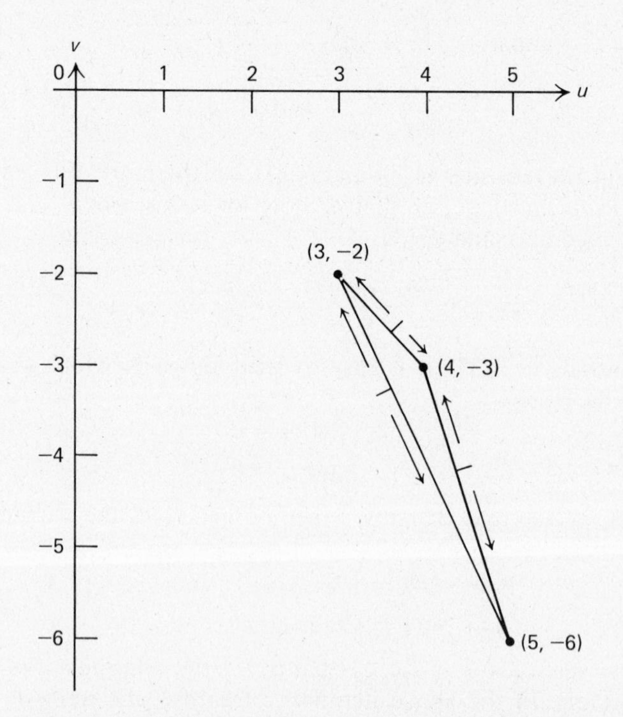


**Fig. 4.10** Successive applications of Bairstow's method.

guesses considered on the line between $(3, -2)$ and $(5, -6)$ converged to either $(3, -2)$ or $(5, -6)$, even though some of them are closer to the other valid point, $(4, -3)$.

Since Bairstow's method uses Newton's method in two variables, it is sensitive to the accuracy of the initial guess, and may fail to converge for that particular initial guess. Thus for an arbitrary polynomial, a reasonable procedure would be to use the localization methods of Section 4.4.3 and find the real roots by Newton's method first. Then use Bairstow's method to find the complex conjugate pairs.

**PROBLEMS**

1. For the polynomial $p(x) = x^3 - 6x^2 + 11x - 6$ explicitly find the functions $b_1(u, v)$, $b_2(u, v)$ and $b_3(u, v)$. Do this by equating like powers of $x$ in (4.25).
2. Verify that $\{b_k\}_{k=0}^{n}$ given by (4.27) satisfy Eq. (4.25) via (4.26), with $p(x)$ given by (4.12).
3. Prove Theorem 4.10, using Theorem 4.9.
4. From (4.28) and (4.30) verify that $d_k = c_k$ for $-2 \leq k \leq n - 2$.
5. Let $p(x) = x^3 - 4x^2 + 6x - 4$, and let $(u_0, v_0) = (0, 0)$. Program Bairstow's method and test your program on this polynomial.
6. Continue the numerical experiment conducted in Example 4.9 by refining the partitions of the three line segments and by considering various points within the triangle defined by these line segments in order to attempt to establish regions of convergence for three points.

### 4.4.3 Localization of Polynomial Zeros

There are numerous results in the literature on locating regions in the complex plane which contain zeros of a polynomial. We shall present only a few such results which are easy to implement, and have a direct bearing on the numerical procedures discussed thus far. Since $a_0 \neq 0$ in (4.12) we may as well consider only polynomials of the form

$$p(x) = x^n + a_1 x^n + \cdots + a_{n-1}x + a_n. \tag{4.33}$$

The first localization technique has already been discussed in terms of eigenvalues of a matrix, but has a special modification for our particular problem. (This result can also be proved directly without the use of matrix theory (see Problem 1), but it is most easily seen using the Gerschgorin circle theorem as a tool.) We note that the $(n \times n)$ *companion matrix* for $p(x)$,

$$A = \begin{bmatrix} 0 & 1 & 0 & \cdots & 0 \\ 0 & 0 & 1 & \cdots & 0 \\ \vdots & & & & \vdots \\ 0 & & & & 1 \\ -a_n & -a_{n-1} & & \cdots & -a_1 \end{bmatrix} \tag{4.34}$$

is a matrix which has (4.33) as its characteristic polynomial. Using the Gerschgorin theorem in column form on (4.34) we have the following theorem and its corollary:

**Theorem 4.11.** Let $p(x)$ be given by (4.33), then all of the zeros of $p(x)$ lie in the union of the circular disks $C_1, C_2, \ldots, C_n$ in the complex plane:

$$\begin{aligned} C_n &= \{z: |z| \leq |a_n|\} \\ C_k &= \{z: |z| \leq 1 + |a_k|\} \qquad 2 \leq k \leq n-1 \\ C_1 &= \{z: |z + a_1| \leq 1\} \end{aligned}$$

**Corollary.** Let $p(x)$ be given by (4.33), and

$$r = 1 + \max_{1 \leq j \leq n} |a_j|.$$

Then every zero of $p(x)$ lies in the circular disk, $C = \{z: |z| \leq r\}$.

Although this corollary follows immediately from Theorem 4.11, it can also be proved by induction using synthetic division (Problem 1), yielding the further information that if $\alpha$ is any number satisfying $|\alpha| \geq r$, then $|p(\alpha)| \geq 1$.

We note that all of the circular disks but one in Theorem 4.11 have centers at the origin, that is, at $z = 0$. Thus if our polynomial has some large coefficients, then Theorem 4.11 will yield large circular disks about the origin. Also, the theorem will not say anything about the location of the zeros within these disks. This is where it becomes valuable to change the variable $x$ to $t = x - \alpha$ (which we have seen can be done quite efficiently using synthetic division) in order to write $p(x)$ in the form of Eq. (4.18). Application of Theorem 4.11 to $p(x)$ rewritten in this form yields a set of circular disks whose center is $z = \alpha$. We illustrate this with the following example.

**Example 4.10.** Let $p(x) = x^4 + 8x^3 - 8x^2 - 200x - 425$. Then

$$p(x) = (x - 5)(x + 5)(x + 4 + i)(x + 4 - i).$$

Using Theorem 4.11, $C_4 = \{z: |z| \leq 425\}$ contains all of the other disks, and so the best we can say is that all of the zeros lie in $C_4$. If we let $x = -4$ and generate $p^{(m)}(-4)/m!$, $1 \leq m \leq 4$, we can write

$$\begin{aligned} p(x) &= (x + 4)^4 - 8(x + 4)^3 - 8(x + 4)^2 - 8(x + 4) - 9 \\ &= t^4 - 8t^3 - 8t^2 - 8t - 9. \end{aligned}$$

Using Theorem 4.11 with respect to the variable $t$, we see that $C_4 = \{z: |z| \leq 9\}$ contains the other disks. By this we know that all of the zeros of $p(x)$ lie in a circular disk of radius 9 with center $\alpha = -4$, a considerable improvement.

We next state, without proof, a result which is helpful in determining the number of real zeros of a polynomial with real coefficients, known as *Descartes rule of signs*. The procedure of this rule is to simply write down the signs (in their respective order) of the nonzero coefficients of $p(x)$, and count the number of sign changes. (We call this number "$v$".) For example, let $p_1(x) = x^3 + x - 2$. Then the coefficient signs (in order) are $+, +, -$; and thus $v = 1$. For $p(x)$ given in Example 4.10, we get: $+, +, -, -, -$; and so $v = 1$ again. The statement of the rule and how to use it is as follows:

> Let $k$ be the number of positive real roots of $p(x)$. Then $k \leq v$, and $(v - k)$ is a nonnegative *even* integer.

For $p_1(x)$, we have $v = 1$. Since $v - k$ is nonnegative and even, we see that $v - k = 0$, so $k$ must be 1. The same analysis holds for the second polynomial as well, so each of these polynomials has one positive real root.

It is easily verified (Problem 3), that if $r$ is a zero of $p(x)$, then $-r$ is a zero of $p(-x)$. Thus we can obtain information on the number of negative real roots of $p(x)$ by using Descartes' rule of signs on $p(-x)$. For example, $p_1(-x) = -x^3 - x - 2$, and its coefficient signs are $-, -, -$; so $v = 0$. Therefore $p_1(x)$ has no negative real roots, and by combining these results we see that $p_1(x)$ has one positive real root and a pair of complex conjugate roots. From Example 4.10, $p(-x) = x^4 - 8x^3 - 8x^2 + 200x - 425$, and its sign pattern is $+, -, -, +, -$. Here $v = 3$, and so $v - k = 2$ or 0. Thus $p(x)$ has either 1 or 3 negative real roots, but it is not possible to determine which from this analysis.

The next root location result we shall derive is directly related to the Newton method iteration and provides some information as to when to terminate the iteration of Newton's method.

**Theorem 4.12.** Given any number $\alpha$ such that $p'(\alpha) \neq 0$, where $p(x)$ is given by (4.33), then there exists at least one zero of $p(x)$ in the circular disk centered at $\alpha$ given by:

$$C = \left\{ z: |z - \alpha| \leq n \left| \frac{p(\alpha)}{p'(\alpha)} \right| \right\}. \tag{4.35}$$

*Proof*: Let $p(x) = (x - r_1)(x - r_2) \cdots (x - r_n)$, then

$$p(x) = (x - \alpha)^n + \beta_1 (x - \alpha)^{n-1} + \cdots + \beta_{n-1}(x - \alpha) + \beta_n,$$

where $\beta_j = p^{(n-j)}(\alpha)/(n - j)!$, $1 \leq j \leq n$. Now we let $t \equiv (x - \alpha)$ and define $s(t) = t^n + \beta_1 t^{n-1} + \cdots + \beta_{n-1} t + \beta_n$. Obviously $s(t_j) = 0$ if and only if $t_j = r_j - \alpha$, $1 \leq j \leq n$. We next define the $n$th degree polynomial, $q(t) \equiv t^n s(1/t) = 1 + \beta_1 t + \cdots + \beta_{n-1} t^{n-1} + \beta_n t^n$. By Problem 3, the zeros of $q(t)$ are

$$u_j = \frac{1}{t_j} = \frac{1}{r_j - \alpha}, \qquad 1 \leq j \leq n.$$

Writing $q(t) = \beta_n (t - u_1)(t - u_2) \cdots (t - u_n)$, we can easily see (Problem 4) that

$$\frac{\beta_{n-1}}{\beta_n} = -(u_1 + u_2 + \cdots + u_n),$$

and so

$$\left|\frac{p'(\alpha)}{p(\alpha)}\right| \equiv \left|\frac{\beta_{n-1}}{\beta_n}\right| \leq n \max_{1 \leq j \leq n} |u_j| \equiv \frac{n}{\min_{1 \leq j \leq n} |r_j - \alpha|}.$$

Therefore

$$\min_{1 \leq j \leq n} |r_j - \alpha| \leq n \left|\frac{p(\alpha)}{p'(\alpha)}\right|, \tag{4.36}$$

and so at least one $r_j$ lies in $C$ given by (4.35). ■

We again consider the polynomial, $p(x)$, of Example 4.10. If $\alpha = -4$, then $p(-4) = -9$, $p'(-4) = -8$, and $n = 4$. By Theorem 4.12, at least one zero lies in $C = \{z: |z + 4| \leq 4(9/8) = 4.5\}$.

To use this theorem, we note that if $\{x_m\}_{m=0}^{\infty}$ are the iterates of the Newton method, then $p(x_m)/p'(x_m)$ is already calculated at each step. Therefore we can easily calculate $n|p(x_m)/p'(x_m)|$ and test to see if it is less than some prescribed tolerance. If this is true, then (4.36) guarantees that $x_m$ is within that tolerance of some root, $r$, of $p(x)$.

The above procedure yields information as to when to terminate a Newton iteration in terms of the absolute error. As we have noted previously, a measure of the relative error is usually of more practical significance than a measure of the absolute error. We now give a procedure for estimating the relative error, which is applicable to any iteration technique. In this analysis we return to formula (4.12) for $p(x)$ and write

$$p(x) = a_0 x^n + a_1 x^{n-1} + \cdots + a_n = a_0(x - r_1)(x - r_2)\cdots(x - r_n),$$

where $a_0 \neq 0$. It is easily seen (Problem 4) that

$$a_n/a_0 = (-1)^n r_1 r_2 \cdots r_n \tag{4.37}$$

where we assume that $a_n \neq 0$, that is, $r_j \neq 0$, $1 \leq j \leq n$. Let $s$ (real or complex) be an approximation for a zero of $p(x)$ that satisfies $|p(s)| \leq \varepsilon$, for some prescribed tolerance, $\varepsilon > 0$. Using (4.37) we see that

$$\frac{p(s)}{a_n} = \frac{a_0(s - r_1)(s - r_2)\cdots(s - r_n)}{a_0(-1)^n r_1 r_2 \cdots r_n} = (-1)^n\left(\frac{s}{r_1} - 1\right)\left(\frac{s}{r_2} - 1\right)\cdots\left(\frac{s}{r_n} - 1\right).$$

Since $|p(s)| \leq \varepsilon$,

$$\left|\frac{p(s)}{a_n}\right| = \left|1 - \frac{s}{r_1}\right|\left|1 - \frac{s}{r_2}\right|\cdots\left|1 - \frac{s}{r_n}\right| \leq \frac{\varepsilon}{a_n}.$$

Therefore

$$\left|\frac{p(s)}{a_n}\right| \geq \min_{1 \leq j \leq n}\left|1 - \frac{s}{r_j}\right|^n,$$

and thus,

$$\min_{1 \leq j \leq n} \left|1 - \frac{s}{r_j}\right| \leq \left(\frac{\varepsilon}{|a_n|}\right)^{1/n}. \tag{4.38}$$

In (4.38) we have the answer to the question: "If $|p(s)| \leq \varepsilon$, then what is an upper bound on the relative error of $s$ with respect to some zero of $p(x)$?" Since most iterative methods will have already calculated $p(s)$, (4.38) is quite simple to use as a test for termination of the iteration.

The mathematical theory becomes deeper and deeper, as we try to develop numerical root-finding techniques that work for larger and larger classes of polynomials. Finally we reach a point where the theory behind a method is so extensive, that it sometimes becomes more practical, for a particular polynomial, to try several simple approaches than to use one general "all-purpose" method. Of course, an efficient all-purpose method is preferable if one is an expert, or if the method is preprogrammed and part of the library of subroutines associated with the computing facility. We have found that the techniques presented here, along with a judicious use of the localization results, are usually adequate for polynomials that arise in actual practical problems. There are quite a few methods that we have not included in these sections. Some of these methods that the interested reader may want to investigate are designed especially for polynomials. These methods include: Bernoulli's method, Graeffe's root-squaring method, Lin's method, and Laguerre's method. Some more-sophisticated methods (some of which are based on localization procedures) include: the use of Sturm sequences, the Lehmer-Schur method, the quotient-difference algorithm and some recent procedures investigated by Henrici. In addition, there are a number of methods we have not discussed, such as Muller's method and those based on inverse interpolation, that are applicable to the problem $f(x) = 0$ even when $f(x)$ is not a polynomial. We refer the interested reader to Householder (1970) and Henrici (1974).

## PROBLEMS

1. a) Prove the corollary of Theorem 4.11 by direct use of Theorem 4.11.
   b) Prove the corollary of Theorem 4.11 by using induction on the synthetic division algorithm. (Hint: For $|\alpha| \geq r$, show that $|b_j| \geq 1$ for all $j$, where $b_0 = 1, b_j = \alpha b_{j-1} + a_j$, $j = 1, 2, \ldots, n$.)
2. Use the Gerschgorin Theorem in row form on the companion matrix (4.34) to derive other localization results.
3. Let $p(x)$ be given by (4.33) with zeros $\{r_j\}_{j=1}^{n}$. Find the zeros of $p(-x)$ and $x^n p(1/x)$.
4. Let $p(x) = a_0x^n + a_1x^{n-1} + \cdots + a_{n-1}x + a_n = a_0(x - r_1)(x - r_2)\cdots(x - r_n)$.
   a) Prove that $(a_1/a_0) = -(r_1 + r_2 + \cdots + r_n)$.
   b) Prove that $(a_n/a_0) = (-1)^n r_1 r_2 \cdots r_n$.

5. Program Newton's method and use (4.36) to include the provision that the iteration terminates when the absolute error is less than $10^{-5}$. Apply this program to the polynomial of Example 4.10 using $x_0 = -4$.
6. In a Bairstow's method program, use (4.38) to include the provision that the iteration terminates when the relative error is less than $10^{-5}$. Apply this program to the polynomial of Example 4.10 using $(u_0, v_0) = (6, 20)$.
7. Show that (4.33) is the characteristic polynomial of (4.34).

## *4.5 NEWTON'S METHOD IN SEVERAL VARIABLES

We wish to consider the problem of simultaneously solving a system of $n$ nonlinear equations in $n$ unknowns. Let $x_1, x_2, \ldots, x_n$ be real variables, and let $f_i(x_1, x_2, \ldots, x_n)$ be a real-valued function for each $i$, $1 \le i \le n$. Then we wish to find values $s_1, s_2, \ldots, s_n$ such that $f_i(s_1, s_2, \ldots, s_n) = 0$ simultaneously, for $1 \le i \le n$. We may simplify the statement of the problem by using vector notation. Let $\mathbf{x} = (x_1, x_2, \ldots, x_n)$ be a vector of $n$ variables, then for each $i$, $f_i(x_1, x_2, \ldots, x_n)$ can be written as $f_i(\mathbf{x})$. Furthermore we let

$$\mathbf{F}(\mathbf{x}) = \begin{bmatrix} f_1(\mathbf{x}) \\ f_2(\mathbf{x}) \\ \vdots \\ f_n(\mathbf{x}) \end{bmatrix}. \tag{4.39}$$

Now our problem can be restated in vector form as: Find a vector $\mathbf{s} = (s_1, s_2, \ldots, s_n)$ such that $\mathbf{F}(\mathbf{s}) = \mathbf{\theta}$, where $\mathbf{\theta}$ is the $(n \times 1)$ vector with all components equal to zero. For a thorough understanding of the material in this section, the reader should have some familiarity with vector and matrix norms (see Chapter 2).

**Example 4.11.** Let $n = 3$, and $f_1(x_1, x_2, x_3) = x_1 \cos(x_2) - x_3$, $f_2(x_1, x_2, x_3) = x_1^2 + x_3$, and $f_3(x_1, x_2, x_3) = e^{x_1+x_3} \sin(x_2/2) + x_3$. Then

$$\mathbf{F}(\mathbf{x}) = \begin{bmatrix} x_1 \cos(x_2) - x_3 \\ x_1^2 + x_3 \\ e^{x_1+x_3} \sin(x_2/2) + x_3 \end{bmatrix}$$

Obviously, $\mathbf{s} = (1, \pi, -1)$ satisfies $\mathbf{F}(\mathbf{s}) = \mathbf{\theta}$. Since the equations are nonlinear, however, we can probably expect other solutions to exist, for example, $\mathbf{s} = (0, 0, 0)$.

In problems of this form, we lose the geometric simplicity that led us to the derivation of simple methods for one variable. This is where the theory of the fixed-point problem becomes invaluable. As we shall see, almost all of the theorems about the fixed-point iteration are true in this setting, and their proofs are almost identical to the proofs in the case of a single variable. Using the same vector notation as above, we construct a vector-valued function

$$\mathbf{g}(\mathbf{x}) = \begin{bmatrix} g_1(x_1, x_2, \ldots, x_n) \\ g_2(x_1, x_2, \ldots, x_n) \\ \vdots \\ g_n(x_1, x_2, \ldots, x_n) \end{bmatrix} \qquad \textbf{(4.40)}$$

such that if $\mathbf{g}(\mathbf{s}) = \mathbf{s}$, then $\mathbf{F}(\mathbf{s}) = \mathbf{\theta}$. For example, given any such $\mathbf{F}(\mathbf{x})$, one choice for $\mathbf{g}(\mathbf{x})$ would be $\mathbf{g}(\mathbf{x}) = \mathbf{x} + A\mathbf{F}(\mathbf{x})$, where $A$ is a non-singular $(n \times n)$ matrix.

If $\mathbf{g}(\mathbf{x})$ has a fixed-point $\mathbf{s}$ (actually a "fixed-vector"), then we define the fixed-point iteration analogously to be:

$$\mathbf{x}_{k+1} = \mathbf{g}(\mathbf{x}_k). \qquad \textbf{(4.41)}$$

Before saying anything about the convergence of the fixed-point iteration, we must pause to introduce some vector concepts. Let $\mathscr{R}$ be a region in $R^n$ (the space of all real $n$-tuples) such that $\mathbf{x} = (x_1, x_2, \ldots, x_n)$ belongs to $\mathscr{R}$ if and only if $a_1 \le x_1 \le b_1$, $a_2 \le x_2 \le b_2, \ldots, a_n \le x_n \le b_n$; where $\{a_i\}_{i=1}^n$ and $\{b_i\}_{i=1}^n$ are specified real numbers. (Thus $\mathscr{R}$ is the generalization of a closed interval. For example, $\mathscr{R}$ is a rectangle in $R^2$.) A *line segment* from the vector $\mathbf{x}$ to the vector $\mathbf{y}$ is defined to be the set of all vectors of the form $\mathbf{w} = \lambda\mathbf{x} + (1 - \lambda)\mathbf{y}$, where $\lambda$ is any real number satisfying $0 \le \lambda \le 1$. Note that if $\lambda = 0$, $\mathbf{w} = \mathbf{y}$, and if $\lambda = 1$, $\mathbf{w} = \mathbf{x}$. It is left to the reader to show that if $\mathbf{x}$ and $\mathbf{y}$ belong to $\mathscr{R}$, then $\mathbf{w} \in \mathscr{R}$, for all $\lambda$, $0 \le \lambda \le 1$.

In order to talk about the rate of convergence of a fixed-point iteration, we must introduce the concept of a *derivative of a vector-valued function.* For a function of a real variable, $g$, the derivative of $g$ at $x$ is simply

$$\lim_{h \to 0} \frac{g(x + h) - g(x)}{h} = g'(x) \qquad \textbf{(4.42)}$$

Equivalently, Eq. (4.42) means:

For every $\varepsilon > 0$, there exists a $\delta > 0$, such that

$$|(g(x + h) - g(x))/h - g'(x)| \le \varepsilon \quad \text{whenever} \quad 0 < |h| \le \delta. \qquad \textbf{(4.43)}$$

For $\mathbf{h} = (h_1, h_2, \ldots, h_n) \in R^n$, division by $\mathbf{h}$ is not defined, so we may not extend (4.42) directly to $R^n$. However, (4.43) does have a natural extension to $R^n$. Let $\|\cdot\|$ be any of the three norms introduced in Chapter 2. Then the derivative of $\mathbf{g}$ at $\mathbf{x}$ is defined to be the $(n \times n)$ matrix, $A_\mathbf{x}$, that satisfies:

For every $\varepsilon > 0$, there exists a $\delta > 0$ such that

$$\|\mathbf{g}(\mathbf{x} + \mathbf{h}) - \mathbf{g}(\mathbf{x}) - A_\mathbf{x}\mathbf{h}\| < \varepsilon\|\mathbf{h}\|, \quad \text{whenever} \quad 0 < \|\mathbf{h}\| \le \delta. \qquad \textbf{(4.44)}$$

We shall write $A_\mathbf{x} \equiv g'(\mathbf{x})$, and note that the matrix $A_\mathbf{x}$ changes as $\mathbf{x}$ changes. This definition also applies if $\mathbf{g}\colon R^n \to R^p$, in which case $A_\mathbf{x}$ is $(p \times n)$.

It is cumbersome, but not difficult (see Problem 2), to verify that $g'(\mathbf{x})$ is the familiar Jacobian matrix whose $(i, j)$th element is

$$\frac{\partial g_i(\mathbf{x})}{\partial x},$$

that is,

$$g'(\mathbf{x}) \equiv J(\mathbf{x}) = \left[\frac{\partial g_i(\mathbf{x})}{\partial x_j}\right] \tag{4.45}$$

**Example 4.12.** For the function $\mathbf{F}(\mathbf{x})$ of Example 4.11, and $\mathbf{x} = (1, 2, 3)$,

$$F'(\mathbf{x}) = \begin{bmatrix} \cos(2) & -\sin(2) & -1 \\ 2 & 0 & 1 \\ e^4 \sin(1) & (1/2)e^4 \cos(1) & e^4 \sin(1) + 1 \end{bmatrix}$$

The concept of the second derivative can also be defined, but it is not necessary for us to do so now. It will appear later in a Taylor expansion, where it will simply be denoted by $g''(\mathbf{x}, \mathbf{h}, \mathbf{h})$. The only thing we will need to know about the second derivative is that if it is continuous for all $\mathbf{x} \in \mathcal{R}$, then it satisfies

$$\|g''(\mathbf{x}, \mathbf{h}, \mathbf{h})\| \leq K\|\mathbf{h}\|^2, \qquad \text{for all } \mathbf{x}, \tag{4.46}$$

where $K$ is a constant independent of $\mathbf{x}$.

The last preliminary result we need is a generalization of the mean-value theorem, which we state without proof.

**Mean-Value Theorem.** Given any two vectors $\mathbf{x}$ and $\mathbf{y}$ in $\mathcal{R}$, assume that for all $\mathbf{w}$ on the line segment between $\mathbf{x}$ and $\mathbf{y}$ that

$$\left|\frac{\partial g_i(\mathbf{w})}{\partial x_j}\right| \leq L_{ij}.$$

Let

$$L_1 = \max_{1 \leq j \leq n} \left\{\sum_{i=1}^{n} L_{ij}\right\}, \qquad L_\infty = \max_{1 \leq i \leq n} \left\{\sum_{j=1}^{n} L_{ij}\right\},$$

and

$$L_2 = \left[\sum_{i=1}^{n} \sum_{j=1}^{n} L_{ij}^2\right]^{1/2}$$

Then

$$\|\mathbf{g}(\mathbf{x}) - \mathbf{g}(\mathbf{y})\|_p \leq L_p\|\mathbf{x} - \mathbf{y}\|_p, \qquad p = 1, 2, \text{ or } \infty \tag{4.47}$$

As an example of the above theorem, assume that each $L_{ij} = \lambda/n$, where $0 \leq \lambda < 1$. Then $L_1 = L_2 = L_\infty = \lambda < 1$, and $\|\mathbf{g}(\mathbf{x}) - \mathbf{g}(\mathbf{y})\|_p \leq \lambda\|\mathbf{x} - \mathbf{y}\|_p$; $p = 1, 2$, or $\infty$. Now we are able to prove the following convergence theorem for the fixed-point iteration. (Compare its results and proof with Theorem 4.3.)

**Theorem 4.13.** Let $\mathscr{R}$ be a rectangle in $R^n$ (as above), and assume that $\mathbf{g}(\mathscr{R}) \subseteq \mathscr{R}$. Also assume that for all $\mathbf{w} \in \mathscr{R}$, at least one of the constants $L_1$, $L_2$, $L_\infty$ defined above (say $L_p$) satisfies $0 < L_p < 1$. Then,

a) $\mathbf{x} = \mathbf{g}(\mathbf{x})$ has exactly one solution, $\mathbf{s}$, in $\mathscr{R}$.

b) the sequence (4.41) converges to $\mathbf{s}$.

c) for all $n$, $\|\mathbf{x}_n - \mathbf{s}\|_p \le L_p^n \|\mathbf{x}_1 - \mathbf{x}_0\|_p/(1 - L_p)$.

*Proof*: By repeated use of (4.47),

$$\begin{aligned}\|\mathbf{x}_{n+1} - \mathbf{x}_n\|_p &= \|\mathbf{g}(\mathbf{x}_n) - \mathbf{g}(\mathbf{x}_{n-1})\|_p \le L_p\|\mathbf{x}_n - \mathbf{x}_{n-1}\|_p \\ &\le L_p^2\|\mathbf{x}_{n-1} - \mathbf{x}_{n-2}\|_p \le \cdots \le L_p^n\|\mathbf{x}_1 - \mathbf{x}_0\|_p. \end{aligned} \qquad \textbf{(4.48)}$$

Let $m > n$, then by the triangle inequality and (4.48),

$$\begin{aligned}\|\mathbf{x}_m - \mathbf{x}_n\|_p &\le \|\mathbf{x}_m - \mathbf{x}_{m-1}\|_p + \|\mathbf{x}_{m-1} - \mathbf{x}_{m-2}\|_p + \cdots + \|\mathbf{x}_{n+1} - \mathbf{x}_n\|_p \\ &\le (L_p^{m-1} + \cdots + L_p^n)\|\mathbf{x}_1 - \mathbf{x}_0\|_p \\ &\le L_p^n\|\mathbf{x}_1 - \mathbf{x}_0\|_p/(1 - L_p). \end{aligned} \qquad \textbf{(4.49)}$$

Since $0 \le L_p < 1$, given any $\varepsilon > 0$, there exists an $N$ such that, if $m$ and $n > N$, then $\|\mathbf{x}_m - \mathbf{x}_n\|_p < \varepsilon$, that is, choose $N$ such that $L_p^n\|\mathbf{x}_1 - \mathbf{x}_0\|/(1 - L_p) < \varepsilon$. This property is called the *Cauchy Criterion.* Just as in the case of a sequence, $\{x_n\}_{n=1}^\infty$, of real numbers in a closed interval $[a, b]$, it is also true that if a sequence, $\{\mathbf{x}_n\}_{n=1}^\infty$, in $\mathscr{R}$ satisfies the Cauchy Criterion, then the sequence must converge to some $\mathbf{s} \in \mathscr{R}$. Since $\mathbf{g}(\mathbf{x})$ is continuous,

$$\mathbf{s} = \lim_{n\to\infty} \mathbf{x}_{n+1} = \lim_{n\to\infty} \mathbf{g}(\mathbf{x}_n) = \mathbf{g}(\mathbf{s}).$$

Now, assume that there is another fixed-point $\mathbf{q} \in \mathscr{R}$, that is, $\mathbf{q} = \mathbf{g}(\mathbf{q})$, $\mathbf{q} \neq \mathbf{s}$. Then, $\|\mathbf{s} - \mathbf{q}\|_p = \|\mathbf{g}(\mathbf{s}) - \mathbf{g}(\mathbf{q})\|_p \le L_p\|\mathbf{s} - \mathbf{q}\|_p < \|\mathbf{s} - \mathbf{q}\|_p$, but this is a contradiction and completes the proof. ■

Our analogy to the fixed-point problem of a single variable continues here with the following local convergence theorem.

**Theorem 4.14.** Let $r > 0$ be given such that the set of vectors, $S = \{\mathbf{x}: \|\mathbf{x} - \mathbf{s}\| < r\}$ contains a fixed-point, $\mathbf{s}$, of $\mathbf{g}(\mathbf{x})$. Further, let $S \subseteq \mathscr{R}$, $g'$ be continuous on $S$, and $\|g'(\mathbf{s})\| < 1$. Then there exists an $\varepsilon > 0$ such that the fixed-point iteration is convergent whenever $\|\mathbf{x}_0 - \mathbf{s}\| < \varepsilon$.

The proof is exactly analogous to the proof of Theorem 4.4 and we omit it. We now define the error vector, $\mathbf{e}_n = \mathbf{x}_n - \mathbf{s}$. By formula (4.46) and the remarks preceding it, there is a Taylor's expansion of the form:

$$\begin{aligned}\|\mathbf{e}_{n+1}\| &= \|\mathbf{x}_{n+1} - \mathbf{s}\| = \|\mathbf{g}(\mathbf{x}_n) - \mathbf{g}(\mathbf{s})\| \\ &= \|g'(\mathbf{s})\mathbf{e}_n + g''(\zeta; \mathbf{e}_n, \mathbf{e}_n)\| \le \|g'(\mathbf{s})\mathbf{e}_n\| + K\|\mathbf{e}_n\|^2. \end{aligned} \qquad \textbf{(4.50)}$$

If $g'(\mathbf{s}) \neq \mathcal{O}$ (the zero matrix), then, in general, $g'(\mathbf{s})\mathbf{e}_n \neq \mathbf{0}$. In this case, the best possible rate of convergence we can obtain is linear, that is,

$$\lim_{n\to\infty} \frac{\|\mathbf{e}_{n+1}\|}{\|\mathbf{e}_n\|} \leq \|g'(\mathbf{s})\|.$$

If $g'(\mathbf{s}) = \mathcal{O}$, however, we obtain quadratic convergence,

$$\lim_{n\to\infty} \frac{\|\mathbf{e}_{n+1}\|}{\|\mathbf{e}_n\|^2} \leq K.$$

Here we again return to our principal problem of finding vectors, $\mathbf{s}$, such that $\mathbf{F}(\mathbf{s}) = \mathbf{0}$. Again, if we let $\mathbf{g}(\mathbf{x}) = \mathbf{x} + A\mathbf{F}(\mathbf{x})$, where $A$ is a non-singular $(n \times n)$ matrix, then $\mathbf{s} = \mathbf{g}(\mathbf{s})$ if and only if $\mathbf{F}(\mathbf{s}) = \mathbf{0}$. By considering the Jacobian of $\mathbf{g}(\mathbf{x}) = \mathbf{x} + A\mathbf{F}(\mathbf{x})$ (Problem 2d), we see that $g'(\mathbf{x}) = I + AF'(\mathbf{x})$. Thus $g'(\mathbf{s}) = \mathcal{O}$ only if $A = -[F'(\mathbf{s})]^{-1}$, that is, the inverse of the Jacobian of $\mathbf{F}(\mathbf{x})$ evaluated at $\mathbf{s}$. Once again this choice for $A$ cannot be made without *a priori* knowledge of $\mathbf{s}$. We are then forced to consider the choice, $\mathbf{g}(\mathbf{x}) = \mathbf{x} + A(\mathbf{x})\mathbf{F}(\mathbf{x})$, where $A(\mathbf{x})$ is an $(n \times n)$ matrix whose elements are real-valued functions of $\mathbf{x}$. It is beyond the level of this text to take the derivative of $A(\mathbf{x})\mathbf{F}(\mathbf{x})$ (and very cumbersome, at best, to calculate its Jacobian and arrive at the desired result). In order that $g'(\mathbf{s}) = \mathcal{O}$, however, it can be shown that $A(\mathbf{x}) = -[\mathbf{F}'(\mathbf{x})]^{-1}$. Therefore with this choice of $A(\mathbf{x})$, the fixed-point iteration becomes Newton's method in $R^n$ where $\mathbf{x}_0$ is an initial guess:

$$\mathbf{x}_{n+1} = \mathbf{x}_n - J(\mathbf{x}_n)^{-1}\mathbf{F}(\mathbf{x}_n), \qquad n \geq 0, \tag{4.51}$$

where $J(\mathbf{x}_n)$ is the Jacobian of $\mathbf{F}$ at $\mathbf{x}_n$.

Thus the Newton method in $R^n$ is notationally the same as (4.11e), possesses a local convergence theorem analogous to Theorem 4.6 (via Theorem 4.14), and again has quadratic convergence whenever $J(\mathbf{s})$ is non-singular.

We have noted previously that it is usually computationally inefficient to compute inverses of matrices. Thus (4.51) is usually performed in the following manner: Assume that $\mathbf{x}_n$ has just been calculated, and define $\mathbf{z}_{n+1} \equiv \mathbf{x}_{n+1} - \mathbf{x}_n$. Rewrite (4.51) as $J(\mathbf{x}_n)\mathbf{z}_{n+1} = -\mathbf{F}(\mathbf{x}_n)$. Solve this linear system for $\mathbf{z}_{n+1}$ by one of the methods of Chapter 2 (usually Gauss elimination), and then find $\mathbf{x}_{n+1}$ by $\mathbf{x}_{n+1} = \mathbf{z}_{n+1} + \mathbf{x}_n$.

**Example 4.13.** Let $f_1(x_1, x_2, x_3) = x_1^5 + x_2^3 x_3^4 + 1$, $f_2(x_1, x_2, x_3) = x_1^2 x_2 x_3$, and $f_3(x_1, x_2, x_3) = x_3^4 - 1$. It can be easily shown algebraically that there are exactly four real solutions:

$$\mathbf{s}_1 = (0, -1, -1), \qquad \mathbf{s}_2 = (0, -1, 1), \qquad \mathbf{s}_3 = (-1, 0, 1), \qquad \text{and}$$
$$\mathbf{s}_4 = (-1, 0, -1).$$

We list in Table 4.8 the results of several trials of Newton's iteration carried out to six-place accuracy, where $\mathbf{x}_0$ is the initial iteration vector and $k$ represents the $k$th iteration.

**Table 4.8**

(1) $\mathbf{x}_0 = (-100, 0, -100)$: (Yields $\mathbf{s}_4$)

| $k$ | $x_1(k)$ | $x_2(k)$ | $x_3(k)$ |
|---|---|---|---|
| 10 | −10.73743 | 0 | −5.63222 |
| 20 | −1.21805 | 0 | −1.00000 |
| 24 | −1.000000 | 0 | −1.00000 |

(2) $\mathbf{x}_0 = (-1000, -1000, -1000)$: (Yields $\mathbf{s}_1$)

| $k$ | $x_1(k)$ | $x_2(k)$ | $x_3(k)$ |
|---|---|---|---|
| 10 | −9.08509 | −1000.00000 | −56.31350 |
| 20 | −0.08265 | −998.47800 | −3.17604 |
| 30 | −0.00110 | −96.58270 | −1.00000 |
| 40 | −0.00002 | −1.74844 | −1.00000 |
| 45 | −0.00000 | −1.00000 | −1.00000 |

(3) $\mathbf{x}_0 = (-0.01, -0.01, -0.01)$: (Yields $\mathbf{s}_1$)

| $k$ | $x_1(k)$ | $x_2(k)$ | $x_3(k)$ |
|---|---|---|---|
| 10 | −149328 | 40254800. | −18771.2 |
| 20 | 154152 | 40255700. | −1057.07 |
| 30 | 42024.2 | 40255700. | −59.5274 |
| 40 | 438.705 | 40206600. | −3.35629 |
| 50 | 4.64666 | 4204370. | −1.00000 |
| 60 | 0.08058 | 72910.2 | −1.00000 |
| 70 | 0.00140 | 1264.37 | −1.00000 |
| 80 | 0.00002 | 21.9258 | −1.00000 |
| 90 | 0 | −1.00718 | −1.00000 |
| 92 | 0 | −1.00000 | −1.00000 |

(4) $\mathbf{x}_0 = (0.1, 0.1, 0.1)$: (Yields $\mathbf{s}_2$)

| $k$ | $x_1(k)$ | $x_2(k)$ | $x_3(k)$ |
|---|---|---|---|
| 10 | −47.1446 | 3799.41 | 25.0358 |
| 20 | −0.43201 | 3657.72 | 1.46340 |
| 30 | −0.00685 | 117.065 | 1.00000 |
| 40 | −0.00012 | 1.97841 | 1.00000 |
| 48 | −0 | −1.00000 | 1.00000 |

(5) $\mathbf{x}_0 = (-100, 0, 100)$: (Yields $\mathbf{s}_3$)

| $k$ | $x_1(k)$ | $x_2(k)$ | $x_3(k)$ |
|---|---|---|---|
| 10 | −10.7374 | 0 | 5.63222 |
| 20 | −1.21805 | 0 | 1.00000 |
| 24 | −1.00000 | 0 | 1.00000 |

## PROBLEMS

*1. Let $\mathscr{R}$ be a rectangle in $R^n$. Let $\mathbf{x}$ and $\mathbf{y}$ be any two vectors in $\mathscr{R}$, and let $\mathbf{w}$ be any vector of the form, $\mathbf{w} = \lambda\mathbf{x} + (1 - \lambda)\mathbf{y}$, where $0 \leq \lambda \leq 1$.

a) What is the geometrical interpretation of $\mathbf{w}$ if $\lambda = 1/2$? (Consider $\|\mathbf{w} - \mathbf{x}\|$ and $\|\mathbf{w} - \mathbf{y}\|$ with respect to $\|\mathbf{x} - \mathbf{y}\|$ for any norm.)

b) Show that if $\mathbf{x}$ and $\mathbf{y}$ are in $\mathscr{R}$, then $\mathbf{w} \in \mathscr{R}$ for any $\lambda$, $0 \leq \lambda \leq 1$.

c) The vector-valued function $\mathbf{F}$, as given by (4.39), is continuous at a vector $\mathbf{z} \in \mathscr{R}$ if: For all $\varepsilon > 0$, there exists a $\delta > 0$, such that $\|\mathbf{F}(\mathbf{z} + \mathbf{h}) - \mathbf{F}(\mathbf{z})\| \leq \varepsilon$ whenever $0 < \|\mathbf{h}\| \leq \delta$. Let $\mathbf{x}$ and $\mathbf{y}$ be any two points in $\mathscr{R}$ satisfying $f_i(\mathbf{x}) < 0$ and $f_i(\mathbf{y}) > 0$ for all $i$, $1 \leq i \leq n$. Assume that $\mathbf{F}$ is continuous at all $\mathbf{z} \in \mathscr{R}$. Does it necessarily follow that there exists an $\mathbf{s}$ on the line segment between $\mathbf{x}$ and $\mathbf{y}$ satisfying $\mathbf{F}(\mathbf{s}) = \mathbf{0}$? (Note that if this result were true, then the bisection method could be extended to this problem. Hint: Consider $R^2$.)

d) Use the definition in Part (c) to prove that if $F'(\mathbf{z})$ exists, then $\mathbf{F}$ is continuous at $\mathbf{z}$. (Remember, $F'(\mathbf{z})$ is an $n \times n$ matrix).

*2. Again let $\mathbf{e}_j \in R^n$ be 1 in the $j$th component and zero elsewhere. Also recall from calculus that

$$\frac{\partial f_i(\mathbf{z})}{\partial x_j} = \lim_{t \to 0} \left( \frac{f_i(\mathbf{z} + t\mathbf{e}_j) - f_i(\mathbf{z})}{t} \right). \tag{4.52}$$

a) Verify that $\|\mathbf{e}_j\|_1 = \|\mathbf{e}_j\|_2 = \|\mathbf{e}_j\|_\infty = 1$, $1 \leq j \leq n$, and let $\mathbf{h} = t\mathbf{e}_j$, where $|t| \leq \delta$. Thus $0 < \|\mathbf{h}\| \leq \delta$ for any of the three norms.

b) Verify, that if $A$ is any $(n \times n)$ matrix, then $A(t\mathbf{e}_j) = t\mathbf{A}_j$, where $\mathbf{A}_j$ is the $j$th column of $A$, $1 \leq j \leq n$.

c) Assume that $\mathbf{F}$ is given by (4.39), and that $F'(\mathbf{z})$ exists, that is, given any $\varepsilon$ there exists a $\delta$ by (4.44) such that

$$\|\mathbf{F}(\mathbf{z} + \mathbf{h}) - \mathbf{F}(\mathbf{z}) - F'(\mathbf{z})\mathbf{h}\| < \varepsilon\|\mathbf{h}\|, \tag{4.53}$$

whenever

$$0 < |t| \leq \delta$$

and let

$$\mathbf{h}_j = t\mathbf{e}_j$$

for a fixed $j$, $1 \leq j \leq n$.

Assuming that $\partial f_i(\mathbf{z})/\partial x_j$ exists, $1 \leq i \leq n$, prove by use of (4.52) that (4.53) can be true for this particular $\mathbf{h}_j$ only if the $j$th column of $F'(\mathbf{z})$ equals:

$$\begin{bmatrix} \dfrac{\partial f_1(\mathbf{z})}{\partial x_j} \\ \dfrac{\partial f_2(\mathbf{z})}{\partial x_j} \\ \vdots \\ \dfrac{\partial f_n(\mathbf{z})}{\partial x_j} \end{bmatrix}$$

(Use any of the three norms that you wish.) Thus we argue that if (4.53) holds for all $\mathbf{h}$ satisfying $0 < \|\mathbf{h}\| \leq \delta$, then it must hold for each $\mathbf{h}_j$, $1 \leq j \leq n$. But if all partials $\partial f_i(\mathbf{z})/\partial x_j$ exist, this implies that $F'(\mathbf{z}) = J(\mathbf{z})$, the Jacobian of $\mathbf{F}$ at $\mathbf{z}$.

d) Let $\mathbf{F}(\mathbf{x})$ be given by (4.39) and let $A$ be a non-singular $(n \times n)$ matrix whose elements are constants. If $\mathbf{g}(\mathbf{x}) = \mathbf{x} + A\mathbf{F}(\mathbf{x})$, show that $g'(\mathbf{x}) = I + AF'(\mathbf{x})$.

3. Verify that the equations of Example 4.13 have only the four real solutions that are given. Find one complex solution. (Recall $e^{i\theta} = \cos\theta + i\sin\theta$, and $(e^{i\theta})^m = e^{im\theta} = \cos m\theta + i\sin m\theta$.)

4. One modification of Newton's method is as follows. Let $\mathbf{x}_0$ be an initial guess for $\mathbf{s}$, where $\mathbf{F}(\mathbf{s}) = \boldsymbol{\theta}$, and define $\{\mathbf{x}_n\}_{n=0}^{\infty}$ by

$$\mathbf{x}_{n+1} = \mathbf{x}_n - J(\mathbf{x}_0)^{-1}\mathbf{F}(\mathbf{x}_n), \tag{4.54}$$

where $J(\mathbf{x}_0)$ is the Jacobian of $\mathbf{F}$ at $\mathbf{x}_0$. This method is still of the fixed-point type, but its convergence is linear. Apply this algorithm to one of the initial guesses in Example 4.13 and compare their rates of convergence. This method is obviously easier to perform since $J(\mathbf{x}_0)$ remains stationary. Can you develop a different algorithm along these lines which should be faster than (4.54), but still not as fast as (4.51)?

5. Let $\mathbf{F}(\mathbf{x})$ be given as in Example 4.11. Write a program for Newton's method to approximate solutions of $\mathbf{F}(\mathbf{x}) = \boldsymbol{\theta}$, with $\mathbf{x}_0 = (1/4, \pi/2, -1/2)$. Continue the iteration until $\|\mathbf{F}(\mathbf{x}_n)\|_2 < 10^{-5}$.

# 5

# INTERPOLATION AND APPROXIMATION

## 5.1 INTRODUCTION

An important problem often encountered in scientific work is that of approximating some very "complicated" function, $f(x)$, by a "simpler" function, $p(x)$. The reader has already seen one form of this problem in calculus where if $f(a)$, $f'(a)$, $f''(a), \ldots, f^{(k)}(a)$ are known at some point $a$, then the truncated Taylor's expansion,

$$p(x) = f(a) + f'(a)(x - a) + \frac{f''(a)}{2!}(x - a)^2 + \cdots + \frac{f^{(k)}(a)}{k!}(x - a)^k,$$

is a $k$th degree polynomial which approximates $f(x)$ for $x$ near $a$. If $f^{(k+1)}(x)$ is continuous then the error of the approximation at $x$ is given by

$$\frac{f^{(k+1)}(\xi)}{(k + 1)!}(x - a)^{k+1},$$

where $\xi$ lies between $x$ and $a$. There are, however, several obvious drawbacks to this type of an approximation—the derivatives may not exist, they may be very difficult to calculate (or even impossible if, for instance, $f(x)$ is given in tabular form instead of in terms of a formula), $f^{(k+1)}(\xi)$ may become quite large and thus make the error large, etc. Even for a "nice" function such as $f(x) = \cos(x)$ with $a = 0$, where the Taylor (Maclaurin) expansion converges to $f(x)$ for any $x$ and $f^{(k)}(0)$ is known for any $k$, this type of approximation is not usually practical since many terms may be required to maintain accuracy for values of $x$ somewhat removed from 0 (see Problem 2).

Thus we see that the problem of approximation of functions must be analyzed quite carefully to obtain practical computational procedures. In this chapter we shall be mainly concerned with polynomial interpolation, spline interpolation, and Fourier approximations. In the following chapter, we will study their use in such problems as numerical integration and numerical differentiation. For example, if $f(x)$ is approximated by $p(x)$ and we wish to find a numerical approximation for $\int_a^b f(x)\,dx$ or for $f'(\alpha)$ for some value $\alpha$, then we could use $\int_a^b p(x)\,dx$ or $p'(\alpha)$ as these approximations, respectively. However, it may be that while $\int_a^b p(x)\,dx$ is a good approximation for $\int_a^b f(x)\,dx$, $p'(\alpha)$ is a poor approximation for $f'(\alpha)$ (and vice versa). Therefore an integral part of designing an approximation for a function is the knowledge of how the approximation is to be subsequently used. Thus Chapters 5 and 6 are intrinsically related, in that Chapter 5 develops some approximation techniques and the main part of Chapter 6 discusses their utilization in integration and differentiation. Then with this background, Chapter 6 concludes by discussing some further approximation techniques.

We shall be principally interested in approximating continuous real-valued functions, $f(x)$, where $x$ belongs to the closed finite interval $[a, b]$. We use the standard notation, $C[a, b]$, to denote the set of real-valued functions which are continuous on $[a, b]$. If $p(x)$ is an approximation to $f(x)$, where $f \in C[a, b]$, then we need some way of measuring how good this approximation is. That is, we must be able to answer the question: How "close" is $p(x)$ to $f(x)$? We will naturally say that $p(x)$ is a good approximation to $f(x)$ if the function $f(x) - p(x)$ is small in some sense. So to measure closeness we need some sort of a measure of size for functions in $C[a, b]$. In (5.1), we have listed three commonly used measures for the size of the function $(f - p)$. These measures are called *norms* and the magnitude of the number $\|f - p\|$ provides us with a quantitative way of gauging how good an approximation $p(x)$ is to $f(x)$. (The reader who has covered the material on vector norms in Chapter 2 will recognize the norms defined in (5.1) as natural extensions of the $\ell_p$ vector norms, where $(f - p)$ is used in place of the vector $(x - y)$ and integration is used in place of summation.)

$$\|f - p\|_1 \equiv \int_a^b |f(x) - p(x)|\, w(x)\,dx \tag{5.1a}$$

$$\|f - p\|_2 \equiv \left(\int_a^b (f(x) - p(x))^2 w(x)\,dx\right)^{1/2} \tag{5.1b}$$

$$\|f - p\|_\infty \equiv \max_{a \le x \le b} |f(x) - p(x)|. \tag{5.1c}$$

In (5.1a) and (5.1b) the function $w(x)$ is a fixed weighting function which provides us with some flexibility in measuring closeness. In all the cases we consider, $w(x)$ is continuous and nonnegative on $(a, b)$, $\int_a^b w(x)\,dx$ exists and $\int_a^b w(x)\,dx > 0$. Finally, by way of notation, we shall denote the "zero function" as $\theta(x)$, where $\theta(x) \equiv 0$ for all $x$ in $[a, b]$ and we will let $\mathscr{P}_n$ denote the set of all polynomials of degree $n$ or less. The following example illustrates that two functions can be "close" in one norm but *not* in another.

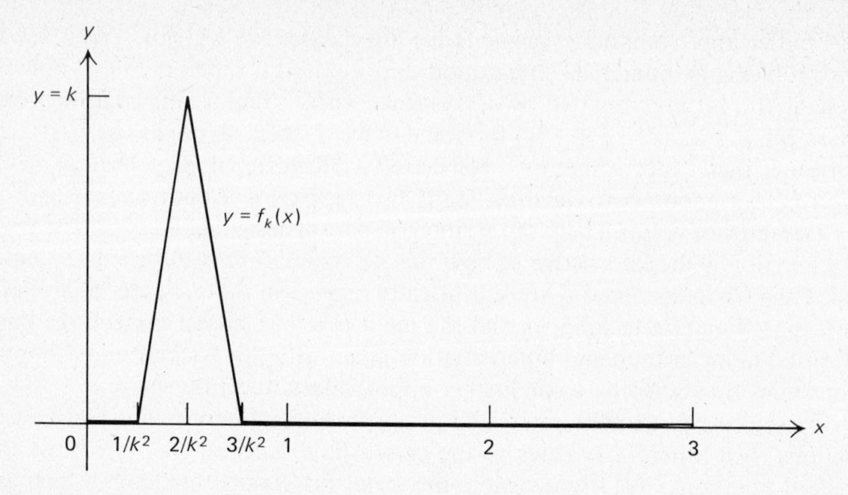


**Fig. 5.1** Graph of $y = f_k(x)$; $k \geq 1$.

**Example 5.1.** Let $f(x) = \theta(x)$ and let $w(x) \equiv 1$ for $x$ in $[a, b]$, with $a = 0$ and $b = 3$. For any positive integer $k$, let $f_k(x)$ be given by (see Fig. 5.1):

$$f_k(x) = \begin{cases} k(k^2x - 1), & \text{for} \quad 1/k^2 \leq x \leq 2/k^2 \\ -k(k^2x - 3), & \text{for} \quad 2/k^2 \leq x \leq 3/k^2 \\ 0, & \text{otherwise.} \end{cases}$$

Using these three formulas, we obtain

$$\|f - f_k\|_1 = 1/k, \qquad \|f - f_k\|_2 = \sqrt{2}/\sqrt{3}, \qquad \|f - f_k\|_\infty = k.$$

Thus, in the sense of (5.1a), the distance between $f(x)$ and $f_k(x)$ becomes small for large $k$, for (5.1b) the distance is constant for any $k$, and for (5.1c) the distance is large for large $k$. Now we consider $\{\|f - f_k\|\}_{k=1}^{\infty}$ as a sequence of real numbers and see that

$$\lim_{k \to \infty} \|f - f_k\|_1 = 0,$$

$$\lim_{k \to \infty} \|f - f_k\|_2 = \sqrt{2}/\sqrt{3},$$

and

$$\lim_{k \to \infty} \|f - f_k\|_\infty = \infty.$$

In this context we therefore say that the sequence of functions, $\{f_k(x)\}_{k=1}^{\infty}$, *converges to* $f(x)$ only in terms of the $\|\cdot\|_1$ distance measurement. (A sequence of functions, $\{q_k(x)\}_{k=1}^{\infty}$, is said to converge to a function $g(x)$ with respect to a given norm, $\|\cdot\|$, if and only if

$$\lim_{k \to \infty} \|q_k - g\| = 0.)$$

The preceding example illustrates that the quality of an approximation is completely dependent on how we choose to measure distance between functions.

In most practical problems we would probably accept $f_k(x)$ for large $k$ as a "good" approximation for $f(x) = \theta(x)$, since it is only "bad" in a small neighborhood of a single point. However, there are practical problems where even moderate errors in a small neighborhood are unacceptable. For example, the cosine routine in a computer must provide *uniformly* good approximations for all $x$ in $[0, \frac{\pi}{2}]$. This illustrates the fact that the choice of a distance measurement (norm) is dependent on the underlying physical or mathematical problem. We note here that

$$\|f - p\|_1 = \int_a^b |f(x) - p(x)|\, w(x)\, dx \le \max_{a \le x \le b} |f(x) - p(x)| \int_a^b w(x)\, dx$$

$$= \|f - p\|_\infty \left(\int_a^b w(x)\, dx\right),$$

and similarly,

$$\|f - p\|_2 \le \|f - p\|_\infty \left(\int_a^b w(x)\, dx\right)^{1/2}.$$

The number $\int_a^b w(x)\, dx$ is a constant, so if $\|f - p\|_\infty$ is "small," then both $\|f - p\|_1$ and $\|f - p\|_2$ are also "small." Thus the $\|\cdot\|_\infty$ norm is "stronger" than the other two norms, and we strive for goodness of approximation with respect to the $\|\cdot\|_\infty$ norm whenever possible. (The reader with some advanced calculus background will easily recognize that convergence of a sequence of functions in the $\|\cdot\|_\infty$ norm is equivalent to *uniform convergence*.)

As we shall see shortly, we will normally be approximating functions with polynomials. Two important reasons for this are that polynomials are easy to use (for example, in integration and differentiation, see Problem 1) and that polynomials can be used to provide very good approximations for functions in $C[a, b]$. The first reason is obvious to the reader and evidence for the second is provided by the following classical result of Weierstrass (which we give without proof).

**Theorem 5.1. Weierstrass.** Let $f \in C[a, b]$. For each $\varepsilon > 0$ there exists a polynomial $p(x)$ of degree $N_\varepsilon$ ($N_\varepsilon$ depends on $\varepsilon$), such that $\|f - p\|_\infty < \varepsilon$.

This theorem says that any continuous function on the finite interval $[a, b]$ may be uniformly approximated by some polynomial. We conclude this section with another classical result (again without proof).

**Theorem 5.2.** Let $f(x)$ be given in $C[a, b]$ and let $n$ be a fixed positive integer. If $\|\cdot\|$ is any one of the three norms given above, then there exists a unique polynomial $p^*(x)$ of degree $n$ or less, such that $\|f - p^*\| \le \|f - p\|$ for all $p(x) \in \mathscr{P}_n$.

This theorem tells us that there is a unique "best" $n$th degree polynomial approximation to $f(x)$ with respect to any of the above norms. The reader should be warned that the polynomial which is best with respect to one norm is usually *not* the same polynomial which is best with respect to another. We shall see in Section 5.3.1 that the polynomial which minimizes $\|f - p\|_2$, for $p(x) \in \mathscr{P}_n$, can be explicitly constructed, whereas this is not usually true with respect to the other two norms. However, Theorem 5.2 still has important theoretical implications with respect to these norms. Finally, let us define $E_n(f) \equiv \|f - p_n^*\|_\infty$, where $p_n^*(x) \in \mathscr{P}_n$ satisfies $\|f - p_n^*\|_\infty \leq \|f - p\|_\infty$, for all $p(x) \in \mathscr{P}_n$. (Usually, $E_n(f)$ is called the *degree of approximation for* $f(x)$ and $p_n^*(x)$ is called the *best nth degree uniform approximation to* $f(x)$.) It is easily seen that if $m > n$, then $E_m(f) \leq E_n(f)$. So, by Theorem 5.1, we obtain the result that $\lim_{n\to\infty} E_n(f) = 0$ for any $f(x)$ in $C[a, b]$.

## PROBLEMS

1. One argument for using polynomials $p(x)$ to approximate complicated functions $f(x)$ is that polynomials are easy to integrate, evaluate, and differentiate on the computer. Write a program that accepts the coefficients of any 20th-degree or less polynomial $p(x)$ as input, together with either an interval $[a, b]$ or a number $c$. Develop the program to have the capability of calculating $\int_a^b p(x)\,dx$ and the $i$th derivative of $p(x)$ evaluated at $x = c$, for any $i$, $0 \leq i \leq 20$.
2. Find the truncated $k$th degree Taylor's series expansion $p_k(x)$ for $f(x) = \cos(x)$ with $a = 0$ and $k$ arbitrary. Show that a bound for the error $|f(x) - p_k(x)|$ is given by $|x^{k+1}|/(k + 1)!$. Given this bound, how large must $k$ be in order that $|p_k(x) - f(x)| \leq 10^{-6}$ for $x = 1$, for $x = 2$, and for $x = 3$? For $k = 6$, 8, and 10, write a short program that calculates and lists $p_k(x)$, $\cos(x)$, and $\cos(x) - p_k(x)$ for $x$ varying between 0 and 3 in steps of 0.1.
3. Verify that the function $\|g\|_\infty$ defined on $C[a, b]$ by (5.1c) satisfies the following three conditions (where $f$ and $g$ are in $C[a, b]$):
   a) $\|g\|_\infty \geq 0$ and $\|g\|_\infty = 0$ if and only if $g(x) \equiv \theta(x)$
   b) $\|\alpha g\|_\infty = |\alpha|\,\|g\|_\infty$, for any scalar $\alpha$
   c) $\|f + g\|_\infty \leq \|f\|_\infty + \|g\|_\infty$.

   To do this problem, recall that a continuous function defined on $[a, b]$ attains its maximum at some point $x_0$ in $[a, b]$.
4. The *sine-integral*, Si($x$), occurs frequently in certain applied problems, where $x > 0$ and Si($x$) is defined by

$$\mathrm{Si}(x) = \int_0^x \frac{\sin(t)}{t}\,dt.$$

   One way to estimate Si($x$) is to take the truncated $k$th-degree Taylor's series expansion $p_k(t)$, for $f(t) = \sin(t)$ with $a = 0$, and use

$$\int_0^x \frac{p_k(t)}{t}\,dt$$

as an approximation to Si($x$). How large must $k$ be in order that

$$\left|\int_0^4 \frac{p_k(t)}{t}\,dt - \text{Si}(4)\right| \le 10^{-6}?$$

To bound the error, use the fact that if $|h(t)| \le |q(t)|$, for $0 \le t \le x$, then $\int_0^x |h(t)|\,dt \le \int_0^x |q(t)|\,dt$.

## 5.2 POLYNOMIAL INTERPOLATION

Perhaps the simplest and best known way to construct an $n$th-degree polynomial approximation, $p(x)$, to a function $f(x)$ in $C[a, b]$ is by interpolation. Let $x_0, x_1, \ldots, x_n$ be $(n + 1)$ distinct points in the interval $[a, b]$. Then $p(x) \in \mathscr{P}_n$ is said to *interpolate* $f(x)$ at these points if $p(x_j) = f(x_j)$ for $0 \le j \le n$. For example, the 2nd-degree polynomial $p(x) = -(4/\pi^2)x^2 + (4/\pi)x$ interpolates $f(x) = \sin(x)$ at the points $x_0 = 0$, $x_1 = \pi/2$ and $x_2 = \pi$. We must first show that such an interpolating polynomial always exists and is unique. In doing so we shall present two separate proofs since they both illustrate ways of constructing the interpolating polynomial.

**Theorem 5.3.** Let $\{x_j\}_{j=0}^n$ be $(n + 1)$ distinct points in the interval $[a, b]$ and let $\{y_j\}_{j=0}^n$ be any set of $(n + 1)$ real numbers. Then there exists a unique polynomial $p(x)$ in $\mathscr{P}_n$ such that $p(x_j) = y_j$ for $0 \le j \le n$. (Often the $y_j$'s are determined by some function, $f(x)$, so we have the property that $p(x_j) = f(x_j)$ for $0 \le j \le n$.)

*Proof* 1: For each $j$, $0 \le j \le n$, let $\ell_j(x)$ be the $n$th-degree polynomial defined by

$$\ell_j(x) \equiv \frac{(x - x_0)(x - x_1)\cdots(x - x_{j-1})(x - x_{j+1})\cdots(x - x_n)}{(x_j - x_0)(x_j - x_1)\cdots(x_j - x_{j-1})(x_j - x_{j+1})\cdots(x_j - x_n)}$$

$$\equiv \prod_{\substack{i=0 \\ i \ne j}}^{n} \frac{(x - x_i)}{(x_j - x_i)}. \tag{5.2}$$

Then it follows for each $i$ and $j$, that $\ell_i(x_j) = \delta_{ij}$. (The symbol $\delta_{ij}$ is defined to be 1 when $i = j$ and 0 otherwise. The symbol $\delta_{ij}$ is called the Kronecker delta.) Since the sum of $n$th-degree polynomials is again a polynomial of at most $n$th degree, the polynomial $p(x)$ defined by (5.3) is in $\mathscr{P}_n$:

$$p(x) \equiv \sum_{j=0}^{n} y_j \ell_j(x). \tag{5.3}$$

Moreover, for $0 \le i \le n$, we have

$$p(x_i) = y_0\ell_0(x_i) + \cdots + y_n\ell_n(x_i) = y_i\ell_i(x_i) = y_i.$$

If $f(x)$ is a function such that $f(x_i) = y_i$, $0 \le i \le n$, then it has an interpolating polynomial in $\mathscr{P}_n$ of the form

$$p(x) = \sum_{j=0}^{n} f(x_j)\ell_j(x). \tag{5.4}$$

To establish uniqueness, let us assume that there are two different polynomials, $p(x)$ and $q(x)$ in $\mathscr{P}_n$, such that $p(x_j) = q(x_j) = y_j$ for $0 \le j \le n$. If $r(x) \equiv p(x) - q(x)$, then $r(x) \in \mathscr{P}_n$ and furthermore $r(x_j) = p(x_j) - q(x_j) = 0$ for $0 \le j \le n$. By the Fundamental Theorem of Algebra, $r(x) \equiv 0$, so $p(x) \equiv q(x)$ which contradicts our assumption.

*Proof* 2: Let $p(x) = a_0 + a_1x + a_2x^2 + \cdots + a_nx^n$ where the coefficients, $a_j$, are to be determined. Consider the $(n + 1)$ equations

$$p(x_j) = a_0 + a_1x_j + a_2x_j^2 + \cdots + a_nx_j^n = y_j, \qquad 0 \le j \le n. \tag{5.5}$$

In matrix form, Eqs. (5.5) become

$$\begin{bmatrix} 1 & x_0 & x_0^2 & \cdots & x_0^n \\ 1 & x_1 & x_1^2 & \cdots & x_1^n \\ \vdots & & & & \\ 1 & x_n & x_n^2 & \cdots & x_n^n \end{bmatrix} \begin{bmatrix} a_0 \\ a_1 \\ \vdots \\ a_n \end{bmatrix} = \begin{bmatrix} y_0 \\ y_1 \\ \vdots \\ y_n \end{bmatrix} \tag{5.6}$$

or $V\mathbf{a} = \mathbf{y}$, where $V$ is the coefficient matrix in (5.6), $\mathbf{a} = [a_0, a_1, \ldots, a_n]^T$ and $\mathbf{y} = [y_0, y_1, \ldots, y_n]^T$. The matrix $V$ is called a *Vandermonde matrix*, and it is easy to see that $V$ is non-singular when $x_0, x_1, \ldots, x_n$ are distinct. To see this, recall from Section 2.1 that $V$ is non-singular if and only if $\boldsymbol{\theta}$ is the only solution of $V\mathbf{a} = \boldsymbol{\theta}$. So, if the vector $\mathbf{a}$ is any solution of $V\mathbf{a} = \boldsymbol{\theta}$, then $p(x) = a_0 + a_1x + \cdots + a_nx^n$ is an $n$th-degree polynomial such that $p(x_j) = 0$, for $j = 0, 1, \ldots, n$. Since the only $n$th-degree polynomial with $n + 1$ zeros is the zero polynomial, it must be $\mathbf{a} = \boldsymbol{\theta}$ and hence $V$ is non-singular. Thus, Eqs. (5.5) have a unique solution for the $a_i$'s and the polynomial $p(x)$ found by solving (5.6) is the unique interpolating polynomial in $\mathscr{P}_n$. ■

From the uniqueness of the interpolating polynomial in $\mathscr{P}_n$, both (5.4) and (5.6) must yield the same polynomial, even though it is written in different forms. The form given by (5.4) is called the *Lagrange form*, whereas the construction of $p(x)$ by (5.6) is called the *method of undetermined coefficients*. We should note three obvious facts here. First, if $f(x)$ is itself in $\mathscr{P}_n$, then $f(x) \equiv p(x)$ for all $x$ by the uniqueness property. Secondly, it is possible that the unique solution of (5.6) may yield $a_n = 0$ (and possibly other coefficients may be zero also), so $p(x)$ may be a polynomial of degree strictly less than $n$. Thirdly, if $m > n$, there are an infinite number of polynomials, $q(x)$, in $\mathscr{P}_m$ which satisfy $q(x_j) = y_j$, $0 \le j \le n$. (Note that we can arbitrarily specify another point, $x_{n+1}$, in $[a, b]$ and any other value $y_{n+1}$, and then construct $q(x) \in \mathscr{P}_{n+1}$ such that $q(x_j) = y_j$, $0 \le j \le n+1$. Thus the uniqueness only holds for $\mathscr{P}_n$.)

**Example 5.2.** As a simple example of a problem involving data fitting, suppose we want a 2nd-degree polynomial, $p(x)$, such that $p(0) = -1$, $p(1) = 2$, and $p(2) = 7$. Using the Lagrange form, we have

$$\ell_0(x) = \frac{(x-1)(x-2)}{2}, \qquad \ell_1(x) = -x(x-2), \qquad \text{and} \qquad \ell_2(x) = \frac{x(x-1)}{2}.$$

Thus $p(x)$ is given by the formula $p(x) = -\ell_0(x) + 2\ell_1(x) + 7\ell_2(x)$ or, upon simplification, $p(x) = x^2 + 2x - 1$. Alternatively, using the method of undetermined coefficients with $x_0 = 0$, $x_1 = 1$ and $x_2 = 2$ in (5.5), we obtain

$$\begin{aligned} a_0 \qquad\qquad\quad &= -1 \\ a_0 + a_1 + a_2 &= 2 \\ a_0 + 2a_1 + 4a_2 &= 7. \end{aligned}$$

Solving this system quickly gives $p(x) = x^2 + 2x - 1$.

The method of undetermined coefficients is a procedure which has wide application to other types of interpolation problems. For example, let $x_0 = 0$, $x_1 = \pi/6$, $x_2 = \pi/4$, and $x_3 = \pi/3$; with $y_0 = 0$, $y_1 = 2 - \sqrt{3}/2$, $y_2 = 1 + \sqrt{2}/2$, and $y_3 = 3/2$. Suppose we wish to find a *trigonometric* polynomial, $p(x)$, of the form $p(x) = C_0 + C_1 \cos(x) + C_2 \cos(2x) + C_3 \cos(3x)$ such that $p(x_j) = y_j$ for $0 \le j \le 3$. These four equations yield the matrix equation (as in the derivation of (5.6))

$$\begin{bmatrix} 1 & 1 & 1 & 1 \\ 1 & \sqrt{3}/2 & 1/2 & 0 \\ 1 & \sqrt{2}/2 & 0 & -\sqrt{2}/2 \\ 1 & 1/2 & -1/2 & -1 \end{bmatrix} \begin{bmatrix} C_0 \\ C_1 \\ C_2 \\ C_3 \end{bmatrix} = \begin{bmatrix} 0 \\ 2 - \sqrt{3}/2 \\ 1 + \sqrt{2}/2 \\ 3/2 \end{bmatrix}. \tag{5.7}$$

The solution of this system by Gauss elimination yields $C_0 = 1$, $C_1 = -1$, $C_2 = 2$, $C_3 = -2$. Therefore $p(x) = 1 - \cos(x) + 2\cos(2x) - 2\cos(3x)$ is a trigonometric polynomial satisfying $p(x_i) = y_i$, $i = 0, 1, 2, 3$.

### 5.2.1 Efficient Evaluation of the Interpolating Polynomial*

We have seen that either (5.4) or (5.6) may be used to construct $p(x) \in \mathscr{P}_n$ which interpolates $f(x)$ at the distinct points $\{x_j\}_{j=0}^n$. Since $p(x)$ is unique, we may construct $p(x)$ by using any technique whatsoever that generates a polynomial interpolating $f(x)$ at these points. In this section we shall introduce a technique due to Aitken (not to be confused with the Aitken's $\Delta^2$-method of Chapter 4), which is usually more efficient than either (5.4) or (5.6) when we wish to evaluate $p(x)$ at $x = \alpha$, where $\alpha$ is some fixed number. This technique has the additional important feature that it can easily be extended to yield the polynomial in $\mathscr{P}_{n+1}$

* The material in this section may be omitted without loss of continuity. However, the problems deal with the first two sections as well.

which interpolates $f(x)$ on the larger set $\{x_j\}_{j=0}^{n+1}$, where $x_{n+1}$ is some point adjoined to the original set in order to achieve greater accuracy in the interpolation. (The reader can readily see that if an additional interpolating point is used in the constructions of Eq. (5.4) or (5.6), then one must essentially start again from the beginning to obtain the new interpolating polynomial.)

The basis for this technique (also known as *iterative linear interpolation*) is Theorem 5.4 below. (This theorem is often given in a more generalized form such as Neville's method, which yields other ways of evaluating the interpolating polynomial. See Isaacson and Keller (1966) or Problem 7 below. However, these other techniques are comparable in efficiency to the Aitken technique, so we only present the theorem in this simple form.)

**Theorem 5.4. Aitken.** Let $f(x)$ be a given function and let $\{x_j\}_{j=0}^{n}$ be distinct points. Let $p_{j,k}(x)$ be in $\mathscr{P}_j$ and interpolate $f(x)$ at the $(j+1)$ points $\{x_0, x_1, \ldots, x_{j-1}, x_k\}$, where $0 \le j \le n$ and $j \le k \le n$ (if $j = 0$ we denote this set simply as the single point, $\{x_k\}$). Then for $j+1 \le k \le n$,

$$p_{j+1,k}(x) = \frac{(x - x_k)p_{j,j}(x) - (x - x_j)p_{j,k}(x)}{(x_j - x_k)}, \qquad 0 \le j \le n-1. \qquad \textbf{(5.8)}$$

*Proof*: We first pause to make sure our notation is clear. For example, for any fixed value of $k$, $p_{0,k}(x)$ is of degree zero, so, for all $x$, $p_{0,k}(x) \equiv f(x_k)$, a constant. The polynomial $p_{2,5}(x)$ is in $\mathscr{P}_2$ and interpolates $f(x)$ at $\{x_0, x_1, x_5\}$. Lastly, $p_{n,n}(x) \in \mathscr{P}_n$ and interpolates $f(x)$ at $\{x_0, x_1, \ldots, x_{n-1}, x_n\}$ (our ultimate goal is to calculate $p_{n,n}(x)$). Now we proceed with the proof. If $x = x_i$ for $0 \le i \le j-1$, then both $p_{j,j}(x)$ and $p_{j,k}(x)$ interpolate $f(x)$ at $x_i$, so from (5.8), $p_{j+1,k}(x_i) = [(x_i - x_k)f(x_i) - (x_i - x_j)f(x_i)]/(x_j - x_k) = f(x_i)$. If $x = x_j$, then $p_{j+1,k}(x_j) = [(x_j - x_k)f(x_j) - 0]/(x_j - x_k) = f(x_j)$, and, similarly, $p_{j+1,k}(x_k) = f(x_k)$. Obviously $p_{j+1,k}(x) \in \mathscr{P}_{j+1}$, and we have just shown that it interpolates $f(x)$ at $\{x_0, x_1, \ldots, x_j, x_k\}$. ■

Essentially what Theorem 5.4 says is that two $j$th-degree interpolating polynomials can be combined via (5.8) to form a $(j+1)$st-degree polynomial which interpolates $f(x)$ at the points $x_0, x_1, \ldots, x_j, x_k$. The basic idea is easy to extend to the general case where the two $j$th-degree polynomials have any $j$ interpolation points in common (see Problem 7). Aitken's technique of iterated interpolation can probably be best understood by considering Table 5.1. In Table 5.1, our objective is to determine $p_{n,n}(\alpha)$, that is, to evaluate the $n$th-degree interpolating polynomial for $f(x)$ at $x = \alpha$. This determination is made by setting $x = \alpha$ in (5.8).

For any fixed value of $\alpha$, Table 5.1 is generated one column at a time by (5.8) for $j = 1, 2, \ldots, n$, respectively, with $p_{0,k}(x) \equiv f(x_k)$, for the $(j = 0)$ column. For example, $p_{2,n}(\alpha) = [(\alpha - x_n)p_{1,1}(\alpha) - (\alpha - x_1)p_{1,n}(\alpha)]/(x_1 - x_n)$. If we are not satisfied with how well $p_{n,n}(\alpha)$ approximates $f(\alpha)$ (see Section 5.2.3 on error estimation), we can take an additional point of interpolation, $x_{n+1}$, and easily

**Table 5.1** Aitken's iterated interpolation.

| | $(j = 0)$ | $(j = 1)$ | $(j = 2)$ | $\cdots$ | $(j = n)$ |
|---|---|---|---|---|---|
| $x_0$ | $p_{0,0}(\alpha)$ | | | | |
| $x_1$ | $p_{0,1}(\alpha)$ | $p_{1,1}(\alpha)$ | | | |
| $x_2$ | $p_{0,2}(\alpha)$ | $p_{1,2}(\alpha)$ | $p_{2,2}(\alpha)$ | | |
| $\vdots$ | $\vdots$ | $\vdots$ | $\vdots$ | $\ddots$ | |
| $x_n$ | $p_{0,n}(\alpha)$ | $p_{1,n}(\alpha)$ | $p_{2,n}(\alpha)$ | $\cdots$ | $p_{n,n}(\alpha)$ |

append another row to Table 5.1. Note that $p_{0,n+1}(\alpha) = f(x_{n+1})$ and we use formula (5.8), along with the diagonal elements $p_{j,j}(\alpha)$, to generate $p_{j,n+1}(\alpha)$ for $1 \le j \le n + 1$. Thus $p_{n+1,n+1}(\alpha)$ is the value of the interpolating polynomial in $\mathscr{P}_{n+1}$ at $\{x_j\}_{j=0}^{n+1}$ for $f(x)$ at $x = \alpha$. Obviously this procedure can be repeated for the addition of other interpolating points. We also note that we only need to retain the values $p_{j,j}(\alpha)$, $0 \le j \le n$, in order to generate the new row, $p_{j,n+1}(\alpha)$, $0 \le j \le n + 1$. Thus, to conserve storage, the other values in the table may be discarded.

In many practical situations, we are not really interested in finding the interpolating polynomial explicitly, but we instead want various values that the interpolating polynomial assumes. This is where iterated interpolation is quite efficient. For example, the method of undetermined coefficients can be used to actually find the interpolating polynomial $p(x)$, but if we want to evaluate $p(x)$ at $x = \alpha$, further calculations are required. On the other hand, iterated interpolation gives the value $p(\alpha)$ without our actually having to find the formula for $p(x)$. Aitken's method of iterated interpolation is fairly simple to program, as subroutine AITINT in Fig. 5.2 demonstrates. For a number $\alpha$, this subroutine calculates

```
      SUBROUTINE AITINT(X,Y,ALPHA,PALPHA,N)
      DIMENSION X(20),Y(20),P(20),Q(20)
C
C   SUBROUTINE AITINT USES AITKEN'S ITERATED INTERPOLATION TO EVALUATE
C   THE INTERPOLATING POLYNOMIAL, P(X), AT A POINT X=ALPHA, WHERE
C   P(X(I))=Y(I), I=1,2,...,N.  THE CALLING PROGRAM MUST SUPPLY THE
C   ARRAYS X(I) AND Y(I), THE POINT ALPHA AND AN INTEGER N.  THE
C   SUBROUTINE RETURNS P(ALPHA) AS PALPHA.
C
      DO 1 K=1,N
    1 P(K)=Y(K)
      DO 3 J=2,N
      F=ALPHA-X(J-1)
      DO 2 K=J,N
    2 Q(K)=((ALPHA-X(K))*P(J-1)-F*P(K))/(X(J-1)-X(K))
      DO 3 K=J,N
    3 P(K)=Q(K)
      PALPHA=P(N)
      RETURN
      END
```

**Figure 5.2**

the value $p(\alpha)$, where $p(x)$ is the $(n - 1)$st-degree polynomial satisfying $p(x_i) = y_i$, $i = 1, 2, \ldots, n$.

**Example 5.3.** To illustrate the use of iterated interpolation, consider a problem of interpolating a function given in tabular form. Suppose $f(x) = \cos(x)$ is given (with $x$ in radians) at $x_0 = 0.3$, $x_1 = 0.4$, $x_2 = 0.5$, and $x_3 = 0.6$. If we want to estimate $\cos(\alpha)$ for $\alpha = 0.44$, we could directly find the 3rd-degree polynomial $p(x)$ that interpolates $\cos(x)$ at $\{x_0, x_1, x_2, x_3\}$ and then evaluate $p(x)$ at $x = 0.44$. But, since all we want is $p(\alpha)$, we can use iterated interpolation, as in Table 5.1, generating:

| | $(j = 0)$ | $(j = 1)$ | $(j = 2)$ | $(j = 3)$ |
|---|---|---|---|---|
| 0.3 | 0.955336 | | | |
| 0.4 | 0.921061 | 0.907351 | | |
| 0.5 | 0.877583 | 0.900909 | 0.904774 | |
| 0.6 | 0.825336 | 0.894669 | 0.904815 | 0.904749 |

To 6 places, $\cos(0.44) = 0.904752$ and our estimate is $p_{3,3}(0.44) = 0.904749$. In the table above, the $(j = 0)$ column lists the values $\cos(x_i)$ for $i = 0, 1, 2, 3$. Most readers have done first-degree (or linear) interpolation when they "interpolated" between values in a table of logarithms or trigonometric functions. Linear interpolation, between $x = 0.4$ and $x = 0.5$, yields $\cos(0.44) \approx \cos(0.4) + (0.4)(\cos(0.5) - \cos(0.4)) = 0.903670$ which is not nearly as accurate. For reasonably smooth functions, higher-order interpolation, as in this example, usually gives better results than linear interpolation.

## PROBLEMS

1. Given $x_0 = -1$, $x_1 = 0$, $x_2 = 2$, and $x_3 = 3$, let $p(x)$ be the cubic polynomial such that $p(x_i) = y_i$, where $y_0 = -1$, $y_1 = 3$, $y_2 = 11$, and $y_3 = 27$. Find $p(1)$, using the Lagrange form, the method of undetermined coefficients and iterated interpolation. (Note that to evaluate the Lagrange form at $x = 1$, you need not find the coefficients of $x^j$ in $p(x)$. Instead, use (5.4) directly, by evaluating $\ell_j(1)$, for $j = 0, 1, 2, 3$.)
2. Use the method of undetermined coefficients to find a quadratic polynomial, $p(x)$, such that $p(1) = 0$, $p'(1) = 7$, and $p(2) = 10$. (Note the derivative evaluation. You must modify (5.6) slightly.)
*3. Suppose we know $f(t)$ has the form $f(t) = ae^{2t} + be^{-t}$, where $a$ and $b$ are unknown. Use the method of undetermined coefficients to find $a$ and $b$, given that $f(t_0) = 1$ and $f(t_1) = 7.5$, where $t_0 = 0$ and $t_1 = \ln(2)$. Suppose next, that $f(t)$ is to be determined by some experimental observations, where we observe $f(t_0) = 1$, $f(t_1) = 7$, and $f(t_2) = 9$, with $t_2 = \ln(3)$. Since we have three data points, and only two parameters ($a$ and $b$), we do not expect to be able to find $a$ and $b$ by the method of undetermined coefficients. Use the procedure of Section 2.5 to find the best least-squares solution for $a$ and $b$.
4. a) Using a given value of $n$, $n \le 20$, interpolating points $\{x_j\}_{j=1}^{n}$, and interpolation values $\{y_j\}_{j=1}^{n}$ as input, write a computer program which uses subroutine AITINT

to evaluate the interpolating polynomial, $p(x)$. Test your program on the data $y_i = \ln(1 + x_i)$, where $x_i = (i - 1)/10$, $i = 1, 2, \ldots, 11$. Print $\ln(1 + \alpha_i)$ and $p(\alpha_i)$ for $\alpha_i = (i - 1)/100$, $1 \leq i \leq 100$.

b) Although Aitken's method is primarily used to evaluate $p(\alpha)$, one can easily use (5.8) to also generate the coefficients of $x^j$, $0 \leq j \leq n$, for $p(x)$. Modify or rewrite subroutine AITINT to accomplish this task. Check your program using the data in Problem 1.

(Hint: Knowing the coefficients of all of the polynomials in the $j$th column enables us via (5.8) to find the coefficients of all of the polynomials in the $(j + 1)$st column, $0 \leq j \leq n - 1$.)

*5. Suppose $p(x)$ is the $n$th-degree interpolating polynomial such that $p(x_i) = y_i$, $i = 0, 1, \ldots, n$. If it is necessary to evaluate $p(x)$ at many different points $\alpha$, it may be that Aitken's method is less efficient than using (5.4) or (5.6) to find a formula for $p(x)$ and then using the formula to find each $p(\alpha)$. Thus we wish to compare the relative computational efficiency of the three methods in this setting. This suggests an operations count for each of the three procedures. To illustrate the ideas, we count only multiplications (where, again, we consider division as a multiplication).

a) Let $W(x) = (x - x_0)(x - x_1) \cdots (x - x_n) = \prod_{i=0}^{n} (x - x_i)$. Show that an alternative expression for $\ell_j(x)$ in (5.2) is

$$\ell_j(x) = \frac{W(x)}{(x - x_j)W'(x_j)}.$$

Thus for a fixed number $\alpha$, $p(\alpha) = W(\alpha) \sum_{i=0}^{n} \beta_i/(\alpha - x_i)$ where $\beta_i = y_i/W'(x_i)$. The numbers $\beta_i$ can be computed and stored, since they do not depend on $\alpha$.

b) Using (a), show that $n(n + 1)$ multiplications are required to find all the numbers $\beta_0, \beta_1, \ldots, \beta_n$. Thus, show that $n(n + 1) + 2(n + 1)$ multiplications are needed to find $p(\alpha)$. If we want to evaluate $p(x)$ at $k$ different points $\alpha_1, \alpha_2, \ldots, \alpha_k$, show that $(n + 1)(n + 2k)$ multiplications are required.

c) Show that $3(n + 1)n/2$ multiplications are needed to find $p(\alpha)$ by Aitken's iterated interpolation. Consequently, $3k(n + 1)n/2$ multiplications are needed to evaluate $p(x)$ at $\alpha_1, \alpha_2, \ldots, \alpha_k$.

d) Show that $(n - 1)(n + 1)$ multiplications are needed to form the Vandermonde matrix in (5.6). Using the operations counts of Gauss elimination in Chapter 2 and nested multiplication (synthetic division) in Chapter 4, show that $(n - 1)(n + 1) + (n + 1)n(2n + 7)/6 + (n + 1)(n + 2)/2 + kn$ multiplications are needed to evaluate $p(x)$ at $\alpha_1, \alpha_2, \ldots, \alpha_k$ with the method of undetermined coefficients.

e) For $n = 5$ and $n = 10$, determine the values $k$ for which a particular method is most efficient (in terms of fewest multiplications).

*6. Referring to Problem 4b, show that $n(n + 1)(n + 3)/2$ multiplications are necessary to find the coefficients of an $n$th-degree interpolating polynomial, using iterated interpolation as in (5.8). Thus $n[(n + 1)(n + 3)/2 + k]$ multiplications are required to find $p(\alpha_i)$ for $i = 1, 2, \ldots, k$. How does this compare with the results of Problem 5?

7. Develop a more inclusive iterated interpolation formula than that given in (5.8). In particular, suppose $P$ and $Q$ are both sets of $(k + 1)$ points, where $P \cap Q$ has $k$ points. Suppose $p(x) \in \mathscr{P}_k$ interpolates $f(x)$ at each of the points of $P$, and $q(x) \in \mathscr{P}_k$ interpolates

$f(x)$ at each of the points of $Q$. The proof of Theorem 5.4 shows how to find $r(x) \in \mathscr{P}_{k+1}$ such that $r(x)$ interpolates $f(x)$ at each of the $(k + 2)$ points of $P \cup Q$. You should obtain a formula similar to (5.8).

### 5.2.2 Interpolation at Equally Spaced Points

In applications where functions are given in tabular form, the tabular points are frequently equally spaced (as, for example, in tables of experimental data or in trigonometric and logarithmic tables). For these problems of interpolating functions or data at equally spaced points, several efficient procedures are available. To examine some of these procedures, let us suppose that the interpolating points $x_0, x_1, \ldots, x_n$ are equally spaced in $[a, b]$, with $x_0 = a$ and $x_n = b$. Then we have

$$x_i = x_0 + ih; \qquad h = (b - a)/n; \qquad i = 0, 1, \ldots, n \tag{5.9}$$

(or equivalently, $x_{j+1} - x_j = h$ for $j = 0, 1, \ldots, n - 1$).

If $x$ is *any* point in $[a, b]$, then $x$ can be written as

$$x = x_0 + rh \tag{5.10}$$

where $r$ is a number (not necessarily an integer) between 0 and $n$. From (5.3), we recall that the Lagrange form of the interpolating polynomial is given by $p(x) = \sum_{j=0}^{n} \ell_j(x)y_j$, where $\ell_j(x)$ is given by (5.2). Since $x_i = x_0 + ih$ and $x = x_0 + rh$, we can write $\ell_j(x)$ in a reduced form as

$$\ell_j(x) = \prod_{\substack{i=0 \\ i \neq j}}^{n} \frac{(r - i)}{(j - i)} \equiv L_j(r), \qquad 0 \leq j \leq n. \tag{5.11}$$

Note that (5.11) defines a function of $r$, $L_j(r)$, where $L_j(r)$ is a result of the simplification of (5.2) provided by equally spaced points. With this definition of $L_j(r)$, the Lagrange form of the interpolating polynomial becomes

$$p(x) = p(x_0 + rh) = \sum_{j=0}^{n} L_j(r)y_j. \tag{5.12}$$

The important feature of (5.12) is that the polynomials $L_j(r)$ do not depend on $[a, b]$, the interpolating points $x_0, x_1, \ldots, x_n$, or even the "step-size" $h$. Thus, for a given $n$, the $n$th-degree polynomials $L_0(r), L_1(r), \ldots, L_n(r)$ can be calculated once and for all, and used in any instance where interpolation at $(n + 1)$ equally spaced points is desired. There are extensive listings of these polynomials $L_j(r)$, as for example, those tables prepared by the National Bureau of Standards.

**Example 5.4.** For $n = 3$, we obtain $L_0(r)$, $L_1(r)$, $L_2(r)$, and $L_3(r)$ by (5.11):

$$L_0(r) = -\frac{(r-1)(r-2)(r-3)}{6} \qquad L_1(r) = \frac{r(r-2)(r-3)}{2}$$

$$L_2(r) = -\frac{r(r-1)(r-3)}{2} \qquad L_3(r) = \frac{r(r-1)(r-2)}{6}.$$

Recall from Example 5.3, that we were interested in approximating cos(0.44) given $\cos(x_i)$ for $x_0 = 0.3$, $x_1 = 0.4$, $x_2 = 0.5$, and $x_3 = 0.6$. For these equally spaced points, the step-size $h$ is 0.1 and $r = 1.4$ (since $0.44 = 0.3 + 1.4h$). Thus if $p(x)$ is the cubic polynomial interpolating $\cos(x)$ at these four points, then we can write $p(x)$ in the form of (5.12) and find

$$p(0.44) = \sum_{j=0}^{3} L_j(1.4)\cos(x_j). \tag{5.13}$$

As a completely different approach, there are a number of interpolation formulas for equally spaced points based on the idea of *differences*. Differences were hinted at in the development of Aitken's $\Delta^2$-process in Section 4.3.2 (see also Problem 2 of that section). Since the theory of differences plays an important role in numerical solutions of differential equations, as well as in interpolation and numerical integration, we will outline some of the basic ideas here. For any function $f(x)$ and for a fixed step-size $h$, we define the *forward difference operator* $\Delta$ by

$$\Delta f(x) = f(x + h) - f(x). \tag{5.14}$$

Thus, for example, if $f(x) = \cos(x)$ and $h = 0.1$, then $\Delta f(x)$ is the function: $\Delta f(x) = \cos(x + 0.1) - \cos(x)$. Higher-order forward differences are defined recursively by

$$\Delta^k f(x) \equiv \Delta(\Delta^{k-1} f(x)), \qquad k = 1, 2, \ldots \tag{5.15}$$

where we set $\Delta^0 f(x) \equiv f(x)$. For example,

$$\begin{aligned}\Delta^2 f(x) &= \Delta(\Delta f(x)) = \Delta(f(x + h) - f(x)) \\ &= (f(x + 2h) - f(x + h)) - (f(x + h) - f(x)) \\ &= f(x + 2h) - 2f(x + h) + f(x).\end{aligned} \tag{5.16}$$

Similarly, we can easily verify that

$$\Delta^3 f(x) = f(x + 3h) - 3f(x + 2h) + 3f(x + h) - f(x).$$

A pattern soon emerges for higher-order forward differences and we can easily prove that

$$\Delta^k f(x) = \sum_{i=0}^{k} \binom{k}{i} (-1)^i f(x + (k - i)h) \tag{5.17}$$

where $\binom{k}{i}$ is the usual symbol for binomial coefficients,

$$\binom{k}{i} = \frac{k!}{i!(k - i)!}$$

and, where $\binom{k}{0} \equiv \binom{k}{k} \equiv 1$. We establish the identity in (5.17) in order to demonstrate some of the manipulations that are useful in the theory of differences. Proceeding by induction, the identity (5.17) is clearly correct for $k = 1$. Supposing (5.17) is correct for an arbitrary $k$, consider

$$\begin{aligned}\Delta^{k+1}f(x) &= \Delta(\Delta^k f(x))\\ &= \sum_{i=0}^{k}\binom{k}{i}(-1)^i\,\Delta f(x+(k-i)h)\\ &= \sum_{i=0}^{k}\binom{k}{i}(-1)^i[f(x+(k-i+1)h)-f(x+(k-i)h)].\end{aligned}$$

Splitting this sum into two parts and changing the limits on the second sum, we have

$$\begin{aligned}\Delta^{k+1}f(x) &= \sum_{i=0}^{k}\binom{k}{i}(-1)^i f(x+(k-i+1)h)\\ &\quad+\sum_{i=1}^{k+1}\binom{k}{i-1}(-1)^i f(x+(k-i+1)h)\\ &= \binom{k}{0}f(x+(k+1)h)+\sum_{i=1}^{k}\left[\binom{k}{i}+\binom{k}{i-1}\right](-1)^i f(x+(k-i+1)h)\\ &\quad+\binom{k}{k}(-1)^{k+1}f(x).\end{aligned}$$

Since $\binom{k}{0}=\binom{k}{k}=\binom{k+1}{0}=\binom{k+1}{k+1}=1$ and since $\binom{k}{i}+\binom{k}{i-1}=\binom{k+1}{i}$ (see Problem 2), we have shown the validity of formula (5.17).

Continuing with differences, we note that $\Delta f(x)=f(x+h)-f(x)$ is the same as

$$f(x+h)=f(x)+\Delta f(x).$$

Similarly, since $\Delta^2 f(x)=f(x+2h)-2f(x+h)+f(x)$, we see that $f(x+2h)=2f(x+h)-f(x)+\Delta^2 f(x)$. Since $f(x+h)=f(x)+\Delta f(x)$ from above, we have

$$f(x+2h)=f(x)+2\Delta f(x)+\Delta^2 f(x).$$

These observations suggest that we should be able to write $f(x+kh)$ in terms of $f(x), \Delta f(x), \ldots, \Delta^k f(x)$. This can be done and we leave it to the reader (Problem 3) to establish the identity

$$f(x+kh)=\sum_{i=0}^{k}\binom{k}{i}\Delta^i f(x). \tag{5.18}$$

Experimenting with differences soon leads us to realize that if $p(x)$ is a polynomial of degree $n$, then $\Delta p(x)$ is a polynomial of degree $n-1$. That is, differencing behaves like differentiation in the sense of reducing the degree of a polynomial.

For example,

$$\Delta(x^2) = (x + h)^2 - x^2 = 2xh + h^2.$$

Similarly,

$$\Delta^2(x^2) = \Delta(\Delta(x^2)) = (2(x + h)h + h^2) - (2xh + h^2) = 2h^2$$

and finally,

$$\Delta^3(x^2) \equiv 0.$$

We leave as a problem for the reader to verify that $\Delta^m p(x) \equiv 0$ whenever $p(x)$ is a polynomial of degree $n$ and $m \geq n + 1$. This observation, together with the identity (5.18), means that when $p(x)$ is a polynomial of degree $n$ and when $k \geq n + 1$, then (5.18) becomes

$$p(x + kh) = \sum_{i=0}^{n} \binom{k}{i} \Delta^i p(x). \tag{5.19}$$

Thus, for any fixed $x$, the value of the polynomial at any of the points $x + kh$, for $k = n + 1, n + 2, \ldots$, is completely determined by $p(x), \Delta p(x), \ldots, \Delta^n p(x)$ and the appropriate binomial coefficients. This agrees with our intuition, since if we know $p(x)$ at $n + 1$ points, we should be able to evaluate $p(x)$ at any point.

In particular, set $x = x_0$ in (5.19). Then since $x_0 + kh = x_k$, we have

$$p(x_0 + kh) = p(x_k) = \sum_{i=0}^{n} \binom{k}{i} \Delta^i p(x_0), \qquad k = n + 1, \quad n + 2, \ldots. \tag{5.20}$$

The next question that comes to mind is how can we use an identity like (5.20) to evaluate $p(x)$ for an arbitrary point $x = x_0 + rh$, where $r$ is not necessarily an integer. Since

$$\binom{k}{i} = \frac{k!}{i!(k - i)!} = \frac{k(k - 1) \cdots (k - i + 1)}{i!},$$

we are led to try the definition

$$\binom{r}{i} \equiv \frac{r(r - 1) \cdots (r - i + 1)}{i!}, \qquad i \geq 1, \qquad \binom{r}{0} \equiv 1. \tag{5.21}$$

Thus

$$\binom{r}{1} = r, \quad \binom{r}{2} = \frac{r(r - 1)}{2},$$

etc. Next, we rewrite (5.20) in a formal fashion, defining $Q(x)$ by

$$Q(x) = Q(x_0 + rh) = \sum_{i=0}^{n} \binom{r}{i} \Delta^i p(x_0). \tag{5.22}$$

The new function $Q(x)$ is introduced to emphasize that (5.22) was arrived at in a completely formal way by substituting $\binom{r}{i}$ for $\binom{k}{i}$. We have yet to show that the function $Q(x)$, defined by the summation in (5.22), is, in fact, the polynomial $p(x)$. To show that $Q(x)$ is $p(x)$, we note first that $r = (x - x_0)/h$, so it follows from (5.21) that $\binom{r}{i}$ is a polynomial (in $x$) of degree $i$. Thus, $Q(x)$ is indeed a polynomial of degree $n$. All that remains to be shown is that $Q(x)$ is precisely the $n$th-degree polynomial $p(x)$. We do this by showing that $Q(x_i) = p(x_i)$ for $i = 0, 1, \ldots, n$. For $x = x_j$, we have $r = j$ for $j = 0, 1, \ldots, n$. Using Eq. (5.21), we note that $\binom{r}{i} = 0$ for $r = 0, 1, \ldots, i - 1$ (or, equivalently, $\binom{j}{i} = 0$ when $i > j$). Thus, for $0 \leq j \leq n$, (5.22) reduces to

$$Q(x_j) = Q(x_0 + jh) = \sum_{i=0}^{j} \binom{j}{i} \Delta^i p(x_0).$$

From (5.18), we have that $Q(x_j) = p(x_j)$ for $j = 0, 1, \ldots, n$ and therefore $Q(x) \equiv p(x)$.

Since any $n$th-degree polynomial $p(x)$ can be written in the form of the summation in (5.22), the connection between this form and the problem of interpolation is now easy to make. Suppose $p(x)$ interpolates $f(x)$ at the $(n + 1)$ equally spaced points $x_0, x_1, \ldots, x_n$. Since $p(x_j) = f(x_j)$, $0 \leq j \leq n$, we see from (5.17) that $\Delta^i f(x_0) = \Delta^i p(x_0)$, for $i = 0, 1, \ldots, n$. Consequently, since we know $Q(x) = p(x)$ in (5.22), the $n$th-degree interpolating polynomial for $f(x)$ can be written as

$$p(x) = \sum_{i=0}^{n} \binom{r}{i} \Delta^i f(x_0) \tag{5.23}$$

where $r = (x - x_0)/h$ and where the numbers $\Delta^i f(x_0)$ can be calculated from the known values of $f(x)$ at $x_0, x_1, \ldots, x_n$. The expression (5.23) is known as *Newton's forward formula* for the interpolating polynomial. Newton's forward formula has the important feature that more interpolation points can easily be added. If we wish to interpolate $f(x)$ at $x_0, x_1, \ldots, x_n, x_{n+1}$ by a polynomial $q(x)$ of degree $n + 1$, then clearly $q(x) = p(x) + \binom{r}{n+1} \Delta^{n+1} f(x_0)$. Hence, the work done in finding $p(x)$ is not lost.

There are efficient ways to evaluate the Newton forward formula, given the numbers $\Delta^i f(x_0)$ (see Problem 6). In addition, the required differences, $\Delta^i f(x_0)$, are easily generated from a table (see Table 5.2).

Note that the table of forward differences is generated one column at a time, from left to right. For example, $\Delta f(x_1) = f(x_2) - f(x_1)$ and $\Delta^2 f(x_0) = \Delta f(x_1) - \Delta f(x_0)$. Note also, that in order to compute $\Delta^{n+1} f(x_0)$, we need not start over. Instead, we merely compute one more entry in each column, that is, $x_{n+1}$, $f(x_{n+1})$, $\Delta f(x_n)$, $\Delta^2 f(x_{n-1}), \ldots, \Delta^n f(x_1)$ and $\Delta^{n+1} f(x_0)$. The subroutine DIFINT shown in Fig. 5.3 implements the ideas outlined above. This subroutine first calculates the forward difference table, as in Table 5.2, for the interpolation points $x_1, x_2, \ldots, x_n$. When these differences are obtained, the interpolating polynomial is evaluated

**Table 5.2** The forward difference table for $f(x_0), f(x_1), \ldots, f(x_n)$.

| $x$ | $f(x)$ | $\Delta f(x)$ | $\Delta^2 f(x)$ | $\Delta^3 f(x)$ | $\cdots$ | $\Delta^n f(x)$ |
|---|---|---|---|---|---|---|
| $x_0$ | $f(x_0)$ | | | | | |
| | | $\Delta f(x_0)$ | | | | |
| $x_1$ | $f(x_1)$ | | $\Delta^2 f(x_0)$ | | | |
| | | $\Delta f(x_1)$ | | $\Delta^3 f(x_0)$ | | |
| $x_2$ | $f(x_2)$ | | $\Delta^2 f(x_1)$ | | $\ddots$ | |
| | | $\Delta f(x_2)$ | | | | |
| $x_3$ | $f(x_3)$ | | | | | $\Delta^n f(x_0)$ |
| $\vdots$ | $\vdots$ | | | | | |
| $x_{n-1}$ | $f(x_{n-1})$ | | | $\cdot^{\cdot^{\cdot}}$ | | |
| | | $\Delta f(x_{n-1})$ | | | | |
| $x_n$ | $f(x_n)$ | | | | | |

```
      SUBROUTINE DIFINT(X,Y,R,PXPRH,N)
      DIMENSION X(20),Y(20),P(20),Q(20)
C
C   SUBROUTINE DIFINT CONSTRUCTS AND EVALUATES THE NEWTON FORWARD FORM OF
C   THE INTERPOLATING POLYNOMIAL, P(X), AT X=X(1)+R*H, WHERE P(X(I))=Y(I)
C   AND WHERE X(I)=X(1)+(I-1)*H, I=1,2,...,N.  THE CALLING PROGRAM MUST
C   SUPPLY THE ARRAYS X(I) AND Y(I), THE NUMBER R AND THE INTEGER N.
C   THE SUBROUTINE RETURNS P(X(1)+R*H) AS PXPRH.
C
      DO 1 I=1,N
    1 P(I)=Y(I)
C
C   SET UP THE FORWARD DIFFERENCE TABLE
C
      DO 3 J=2,N
      DO 2 K=J,N
    2 Q(K)=P(K)-P(K-1)
      DO 3 K=J,N
    3 P(K)=Q(K)
C
C   EVALUATE THE FORWARD FORM OF THE INTERPOLATING
C   POLYNOMIAL BY NESTED MULTIPLICATION
C
      C=P(N)
      DO 4 J=2,N
      I=N-J
    4 C=P(I+1)+(R-I)*C/(I+1)
      PXPRH=C
      RETURN
      END
```

**Figure 5.3**

at $x = x_1 + rh$ and the nested multiplication algorithm given in Problem 6 is used. For ease of programming, the range on the subscript $i$ is $1 \le i \le n$ instead of $0 \le i \le n$.

**Example 5.5.** We return to the function $f(x) = \cos(x)$, as in Examples 5.3 and 5.4. Following Table 5.2, we construct the table

| $x$ | $f(x)$ | $\Delta f(x)$ | $\Delta^2 f(x)$ | $\Delta^3 f(x)$ |
|---|---|---|---|---|
| 0.3 | 0.955336 | | | |
| | | −0.034275 | | |
| 0.4 | 0.921061 | | −0.009203 | |
| | | −0.043478 | | 0.000434 |
| 0.5 | 0.877583 | | −0.008769 | |
| | | −0.052247 | | |
| 0.6 | 0.825336 | | | |

Therefore $p(x) = 0.955336 - 0.034275\binom{r}{1} - 0.009203\binom{r}{2} + 0.000434\binom{r}{3}$. To find $p(0.44)$, since $0.44 = 0.3 + (1.4)h$, we set $r = 1.4$ in the above expression, obtaining $p(0.44) = 0.904750$. The small discrepancy between this result and that of Example 5.3 is due to round-off error.

In many applications, such as those involving numerical solutions of differential equations, we are interested in interpolating at points labeled $x_m, x_{m+1}, \ldots, x_{m+n}$ instead of at points labeled $x_0, x_1, \ldots, x_n$. While we could rename the points $x_m, x_{m+1}, \ldots, x_{m+n}$ and then use (5.23) to find the interpolating polynomial this is both inconvenient and unnecessary. Instead, we can use the formula

$$p(x) = p(x_m + rh) = \sum_{j=0}^{n} \binom{r}{j} \Delta^j f(x_m) \tag{5.24}$$

where now $r = (x - x_m)/h$. Clearly, (5.24) is a valid formula for the interpolating polynomial, since it can be gotten from (5.23) by relabeling points (for example, set $z_0 = x_m$, $z_j = z_0 + jh$ and (5.24) reduces directly to (5.23)).

A somewhat different labeling problem occurs when, as often happens, we wish to interpolate at the points $x_m, x_{m-1}, \ldots, x_{m-n}$. This is a problem of "backwards interpolation," where we again want a formula in which we can add an additional point $x_{m-n-1}$ without losing the effort we expended to interpolate at $x_m, x_{m-1}, \ldots, x_{m-n}$. The formula needed to do this is known as *Newton's backward formula* for the interpolating polynomial, and is given by

$$p(x) = p(x_m - rh) = \sum_{j=0}^{n} \binom{r}{j} (-1)^j \Delta^j f(x_{m-j}). \tag{5.25}$$

Notice that each difference $\Delta^j f(x_{m-j})$ used in (5.25) can be obtained from a difference table like that displayed in Table 5.2. In fact, when $m = n$, each difference $\Delta^j f(x_{n-j})$ is found as the *last entry* of the $j$th column of Table 5.2.

To see how formula (5.25) is derived, let us rename the points $\{x_m, x_{m-1}, \ldots, x_{m-n}\}$ as $\{t_m, t_{m+1}, \ldots, t_{m+n}\}$. Then $t_{m+i} = t_m + ih'$, where $h' = -h$ and where, therefore, $t_{m+i} = x_{m-i}$, $i = 0, 1, \ldots, n$. Using (5.24), with $x_m = t_m$, we obtain

$$p(x) = p(t_m + rh') = \sum_{j=0}^{n} \binom{r}{j} \Delta^j f(t_m). \qquad \textbf{(5.26)}$$

To translate $\Delta^j f(t_m)$ in terms of our original labeling, note by (5.17)

$$\begin{aligned} \Delta^j f(t_m) &= \sum_{i=0}^{j} \binom{j}{i} (-1)^i f(t_{m+j-i}) \\ &= \sum_{i=0}^{j} \binom{j}{i} (-1)^i f(x_{m+i-j}). \end{aligned}$$

Next, we change the index of the summation by setting $k = j - i$, and observe that $\binom{j}{i} = \binom{j}{j-i}$ and that $(-1)^i = (-1)^j(-1)^k$. With these changes, we find

$$\Delta^j f(t_m) = (-1)^j \sum_{k=0}^{j} \binom{j}{k} (-1)^k f(x_{m-k}) = (-1)^j \, \Delta^j f(x_{m-j})$$

where the last equality comes from using (5.17) again. This identity, when inserted into (5.26), gives the Newton backward formula of (5.25).

**Example 5.6.** As an illustration of how the backwards and forwards formulas can be set up, we consider interpolating $f(x) = e^{3x}$ at five points, with $x_0 = -1$ and $h = 0.5$. As in Table 5.2, we obtain the following difference table:

| $x$ | $f(x)$ | $\Delta f(x)$ | $\Delta^2 f(x)$ | $\Delta^3 f(x)$ | $\Delta^4 f(x)$ |
|---|---|---|---|---|---|
| −1. | 0.049787 | | | | |
| | | 0.173343 | | | |
| −0.5 | 0.223130 | | 0.603527 | | |
| | | 0.776870 | | 2.10129 | |
| 0. | 1. | | 2.70482 | | 7.31599 |
| | | 3.48169 | | 9.41728 | |
| 0.5 | 4.48169 | | 12.1221 | | |
| | | 15.6038 | | | |
| 1. | 20.0855 | | | | |

Using the backwards formula (5.25), with $m = 4$ and $n = 2$, the 2nd-degree polynomial interpolating $f(x)$ at $x_4$, $x_3$, and $x_2$ is

$$p(x) = p(x_4 - rh) = f(x_4) - \binom{r}{1} \Delta f(x_3) + \binom{r}{2} \Delta^2 f(x_2).$$

For example, for $x = 0.8$, we have $r = 0.4$ and $p(0.8) = 20.0855 - (0.4)(15.6038) + (-0.12)(12.1221) = 12.3893$. The forward formula (5.24), using the same three points, has $m = 2$ and $n = 2$, giving

$$p(x) = p(x_2 + sh) = f(x_2) + \binom{s}{1} \Delta f(x_2) + \binom{s}{2} \Delta^2 f(x_2).$$

In this case, for $x = 0.8$ we have $s = 1.6$ and $p(0.8) = 1.0 + (1.6)(3.48169) + (0.48)(12.1221) = 12.3893$.

To include the tabulated information at $x_1$ and $x_0$ in our approximation to $f(0.8)$, we merely have to add the appropriate terms to the backwards form of the interpolating polynomial. Thus the 3rd-degree polynomial $p(x_4 - rh) - \binom{r}{3} \Delta^3 f(x_1)$ and the 4th-degree polynomial $p(x_4 - rh) - \binom{r}{3} \Delta^3 f(x_1) + \binom{r}{4} \Delta^4 f(x_0)$ interpolate $f(x)$ at $\{x_4, x_3, x_2, x_1\}$ and $\{x_4, x_3, x_2, x_1, x_0\}$, respectively. The respective estimates for $f(0.8)$ are $12.3893 - (0.064)(9.41728) = 11.7866$ and $11.7866 + (-0.0416)(7.31599) = 11.4823$, whereas the correct value of $f(0.8)$ is 11.0232.

## PROBLEMS

1. For $n = 3$, find the 2nd-degree polynomials $L_0(r)$, $L_1(r)$ and $L_2(r)$ defined by Eq. (5.11). Use these, as in (5.12), to construct the 2nd-degree interpolating polynomial to $f(x) = e^{3x}$ at 0, 0.5, and 1. Evaluate this form of the interpolating polynomial for $x = 0.8$, as in Example 5.6.
2. Verify the identities $\binom{k}{i} + \binom{k}{i-1} = \binom{k+1}{i}$ and $\binom{k}{i} = \binom{k}{k-i}$.
3. Establish formula (5.18). (Hint: By Eq. (5.18), $f(x + (k + 1)h) - f(x + kh) = \Delta f(x + kh) = \sum_{i=0}^{k} \binom{k}{i} \Delta^{i+1} f(x)$, so induction is indicated.)
4. Show that $\Delta^m p(x) \equiv 0$ when $p(x) \in \mathscr{P}_n$ and $m \geq n + 1$. (Hint: If $p(x) \in \mathscr{P}_n$, show that $\Delta p(x)$ is a polynomial in $\mathscr{P}_{n-1}$, thus concluding that $\Delta^{n-1} p(x)$ is a linear polynomial, $\Delta^n p(x)$ is a constant, and $\Delta^{n+1} p(x) \equiv 0$.)
5. Continue Example 5.6, by adding the interpolation point $x_5 = 1.5$ and forming $\Delta^1 f(x_4)$, $\Delta^2 f(x_3), \ldots, \Delta^5 f(x_0)$. Knowing the value of the 4th-degree interpolating polynomial at $x = 0.8$ (see Example 5.6), use the forward form to evaluate the 5th-degree polynomial interpolating $f(x)$ at $x_0, x_1, \ldots, x_5$ for $x = 0.8$. Next add the point $x_{-1} = -1.5$ and use the backwards form to evaluate (at $x = 0.8$) the 6th-degree polynomial interpolating $f(x)$ at $x_{-1}, x_0, \ldots, x_5$. Compare all these results with the estimates provided by the truncated Taylor's series of degrees 4, 5, and 6.
6. For $n = 1$, 2, and 3, verify that the following version of nested multiplication can be used to evaluate Newton's forward formula for the interpolating polynomial (5.23): Given any number $r$, form the numbers $C_n, C_{n-1}, \ldots, C_1$ and $C_0$ by

$$C_n = \Delta^n f(x_0)$$
$$C_i = \Delta^i f(x_0) + (r - i)C_{i+1}/(i + 1), \qquad i = n - 1, n - 2, \ldots, 1, 0.$$

Then $C_0 = p(x_0 + rh)$. Use this iteration to evaluate $p(0.44)$ in Example 5.5. (Note: This form of nested multiplication is valid for all $n$, but, in general, verification is a somewhat difficult exercise.)

7. Write a program that calculates the difference table (Table 5.2) and then evaluates Newton's backward form of the interpolating polynomial. Test your program on $f(x) = \ln(x)$, where the interpolation points are $x_i = i/10$, $i = 1, 2, \ldots, 10$. Print out $\ln(x)$, the value of the interpolating polynomial at $x$ and the *relative* error for $x = j/100$, $j = 1, 2, \ldots, 100$.

### 5.2.3 Error of Polynomial Interpolation

We know that given a function, $f(x)$, and $(n + 1)$ distinct points $\{x_j\}_{j=0}^{n}$ in $[a, b]$, we can construct a polynomial, $p(x)$, in $\mathscr{P}_n$ which passes through the points $(x_j, f(x_j))$, $0 \le j \le n$. The question still remains as to how well $p(x)$ approximates $f(x)$ for other values of $x$ in $[a, b]$. (See Fig. 5.4.) For example, as in Fig. 5.4, $f(x)$ may be very ill-behaved at one or more points, $\alpha$, and $p(x)$ may not be a good approximation to $f(x)$ in a neighborhood of $\alpha$. Thus if $e(x) \equiv f(x) - p(x)$ is defined to be the error function, a natural question is: How large can $e(x)$ be for any $x$ in $[a, b]$? If $f^{(n+1)}(x)$ is continuous on $[a, b]$, then the following theorem provides an answer to this question. Before presenting this theorem, it is convenient to give some notation: We let $C^{n+1}[a, b]$ denote the set of functions defined on $[a, b]$ which have a continuous $(n + 1)$st derivative on $[a, b]$ and use $\mathrm{Spr}\{y_1, y_2, \ldots, y_n\}$ to denote the smallest interval containing $y_1, y_2, \ldots, y_n$ (thus $\mathrm{Spr}\{1, -3, 2\} = [-3, 2]$).

**Theorem 5.5.** If $p(x) \in \mathscr{P}_n$ interpolates $f(x) \in C^{n+1}[a, b]$ at $(n + 1)$ distinct points $\{x_j\}_{j=0}^{n}$ in $[a, b]$, then

$$e(x) = f(x) - p(x) = \frac{f^{(n+1)}(\xi)}{(n + 1)!} W(x) \tag{5.27}$$

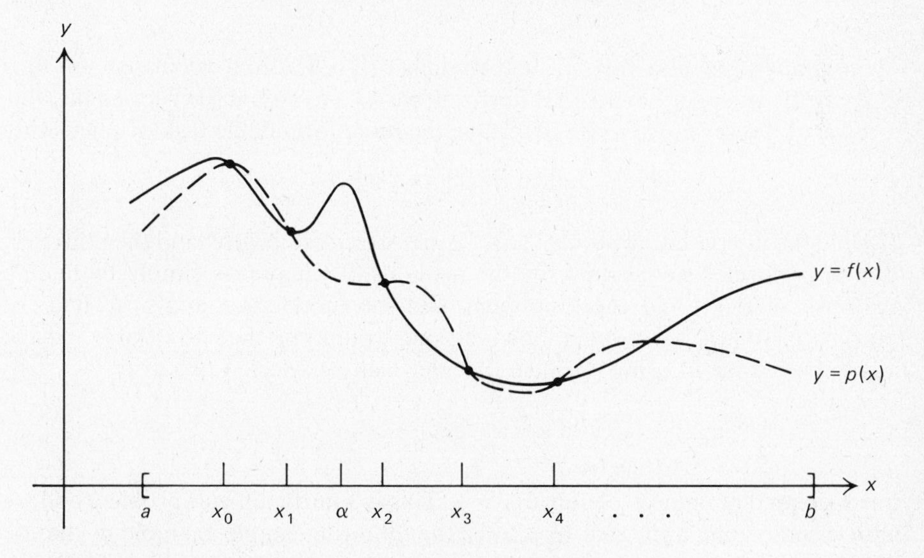


**Fig. 5.4** The graphs of $f(x)$ and its interpolating polynomial, $p(x)$.

where $W(x) \equiv (x - x_0)(x - x_1)\cdots(x - x_n)$ and where $\xi$ is some point lying in $\operatorname{Spr}\{x, x_0, x_1, \ldots, x_n\}$.

*Proof*: Let $x$ be arbitrary but *fixed* in $[a, b]$. If $x = x_j$ for $0 \le j \le n$ then the theorem is trivially true. Thus we assume that $x \ne x_j$ for any $j$, and define the function $F(t) = f(t) - p(t) - CW(t)$, where the constant $C$ is given by $C = (f(x) - p(x))/W(x)$. Now $F^{(n+1)}(t)$ is continuous on $[a, b]$ and furthermore $F(x) = F(x_0) = F(x_1) = \cdots = F(x_n) = 0$. By Rolle's theorem, $F'(t)$ has at least one zero between each distinct zero of $F(t)$, and thus $F'(t)$ has at least $(n + 1)$ distinct zeros in $\operatorname{Spr}\{x, x_0, x_1, \ldots, x_n\}$. Using Rolle's theorem for $F'(t)$, we can reason that $F''(t)$ has at least $n$ distinct zeros. Rolle's theorem applied to $F''(t)$ shows that $F'''(t)$ has at least $(n - 1)$ distinct zeros, etc. Finally we conclude that $F^{(n+1)}(t)$ has at least one zero, $\xi$, and $\xi \in \operatorname{Spr}\{x, x_0, x_1, \ldots, x_n\}$. From the definition of $F(t)$ we see that $F^{(n+1)}(t) = f^{(n+1)}(t) - 0 - C(n + 1)!$. Thus, $0 = F^{(n+1)}(\xi) = f^{(n+1)}(\xi) - C(n + 1)!$. Substituting $(f(x) - p(x))/W(x)$ for $C$ in this expression we obtain (5.27). ■

The reader should note that the value $\xi$ in (5.27) changes as $x$ changes, so $\xi$ can be regarded as a function of $x$. This is one of the drawbacks in using (5.27); since $\xi$ is an unknown function of $x$ it is impossible to evaluate $f^{(n+1)}(\xi)$ exactly.

However, we have assumed in Theorem 5.5 that $f^{(n+1)}(x)$ is continuous, so there is a number $K_n$ such that $|f^{(n+1)}(\xi)| \le K_n$ for any $\xi$ in $[a, b]$. Consequently, a more useful form of (5.27), which we can use to bound the error for any $x$ in $[a, b]$, is

$$|e(x)| = |f(x) - p(x)| \le \frac{K_n}{(n + 1)!}|W(x)|. \tag{5.28}$$

For any particular value of $x$, the number $|W(x)|$ can be calculated and thus (5.28) yields an upper bound on the error. If we wish to bound $|e(x)|$ for all possible $x$ in $[a, b]$, however, we must undertake the more formidable task of calculating

$$\max_{a \le x \le b} |W(x)| \equiv \|W\|_\infty.$$

If $n$ and the interpolating points, $\{x_j\}_{j=0}^n$, are prescribed beforehand then this task can be performed numerically by the methods of Chapter 4, simply by finding the zeros of $W'(x)$ and checking them (and the endpoints $a$ and $b$) to find the maxima of $|W(x)|$. Most often, however, one would like the choice of $n$ and the $x_j$'s to be flexible in using (5.28) so that they can be selected to satisfy

$$\frac{K_n}{(n + 1)!}\|W\|_\infty < \varepsilon$$

for some predetermined tolerance, $\varepsilon > 0$. This is a fairly difficult problem and we shall examine one approach to it after the following simple example of the use of (5.28).

**Example 5.7.** We shall use two previous examples involving interpolation which will show that sometimes (5.28) is a very good bound while on other occasions it may be overly pessimistic. In Example 5.5, $f(x) = \cos(x)$ was interpolated at four points, with $W(x) = (x - 0.3)(x - 0.4)(x - 0.5)(x - 0.6)$. In this case $n = 3$ and $f^{(4)}(x) = \cos(x)$. In the notation of (5.28), we have

$$K_3 = \max_{0.3 \le x \le 0.6} |\cos(x)| = |\cos(0.3)| \le 0.999989$$

and

$$|W(0.44)| = 0.5376 \times 10^{-4}.$$

Thus

$$|e(0.44)| \le K_3 |W(0.44)|/4! \le 0.224 \times 10^{-5}$$

whereas, in fact, $|e(0.44)| = 0.2 \times 10^{-5}$.

In Example 5.6, $f(x) = e^{3x}$ was interpolated at five points, with

$$W(x) = (x + 1)(x + 0.5)x(x - 0.5)(x - 1).$$

Thus $n = 4$ and $f^{(5)}(x) = 243e^{3x}$. Since the interpolation points are in $[-1, 1]$, $K_4 = 243e^3 \le 4880.79$ and $|W(0.8)| = 0.11232$. Thus,

$$|e(0.8)| \le K_4 |W(0.8)|/5! \le 4.57$$

whereas, in fact, $|e(0.8)| = 0.4591$. In this second case, the error bound has overestimated the true error by a factor of about 10.

When using the error bound given by (5.28), it is apparent that the function $|W(x)|$ plays an important role in determining the size of the bound. Clearly, $|W(x)|$ will be small when $x$ is near an interpolation point $x_j$, but more can be said about the location of values $x$ which make $|W(x)|$ small (or large). For instance, in the second case of Example 5.7 above, $W(x) = (x + 1)(x + 0.5)x(x - 0.5)(x - 1)$ and $|W(0.8)| = 0.11232$. Note that $x = 0.2$ is no nearer an interpolation point than is $x = 0.8$, but a simple calculation shows that $|W(0.2)| = 0.04032$. Intuitively, this makes sense, since $W(0.2)$ has more "small" factors than does $W(0.8)$. For equally spaced interpolation points in an interval $[a, b]$, it can be shown that the relative maxima of $|W(x)|$ decrease as $x$ approaches the midpoint $(a + b)/2$ of $[a, b]$ (see Problems 1, 2, and 4). As a general rule, (5.28) shows that interpolation at equally spaced points is more accurate near the center of the interval than near the endpoints. (However, the error is also influenced by $f^{(n+1)}(\xi)$ and even though $|W(x)|$ is small, $|f^{(n+1)}(\xi)|\,|W(x)|$ can be large.) Figure 5.5 shows the graph of $W(x)$ as given in the second case of Example 5.7.

Examination of the derivation of (5.28) reveals that we can do nothing to improve the value of $K_n$, since we do not know the location of the mean-value point $\xi$. However, for $\|W\|_\infty = \max_{a \le x \le b} |W(x)|$, (5.28) tells us that the inequality $|e(x)| \le K_n \|W\|_\infty/(n + 1)!$ holds for all $x$ in $[a, b]$. Thus if we wish to make the error bound as small as possible *for all* $x$ in $[a, b]$, we should concentrate on the problem of choosing the interpolating points, $\{x_j\}_{j=0}^n$, so that $\|W\|_\infty$ is minimized. On the surface this seems to be a very difficult task. This problem was investigated

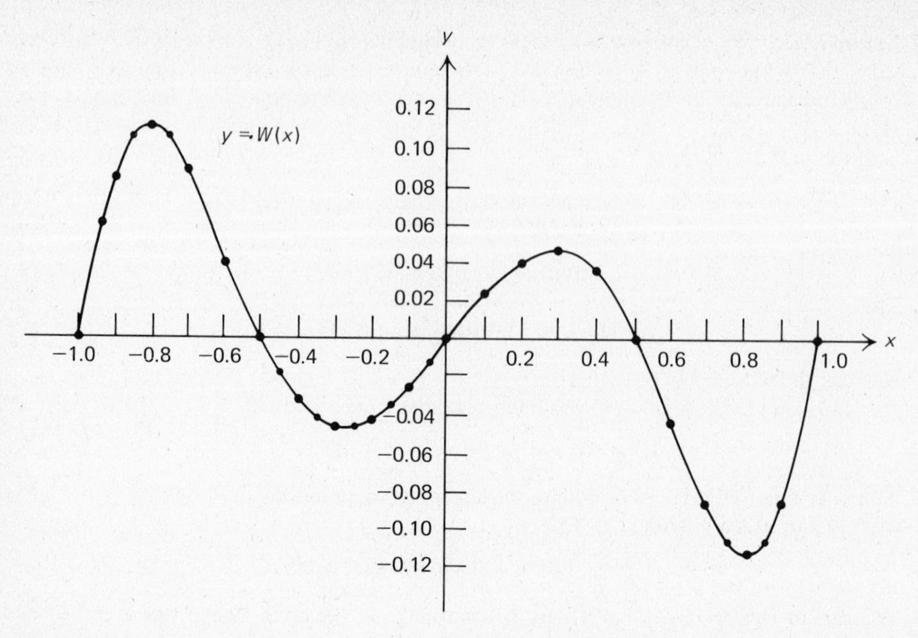


**Fig. 5.5** The graph of $W(x) = (x + 1)(x + 0.5)x(x - 0.5)(x - 1)$

by P. L. Chebyshev in the 19th century for the interval $[-1, 1]$ and the solution involves polynomials which were named after him. The solution is not too hard to obtain and can even be extended to an arbitrary interval, $[a, b]$, by a simple change of variable.

Before giving the result, we must introduce and investigate the *Chebyshev polynomials of the first kind*, $T_k(x) \equiv \cos(k \cos^{-1}(x))$, $k = 0, 1, 2, \ldots$. At first glance it probably seems difficult to believe that each $T_k(x)$ is actually an algebraic polynomial. However, we easily see that $T_0(x) \equiv 1$ and $T_1(x) \equiv x$. Now we make the variable substitution, $x = \cos(\theta), 0 \leq \theta \leq \pi$. Then $T_k(x) = T_k(\cos(\theta)) = \cos(k\theta)$. By elementary trigonometric identities, we can easily verify that

$$\begin{aligned} T_{k+1}(x) &= \cos((k + 1)\theta) = 2\cos(\theta)\cos(k\theta) - \cos((k - 1)\theta) \\ &= 2xT_k(x) - T_{k-1}(x), \qquad \text{for } k \geq 1. \end{aligned} \tag{5.29}$$

Therefore $T_2(x) = 2xT_1(x) - T_0(x) = 2x^2 - 1$ and $T_3(x) = 2xT_2(x) - T_1(x) = 4x^3 - 3x$, etc. We can easily verify inductively that $T_k(x)$ is a $k$th-degree polynomial, for $k = 0, 1, 2, \ldots$. Furthermore, we see that the leading coefficient of $T_k(x)$ equals $2^{k-1}$, and that $T_k(x_i) = 0, 0 \leq i \leq k - 1$, when

$$x_i = \cos\frac{(2i + 1)\pi}{2k}.$$

(Note: $\cos(kt) = 0$ whenever $t$ is an odd multiple of $\pi/(2k)$.) Lastly we see that $|T_k(x)|$ is never larger than 1 for $x$ in $[-1, 1]$, that is, $\|T_k\|_\infty \leq 1$. Moreover, for $y_i = \cos(i\pi/k)$, we have $T_k(y_i) = \cos(i\pi) = (-1)^i$, so $\|T_k\|_\infty = |T_k(y_i)| = 1$ at each of the $(k + 1)$ distinct points $y_0, y_1, \ldots, y_k$. Chebyshev polynomials are very important in many types of numerical approximations as we shall see in subsequent sections. The graphs of the first five Chebyshev polynomials are given in Fig. 5.6.

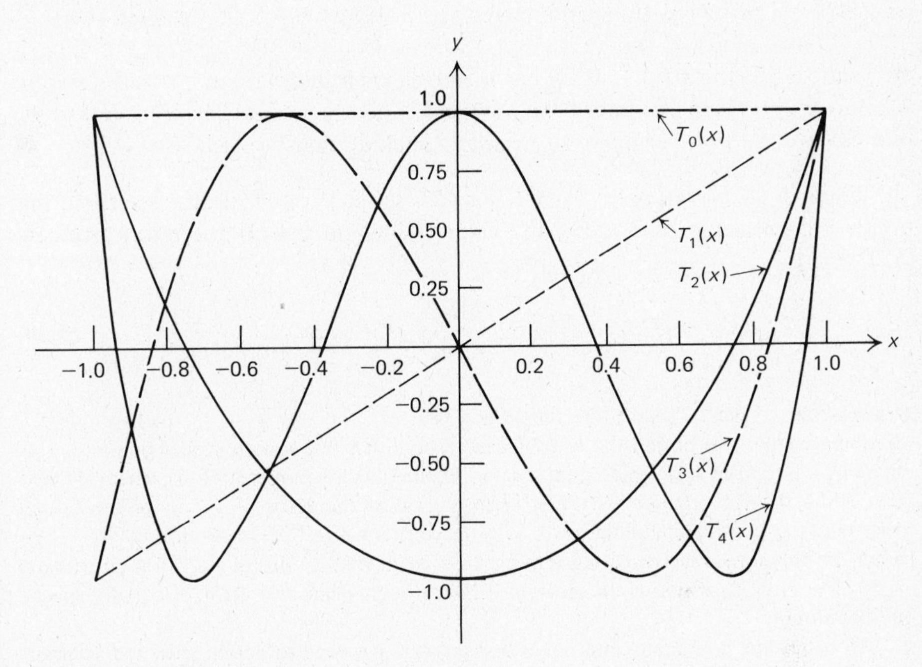


**Fig. 5.6** The Chebyshev polynomials, $\{T_k(x)\}_{k=0}^4$.

Now we are ready to prove the famous result for minimizing

$$\max_{-1 \leq x \leq 1} |W(x)| \equiv \|W\|_\infty.$$

**Theorem 5.6.** Let $W(x) = (x - x_0)(x - x_1) \cdots (x - x_n) \in \mathscr{P}_{n+1}$. Among all possible choices for distinct $\{x_j\}_{j=0}^n$ in $[-1, 1]$, $\|W\|_\infty \equiv \max_{-1 \leq x \leq 1} |W(x)|$ is minimized if $W(x) = (1/2^n)T_{n+1}(x)$ (i.e., the $x_j$'s are the zeros of $T_{n+1}(x)$).

*Proof*: By the above remarks on Chebyshev polynomials, the leading coefficient of the $(n + 1)$st-degree polynomial, $(1/2^n)T_{n+1}(x)$, is 1. Therefore, $(1/2^n)T_{n+1}(x) = (x - x_0)(x - x_1) \cdots (x - x_n)$, $x_i = \cos[(2i + 1)\pi/2(n + 1)]$, and hence $W(x) = (1/2^n)T_{n+1}(x)$ is a candidate for the minimum $W$. Also, from the remarks above:

$\|W\|_\infty = \|(1/2^n)T_{n+1}\|_\infty = 1/2^n$, $W(y_i) = (1/2^n)(-1)^i$, $0 \le i \le n+1$, and $W(y_i) = -W(y_{i+1})$, $0 \le i \le n$, where

$$y_i = \cos\frac{i\pi}{n+1}, \qquad 0 \le i \le n+1.$$

Now assume there exists a polynomial, $V(x) \in \mathscr{P}_{n+1}$, where $V(x)$ is monic (leading coefficient equals 1) and $\|V\|_\infty < \|W\|_\infty$. If $i$ is even, then $V(y_i) < W(y_i) = 1/2^n$, or else $\|V\|_\infty \ge \|W\|_\infty$. If $i$ is odd, then $V(y_i) > W(y_i) = -1/2^n$, or again $\|V\|_\infty \ge \|W\|_\infty$. Thus $V(y_0) < W(y_0)$, $V(y_1) > W(y_1)$, $V(y_2) < W(y_2)$, $V(y_3) > W(y_3)$, etc., and so $H(x) \equiv V(x) - W(x)$ has a zero in each interval $(y_i, y_{i+1})$, $0 \le i \le n$. Moreover, $H(x)$ is in $\mathscr{P}_n$, since both $V(x)$ and $W(x)$ are monic. Therefore $H(x) \in \mathscr{P}_n$ and has at least $(n+1)$ zeros, so we may conclude that no such $V(x)$ exists. ■

Thus if $[a, b] = [-1, 1]$ with $f \in C^{n+1}[-1, 1]$, and if the interpolating points are the zeros of $T_{n+1}(x)$, the error bound on the interpolation given by (5.28) yields

$$|e(x)| = |f(x) - p(x)| \le \frac{K_n}{(n+1)!}\|W\|_\infty = \frac{K_n}{(n+1)!2^n}. \tag{5.30}$$

**Example 5.8.** Again, consider the function $f(x) = e^{3x}$, for $x$ in $[-1, 1]$. In Example 5.7, we computed the error bound at $x = 0.8$ for interpolation at five equally spaced points ($n = 4$), obtaining a bound of 4.57. Finding this bound required us to compute $W(0.8)$, where $W(x) = (x+1)(x+0.5)x(x-0.5)(x-1)$. For interpolation at the zeros of $T_5(x)$, we can bound the error at any point $x$ (including $x = 0.8$) by the constant $K_4/(2^4 5!) = 4880.79/(2^4 5!) = 2.54$. In fact, the polynomial interpolating at the zeros of $T_5(x)$ gives a value of 11.2776 as an estimate to $f(0.8) = 11.0232$, which is an error of 0.2544 (roughly half the error of equally spaced interpolation).

To determine how many interpolation points are required to obtain a desired accuracy when interpolating $f(x)$ at the zeros of $T_{n+1}(x)$, we note that $f^{(n+1)}(x) = 3^{n+1}e^{3x}$, so $K_n = 3^{n+1}e^3$. Using (5.30), we have

$$|e(x)| \le \left(\frac{3}{2}\right)^n \frac{3e^3}{(n+1)!}$$

which can be shown to go to zero as $n \to \infty$. For example, with $n = 10$, $|e(x)| \le 0.000087$ and with $n = 20$, $|e(x)| \le 0.39 \times 10^{-16}$. Using 11 equally spaced interpolation points ($n = 10$) and evaluating the error bound of formula (5.28) at $x = 0.9$ (a noninterpolation point), we find $|W(0.9)| = 0.0065$ and hence we have an error bound of 0.00058 for the equally spaced case, which is more than six times as large as the Chebyshev interpolation error bound with $n = 10$.

Obviously, we do not expect the error of Chebyshev interpolation, $|e_T(x)|$, to be smaller than the error for equally spaced interpolation, $|e_h(x)|$, for every $x$ in $[a, b]$. (A trivial example of this is given when $x = x_j = a + jh$, but $x_j$ is not a Chebyshev zero. For then, $0 = |e_h(x_j)| \leq |e_T(x_j)|$.) Usually, however, if we are free to select the interpolating points at our own discretion, we prefer to use Chebyshev interpolation due to Theorem 5.6 which allows us to determine an error bound *a priori* for all values of $x$.

There are other ways of using (5.27) to estimate the error of interpolation. For example, let $w(x)$ be a nonnegative weighting function as given in (5.1b). Multiplying the squares of both sides of (5.27) by $w(x)$ and integrating from $a$ to $b$ yields

$$
\begin{aligned}
\|e(x)\|_2 &= \left(\int_a^b (f(x) - p(x))^2 w(x)\,dx\right)^{1/2} \\
&= \left(\int_a^b \left(\frac{f^{(n+1)}(\xi)}{(n+1)!} W(x)\right)^2 w(x)\,dx\right)^{1/2} \\
&\leq \frac{\max\limits_{a \leq x \leq b} |f^{(n+1)}(x)|}{(n+1)!} \left(\int_a^b (W(x))^2 w(x)\,dx\right)^{1/2} \\
&= \frac{K_n}{(n+1)!} \|W\|_2. \qquad \textbf{(5.31)}
\end{aligned}
$$

After we investigate orthogonal polynomials in a later section, we will see how to choose the interpolating points, $\{x_j\}_{j=0}^n$, in order to minimize $\|W\|_2$. We merely state here that they are not, in most cases, the zeros of a Chebyshev polynomial.

### 5.2.4 Translating the Interval

Often it is necessary for practical purposes to take a problem stated on an interval $[c, d]$ and reformulate the same problem on a different interval $[a, b]$. This is especially true in problems which require the numerical approximation of integrals or problems concerning orthogonal polynomials. We have already come across this problem in Chebyshev interpolation, where Theorem 5.6 is applicable to the specific interval $[-1, 1]$ and says nothing about the optimal choice of interpolation points in an arbitrary interval $[a, b]$. The change of variable which usually achieves our purposes of reformulating a problem on $[c, d]$ to one on $[a, b]$, is the "straight line" transformation $x = mt + \beta$, for $t \in [c, d]$ and $x \in [a, b]$. Here $m$ and $\beta$ are chosen such that $x = a$ when $t = c$ and $x = b$ when $t = d$. A simple algebraic computation will show that this transformation is given by

$$x = \left(\frac{b - a}{d - c}\right) t + \left(\frac{ad - bc}{d - c}\right) \qquad \textbf{(5.32)}$$

To illustrate this procedure, let us suppose that we wish to perform Chebyshev interpolation on an arbitrary interval $[a, b]$ instead of the interval $[-1, 1]$.

Then $t \in [-1, 1] \equiv [c, d]$ and $x \in [a, b]$, so by (5.32),

$$x = \left(\frac{b-a}{2}\right)t + \left(\frac{a+b}{2}\right) \quad \text{or} \quad t = \frac{2}{b-a}\left(x - \frac{a+b}{2}\right) = \frac{2(x-a)}{(b-a)} - 1. \qquad \textbf{(5.33)}$$

We now define the "shifted" Chebyshev polynomial of degree $k$ for $x \in [a, b]$, as

$$\tilde{T}_k(x) \equiv T_k(t) = T_k\left(\frac{2(x-a)}{(b-a)} - 1\right) = \cos\left[k \cos^{-1}\left(\frac{2(x-a)}{(b-a)} - 1\right)\right]. \qquad \textbf{(5.34)}$$

Since $T_k(t_i) = 0$ for $t_i = \cos((2i+1)\pi)/2k$, $0 \le i \le k-1$, then $x_i = ((b-a)/2)t_i + (b+a)/2$, $0 \le i \le k-1$, are the zeros of $\tilde{T}_k(x)$. Furthermore, $\tilde{T}_0(x) = 1$, $\tilde{T}_1(x) = 2(x-a)/(b-a) - 1$, and so from (5.29)

$$\begin{aligned}\tilde{T}_{k+1}(x) &= T_{k+1}(t) = 2tT_k(t) - T_{k-1}(t)\\ &= 2\left(\frac{2(x-a)}{(b-a)} - 1\right)\tilde{T}_k(x) - \tilde{T}_{k-1}(x), \qquad k \ge 1. \qquad \textbf{(5.35)}\end{aligned}$$

We leave it to the reader to verify that the leading coefficient of $\tilde{T}_k(x)$ is $2^{k-1}(2/(b-a))^k$, $k \ge 1$. Therefore, if we interpolate in $[a, b]$ at the zeros $\{x_i\}_{i=0}^n$ of $\tilde{T}_{n+1}(x)$, we have $W(x) = 2^{-n}(2/(b-a))^{-n-1}\,\tilde{T}_{n+1}(x)$. Then,

$$\max_{a \le x \le b} |W(x)| = 2^{-n}\left(\frac{2}{b-a}\right)^{-n-1},$$

since

$$\max_{a \le x \le b} |\tilde{T}_{n+1}(x)| = \max_{-1 \le t \le 1} |T_{n+1}(t)| = 1.$$

Thus the error bound, (5.30), becomes

$$|e(x)| = |f(x) - p(x)| \le \frac{K_n}{(n+1)!}\max_{a \le x \le b}|W(x)| = \frac{K_n}{2^n(n+1)!}\left(\frac{b-a}{2}\right)^{n+1}. \qquad \textbf{(5.36)}$$

It is easily seen that Theorem 5.6 is valid on $[a, b]$ with respect to this $W(x)$, and so $W(x)$ is the smallest monic polynomial of degree $(n + 1)$ with respect to

$$\|W\|_\infty \equiv \max_{a \le x \le b} |W(x)|.$$

**Example 5.9.** Suppose $f(x) = \cos(x)$ and $[a, b] = [0, \pi/2]$, and suppose we wish to find a value of $n$ and interpolating points $\{x_j\}_{j=0}^n$, such that the error of the $n$th-degree interpolation is less than $10^{-6}$ for all $x$ in $[0, \pi/2]$. Now $K_n = 1$ for all $n$, so considering Eq. (5.36), the smallest value of $n$ for which

$$\frac{1}{2^n(n+1)!}\left(\frac{b-a}{2}\right)^{n+1} = \frac{1}{2^n(n+1)!}(\pi/4)^{n+1} < 10^{-6},$$

is $n = 6$. Thus the interpolating points are the zeros of $\tilde{T}_7(x)$, which are the points

$$x_j = \frac{\pi}{4}(\cos[(2j+1)\pi/14] + 1), \qquad 0 \le j \le 6.$$

From (5.36),

$$|e(x)| \leq \frac{1}{2^6 7!}(\pi/4)^7 = 5.71 \times 10^{-7}, \qquad x \in [0, \pi/2].$$

The transformation technique of this section is quite simple, but it is used often, and we shall refer to it in later sections.

## PROBLEMS

1. To use the error estimate (5.28), we need some way of estimating $|W(x)|$. For equally spaced interpolation points, there are a number of ways to do this. For simplicity, suppose we are interpolating in $[-1, 1]$ with an odd number of equally spaced points. Thus suppose $n$ is even, $W(x) = (x - x_0)(x - x_1)\cdots(x - x_n)$ where $h = 2/n$, $x_j = x_0 + jh$ for $j = 0, 1, \ldots, n$ and where $x_0 = -1$, $x_n = 1$. Let $N = n/2$ and show

   $$W(x) = (x + Nh)(x + (N - 1)h)\cdots(x + h)x(x - h)\cdots(x - (N - 1)h)(x - Nh).$$

   For $x \in [-1, 1]$, write $x$ as $x = rh$, where $r$ is a number such that $-N \leq r \leq N$. Suppose next that $x_{n-1} < x < x_n$ so that $N - 1 < r < N$. Show that:

   $$(n - 1)!h^{n-1}|(x - x_{n-1})(x - x_n)| \leq |W(x)| \leq n!h^{n-1}|(x - x_{n-1})(x - x_n)|.$$

2. Show that the maximum value of the factor $|(x - x_{n-1})(x - x_n)|$ in Problem 1, for $x_{n-1} \leq x \leq x_n$, is $h^2/4$. Thus conclude that $|W(x)| \leq n!2^{n-1}/n^{n+1}$. Also conclude, for $x = (x_n + x_{n-1})/2$, that $(n - 1)!2^{n-1}/n^{n+1} \leq |W(x)|$.

3. In Problem 2, for $n = 5$, 10, and 20, evaluate the upper and lower bounds for $|W(x)|$. Contrast the upper bounds with $1/2^n$, which is the bound on $|(x - t_0)(x - t_1)\cdots(x - t_n)|$, where $t_0, t_1, \ldots, t_n$ are the zeros of $T_{n+1}(x)$.

4. Given $n$ equally spaced interpolation points in $[-1, 1]$, where $n$ is even as in Problem 1, show that $W(x)$ is an odd function (i.e., $W(x) = -W(-x)$). Thus, if we can find the maximum of $|W(x)|$ for $0 \leq x \leq 1$, we have a maximum for $|W(x)|$ for $-1 \leq x \leq 1$. Show that the maximum of $|W(x)|$ occurs somewhere in $(x_{n-1}, x_n)$. (Hint: Show that

   $$\left|\frac{W(x + h)}{W(x)}\right| > 1,$$

   for $0 < x < x_{n-1}$ and for $x$ not an interpolation point. Use the representation of $W(x)$ in Problem 1 to establish this inequality. This result means that the upper and lower bounds established in Problem 2 are upper and lower bounds for $\|W\|_\infty$. The inequality in the hint also shows why interpolation at equally spaced points is usually more reliable near the center of the entire interval.

5. Let $f(x) = 1/(x + 2)$. Using (5.28) and the results of Problems 2 and 4, give an error estimate for interpolating $f(x)$ at $n$ equally spaced points in $[-1, 1]$, where $n$ is even. What is the corresponding error estimate if $f(x)$ is interpolated at the zeros of $T_{n+1}(x)$?

*6. Let $n$ be odd and let $x_0, x_1, \ldots, x_n$ be equally spaced in $[-1, 1]$, $x_0 = -1$, $x_n = 1$. The length of each subinterval is $2h$, where $h = 1/n$. Show that

   $$W(x) = (x + nh)(x + (n - 2)h)\cdots(x + h)(x - h)\cdots(x - (n - 2)h)(x - nh).$$

Show that the maximum value of $W(x)$ for $-h \le x \le h$ occurs at $x = 0$. (Hint: Group the factors of $W(x)$ as $(x^2 - n^2h^2)\cdots(x^2 - h^2)$ and show that $W'(x)$ and $W'(-x)$ have opposite signs for any $x$ in $(-h, h)$.) Evaluate $W(0)$ and determine the maximum interpolation error for $f(x) = 1/(x + 2)$, with $x \in (-h, h)$.

7. Verify (5.29) by showing that

$$\cos((k + 1)\theta) = 2\cos(\theta)\cos(k\theta) - \cos((k - 1)\theta).$$

8. Show by induction, using Eq. (5.29) that:

   a) $T_k(x)$ is a $k$th-degree polynomial with leading coefficient $2^{k-1}$;

   b) $T_k(x)$ is an even (odd) function if $k$ is even (odd) (recall that $f(x)$ is an even function if $f(x) = f(-x)$ and that $f(x)$ is an odd function if $f(x) = -f(-x)$).

9. a) In Section 5.2.3 we argued that

$$x_i = \cos\frac{(2i + 1)\pi}{2k}, \qquad 0 \le i \le k - 1,$$

   were zeros of $T_k(x)$ by use of the definition $T_k(x) = \cos(k\cos^{-1}(x))$. Now the cosine function has an infinite number of zeros, thus explain why $T_k(x)$ does not also have an infinite number of zeros. For instance, what about

$$x_i = \cos\frac{(2i + 1)\pi}{2k}$$

   for $i < 0$ or $i \ge k$?

   b) If $|x| > 1$, (say $x = 2$), then $\cos^{-1}(x)$ is not defined since there is no value $\theta$ such that $\cos(\theta) = x$. However, since $T_k(x)$ is a polynomial it must be true that $T_k(x)$ is defined for all $x$. How must the definition of $T_k(x)$ be interpreted to do this? Hint: What are the values of $T_0(2)$, $T_1(2)$, $T_2(2)$, $T_3(2)$, etc.?

10. Show that $\tilde{T}_k(x)$ defined by (5.35) has leading coefficients equal to

$$2^{k-1}\left(\frac{2}{b - a}\right)^k, \qquad k \ge 1.$$

   What are the polynomials $\tilde{T}_k(x)$ in the special case when $[a, b] = [0, 1]$?

11. How large should $k$ be if we wish to obtain an interpolation error of $10^{-6}$ or less throughout $[a, b]$, by interpolating $f(x)$ at the zeros of $\tilde{T}_k(x)$, for

   a) $f(x) = \cos(x)$, $0.3 \le x \le 0.6$

   b) $f(x) = 1/(2 + x)$, $-1 \le x \le 1$

   c) $f(x) = \ln(x)$, $0.1 \le x \le 1$

   d) $f(x) = e^{3x}$, $-1 \le x \le 1$.

### *5.2.5 Polynomial Interpolation with Derivative Data

Examination of (5.5) and (5.6) reveals that the type of $n$th-degree polynomial interpolation that we have studied up to this point is based on the fact that there exists a unique solution to the $(n + 1)$ linear equations, $p(x_j) = f(x_j)$, $0 \le j \le n$, in the $(n + 1)$ unknowns, $a_j$ (the coefficients of $p(x)$). Thus $p(x)$ "matches" $f(x)$ at the interpolating points. It is natural to ask if we might not get a better polynomial approximation to $f(x)$ if we forced some derivatives of $p(x)$ at various points to match the respective derivatives of $f(x)$ at these points (for then we would be using more information about $f(x)$ and would expect a better approximation). The extreme example of this is to take only one point, $x_0$, and choose the coefficients of $p(x)$ such that $p^{(j)}(x_0) = f^{(j)}(x_0)$ for $0 \le j \le n$. This yields the following $(n + 1)$ equations in $\{a_j\}_{j=0}^n$:

$$\begin{aligned}
a_0 + a_1 x_0 + a_2 x_0^2 + \cdots + a_n x_0^n &= f(x_0) \\
a_1 + 2a_2 x_0 + \cdots + n a_n x_0^{n-1} &= f'(x_0) \\
2a_2 + \cdots + n(n-1) a_n x_0^{n-2} &= f''(x_0) \\
\ddots \qquad \vdots \quad &\quad \vdots \\
n! a_n &= f^{(n)}(x_0)
\end{aligned}$$

We note that this triangular system is non-singular (the coefficient matrix has positive diagonal elements), and so $p(x)$ exists and is unique. The reader will recall from calculus that the Taylor polynomial,

$$q(x) = \sum_{j=0}^{n} \frac{f^{(j)}(x_0)}{j!}(x - x_0)^j,$$

also has the property that $q^{(j)}(x_0) = f^{(j)}(x_0)$, $0 \le j \le n$, and so, by the uniqueness of $p(x)$, $q(x) = p(x)$. Thus truncated Taylor's expansions are actually special cases of this generalized interpolation problem. Since the error of the Taylor approximation at $x$ is

$$\frac{f^{(n+1)}(\xi)}{(n+1)!}(x - x_0)^{n+1},$$

it is seen that this approximation is quite good if $x$ is "close" to $x_0$ and usually becomes worse as $x$ moves away from $x_0$. Note the similarity of this error formula to that of (5.27).

There are obviously many ways of matching derivatives of a polynomial to respective derivatives of $f(x)$ at various points in order to obtain a linear system of $(n + 1)$ equations in $(n + 1)$ unknowns (the coefficients of $p(x)$). This problem in its most general form is known as *Hermite-Birkoff interpolation.* For example, we might ask for a 3rd-degree polynomial, $p(x)$, such that $p(x_0) = f(x_0)$, $p'''(x_0) = f'''(x_0)$, $p'(x_1) = f'(x_1)$ and $p''(x_1) = f''(x_1)$. In this general setting, the problem need not always have a solution, and the question of finding conditions under which solutions exist remains unanswered and the subject of extensive research.

We wish now to consider a special form of this problem known as *Hermite* or *osculatory* interpolation. This form of interpolation is important not only as an approximation technique, but is useful in the derivation of the cubic spline approximations of the next section. For simplicity we assume that $n$ is even and define $N = n/2$. We consider $N + 1$ distinct points $\{x_j\}_{j=0}^N$, and impose the conditions that $p(x_j) = f(x_j)$ *and* $p'(x_j) = f'(x_j)$, $0 \le j \le N$. Thus $p(x)$ matches $f(x)$ and $p'(x)$ matches $f'(x)$ at the interpolating points. Note that we have $2(N + 1) = n + 2$ equations, so we will be searching for $p(x)$ in $\mathscr{P}_{2N+1} = \mathscr{P}_{n+1}$, in order that we have $2N + 2$ unknowns, $\{a_j\}_{j=0}^{2N+1}$. To illustrate the problem, consider the following example.

**Example 5.10.** Let $x_0 = \alpha$ and $x_1 = \beta$, and assume that $f(x)$ is given such that $f(\alpha) = y_0$, $f'(\alpha) = y'_0$, $f(\beta) = y_1$, and $f'(\beta) = y'_1$. Then $N = 1$ and so $2N + 1 = 3$. Find $p(x) \in \mathscr{P}_3$ such that $p(x_j) = f(x_j) = y_j$ and $p'(x_j) = f'(x_j) = y'_j$, for $j = 0$ and 1. If $p(x) = a_0 + a_1x + a_2x^2 + a_3x^3$, then the above equations become (in matrix form)

$$\begin{bmatrix} 1 & \alpha & \alpha^2 & \alpha^3 \\ 0 & 1 & 2\alpha & 3\alpha^2 \\ 1 & \beta & \beta^2 & \beta^3 \\ 0 & 1 & 2\beta & 3\beta^2 \end{bmatrix} \begin{bmatrix} a_0 \\ a_1 \\ a_2 \\ a_3 \end{bmatrix} = \begin{bmatrix} y_0 \\ y'_0 \\ y_1 \\ y'_1 \end{bmatrix}$$

The determinant of the coefficient matrix for this system equals $(\beta - \alpha)^4$. Then, when $\alpha \neq \beta$, $p(x)$ exists and is unique. We could solve this system for the $a_i$'s to obtain $p(x)$, but we shall show for $\alpha < \beta$ that $p(x)$ is given by

$$p(x) = y_0\left[\frac{(x-\beta)^2}{(\beta-\alpha)^2} + 2\frac{(x-\alpha)(x-\beta)^2}{(\beta-\alpha)^3}\right] + y_1\left[\frac{(x-\alpha)^2}{(\beta-\alpha)^2} - 2\frac{(x-\beta)(x-\alpha)^2}{(\beta-\alpha)^3}\right]$$
$$+ y'_0\left[\frac{(x-\alpha)(x-\beta)^2}{(\beta-\alpha)^2}\right] + y'_1\left[\frac{(x-\alpha)^2(x-\beta)}{(\beta-\alpha)^2}\right]. \tag{5.37}$$

We could approach the Hermite interpolation problem in the manner of undetermined coefficients as in Example 5.10, but we choose a Lagrangian-type approach instead. For each $j$, $0 \le j \le N$, let $\ell_j(x)$ be the $N$th-degree polynomial, as defined by (5.2), with $N$ substituted for $n$. Now for each $j$, $0 \le j \le N$ define the $(2N + 1)$st-degree polynomials

$$A_j(x) \equiv [1 - 2(x - x_j)\ell'_j(x_j)]\ell_j^2(x)$$
$$B_j(x) \equiv (x - x_j)\ell_j^2(x).$$

We leave to the reader to verify that $A_j(x_i) = \delta_{ij}$, $B_j(x_i) = 0$, $A'_j(x_i) = 0$, and $B'_j(x_i) = \delta_{ij}$, where $0 \le i, j \le N$ and $\delta_{ij}$ is again the Kronecker delta. Let $\{y_0, y_1, \ldots, y_N, y'_0, y'_1, \ldots, y'_N\}$ be any set of $(2N + 2)$ values, and define $p(x) \in \mathscr{P}_{2N+1}$ by

$$p(x) = \sum_{j=0}^{N} (y_jA_j(x) + y'_jB_j(x)). \tag{5.38a}$$

Because of the properties of $A_j(x)$ and $B_j(x)$, it is easily seen that

$$p(x_i) = \sum_{j=0}^{N} (y_j \cdot \delta_{ij} + y'_j \cdot 0) = y_i, \qquad 0 \le i \le N,$$

and

$$p'(x_i) = \sum_{j=0}^{N} (y_j \cdot 0 + y'_j \cdot \delta_{ij}) = y'_i, \qquad 0 \le i \le N.$$

We can also show that the Hermite interpolating polynomial, $p(x)$, in (5.38a) is unique. Assume that $q(x) \in \mathscr{P}_{2N+1}$, $q(x) \neq p(x)$, and $q(x)$ also has the property that $q(x_j) = y_j$ and $q'(x_j) = y'_j$, $0 \le j \le N$. Let $r(x) \equiv p(x) - q(x)$. Then $r(x) \in \mathscr{P}_{2N+1}$ and $r(x_j) = r'(x_j) = 0$, $0 \le j \le N$. Thus, counting multiplicities, $r(x)$ has $2N + 2$ zeros which contradicts the Fundamental Theorem of Algebra. Therefore no such $q(x)$ exists and $p(x)$ is unique. Hence we have just proved:

**Theorem 5.7.** Let $\{x_j\}_{j=0}^N$ be a set of $(N + 1)$ distinct points and $\{y_j\}_{j=0}^N$ and $\{y'_j\}_{j=0}^N$ be any two sets of $N + 1$ values. Then there exists a unique $p(x) \in \mathscr{P}_{2N+1}$ such that $p(x_j) = y_j$ and $p'(x_j) = y'_j$, $0 \le j \le N$, and $p(x)$ is given by (5.38a).

Usually the $y_j$ and $y'_j$ are given by some function $f(x)$, where $f(x_j) = y_j$ and $f'(x_j) = y'_j$, $0 \le j \le N$. In this case, (5.38a) becomes

$$p(x) = \sum_{j=0}^{N} (f(x_j)A_j(x) + f'(x_j)B_j(x)). \tag{5.38b}$$

We leave as a problem for the reader that (5.38b) reduces to (5.37) in the case where $N = 1$.

In the case where $f'(x)$ does not exist or is unknown, Hermite interpolation is still sometimes used as in (5.38a), where we set $y_j = f(x_j)$ and $y'_j = 0, 0 \le j \le n$. This is known as Hermite-Fejér interpolation, and the following result was proved by Fejér:

> For any $N$, let $\{x_j\}_{j=0}^N$ be the zeros of the Chebyshev polynomial $T_{N+1}(x)$. Let $f(x)$ be any function in $C[-1, 1]$ and let $p_{2N+1}(x)$ be the Hermite-Fejér interpolating polynomial in $\mathscr{P}_{2N+1}$ for $f(x)$. Then
>
> $$\lim_{N\to\infty} \left\{ \max_{-1 \le x \le 1} |f(x) - p_{2N+1}(x)| \right\} = 0.$$

This theorem of Fejér is a curious result since we ask for the condition $p'_{2N+1}(x_j) = 0$ (which is unrelated to $f(x)$) and still obtain uniform convergence of the Hermite interpolating polynomials. The result is doubly curious in that no matter what set of interpolation points are chosen, there are functions $f \in C[-1, 1]$ such that $\lim_{N\to\infty} \|p_N - f\|_\infty = +\infty$, where $p_N(x)$ denotes the Lagrange interpolating polynomial as in (5.4) (c.f., Natanson 1965).

The error analysis for Hermite interpolation, (5.38b), can be carried out in much the same way as in Theorem 5.5. This time, however, we define the auxiliary function as $F(t) = f(t) - p(t) - CW(t)^2$, with $C = (f(x) - p(x))/W(x)^2$, $W(t) = \prod_{j=0}^N (t - x_j)$, and $p(t)$ as in (5.38b). Using Rolle's theorem as before we obtain an expression for the error:

$$f(x) - p(x) = \frac{f^{(2N+2)}(\xi)W(x)^2}{(2N+2)!} \tag{5.39}$$

### 5.2.6 Interpolation by Cubic Splines

In some practical problems, interpolating polynomials are not suitable for use as an approximation. For example, in order to obtain a "good" approximation to a function $f(x)$ by an $n$th-degree interpolating polynomial, it may be necessary to use a fairly large value of $n$. Unfortunately, polynomials of high degree often have a very oscillatory behavior which is not desirable in approximating functions which are reasonably smooth. This disadvantage of polynomial interpolation becomes particularly apparent when interpolating tabular data, as is shown in the example below. Also, computational problems arise when the number of data points is large. For instance, given 100 data points $(x_0, y_0), (x_1, y_1), \ldots, (x_{99}, y_{99})$, it would usually be foolhardy to find the 99th-degree polynomial $p(x)$ such that $p(x_i) = y_i$, $i = 0, 1, \ldots, 99$. One alternative to interpolation, as we shall see in Section 5.3, would be to find a polynomial of a low degree that "best fits" the data. Unfortunately, the polynomial that best fits the data will not generally interpolate the data as well. When it is desirable to have an approximation $q(x)$ such that $q(x_i) = y_i$, $i = 0, 1, \ldots, n$, then *piece-wise* polynomial interpolation is an attractive alternative and the one that we focus on in this section. In piece-wise polynomial interpolation, several lower-degree polynomials are joined together in a continuous fashion so that the resulting piece-wise polynomial, $q(x)$, interpolates the data. The extreme case of piece-wise polynomial interpolation is the "broken line," as in Example 5.11, where the approximating function $q(x)$ is obtained by joining $(x_i, y_i)$ and $(x_{i+1}, y_{i+1})$ by a straight line, for $i = 0, 1, \ldots, n - 1$. In this case, $q(x_i) = y_i$, and $q(x)$ is a linear polynomial in each interval $[x_i, x_{i+1}]$.

**Example 5.11.** The table below represents data taken from a hypothetical experiment.

| $x_i$ | 0 | 1 | 2 | 3 | 4 | 5 | 6 | 7 | 8 | 9 | 10 |
|---|---|---|---|---|---|---|---|---|---|---|---|
| $y_i$ | 3 | 2 | 3 | 5 | 3 | 4 | 3 | 2 | 2 | 3 | 2 |

In Fig. 5.7, we show the piece-wise linear polynomial $q(x)$ (the broken line) and the 10th degree interpolating polynomial, $p(x)$. Clearly, unless the function from which the data was taken is strange indeed, the broken line is a more acceptable approximation than is the interpolating polynomial. (Note that, as mentioned before, the interpolating polynomial appears to be more reasonable near the center of the interval.)

While polynomial interpolation may not give an acceptable approximation, as indicated by Fig. 5.7, the broken line approximation also has its disadvantages.

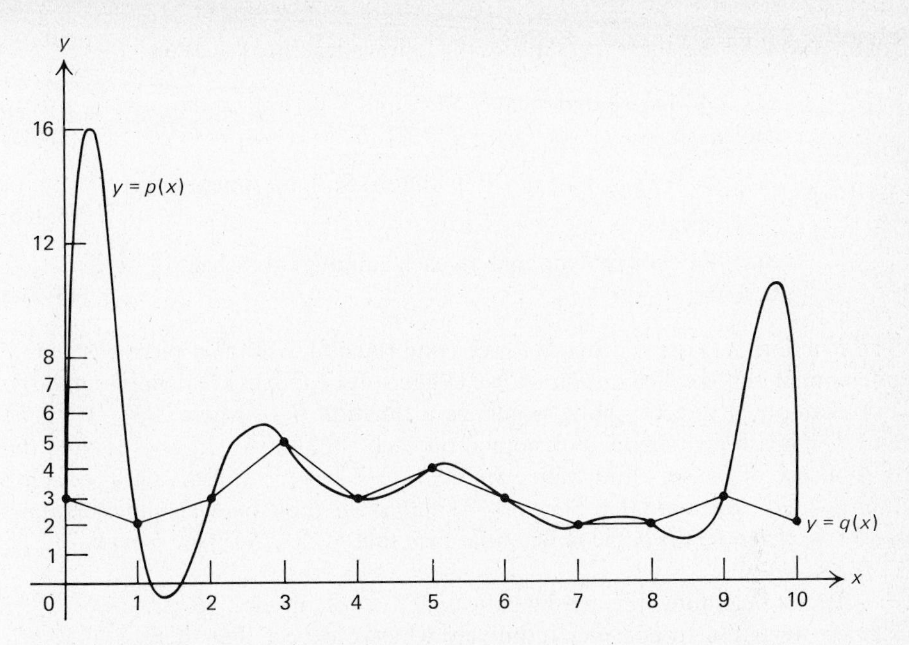


**Fig. 5.7** Piece-wise linear and interpolating approximations of Example 5.11.

In particular, the broken line suffers from a lack of smoothness and has a discontinuous derivative. Thus, except to give a rough idea of the shape of the graph, the broken line is not well suited to approximate most of the functions which arise in physical problems, since such functions are usually fairly smooth. To overcome the oscillatory behavior of polynomials and still provide a smooth approximation, an approximation technique using *splines* was presented in a paper by Schoenberg (1946), and has been the subject of extensive research ever since. This approximation technique resembles a physical process which has been used by draftsmen for many years: Given a set of points, $X_n = \{x_j\}_{j=0}^n$ where $a = x_0 < x_1 < \cdots < x_n = b$, and a set of functional values, $\{f(x_j)\}_{j=0}^n$, a draftsman will plot the data points $P_j = (x_j, f(x_j))$, $0 \le j \le n$. He will then take a thin elastic rod (called a *spline*) and a set of weights and will place the weights on the rod so that the rod must pass over each point, $P_j$, $0 \le j \le n$. The resulting curve traced out by the spline then has the property that it interpolates $f(x)$ at each $x_j$, and furthermore it smooths out as much as possible between the points, due to the elasticity of the rod. Thus the oscillatory behavior of the approximation is minimized as much as possible, but the approximation still retains the interpolation property.

We shall now consider a mathematical procedure which models the draftsman's technique. To do this we first consider the set of all functions, $Sp(X_n)$, such that if $S(x) \in Sp(X_n)$ then $S(x)$ satisfies the following three properties:

$S(x) \in C^2[a, b]$, that is, $S(x)$, $S'(x)$, and $S''(x)$ are continuous on $[a, b]$. **(5.40a)**

$S(x_j) = f(x_j) \equiv f_j$, $0 \le j \le n$, that is, $S(x)$ interpolates $f(x)$ on $[a, b]$. **(5.40b)**

$S(x)$ is a cubic polynomial on each subinterval $[x_j, x_{j+1}]$, $0 \le j \le n - 1$. **(5.40c)**

The function $S(x)$ defined by the three conditions of (5.40) is a piece-wise cubic polynomial and is called a *cubic spline*. Higher-order splines are defined similarly. For example, a quartic spline would be a function $S(x)$, where $S(x) \in C^3[a, b]$ and $S(x)$ is a fourth-degree polynomial on each subinterval $[x_j, x_{j+1}]$. Note the dependence of the set of functions, $Sp(X_n)$, on the particular function, $f(x)$, being interpolated. Also note that $S(x)$ may be a *different* cubic on each subinterval, so we let $S_j(x)$ denote the cubic polynomial such that $S(x) = S_j(x)$ for $x$ in $[x_j, x_{j+1}]$, $0 \le j \le n - 1$.

Before beginning the mathematical derivation needed to construct cubic splines, we pause to consider intuitively why we expect that there is any such function, $S(x)$, satisfying the three conditions of (5.40). For simplicity we let $n = 3$ so that $S(x)$ is represented by the cubic polynomials $S_0(x)$ on $[x_0, x_1]$, $S_1(x)$ on $[x_1, x_2]$, and $S_2(x)$ on $[x_2, x_3]$. Since each $S_j(x) \in \mathscr{P}_3$, it has four coefficients which we would like to choose to satisfy the conditions of (5.40). Therefore, in this case, we have twelve coefficients or unknowns at our disposal. Since each $S_j(x) \in \mathscr{P}_3$, all of its derivatives are continuous for any $x \in (x_j, x_{j+1})$—the *open interval*. Thus we need only worry about the continuity conditions at the interior interpolating points, $x_1$ and $x_2$, where the cubics must "patch together" with second-order continuity. With this in mind, let us work from left to right on $[a, b] \equiv [x_0, x_3]$ and determine the number of equations that must be satisfied. Since $S(x)$ must interpolate $f(x)$ at each $x_j$, and since $S(x)$ must be continuous at $x_1$ and $x_2$, we have the following six equations to be satisfied:

$$S_0(x_0) = f_0; \quad S_1(x_1) = f_1; \quad S_0(x_1) = S_1(x_1); \quad S_2(x_2) = f_2;$$
$$S_1(x_2) = S_2(x_2); \quad \text{and} \quad S_2(x_3) = f_3.$$

Finally since $S'(x)$ and $S''(x)$ must also be continuous at $x_1$ and $x_2$, we must have $S_0'(x_1) = S_1'(x_1)$, $S_0''(x_1) = S_1''(x_1)$, $S_1'(x_2) = S_2'(x_2)$, and $S_1''(x_2) = S_2''(x_2)$—four more equations. Therefore we have a total of ten equations in twelve unknowns. Thus we not only expect just one solution, but hopefully we can specify two of the unknowns to be designated arbitrarily and still have a solution for the remaining ten equations in ten unknowns. So, intuitively, we expect that $Sp(X_n)$ is an infinite two-parameter family of functions. Two logical ways to obtain a good approximation for $f(x)$ would be to choose the two free parameters by specifying either

$S'(x_0)$ and $S'(x_n)$ or $S''(x_0)$ and $S''(x_n)$. We choose the second alternative in the following mathematical derivation, but will also show how the first alternative may be implemented.

To construct a function $S(x)$ that satisfies the three conditions of (5.40), we first introduce some notation and then use the three conditions to obtain a linear system of equations that will enable us to determine the coefficients of each cubic polynomial, $S_j(x)$. We define $h_j \equiv \Delta x_j \equiv x_{j+1} - x_j$ and $\Delta f_j \equiv f(x_{j+1}) - f(x_j) \equiv f_{j+1} - f_j$. We next define $S''(x_j) \equiv y''_j$ and note that the quantities $y''_0, y''_1, \ldots, y''_n$ will appear as *unknowns* in a linear system of equations which will define the cubic spline $S(x)$. (The reader should note that usually $S''(x_j) \neq f''(x_j)$ and even $S'(x_j) \neq f'(x_j)$. All that (5.40) requires is that $S(x_j) = f(x_j)$, $0 \leq j \leq n$.)

The construction of the cubic spline proceeds roughly as follows: Suppose we choose any $n + 1$ values $y''_0, y''_1, \ldots, y''_n$ and then let $q(x)$ be the broken line such that $q(x_j) = y''_j$, $0 \leq j \leq n$. If we integrate $q(x)$ twice, we obtain a function $S(x)$ in $C^2[a, b]$ which is a piece-wise cubic polynomial. The question is: Can we choose $y''_0, y''_1, \ldots, y''_n$ so that $S(x_j) = f_j$, $0 \leq j \leq n$? If we can, then $S(x)$ interpolates $f(x)$, so by (5.40), $S(x) \in Sp(X_n)$. Following this idea, we define the first-degree polynomials, $S''_j(x)$, on $[x_j, x_{j+1}]$ by

$$S''_j(x) = y''_j \frac{x_{j+1} - x}{h_j} + y''_{j+1} \frac{x - x_j}{h_j}, \qquad 0 \leq j \leq n - 1. \tag{5.41}$$

(In (5.41) we have not yet specified $y''_0, y''_1, \ldots, y''_n$; they can be any set of values.) Note that $S''_j(x_j) = y''_j$, $S''_j(x_{j+1}) = y''_{j+1}$, $0 \leq j \leq n-1$. Also, $S''_{j+1}(x_{j+1}) = S''_j(x_{j+1})$, for $0 \leq j \leq n - 2$, so the function $S''(x)$ which has the value $S''_j(x)$ for $x \in [x_j, x_{j+1}]$ is a continuous function defined on $[a, b]$ ($S''(x)$ is a broken line.) Integrating (5.41) twice, we obtain for $x \in [x_j, x_{j+1}]$

$$S_j(x) = \frac{y''_j}{6h_j}(x_{j+1} - x)^3 + \frac{y''_{j+1}}{6h_j}(x - x_j)^3 + c_j(x - x_j) + d_j(x_{j+1} - x), \tag{5.42}$$

where $c_j$ and $d_j$ are constants of integration. To make $S(x)$ continuous and to satisfy the interpolation constraints, we need to choose the constants of integration so that $S_j(x_j) = f_j$, $S_j(x_{j+1}) = f_{j+1}$, for $0 \leq j \leq n - 1$. Substituting these two conditions into (5.42) we obtain

$$c_j = \frac{f_{j+1}}{h_j} - \frac{y''_{j+1}h_j}{6} \qquad \text{and} \qquad d_j = \frac{f_j}{h_j} - \frac{y''_j h_j}{6}$$

so

$$\begin{aligned} S_j(x) = {} & \frac{y''_j}{6h_j}(x_{j+1} - x)^3 + \frac{y''_{j+1}}{6h_j}(x - x_j)^3 + \left(\frac{f_{j+1}}{h_j} - \frac{y''_{j+1}h_j}{6}\right)(x - x_j) \\ & + \left(\frac{f_j}{h_j} - \frac{y''_j h_j}{6}\right)(x_{j+1} - x), \qquad 0 \leq j \leq n - 1. \end{aligned} \tag{5.43}$$

Now, $S_j(x)$ as defined in (5.43), was chosen to match $f(x)$ at $x_j$ and $x_{j+1}$ by adjusting the constants of integration, $c_j$ and $d_j$. Since we do this independently in each interval $[x_j, x_{j+1}]$, we have no guarantee that $S_j'(x_j) = S_{j-1}'(x_j)$, for $j = 1, 2, \ldots, n-1$ (that is, that $S'(x)$ is continuous). Differentiating (5.43) yields

$$S_j'(x) = -\frac{y_j''}{2h_j}(x_{j+1} - x)^2 + \frac{y_{j+1}''}{2h_j}(x - x_j)^2 + \frac{\Delta f_j}{h_j} - \frac{h_j}{6}(y_{j+1}'' - y_j''). \tag{5.44}$$

So the last condition that we must meet to satisfy all of (5.40) is to make $S'(x)$ continuous at the interior interpolating points. We do this by using (5.44) and choosing the values $y_j''$, $1 \le j \le n-1$, so that $S_j'(x_j) = S_{j-1}'(x_j)$ for $1 \le j \le n-1$. This yields the following linear system of $(n-1)$ equations in the unknowns $\{y_j''\}_{j=0}^{n}$:

$$\begin{aligned} h_{j-1}y_{j-1}'' + 2(h_j + h_{j-1})y_j'' + h_j y_{j+1}'' &= b_j \\ b_j \equiv 6\left(\frac{\Delta f_j}{h_j} - \frac{\Delta f_{j-1}}{h_{j-1}}\right), &\qquad 1 \le j \le n-1. \end{aligned} \tag{5.45}$$

Since we have $(n+1)$ unknowns and $(n-1)$ equations in (5.45), we can specify fixed values for $y_0''$ and $y_n''$ and transfer the two terms involving these values to the right-hand side. We then have the following $(n-1) \times (n-1)$ linear system (written in matrix form and denoted by $A\mathbf{y} = \mathbf{b}$):

$$\begin{bmatrix} \gamma_1 & h_1 & 0 & 0 & \cdots & 0 & 0 & 0 \\ h_1 & \gamma_2 & h_2 & 0 & \cdots & 0 & 0 & 0 \\ \vdots & & \ddots & & \ddots & & & \\ 0 & 0 & 0 & 0 & \cdots & h_{n-3} & \gamma_{n-2} & h_{n-2} \\ 0 & 0 & 0 & 0 & \cdots & 0 & h_{n-2} & \gamma_{n-1} \end{bmatrix} \begin{bmatrix} y_1'' \\ y_2'' \\ \vdots \\ y_{n-2}'' \\ y_{n-1}'' \end{bmatrix} = \begin{bmatrix} b_1 - h_0 y_0'' \\ b_2 \\ \vdots \\ b_{n-2} \\ b_{n-1} - h_{n-1} y_n'' \end{bmatrix} \tag{5.46}$$

where $\gamma_i = 2(h_i + h_{i-1})$.

It is clear that the coefficient matrix $A$ is tridiagonal and diagonally dominant and hence non-singular (see Chapter 2). Thus, no matter what values we selected for $y_0''$ and $y_n''$, (5.46) has a unique solution for $\{y_j''\}_{j=1}^{n-1}$ (the solution does depend on our choice of $y_0''$ and $y_n''$, of course). Furthermore the $\{y_j''\}_{j=1}^{n-1}$ can be easily found by the finite recurrence given for the LU decomposition of a tridiagonal matrix. (Refer to formula (2.33).) These values are then substituted into (5.43) and thus yield $S(x)$. It is common, as we shall see shortly, to set $y_0'' = y_n'' = 0$ and the unique cubic spline which results from this choice is called the *natural cubic spline*. A final point that should be made is that once the values $y_1'', y_2'', \ldots, y_{n-1}''$ are determined from (5.46), we then use (5.43) to evaluate $S(x)$ at $x = \alpha$. That is, if $\alpha \in [x_j, x_{j+1}]$, then $S(\alpha) = S_j(\alpha)$.

**Example 5.12.** In this example, we consider approximating $f(x) = |x|$ by a cubic spline and contrast the spline approximation with an interpolating polynomial. Let $x_0 = -2$,

$x_1 = -1$, $x_2 = 0$, $x_3 = 1$, and $x_4 = 2$. In this case, $h_j = 1$ for $j = 0, 1, 2, 3$ and $b_j = 6(f_{j+1} - 2f_j + f_{j-1}) = 6\Delta^2 f_{j-1}$ for $j = 1, 2, 3$. Thus we are led to the linear system (as in Eq. (5.46))

$$\begin{aligned} 4y_1'' + y_2'' \qquad &= 0 \\ y_1'' + 4y_2'' + y_3'' &= 2 \\ y_2'' + 4y_3'' &= 0. \end{aligned}$$

Solving this system, we find $y_1'' = -6/7$, $y_2'' = 24/7$, and $y_3'' = -6/7$. Using these values in (5.43), we can evaluate the cubic spline in any subinterval $[x_j, x_{j+1}]$.

The 4th-degree interpolating polynomial for $f(x)$ at the points $x_j$ above, is easily seen to be given by $p(x) = \frac{7}{6}x^2 - \frac{1}{6}x^4$. We will compare these two approximations in the interval $[1, 2]$, where to evaluate the cubic spline in $[1, 2]$ we use (from (5.43))

$$S_3(x) = -(2 - x)^3/7 + 2(x - 1) + 8(2 - x)/7.$$

Since both of these approximations, and $f(x)$ as well, have such simple forms, it is easy to verify that the maximum value of $p(x) - |x|$ occurs at approximately $x = 1.6$. The maximum value of $(S_3(x) - |x|)$ occurs approximately at $x = 1.423$ and $|S_3(x) - |x|| \le 0.055$ for all $x$ in $[1, 2]$. In Table 5.3, we have listed values of $S_3'(x)$ and $p'(x)$ as well. (As we shall see, the derivative of the cubic spline provides a fairly good approximation to $f'(x)$. On the other hand, derivatives of the interpolating polynomial are not generally good approximations to the derivatives of $f(x)$.)

**Table 5.3**

| $x$ | $f(x)$ | $S_3(x)$ | $S_3'(x)$ | $p(x)$ | $p'(x)$ |
|---|---|---|---|---|---|
| 1.0 | 1.0 | 1.000 | 1.286 | 1.000 | 1.667 |
| 1.2 | 1.2 | 1.241 | 1.131 | 1.334 | 1.648 |
| 1.4 | 1.4 | 1.455 | 1.011 | 1.646 | 1.437 |
| 1.6 | 1.6 | 1.648 | 0.926 | 1.894 | 1.003 |
| 1.8 | 1.8 | 1.827 | 0.874 | 2.030 | 0.312 |
| 2.0 | 2.0 | 2.000 | 0.857 | 2.000 | −0.667 |

Since $f'(x) \equiv 1$ for $1 \le x \le 2$, this example shows how well the cubic spline can duplicate the general shape of $f(x)$. We shall return to this point when we discuss numerical differentiation in the next chapter.

If the derivatives of $f(x)$ at the endpoints are known, that is, if $f'(x_0) \equiv y_0'$ and $f'(x_n) \equiv y_n'$ are given, then we may suspect that we will get a better cubic spline approximation to $f(x)$ if we choose the two free parameters, $y_0''$ and $y_n''$, in a manner such that the resulting cubic spline, $S(x)$, satisfies $S'(x_0) = f'(x_0) = y_0'$ and $S'(x_n) = f'(x_n) = y_n'$ as well as the interpolation property, $S(x_j) = f_j$, $0 \le j \le n$. From Eq. (5.44) we see for $j = n - 1$ that

$$f'(x_n) = S_{n-1}'(x_n) = \frac{y_n''}{2h_{n-1}}(x_n - x_{n-1})^2 + \frac{\Delta f_{n-1}}{h_{n-1}} - \frac{h_{n-1}}{6}(y_n'' - y_{n-1}''),$$

or

$$\frac{h_{n-1}}{3} y''_n + \frac{h_{n-1}}{6} y''_{n-1} = y'_n - \frac{\Delta f_{n-1}}{h_{n-1}}. \tag{5.47a}$$

Similarly, from (5.44) with $j = 0$, we obtain

$$\frac{h_0}{6} y''_1 + \frac{h_0}{3} y''_0 = \frac{\Delta f_0}{h_0} - y'_0. \tag{5.47b}$$

The continuity of $S'(x)$ must be maintained so all of the equations of (5.45) must still hold. If Eqs. (5.45) are written together with (5.47), we obtain an $(n + 1) \times (n + 1)$ tridiagonal diagonally dominant linear system of equations. These may be easily solved as before and again substituted into (5.43) to obtain $S(x)$. (There are other ways of deriving this particular cubic spline, but they involve a different derivation. See Rivlin, 1969.)

For brevity of notation we shall denote the above cubic spline as $S^{(1)}(x)$ and the natural cubic spline as $S^{(2)}(x)$. These two particular cubic spines are the ones that are most often used in practice. The reasons for this are the so-called *extremal properties* that we shall now develop. First we note by a simple algebraic reduction that, for any $g(x)$ and $S(x) \in C^2[a, b]$,

$$\int_a^b [g''(x) - S''(x)]^2 \, dx = \int_a^b g''(x)^2 \, dx - \int_a^b S''(x)^2 \, dx - 2 \int_a^b S''(x)[g''(x) - S''(x)] \, dx. \tag{5.48}$$

We concentrate on the last integral in (5.48) and see that integration by parts yields

$$\int_a^b S''(x)[g''(x) - S''(x)] \, dx = S''(x)[g'(x) - S'(x)]\Big|_a^b - \int_a^b S'''(x)[g'(x) - S'(x)] \, dx. \tag{5.49}$$

For the first extremal property we let $g(x)$ be any function in $C^2[a, b]$ that interpolates $f(x)$ at $\{x_j\}_{j=0}^n$. If $S(x) \in Sp(X_n)$, then $S'''(x)$ is a constant, say $\alpha_j$, on each subinterval $(x_j, x_{j+1})$. Therefore

$$\int_a^b S'''(x)[g'(x) - S'(x)] \, dx = \sum_{j=0}^{n-1} \alpha_j \int_{x_j}^{x_{j+1}} [g'(x) - S'(x)] \, dx = \sum_{j=0}^{n-1} \alpha_j [g(x) - S(x)] \Big|_{x_j}^{x_{j+1}} = 0,$$

since $g(x_j) = f(x_j) = S(x_j)$, $0 \le j \le n$.

Now if $S(x) = S^{(2)}(x)$ (the natural cubic spline), then $S''(b) = S''(a) = 0$, and so the integral on the left-hand side of (5.49) is zero. Using this result in (5.48) yields

$$\int_a^b g''(x)^2 \, dx = \int_a^b S''(x)^2 \, dx + \int_a^b [g''(x) - S''(x)]^2 \, dx \ge \int_a^b S''(x)^2 \, dx. \tag{5.50}$$

We leave it to the reader to show that equality holds in (5.50) if and only if $g(x) \equiv S^{(2)}(x)$. Hence, by (5.50), among all possible functions in $C^2[a, b]$ which interpolate $f(x)$ (including all of $Sp(X_n)$, all interpolating polynomials, and even $f(x)$ itself if $f''(x)$ is continuous), the integral, $\int_a^b g''(x)^2\, dx$, is minimized if and only if $g(x) = S^{(2)}(x)$. This is the first extremal property and it explains the origin of the name "spline" for the mathematical approximation, since the "strain energy" of the draftsman's elastic rod is essentially proportional to the integral of the square of the second derivative, which we see is minimized by $S^{(2)}(x)$. The property is also called the "minimum curvature" property, since the curvature of any approximation is essentially the integral of the square of the second derivative. Thus the oscillatory behavior of the approximation is minimized by $S^{(2)}(x)$. Yet another interpretation of (5.50) is that among all functions in $C^2[a, b]$ that interpolate $f(x)$, the natural cubic spline is closest to being a broken line, since the broken line, $q(x)$, has zero curvature, that is, $\int_a^b q''(x)^2\, dx = 0$.

For the second extremal property we now restrict the function $g(x)$ in (5.49) to be in the smaller set of functions which interpolate $f(x)$ *and* also satisfy $g'(a) = f'(a)$ and $g'(b) = f'(b)$ (such as the Hermite interpolating polynomial, for instance). This time we let $\hat{S}(x) = S^{(1)}(x)$ so $\hat{S}'(a) = f'(a) = g'(a)$ and $\hat{S}'(b) = f'(b) = g'(b)$. Using this, we again see that the integral in (5.49) is zero. Therefore, by (5.48), for all $g(x)$ of this particular form,

$$\int_a^b g''(x)^2\, dx = \int_a^b \hat{S}''(x)^2\, dx + \int_a^b [g''(x) - \hat{S}''(x)]^2\, dx \qquad \textbf{(5.51)}$$

((5.51) should not be confused with (5.50), since $g(x)$ is now more restricted but note that $S^{(1)}(x)$ has minimum curvature in this smaller class of functions.)) Now we let $u(x)$ be any cubic spline on the points $\{x_j\}_{j=0}^n$, whether it interpolates $f(x)$ or not. We then let $g(x) \equiv f(x) - u(x)$ and $S(x) \equiv S^{(1)}(x) - u(x)$. Note that $S(x)$ is still a cubic spline and has the properties that $S(x_j) = g(x_j)$, $0 \le j \le n$, and $S'(a) = g'(a)$ and $S'(b) = g'(b)$. Thus this particular choice of $g(x)$ and $S(x)$ can be used in (5.51) and yields $\int_a^b g''(x)^2\, dx \ge \int_a^b [g''(x) - S''(x)]^2\, dx$, or

$$\int_a^b \left[\frac{d^2f(x)}{dx^2} - \frac{d^2u(x)}{dx^2}\right]^2 dx \ge \int_a^b \left[\frac{d^2f(x)}{dx^2} - \frac{d^2S^{(1)}(x)}{dx^2}\right]^2 dx. \qquad \textbf{(5.52)}$$

(We leave it to the reader to verify that equality holds in (5.52) if and only if $u(x) = S^{(1)}(x) + \alpha x + \beta$.) Formula (5.52) is the second extremal property. It says that if we measure the distance between $f(x)$ and any cubic spline $u(x)$ by the formula $\int_a^b [f''(x) - u''(x)]^2\, dx$, then this distance is minimized when $u(x) = S^{(1)}(x)$ (so $S^{(1)}(x)$ is a "best approximation" to $f(x)$ in this sense). To summarize, among all cubic splines $u(x)$, the *error* $f(x) - u(x)$ has *minimum curvature* when $u(x) = S^{(1)}(x)$. Among all functions $g(x)$ in $C^2[a, b]$ that interpolate $f(x)$, the *function* with minimum curvature is $g(x) = S^{(2)}(x)$. Among all functions $g(x)$ that interpolate $f(x)$ and satisfy $g'(a) = f'(a)$, $g'(b) = f'(b)$, the function with minimum curvature is $g(x) = S^{(1)}(x)$.

We conclude this section by giving error bounds on the approximation of $f(x)$ by $S(x)$ *and* the approximation of $f'(x)$ by $S'(x)$, where $S(x)$ can be taken to be either $S^{(1)}(x)$ or $S^{(2)}(x)$. For the sake of notation we let the error function be given by $E(x) = f(x) - S(x)$, for $x \in [a, b]$. Then $E'(x) = f'(x) - S'(x)$ is the error of the derivative approximation. We also let

$$h \equiv \max_{0 \le j \le n-1} (x_{j+1} - x_j),$$

the maximum step-size. Let $x$ be any arbitrary fixed point in $[a, b]$, then there exists some $j$, $0 \le j \le n - 1$, such that $x \in [x_j, x_{j+1}]$. Since $E(x_j) = E(x_{j+1}) = 0$, by Rolle's theorem there exists a point $c \in [x_j, x_{j+1}]$ such that $E'(c) = 0$. Thus, $\int_c^x E''(t)\,dt = E'(x) - E'(c) = E'(x)$. By the Cauchy-Schwarz inequality (see Problem 9, Section 2.6, or Theorem 5.8, Section 5.3), we have

$$\begin{aligned} |E'(x)|^2 &= \left|\int_c^x E''(t) \cdot 1\,dt\right|^2 \le \left(\int_c^x E''(t)^2\,dt\right)\left(\int_c^x 1^2\,dt\right) \\ &\le \left(\int_c^x E''(t)^2\,dt\right)|x - c| \le h \int_c^x E''(t)^2\,dt. \end{aligned} \tag{5.53}$$

From either (5.50) or (5.51) we find

$$\begin{aligned} \int_c^x E''(t)^2\,dt &\le \int_a^b E''(t)^2\,dt = \int_a^b [f''(t) - S''(t)]^2\,dt \\ &= \int_a^b f''(t)^2\,dt - \int_a^b S''(t)^2\,dt \le \int_a^b f''(t)^2\,dt. \end{aligned}$$

Substituting this expression into (5.53) and taking square roots yields

$$|E'(x)| \equiv |f'(x) - S'(x)| \le h^{1/2}\left(\int_a^b f''(t)^2\,dt\right)^{1/2}, \qquad \text{for all } x \in [a, b]. \tag{5.54}$$

Thus, for $x \in [a, b]$, $|f'(x) - S'(x)|$ is bounded by an expression proportional to $h^{1/2}$.

Now, as above, let $x$ be fixed in $[a, b]$ and thus $x \in [x_j, x_{j+1}]$ for some $j$. Since $E(x) = \int_{x_j}^x E'(t)\,dt = E(x) - E(x_j) = E(x) - 0$, then

$$|E(x)| = \left|\int_{x_j}^x E'(t)\,dt\right| \le \int_{x_j}^x \left(\max_{a \le z \le b} |E'(z)|\right) dt \le h \max_{a \le z \le b} |E'(z)|.$$

Then by (5.54) we have

$$|E(x)| \le h^{3/2}\left(\int_a^b f''(t)^2\,dt\right)^{1/2} \tag{5.55}$$

giving a bound for $|f(x) - S(x)|$ that is proportional to $h^{3/2}$.

Formulas (5.54) and (5.55) are important not only as error bounds, but also since the bounds are independent of $x$ they tell us that if we increase the number of interpolating points in a manner such that $h \to 0$ as $n \to \infty$, then $S(x)$ and

$S'(x)$ converge uniformly to $f(x)$ and $f'(x)$, respectively. The inequality (5.54) is also significant in that it tells us that $S'(\alpha)$ is a good approximation to $f'(\alpha)$ for $\alpha \in [a, b]$. Thus cubic splines can be used as a method to find a numerical approximation not only for $f(x)$, but $f'(x)$ as well (see Example 5.12). This property is not shared by interpolating polynomials, $p_n(x)$, as their oscillatory behavior tends to exaggerate the difference between $f'(x)$ and $p_n'(x)$ and one must be extremely cautious in using interpolating polynomials for the purpose of numerical differentiation.

## PROBLEMS

1. To illustrate what can happen in even the simplest Hermite-Birkhoff interpolation problems, consider the following three problems:
   a) find $p \in \mathscr{P}_3$ such that $p(0) = 1, p'(0) = 1, p'(1) = 2, p(2) = 1$
   b) find $p \in \mathscr{P}_3$ such that $p(-1) = 1, p'(-1) = 1, p'(1) = 2, p(2) = 1$
   c) find $p \in \mathscr{P}_3$ such that $p(-1) = 1, p'(-1) = -6, p'(1) = 2, p(2) = 1$.

   Using the method of undetermined coefficients, show that problem (a) has a unique solution, problem (b) has no solution, and problem (c) has infinitely many solutions.
2. Find the third-degree Hermite interpolating polynomial for $f(x) = \cos(x)$ on $[0.3, 0.6]$ and compare the results with those of Example 5.3 ($\sin(0.3) \approx 0.295520$ and $\sin(0.6) \approx 0.564642$).
3. Find the polynomial $p(x)$ in $\mathscr{P}_4$ that interpolates $f(x) = |x|$ as follows: $p(-2) = f(-2)$, $p'(-2) = f'(-2)$, $p(0) = f(0)$, $p(2) = f(2)$ and $p'(2) = f'(2)$. Compare your results with those of Example 5.12 to see that this polynomial is generally better than the interpolating polynomial but not as good as the cubic spline.
4. Write a computer program to find the natural cubic spline, solving the tridiagonal system (5.46) and using (5.43) to evaluate the spline. Test your program on the data of Example 5.11.
5. Modify the program of Problem 4 to include the calculation of $S^{(1)}(x)$, as well as $S^{(2)}(x)$. Test this program on $f(x) = e^{3x}$, using $n = 5$ and $n = 10$ in (5.40).
6. Graph the derivative of the broken line, $q(x)$, in Example 5.11.
7. Verify that the polynomials $A_j(x)$ and $B_j(x)$ in (5.38a) satisfy the conditions:

$$A_j(x_i) = \delta_{ij} \qquad B_j(x_i) = 0$$
$$A_j'(x_i) = 0 \qquad B_j'(x_i) = \delta_{ij}.$$

8. Verify, for $N = 1$, that (5.38a) reduces to (5.37).
9. In some cases, the error bound (5.55) for cubic spline approximation may be pessimistic. How large must $n$ be in order that $|E(x)| \leq 10^{-3}$ for $f(x) = e^{3x}$, $[a, b] = [-1, 1]$, and $h = 2/n$? Contrast this with the results of Problem 5.

*10. Verify that equality holds in (5.50) if and only if $g(x) = S^{(2)}(x)$. Verify that equality holds in (5.52) if and only if $u(x) = S^{(1)}(x) + \alpha x + \beta$. (Recall that $\int_a^b h(x)^2\,dx > 0$ for any function $h(x)$ that is continuous on $[a, b]$, unless $h(x) \equiv 0$ on $[a, b]$.)

## 5.3 ORTHOGONAL POLYNOMIALS AND LEAST-SQUARES APPROXIMATIONS

Given a function $f(x)$, all of the approximations for $f(x)$ that we have discussed thus far have been interpolatory (in that they matched $f(x)$ and/or its derivatives at a predetermined set of points). We now consider another type of method to approximate $f(x)$: the least-squares or Fourier approach. We shall restrict ourselves to approximation by polynomials (algebraic or trigonometric) although the theory extends to more general approximations. Two simple but important problems serve to illustrate our objective:

*Problem A* Let $w_0, w_1, \ldots, w_m$ be a set of positive constants (weights). Given data points $(x_0, y_0), (x_1, y_1), \ldots, (x_m, y_m)$, $m > n$, find $p^*(x) \in \mathscr{P}_n$ such that $\sum_{i=0}^{m} w_i[p^*(x_i) - y_i]^2$ is minimized.

*Problem B* Let $w(x)$ be a function that is continuous and positive on $(a, b)$; find $p^*(x) \in \mathscr{P}_n$ such that $\int_a^b w(x)[p^*(x) - f(x)]^2\,dx$ is minimized.

Perhaps the most important feature about these two problems is that they are "easy" to solve. In both cases, a unique solution exists and, moreover, the solution can be computed in a finite number of steps by a sequence of formulas, once the underlying mathematical theory is understood. By contrast, the problem of finding $p^*(x) \in \mathscr{P}_n$ to minimize $\|p^* - f\|_\infty$ (see (5.1c)), is not usually solvable explicitly in terms of formulas. (Although $p^*(x)$ can be approximated by various iterative procedures, we cannot normally hope to find $p^*(x)$ precisely.) To give an idea of the sorts of approximations the method of least-squares generates, we present an example below:

**Example 5.13.** In this example, we consider Problem A, with $m = 10$ and $w_i = 1$, $i = 0, 1, \ldots, 10$. Using the data of Example 5.11, we can find polynomials $p(x) \in \mathscr{P}_3$ and $q(x) \in \mathscr{P}_5$ to minimize the summation for Problem A. The procedure we use, in this example, is the procedure outlined in Section 2.5 using the matrix equation (2.74).

Solving (2.74) (in a least-squares sense, using (2.73)), we find:

$$p(x) = 0.010685x^3 - 0.202239x^2 + 0.939874x + 2.34959$$

$$q(x) = -0.002162x^5 + 0.056869x^4 - 0.517548x^3 + 1.82337x^2 - 1.79051x + 2.80705$$

and

$$\sum_{i=0}^{10} [p(x_i) - y_i]^2 = 5.78, \qquad \sum_{i=0}^{10} [q(x_i) - y_i]^2 = 3.08.$$

The graphs of these two approximations are given in Fig. 5.8.

A more realistic example of data fitting is provided in Example 4.0, where the setting is that of finding an approximate aerodynamic model for a ballistic reentry trajectory. In deriving the Allen and Eggers reentry model in Example 4.0, one assumption made was that atmospheric density, $\rho(H)$, was given by $\rho(H) = \rho_0 e^{-\lambda H}$. In reality, atmospheric density is only approximately exponential, so we ask for constants $\rho_0$ and $\lambda$ that minimize

$$\sum_{i=1}^{N} [\rho(H_i) - \rho_0 e^{-\lambda H_i}]^2$$

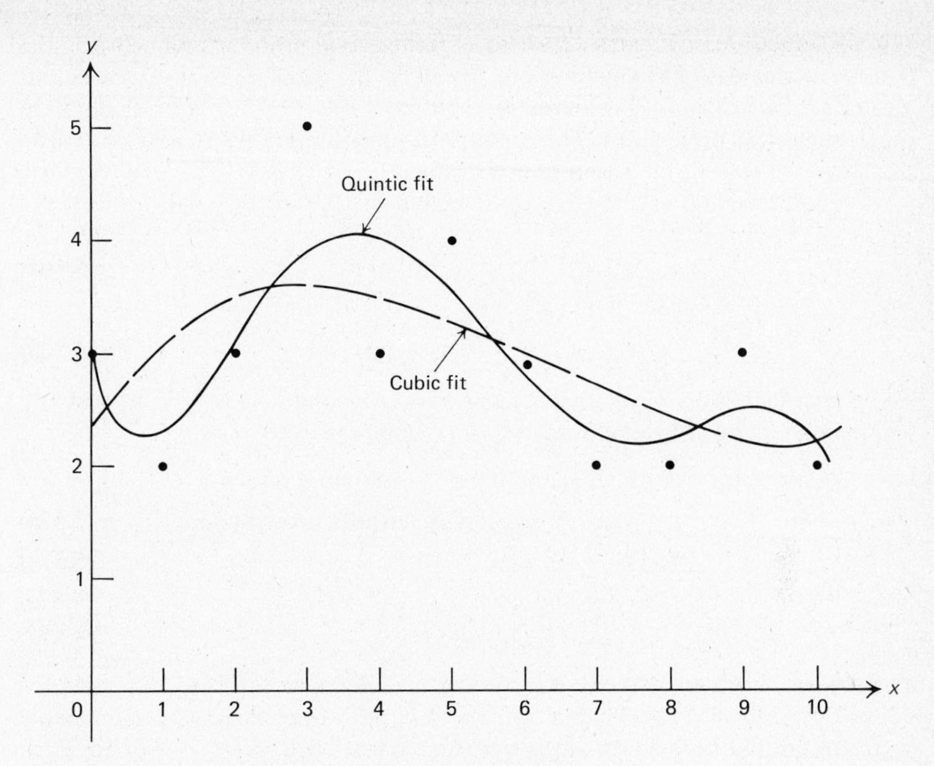


**Fig. 5.8** Cubic and quintic fits of data from Example 5.11.

where the values $\rho(H_i)$ denote the actual atmospheric density at selected altitudes, $H_i$. Solving the problem above gives a reasonable density model for the Allen and Eggers reentry model. The least-squares problem above is an example of a nonlinear least-squares problem, since the parameters $\rho_0$ and $\lambda$ occur in a nonlinear fashion (unlike the coefficients $a_i$ of the polynomial $p^*(x)$ in Problem A or Problem B). Usually such nonlinear problems are quite difficult, but in this case we can convert the problem to one in which the parameters occur linearly by fitting $\ln(\rho_0 e^{-\lambda H_i})$ to the values $\ln(\rho(H_i))$. This gives a minimization problem of the form

$$\sum_{i=1}^{N} [y_i - (a - \lambda H_i)]^2$$

where $y_i = \ln(\rho(H_i))$ and $a = \ln(\rho_0)$. Thus we have reduced the problem to that of fitting a straight line to $N$ data values $y_1, \ldots, y_N$. The resulting problem can be solved and gives an approximation $\rho_0 e^{-\lambda H}$ to $\rho(H)$ which has a relative error of 25 percent or less over an altitude range of 0 to 200,000 feet.

While the method of Section 2.5 serves to solve the problem posed in Example 5.13, it is not the most efficient method computationally. In addition, the matrix equation (2.73) which was used to generate $p(x)$ and $q(x)$, can frequently be ill-conditioned. Finally, the method of Chapter 2 cannot be used to solve Problem B.

For all of these reasons, many modern computational procedures for finding the best least-squares approximation are based on the ideas of *inner-products* and *orthogonal polynomials.* Furthermore these concepts are also fundamental to other numerical procedures. These powerful tools are relatively easy to understand and we present the basics in this section.

For any $f(x)$ and $g(x)$ in $C[a, b]$, we define

$$\langle f, g\rangle_1 = \sum_{j=0}^{m} w_j f(x_j) g(x_j) \tag{5.56a}$$

$$\langle f, g\rangle_2 = \int_a^b w(x) f(x) g(x)\, dx. \tag{5.56b}$$

The reader can easily verify that for any $f(x)$, $g(x)$, and $h(x)$ in $C[a, b]$ and any real number $\alpha$, the following properties for (5.56b) are valid:

$$\langle f, f\rangle_2 \geq 0 \text{ and } \langle f, f\rangle_2 = 0 \text{ if and only if}$$
$$f(x) = \theta(x), \text{ where } \theta(x) \text{ is the zero function} \tag{5.57a}$$
$$\langle f, g\rangle_2 = \langle g, f\rangle_2 \tag{5.57b}$$
$$\langle \alpha f, g\rangle_2 = \alpha\langle f, g\rangle_2 \tag{5.57c}$$
$$\langle f, g + h\rangle_2 = \langle f, g\rangle_2 + \langle f, h\rangle_2. \tag{5.57d}$$

The reader can also easily see that properties (b), (c), and (d) are also valid for $\langle f, g\rangle_1$ given by (5.56a) and that $\langle f, f\rangle_1 \geq 0$. However, there are innumerable nonzero continuous functions, $f(x)$, such that $f(x_j) = 0, 0 \leq j \leq m$. For any such function, we have $\langle f, f\rangle_1 = 0$ but $f(x) \neq \theta(x)$. However, if we only consider functions $f(x)$ which are polynomials of degree $m$ or less, then it is not possible to have $f(x_j) = 0, 0 \leq j \leq m$. Thus, for all $f(x) \in \mathscr{P}_m$, $\langle f, f\rangle_1 = 0$ if and only if $f(x) = \theta(x)$. So the four properties of (5.57) are satisfied by (5.56a) if the functions are restricted to be in $\mathscr{P}_m$. We note that the four properties of (5.57) are also satisfied for all vectors in $R^n$ (that is, all $n$-tuples), where $\langle \mathbf{x}, \mathbf{y}\rangle$ is defined to be the usual inner-product: $\langle \mathbf{x}, \mathbf{y}\rangle \equiv \mathbf{x}^T\mathbf{y}$, for $\mathbf{x}$ any $\mathbf{y}$ in $R^n$ (recall 2.75, Chapter 2). We call both (5.56a and b) *inner-products.* (Equation (5.56a) is often called the *discrete inner-product.*) In the material to follow we will use $\langle f, g\rangle$ to denote both $\langle f, g\rangle_1$ and $\langle f, g\rangle_2$ when we make statements which are valid for both.

We first notice that

$$\langle f - p, f - p\rangle_2^{1/2} = \left(\int_a^b (f(x) - p(x))^2 w(x)\, dx\right)^{1/2} \equiv \|f - p\|_2, \tag{5.58a}$$

which is the norm given by (5.1b). This norm can be used as a way of measuring "distance" between $f(x)$ and $p(x)$, as was discussed in Section 1, and suggests that we can get a different distance measurement between $f(x)$ and $p(x)$ if we define $\|f - p\|_d$ by

$$\langle f - p, f - p\rangle_1^{1/2} = \left(\sum_{j=0}^{m} (f(x_j) - p(x_j))^2 w_j\right)^{1/2} \equiv \|f - p\|_d \tag{5.58b}$$

(We use the subscript "$d$" to denote the discrete form.) Now we can show that $\|f\|_d \equiv \langle f, f\rangle_1^{1/2}$, satisfies the properties of a norm (formula (2.37), Section 2.3.1), *except* it is possible to have a continuous function, $f(x)$, such that $\|f\|_d = 0$ but $f(x) \neq \theta(x)$. (Again this possibility is eliminated if we restrict $f(x)$ to be in $\mathscr{P}_m$.) Both of these norms, or distance measurements, are important in practical computational approximation problems because we are able to develop an efficient technique to find polynomials, $p^*(x)$ and $p_d^*(x)$ in $\mathscr{P}_n$, such that $\|f - p^*\|_2 \leq \|f - p\|_2$ and $\|f - p_d^*\|_d \leq \|f - p\|_d$, for all $p(x) \in \mathscr{P}_n$, that is, we can find the *best* $n$th-degree polynomial approximation to $f(x)$. (We note that it is usually the case that $p^*(x) \neq p_d^*(x)$.)

Both of these inner-products satisfy the Cauchy-Schwarz inequality given by

**Theorem 5.8. Cauchy-Schwarz.** Let $f(x)$ and $g(x) \in C[a, b]$, then

$$|\langle f, g\rangle| \leq (\langle f, f\rangle)^{1/2}(\langle g, g\rangle)^{1/2} \tag{5.59}$$

*Proof*: For either inner-product, if $\langle g, g\rangle = 0$ then $\langle f, g\rangle = 0$, and the theorem is trivially true. Let us assume, then, that $\langle g, g\rangle \neq 0$. Then by (5.57), for *any* scalar $\alpha$,

$$0 \leq \langle f - \alpha g, f - \alpha g\rangle = \langle f, f\rangle - 2\alpha\langle f, g\rangle + \alpha^2\langle g, g\rangle.$$

We use the particular value $\alpha = \langle f, g\rangle/\langle g, g\rangle$ and this expression yields

$$0 \leq \langle f, f\rangle - \langle f, g\rangle^2/\langle g, g\rangle.$$

Transferring $\langle f, g\rangle^2/\langle g, g\rangle$ to the left-hand side and taking square roots yields the desired result. ■

Letting $\|\cdot\|$ denote either $\|\cdot\|_2$ or $\|\cdot\|_d$ we can use the Cauchy-Schwarz inequality to prove that the triangle inequality, $\|f + g\| \leq \|f\| + \|g\|$, holds for both.

**Corollary.** Given $f(x)$ and $g(x) \in C[a, b]$ and $\|\cdot\|$ as above, then

$$\|f + g\| \leq \|f\| + \|g\|$$

*Proof*: $\|f + g\|^2 = \langle f + g, f + g\rangle = \langle f, f\rangle + 2\langle f, g\rangle + \langle g, g\rangle$
$\leq \|f\|^2 + 2\|f\|\,\|g\| + \|g\|^2 = (\|f\| + \|g\|)^2$. ■

To develop our least-squares approximation technique we must introduce the concepts of orthogonality and orthonormality. Since we are primarily interested in polynomial approximation, we shall give these definitions only for polynomials in $\mathscr{P}_m$, although the generalizations to a broader class of functions are obvious. If $p(x)$ and $q(x)$ are in $\mathscr{P}_m$, we say that $p(x)$ and $q(x)$ are *orthogonal* if $\langle p, q\rangle = 0$. Furthermore, if $\{p_j(x)\}_{j=0}^n$ is a set of polynomials in $\mathscr{P}_m$, we say that this set is *orthogonal* if $\langle p_i, p_j\rangle = 0$ for all $i \neq j$. (Thus we see that the concept of orthogonality in $\mathscr{P}_m$ is a direct generalization of the familiar concept of perpendicularity in $R^n$.) We say that the set is *orthonormal* if $\langle p_i, p_j\rangle = \delta_{ij}$. Note: if $p(x) \in \mathscr{P}_m$ and if $q(x) = p(x)/\|p\|$, then $\langle q, q\rangle = \langle p, p\rangle/\|p\|^2 = 1$. Thus to make an orthogonal set orthonormal, all we need do is to divide each element in the set by its norm. Note also that a set can be orthogonal with respect to one inner-product and not to another.

We see by Problem 4 that the set of polynomials, $S = \{1, x, x^2, \ldots, x^n\}$, $n \geq 2$, can never be orthogonal with respect to either of the inner-products (5.56a or b), no matter how the weighting function or the weights are chosen. Yet any polynomial, $p(x) \in \mathscr{P}_n$, can be written as a linear combination of the elements of $S$, that is, there exist constants, $\{a_i\}_{i=0}^n$, such that $p(x) = a_0 + a_1 x + \cdots + a_n x^n$. Given an inner-product we would like to be able to express any $p(x) \in \mathscr{P}_n$ as a linear combination of orthogonal polynomials. The first step in this is to transform $S$ into a different set of polynomials which is orthogonal. This can be done via the Gram-Schmidt algorithm given by the following theorem. (See also Problem 4, Section 3.4.)

**Theorem 5.9. Gram-Schmidt.** Given $S = \{1, x, x^2, \ldots, x^n\}$, let

$$q_0(x) \equiv 1 \qquad \text{and} \qquad p_0(x) \equiv q_0(x)/\|q_0\|$$

and let

$$q_k(x) = x^k - \sum_{j=0}^{k-1} \langle x^k, p_j \rangle p_j(x)$$

and

$$p_k(x) = q_k(x)/\|q_k\|, \qquad \text{for} \qquad 1 \leq k \leq n.$$

Then: $\|q_k\| > 0$ for each $k$, the set $\{q_k(x)\}_{k=0}^n$ is orthogonal, and the set $\{p_k(x)\}_{k=0}^n$ is orthonormal.

*Proof*: We use induction and assume that the set $\{q_k(x)\}_{k=0}^{r-1}$ is orthogonal and that $\|q_k\| > 0$, where $r \leq n$. Note that for $j \leq r - 1$, $j \neq k$, $\langle q_k, p_j \rangle = \langle q_k, q_j/\|q_j\| \rangle = \langle q_k, q_j \rangle / \|q_j\| = 0$ by the induction hypotheses. Now, with $k \leq r - 1$, and using $p_k(x) = q_k(x)/\langle q_k, q_k \rangle^{1/2}$

$$\langle q_k, q_r \rangle = \langle q_k, x^r \rangle - \sum_{j=0}^{r-1} \langle x^r, p_j \rangle \langle p_j, q_k \rangle$$

$$= \langle q_k, x^r \rangle - \langle x^r, q_k \rangle \langle q_k, q_k \rangle / \langle q_k, q_k \rangle = 0.$$

Since $q_r(x)$ is a monic polynomial of degree $r$, $\|q_r\| > 0$, completing the proof. ■

**Example 5.14.** In this example, we illustrate the Gram-Schmidt process for each of the inner-products of (5.56). Let $x_0 = -1$, $x_1 = -1/2$, $x_2 = 0$, $x_3 = 1/2$, $x_4 = 1$ and find $p_0(x)$, $p_1(x)$, $p_2(x)$ for the inner product (5.56a), with $w_i = 1$, $0 \leq i \leq 4$. First, we set $q_0(x) = 1$ and find

$$\langle q_0, q_0 \rangle = \sum_{i=0}^{4} q_0(x_i) q_0(x_i) = 5,$$

so $p_0(x) = 1/\sqrt{5}$. Next, $q_1(x) = x - \langle x, p_0 \rangle p_0(x)$. But

$$\langle x, p_0 \rangle = \frac{1}{\sqrt{5}} \sum_{i=0}^{4} x_i = 0,$$

so $q_1(x) = x$. Since $\langle q_1, q_1 \rangle = 5/2$, we have $p_1(x) = \sqrt{2/5}x$. Finally, $q_2(x) = x^2 - \langle x^2, p_0 \rangle p_0(x) - \langle x^2, p_1 \rangle p_1(x)$; where we see that $\langle x^2, p_0 \rangle = 5/(2\sqrt{5})$ and $\langle x^2, p_1 \rangle = 0$. Thus $q_2(x) = x^2 - 1/2$ and $p_2(x) = \sqrt{8/7}(x^2 - 1/2)$.

Next, let us consider the inner-product $\langle f, g \rangle = \int_{-1}^{1} f(x)g(x)\,dx$ and let $S = \{1, x, x^2\}$. In this case, the Gram-Schmidt process of Theorem 5.9 generates the orthonormal polynomials $p_0(x) = 1/\sqrt{2}$, $p_1(x) = \sqrt{3/2}x$ and $p_2(x) = \sqrt{45/8}(x^2 - 1/3)$. For any integer $n$, we could continue the Gram-Schmidt process and find the orthonormal set $\{p_0(x), p_1(x), \ldots, p_n(x)\}$. The polynomials in this set are known as the *Legendre* polynomials and are widely used for approximation and for numerical integration. The Gram-Schmidt process becomes cumbersome for large values of $n$, and in the next section we will derive a more efficient procedure for generating orthonormal polynomials.

Theorem 5.9 presents a simplified version of the Gram-Schmidt Theorem, but it goes beyond the purposes of our text to introduce the concepts necessary to generalize it. (The reader should be careful not to make the mistake in using the Gram-Schmidt process of assuming that $\|1\| = \|q_0\| = 1$. This is usually *not* true. For instance, in Example 5.14,

$$\langle f, g \rangle = \int_{-1}^{1} f(x)g(x)\,dx, \qquad \text{so} \qquad \|1\| = \left(\int_{-1}^{1} 1^2\,dx\right)^{1/2} = \sqrt{2}.$$

Thus, in this case, $p_0(x) = q_0(x)/\|q_0\| = 1/\sqrt{2}$). We note in this version of Theorem 5.9, that each $q_k(x)$ (and thus each $p_k(x)$), has degree exactly equal to $k$ and $q_k(x)$ has leading coefficient equal to 1, that is, it is monic. We further see that for each $k \geq 1$, $x^k$ can be written as

$$x^k = q_k(x) + \sum_{j=0}^{k-1} \langle x^k, p_j \rangle p_j(x) = \|q_k\| p_k(x) + \sum_{j=0}^{k-1} \langle x^k, p_j \rangle p_j(x).$$

Thus each $x^k$ can be written as a linear combination of the polynomials $\{p_j(x)\}_{j=0}^{k}$. Hence if $p(x) = a_0 + a_1 x + a_2 x^2 + \cdots + a_r x^r = \sum_{k=0}^{r} a_k x^k$, where $r \leq n$, then each $x^k$ may be replaced by its expression in terms of $\{p_j(x)\}_{j=0}^{k}$, and by collecting like coefficients of each $p_j(x)$, we may write $p(x) = \sum_{j=0}^{r} \beta_j p_j(x)$. Now that we know that any $p(x) \in \mathcal{P}_n$ can be expressed in this manner, it is a simple task to determine the value of each $\beta_j$. Let $k$ be fixed, $0 \leq k \leq n$. Then by the orthonormality of the $p_j(x)$'s

$$\langle p_k, p \rangle = \sum_{j=0}^{n} \beta_j \langle p_j, p_k \rangle = \beta_k \langle p_k, p_k \rangle = \beta_k.$$

Therefore, for any $p(x) \in \mathscr{P}_n$, we can represent $p(x)$ in terms of $p_0(x), p_1(x), \ldots, p_n(x)$ by:

$$p(x) = \sum_{j=0}^{n} \langle p, p_j \rangle p_j(x). \tag{5.60}$$

Now if $p(x) \in \mathscr{P}_r$, where $r < n$, we still have, by the above argument, $p(x) = \sum_{j=0}^{r} \langle p, p_j \rangle p_j(x)$. Therefore $\langle p_n, p \rangle = \sum_{j=0}^{r} \langle p, p_j \rangle \langle p_n, p_j \rangle = 0$ (since $r < n$), and we have proved:

**Corollary 1.** If the set $\{p_j(x)\}_{j=0}^{n}$ is orthonormal, as in Theorem 5.9, and if $p(x)$ is *any* polynomial of degree less than $k$, then $\langle p_i, p \rangle = 0$ for $i \geq k$. (This result is obviously also true for the orthogonal polynomials $\{q_j(x)\}_{j=0}^{n}$ of Theorem 5.9.)

We recall that in $R^3$ there are infinitely many orthonormal sets besides the canonical one, $\{\mathbf{e}_1, \mathbf{e}_2, \mathbf{e}_3\} \equiv \{(1, 0, 0)^T, (0, 1, 0)^T, (0, 0, 1)^T\}$ (for example, any constant rotation of $\{\mathbf{e}_1, \mathbf{e}_2, \mathbf{e}_3\}$.) The same is true for $\mathscr{P}_n$. (For example, using the Gram-Schmidt process on $S$ in the order $x^n, x^{n-1}, \ldots, x, 1$ results in an orthonormal set of polynomials, *each* of which has degree $n$.) However, the orthonormal set $\{p_j(x)\}_{j=0}^{n}$ of Theorem 5.9 *is* unique in the following respect:

**Corollary 2.** Let $\{q_j(x)\}_{j=0}^{n}$ be the orthogonal polynomials generated by the Gram-Schmidt algorithm of Theorem 5.9. Let $\{r_j(x)\}_{j=0}^{n}$ be any other set of orthogonal polynomials with respect to the given inner-product which also satisfies: degree $(r_j(x)) = j$ (= degree $(q_j(x))$), $0 \leq j \leq n$. Then for each $j$, $r_j(x)$ is a constant multiple of $q_j(x)$. Furthermore, $\pm p_j(x) = \pm q_j(x)/\|q_j\|$, $0 \leq j \leq n$, is the only orthonormal set that satisfies: degree $(p_j(x)) = j$.

*Proof*: Let $k \leq n$ be fixed and consider the equation

$$q_k(x) = \sum_{j=0}^{k} \alpha_j r_j(x).$$

Then for any $i$ where $0 \leq i \leq k$, the orthogonality of the $r_j(x)$'s yields

$$\langle q_k, r_i \rangle = \sum_{j=0}^{k} \alpha_j \langle r_i, r_j \rangle = \alpha_i \langle r_i, r_i \rangle.$$

By Corollary 1 we have $\langle q_k, r_i \rangle = 0$ for $0 \leq i \leq k - 1$, and so the above equation yields $\alpha_0 = \alpha_1 = \cdots = \alpha_{k-1} = 0$ and $\alpha_k = \langle q_k, r_k \rangle / \langle r_k, r_k \rangle$. Thus $q_k(x) = \alpha_k r_k(x)$ for $0 \leq k \leq n$.

Now suppose that $\{\tilde{p}_j(x)\}_{j=0}^{n}$ is any orthonormal set of polynomials where degree $(\tilde{p}_j(x)) = j$, $0 \leq j \leq n$, and $c_j \neq 0$ is the leading coefficient of each $\tilde{p}_j(x)$. Let $r_j(x) = \tilde{p}_j(x)/c_j$, $0 \leq j \leq n$, then $\{r_j(x)\}_{j=0}^{n}$ is a monic orthogonal set with degree $(r_j(x)) = j$. By the above argument, $q_j(x) = \alpha_j r_j(x)$, $0 \leq j \leq n$. Thus, for each $j$,

$$p_j(x) = q_j(x)/\|q_j\| = \alpha_j r_j(x)/\|\alpha_j r_j\| = \pm r_j(x)/\|r_j\| = \pm \tilde{p}_j(x). \quad \blacksquare$$

Now that we have developed some basic concepts of orthogonal polynomials and how to generate them, we are ready to prove the fundamental theorem of least-squares or Fourier approximation. Before doing so, however, we give a few more examples to illustrate orthogonality in several different situations.

**Example 5.15.**

a) Let $\langle f, g\rangle = \int_{-\pi}^{\pi} f(x)g(x)\,dx$ and consider the set of trigonometric functions, $S_n = \{1, \cos(jx), \sin(jx)\}_{j=1}^{n}$, for any $n$. Using identities such as

$$\cos(mx)\cos(kx) = 1/2(\cos((m+k)x) + \cos((m-k)x)),$$

the reader can easily verify that $S_n$ is an orthogonal set. Furthermore, $\|1\|^2 = \int_{-\pi}^{\pi} 1^2\,dx = 2\pi$ and $\|\cos(jx)\|^2 = \int_{-\pi}^{\pi} \cos^2(jx)\,dx = \|\sin(jx)\|^2 = \int_{-\pi}^{\pi} \sin^2(jx)\,dx = \pi, 1 \le j \le n$. Thus $S_n^* = \{1/\sqrt{2\pi}, (\cos(jx))/\sqrt{\pi}, (\sin(jx))/\sqrt{\pi}\}_{j=1}^{n}$ is an orthonormal set. This example illustrates our previous statement that the concept of orthogonality goes far beyond the simple results we have presented here. This example is also of historical importance in that it was the orthogonal set first used by Fourier when he introduced the concept of orthogonal expansions in 1807.

b) Let $\langle f, g\rangle = \int_{-1}^{1} f(x)g(x)(1-x^2)^{-1/2}\,dx$ and let $S_n = \{T_j(x)\}_{j=0}^{n}$, where $T_j(x) = \cos(j\cos^{-1}(x))$, the $j$th-degree Chebyshev polynomial. If we make the change of variable, $x = \cos(\theta)$, $-1 \le x \le 1$, and $0 \le \theta \le \pi$, then $T_j(x) = \cos(j\theta)$ and we have

$$\langle T_k(x), T_m(x)\rangle = \int_0^{\pi} \cos(k\theta)\cos(m\theta)\,d\theta = 0,$$

for $k \ne m$. Moreover,

$$\|T_0(x)\|^2 = \int_{-1}^{1} 1^2(1-x^2)^{-1/2}\,dx = \pi$$

and

$$\|T_j(x)\|^2 = \int_{-1}^{1} T_j(x)^2(1-x^2)^{-1/2}\,dx = \int_0^{\pi} \cos^2(m\theta)\,d\theta = \frac{\pi}{2}.$$

Thus $S_n^* = \{T_0(x)/\sqrt{\pi}, \sqrt{2/\pi}\,T_j(x)\}_{j=1}^{n}$ is an orthonormal set with respect to this inner-product.

c) Let $x_j = \cos(j\pi/n)$, $0 \le j \le n$, and let $\langle f, g\rangle = \sum_{j=0}^{\prime\prime\, n} f(x_j)g(x_j)$, where the double prime on the summation means to halve the first and last terms, that is, in Eq. (5.56a), $w_0 = w_n = 1/2$ and $w_j = 1, 1 \le j \le n-1$. This example is important not only in terms of practical discrete least-squares approximation, but also in the analysis of certain numerical integration methods. We claim that the Chebyshev polynomials, $T_j(x) = \cos(j\cos^{-1}(x))$, $0 \le j \le n$, are orthogonal with respect to this inner product. To verify this claim we need the following lemma which is also useful in other settings.

**Lemma 5.1.** Let $S_1(x) = \sum_{j=1}^{n} \cos(jx)$ and $S_2(x) = \sum_{j=0}^{\prime\prime\, n} \cos(jx)$. Then,

$$S_1(x) = \frac{1}{2}\left(\frac{\sin((n+1/2)x)}{\sin(x/2)} - 1\right) \quad \text{and} \quad S_2(x) = \frac{\sin(nx)\cos(x/2)}{2\sin(x/2)}, \tag{5.61}$$

where, since $S_1(x)$ and $S_2(x)$ have continuous derivatives, we can use L'Hôpital's rule to evaluate them when $x = k\pi$ for an even integer $k$.

*Proof*: Assume $x \ne k\pi$, then by the trigonometric identity,

$$\sin(x/2)\cos(jx) = 1/2(\sin((j+1/2)x) - \sin((j-1/2)x)),$$

we can write $S_1(x)$ as the telescoping summation:

$$S_1(x) = (2\sin(x/2))^{-1} \sum_{j=1}^{n} (\sin((j + 1/2)x) - \sin((j - 1/2)x))$$
$$= (2\sin(x/2))^{-1}(\sin((n + 1/2)x) - \sin(x/2)).$$

To obtain a closed form for the sum $S_2(x)$, note that

$$S_2(x) = \sum_{j=0}^{n}{}'' \cos(jx) = 1/2 + S_1(x) - (\cos(nx))/2$$
$$= 1/2 + \left(\frac{\sin((n + 1/2)x)}{2\sin(x/2)}\right) - 1/2 - (\cos(nx))/2$$
$$= \frac{\sin(nx)\cos(x/2) + \sin(x/2)\cos(nx) - \cos(nx)\sin(x/2)}{2\sin x/2}$$
$$= \frac{\sin(nx)\cos(x/2)}{2\sin(x/2)}. \quad \blacksquare$$

Now, continuing Example 5.15(c), with $x = \cos(\theta)$, $T_j(x) = \cos(j\theta)$, $0 \le j \le n$. Let $0 \le k$, $m \le n$ (and without loss of generality let $m \ge k$), then

$$\langle T_m, T_k \rangle = \sum_{j=0}^{n}{}'' T_m(x_j)T_k(x_j) = \sum_{j=0}^{n}{}'' \cos\frac{mj\pi}{n}\cos\frac{kj\pi}{n}$$
$$= 1/2 \sum_{j=0}^{n}{}'' \left(\cos\frac{(m + k)j\pi}{n} + \cos\frac{(m - k)j\pi}{n}\right)$$
$$= 1/2(S_2(y) + S_2(z)),$$

where $y \equiv (m + k)\pi/n$ and $z \equiv (m - k)\pi/n$. Then by (5.61),

$$\langle T_m, T_k \rangle = \frac{\sin[(m + k)\pi]\cos[(m + k)\pi/2n]}{4\sin[(m + k)\pi/2n]} + \frac{\sin[(m - k)\pi]\cos[(m - k)\pi/2n]}{4\sin[(m - k)\pi/2n]}.$$

Thus since $\sin[(m + k)\pi] = \sin[(m - k)\pi] = 0$, $\langle T_m, T_k \rangle = 0$, unless $\sin[(m + k)\pi/2n]$ and/or $\sin[(m - k)\pi/2n] = 0$. This happens when (i) $m = k = n$, (ii) $m = k = 0$, or (iii) $m = k$, $1 \le m \le n - 1$. In cases (i) and (ii), we see easily by the definition of $S_2(x)$, that $S_2(y) = S_2(z) = n$. In case (iii), $S_2(y) = 0$ and $S_2(z) = n$. Putting this all together we have

$$\langle T_m, T_k \rangle = \begin{cases} n, & \text{if } m = k = n \quad \text{or} \quad m = k = 0 \\ n/2, & \text{if } m = k, \quad 1 \le m \le n - 1 \\ 0, & \text{if } m \ne k. \end{cases} \tag{5.62}$$

Hence the orthogonality of the Chebyshev polynomials for this discrete inner-product is established.

We now prove a classical theorem which will show us how to construct the best least-squares $n$th-degree polynomial approximation for any continuous function, $f(x)$.

**Theorem 5.10. Best Least-Squares Approximation.** Let $\langle f, g\rangle$ be the inner-product given by either (5.56a or b), and let $\|f\| \equiv \langle f, f\rangle^{1/2}$. If $f(x) \in C[a, b]$, then the polynomial, $p_n^*(x) \in \mathscr{P}_n$, which satisfies $\|f - p_n^*\| \leq \|f - p\|$ for all $p(x) \in \mathscr{P}_n$, is given by

$$p_n^*(x) = \sum_{j=0}^{n} \langle f, p_j\rangle p_j(x), \tag{5.63}$$

where $\{p_j(x)\}_{j=0}^n$ is the *orthonormal* set of polynomials generated by the Gram-Schmidt theorem. (Among all polynomials in $\mathscr{P}_n$, $p_n^*(x)$ is "closest" to $f(x)$ with respect to the given least-squares norm, that is, formula (5.63) provides the solution to Problem A *and* Problem B.)

*Proof*: Let $p(x)$ be any polynomial in $\mathscr{P}_n$. Then by (5.60) $p(x)$ may be written as $p(x) = \sum_{j=0}^n \alpha_j p_j(x)$. Since $\langle p_i, p_j\rangle = \delta_{ij}$, we may write

$$\begin{aligned}
0 \leq \|f - p\|^2 &= \left\langle f - \sum_{j=0}^{n} \alpha_j p_j, \quad f - \sum_{i=0}^{n} \alpha_i p_i \right\rangle \\
&= \langle f, f\rangle - 2\sum_{j=0}^{n} \alpha_j \langle f, p_j\rangle + \sum_{j=0}^{n} \alpha_j^2 \\
&= \langle f, f\rangle - \sum_{j=0}^{n} \langle f, p_j\rangle^2 + \sum_{j=0}^{n} (\alpha_j^2 - 2\alpha_j\langle f, p_j\rangle + \langle f, p_j\rangle^2) \\
&= \langle f, f\rangle - \sum_{j=0}^{n} \langle f, p_j\rangle^2 + \sum_{j=0}^{n} (\alpha_j - \langle f, p_j\rangle)^2.
\end{aligned} \tag{5.64}$$

Since the right-hand sum is the only term containing the $\alpha_j$'s, the above expression is minimized when we choose $\alpha_j = \langle f, p_j\rangle, 0 \leq j \leq n$. Thus $\|f - p\|$ is minimum when

$$p(x) = \sum_{j=0}^{n} \langle f, p_j\rangle p_j(x) = p_n^*(x).$$

(The coefficients, $\langle f, p_j\rangle$, of (5.63) are called *generalized Fourier coefficients.*) ■

Therefore we can find a best least-squares polynomial approximation to $f(x)$ with respect to either the discrete norm or the integral norm by the following steps: (1) Generate the orthonormal set $\{p_j(x)\}_{j=0}^n$ by the Gram-Schmidt algorithm, (2) Calculate $\langle f, p_j\rangle$ for $0 \leq j \leq n$, and (3) Construct $p_n^*(x)$ by formula (5.63). We could give examples of this here, but we defer them to the next section where we develop a more efficient technique. We reiterate here that $p_n^*(x)$ will be different for different inner-products. It is up to the reader to decide how it is best to choose the weighting function of (5.56b) or the weights of (5.56a) to get a best fit for his particular problem. For example, suppose that $\{f(x_j)\}_{j=0}^m$ represent the results of some physical experiment, and suppose that the reader has reason to believe that $f(x_0)$, $f(x_1)$, $f(x_{m-1})$, and $f(x_m)$ are less reliable than the others. Then in using (5.56a) he or she would choose $w_0$, $w_1$, $w_{m-1}$, and $w_m$ to be smaller than the other weights in order that more emphasis be placed on the more reliable results.

Approximation in a least-squares sense also has many "geometric" properties. For instance, suppose that $\Pi$ is a plane in Euclidean 3-dimensional space and $P$ is a point that does not lie in $\Pi$. If $P^*$ is the point in $\Pi$ closest to $P$, then the vector from $P^*$ to $P$ is perpendicular to all vectors in $\Pi$. The following corollary shows that this geometric concept is valid for the Fourier approximation, $p_n^*(x)$ in (5.63), as well.

**Corollary 1.** Under the hypotheses of Theorem 5.10, the remainder function, $(f(x) - p_n^*(x))$, is orthogonal to every polynomial in $\mathscr{P}_n$.

*Proof*: Since any $p(x)$ in $\mathscr{P}_n$ can be written as $p(x) = \sum_{j=0}^n \alpha_j p_j(x)$, where $\{p_j(x)\}_{j=0}^n$ is orthonormal as above, then our result is valid if $(f(x) - p_n^*(x))$ is orthogonal to each $p_k(x)$, $0 \leq k \leq n$. Now,

$$\langle f - p_n^*, p_k \rangle = \langle f, p_k \rangle - \sum_{j=0}^{n} \langle f, p_j \rangle \langle p_j, p_k \rangle = \langle f, p_k \rangle - \langle f, p_k \rangle = 0. \quad \blacksquare$$

Another corollary of Theorem 5.10 is the following classical result which is immediate from formula (5.64).

**Corollary 2. Bessel's Inequality.** Under the hypotheses of Theorem 5.10, for any $f \in C[a, b]$,

$$\|f\|^2 \geq \sum_{j=0}^{n} \langle f, p_j \rangle^2.$$

We shall see in Section 5.3.2 that Theorem 5.10 and its two corollaries lead to some important practical results which can be used to estimate errors and to accelerate convergence. We conclude this section by again emphasizing that the results presented here are only very special cases of the theory of orthogonal expansions, which is fundamental to many areas of mathematical sciences. For example, discrete least-squares estimation is a primary tool in the statistical analysis of data. As another instance, the integral least-squares expansion is often used in the solution of differential equations which model a physical phenomenon. For example, let $L(y)$ be the differential operator $L(y) = (1 - x^2)y''(x) - xy'(x)$, $-1 \leq x \leq 1$, and suppose we wish to solve the differential equation, $L(y) = h(x)$, for some function $h(x)$. The reader can easily verify that $L(T_n) = -n^2 T_n(x)$, where

$T_n(x)$ is the $n$th degree Chebyshev polynomial, $n = 0, 1, 2, \ldots$, that is $T_n(x)$ satisfies $(1 - x^2)T_n''(x) - xT_n'(x) + n^2T_n(x) = 0$. Suppose that $h(x)$ may be expressed as $h(x) = \sum_{j=0}^{\infty} c_jT_j(x)$ where $c_j = \langle h, T_j\rangle/\langle T_j, T_j\rangle$ and where the inner-product is as in Example 5.15(b). If the solution $y(x)$ can be written $y(x) = \sum_{j=0}^{\infty} \alpha_jT_j(x)$ and if the operation $L(y) = \sum_{j=0}^{\infty} \alpha_jL(T_j)$ is valid, then $L(y) = h$ becomes

$$L(y) = \sum_{j=0}^{\infty} \alpha_jL(T_j) = \sum_{j=0}^{\infty} \alpha_j(-j^2T_j(x)) = h(x) = \sum_{j=0}^{\infty} c_jT_j(x).$$

Therefore, $\alpha_j = -c_j/j^2, j = 1, 2, 3, \ldots$ which determines the solution $y(x)$ in terms of the Fourier coefficients of $h(x)$ (note that $L(T_0) = 0$). This type of problem is very common in the solution of partial differential equations by a technique known as separation of variables. The solution $y(x)$ above is not the complete solution of $L(y) = h$, but we shall not finish the solution since our intention here is only to illustrate the utility and importance of integral least-squares approximations.

### 5.3.1 Efficient Computation of Least-Squares Approximations

We have seen by Theorem 5.10 and formula (5.63) that in order to find the best least-squares approximation in $\mathscr{P}_n$ for $f(x)$ we must calculate an orthonormal set, $\{p_j(x)\}_{j=0}^{n}$, and then calculate the Fourier coefficients, $\{\langle f, p_j\rangle\}_{j=0}^{n}$. For $p_n^*(x)$ to be a useful approximation, we should also be able to evaluate $p_n^*(x)$ easily for any value of $x$. We shall now develop efficient and practical computational procedures for each of these steps.

We have seen in Example 5.14 that it can become quite cumbersome to use the Gram-Schmidt process even for relatively small values of $n$. For example, if we wish to find $p_{10}(x)$, we must not only first find $p_j(x)$, $0 \le j \le 9$, but each $p_j(x)$ is actually explicitly used in the computation of $p_{10}(x)$. Thus we first seek an efficient alternative to the Gram-Schmidt process. The key to finding this alternative is to note that the orthogonal polynomials $\{q_j(x)\}_{j=0}^{n}$ of Theorem 5.9 are monic (leading coefficient is 1) and that for either inner product, (5.56a or b), it is true that $\langle xf(x), g(x)\rangle = \langle f(x), xg(x)\rangle$ for any functions $f(x)$ and $g(x)$. This will enable us to write $q_k(x)$ in terms of $q_{k-1}(x)$ and $q_{k-2}(x)$. Let $k$ be any positive integer such that $2 \le k \le n$. Since $q_k(x)$ is monic,

$$\begin{aligned} q_k(x) &= x^k + \alpha_{k-1}x^{k-1} + \alpha_{k-2}x^{k-2} + \cdots + \alpha_1x + \alpha_0 \\ &= x(x^{k-1} + \alpha_{k-1}x^{k-2} + \cdots + \alpha_1) + \alpha_0 \equiv xr_{k-1}(x) + \alpha_0. \end{aligned}$$

Since $r_{k-1}(x)$ and $q_{k-1}(x)$ are monic in $\mathscr{P}_{k-1}$, we see $r_{k-1}(x) - q_{k-1}(x) \equiv u_{k-2}(x)$ is a polynomial of degree $k - 2$. By (5.60), $(xu_{k-2}(x) + \alpha_0)$ can be written as

$$\sum_{j=0}^{k-1} \beta_jq_j(x),$$

so

$$q_k(x) = xr_{k-1}(x) + \alpha_0 = x(q_{k-1}(x) + u_{k-2}(x)) + \alpha_0$$
$$= xq_{k-1}(x) + \sum_{j=0}^{k-1} \beta_j q_j(x) = (x - \beta_{k-1})q_{k-1}(x) + \sum_{j=0}^{k-2} \beta_j q_j(x).$$

For $i < k - 2$,

$$0 = \langle q_k, q_i \rangle = \langle (x - \beta_{k-1})q_{k-1}(x) + \sum_{j=0}^{k-2} \beta_j q_j(x), q_i(x) \rangle$$
$$= \langle xq_{k-1}, q_i \rangle + \beta_i \langle q_i, q_i \rangle.$$

But $\langle xq_{k-1}, q_i \rangle = \langle q_{k-1}, xq_i \rangle = 0$ by Corollary 1, Theorem 5.9, and hence $\beta_i = 0$ for $i < k - 2$. Thus we have just demonstrated for $k \geq 2$ that $q_k(x)$ can be written in the form

$$q_k(x) = (x - a_k)q_{k-1}(x) - b_k q_{k-2}(x), \qquad k \geq 2. \tag{5.65}$$

Therefore, in order to generate $q_k(x)$, all we need use is $q_{k-1}(x)$ and $q_{k-2}(x)$, a considerable savings of computation over the Gram-Schmidt algorithm. Formula (5.65) is called a *three-term recurrence relation* for orthogonal polynomials. Now all we need to do is find the formulas for $a_k$ and $b_k$ in (5.65). This is quite easily done as follows:

$$0 = \langle q_k, q_{k-1} \rangle = \langle xq_{k-1}, q_{k-1} \rangle - a_k \langle q_{k-1}, q_{k-1} \rangle + 0$$

and

$$0 = \langle q_k, q_{k-2} \rangle = \langle xq_{k-1}, q_{k-2} \rangle - a_k \langle q_{k-1}, q_{k-2} \rangle - b_k \langle q_{k-2}, q_{k-2} \rangle$$
$$= \langle q_{k-1}, xq_{k-2} \rangle - 0 - b_k \langle q_{k-2}, q_{k-2} \rangle.$$

Thus, for $k \geq 2$, we have $a_k = \langle xq_{k-1}, q_{k-1} \rangle / \langle q_{k-1}, q_{k-1} \rangle$ and $b_k = \langle q_{k-1}, xq_{k-2} \rangle / \langle q_{k-2}, q_{k-2} \rangle$ in Eq. (5.65). (We leave to the reader to verify that $\langle q_{k-1}, xq_{k-2} \rangle = \langle q_{k-1}, q_{k-1} \rangle$, and so $b_k = \|q_{k-1}\|^2 / \|q_{k-2}\|^2$.) We also leave to the reader to verify from the Gram-Schmidt algorithm that (5.65) is valid for $k = 1$ if we set $a_1 = \langle xq_0, q_0 \rangle / \langle q_0, q_0 \rangle$, and $b_1 = 0$. We wish to make clear that to obtain $q_{10}(x)$, for example, we still need to know $q_j(x)$, $0 \leq j \leq 9$. But in the computation of $q_{10}(x)$, we only need to use $q_8(x)$ and $q_9(x)$ which is the advantage of (5.65) over the Gram-Schmidt process. For many of the widely used sets of orthogonal polynomials, the constants $a_k$ and $b_k$ appearing in (5.65) have been tabulated. These tables further facilitate the use of (5.65) in least-squares approximation.

Efficient computation of $\langle f, p_j \rangle$, $0 \leq j \leq n$, we will defer to the next chapter. (See especially the sections on the Fast Fourier Transform and Gaussian quadrature and the subsequent development of interpolation at the zeros of orthogonal polynomials.) We merely note here that the computation of $\langle f, p_k \rangle$ can often be facilitated by use of (5.65). Since $p_k(x) = q_k(x)/\|q_k\|$, $\langle f, p_k \rangle = \langle f, q_k \rangle / \|q_k\|$, and by (5.65) the term $\langle f, q_k \rangle$ is given by

$$\langle f, q_k \rangle = \langle f, xq_{k-1} \rangle - a_k \langle f, q_{k-1} \rangle - b_k \langle f, q_{k-2} \rangle.$$

The last two terms of this expression can be easily calculated from the previously calculated values of $\langle f, p_{k-1} \rangle$ and $\langle f, p_{k-2} \rangle$, $k \geq 2$. This leaves only the computation of $\langle f, xq_{k-1} \rangle$, which can often be simplified by the use of integration or summation by parts.

We pause to note here that

$$p_n^*(x) = \left( \sum_{j=0}^{n-1} \langle f, p_j \rangle p_j(x) \right) + \langle f, p_n \rangle p_n(x) = p_{n-1}^*(x) + \langle f, p_n \rangle p_n(x).$$

Thus for any $n \geq 1$, the best $n$th degree approximation may be obtained by adding one term to the best $(n - 1)$st degree approximation. So if we wish to increase our accuracy by going from $n - 1$ to $n$, the previous work in computing $p_{n-1}^*(x)$ is not wasted. When actually calculating $p_n^*(x)$ on a computer, rounding errors will occur in the computation of the Fourier coefficients. Hence, if we compute $p_n^*(x)$ from $p_n^*(x) = p_{n-1}^*(x) + \langle f, p_n \rangle p_n(x)$, $p_n^*(x)$ will inherit the accumulated errors of $p_{n-1}^*(x)$ as well as the error introduced by calculating $\langle f, p_n \rangle$. Computational experience and some theoretical results have led many analysts to suggest (particularly for the inner-product (5.56a)) that $p_n^*(x)$ be computed from the formula

$$p_n^*(x) = \sum_{j=0}^{n} \langle f - p_{j-1}^*, p_j \rangle p_j(x)$$

where $p_{j-1}^*(x)$ is the *computed* $(j - 1)$st degree approximation to $f(x)$ with $p_{-1}^* = 0$ in the above formula. Since (theoretically), $\langle p_{j-1}^*, p_j \rangle = 0$ by Corollary 1 of Theorem 5.9, this method of finding $p_n^*(x)$ is equivalent to (5.63) under the assumption that no errors are made in calculating $\langle f - p_{j-1}^*, p_j \rangle$.

From (5.63) we know that the best least-squares approximation is given by

$$p_n^*(x) = \sum_{j=0}^{n} \langle f, p_j \rangle p_j(x) = \sum_{j=0}^{n} \frac{\langle f, q_j \rangle}{\langle q_j, q_j \rangle} q_j(x) \equiv \sum_{j=0}^{n} c_j q_j(x).$$

In practice we usually wish to evaluate $p_n^*(x)$ at one or more values, $x = \alpha$. Assuming that $c_j$ and the values $a_j$ and $b_j$ of the three-term recurrence (5.65) are known quantities, we may take advantage of (5.65) and calculate $p_n^*(\alpha)$ using only $(2n - 1)$ multiplications. The evaluation algorithm proceeds as follows: Set $d_{n+2} = d_{n+1} = 0$, and compute

$$d_k = c_k + (\alpha - a_{k+1})d_{k+1} - b_{k+2}d_{k+2}; \qquad k = n, n-1, \ldots, 1, 0. \qquad \textbf{(5.66)}$$

(Note that although $a_{n+1}$, $b_{n+1}$, and $b_{n+2}$ may not be known, they are always multiplied by zero when they occur in (5.66). Thus each $d_k$ in this algorithm is well-defined.) We now claim that $d_0 = p_n^*(\alpha)$. To see that this is true consider the following, using (5.66) for each $c_k$

$$p_n^*(\alpha) = \sum_{k=0}^{n} c_k q_k(\alpha) = \sum_{k=0}^{n} [d_k - (\alpha - a_{k+1})d_{k+1} + b_{k+2}d_{k+2}]q_k(\alpha).$$

Now, collecting like coefficients of each $d_k$ we obtain,

$$p_n^*(\alpha) = d_0 q_0(\alpha) + d_1[q_1(\alpha) - (\alpha - a_1)q_0(\alpha)] + \sum_{k=2}^{n} d_k[q_k(\alpha) - (\alpha - a_k)q_{k-1}(\alpha) + b_k q_{k-2}(\alpha)].$$

But, by (5.65), all the coefficients of $d_k$, $1 \le k \le n$, are $[q_k(\alpha) - q_k(\alpha)] = 0$. So, since $q_0(\alpha) = 1$ for any $\alpha$, we have $p_n^*(\alpha) = d_0$.

**Example 5.16.** To see how the three-term recurrence relation in (5.65) simplifies the computation of orthogonal polynomials, we calculate the first few Legendre polynomials (i.e., we use the inner-product $\langle f, g\rangle = \int_{-1}^{1} f(x)g(x)\,dx$). As suggested by (5.65), we let $q_0(x) \equiv 1$ and find $\langle q_0, q_0\rangle = \int_{-1}^{1} dx = 2$. Next $q_1(x) = (x - a_1)q_0(x)$, where $a_1 = \langle xq_0, q_0\rangle/\langle q_0, q_0\rangle$. But $\langle xq_0, q_0\rangle = \int_{-1}^{1} x\,dx = 0$, so $q_1(x) = x$ and $\langle q_1, q_1\rangle = \int_{-1}^{1} x^2\,dx = 2/3$. The next step is to find $q_2(x)$ from the formula $q_2(x) = (x - a_2)q_1(x) - b_2 q_0(x)$ where $a_2 = \langle xq_1, q_1\rangle/\langle q_1, q_1\rangle$ and $b_2 = \langle q_1, xq_0\rangle/\langle q_0, q_0\rangle$. Again $a_2 = 0$, since $\langle xq_1, q_1\rangle = \int_{-1}^{1} x^3\,dx = 0$, so $q_2(x) = xq_1(x) - b_2 q_0(x)$.

The pattern soon emerges and we suspect that $a_i = 0$ for all $i$. This is easy to show, for $a_i = \langle xq_{i-1}, q_{i-1}\rangle/\langle q_{i-1}, q_{i-1}\rangle$, where $\langle xq_{i-1}, q_{i-1}\rangle = \int_{-1}^{1} x[q_{i-1}(x)]^2\,dx$. Recalling the elementary result that $\int_{-a}^{a} h(x)\,dx = 0$ whenever $h(x)$ is an odd function, we see $a_i = 0$, since $h(x) = x[q_{i-1}(x)]^2$ is an odd function. Thus, for all $i$, $q_i(x) = xq_{i-1}(x) - b_i q_{i-2}(x)$. By Problem 8, we note that $b_i = \langle q_{i-1}, q_{i-1}\rangle/\langle q_{i-2}, q_{i-2}\rangle$ for $i \ge 2$, so the computation of the Legendre polynomials is easily systematized as follows:

$$q_0(x) = 1 \qquad \langle q_0, q_0\rangle = 2$$

$$q_1(x) = x \qquad \langle q_1, q_1\rangle = \frac{2}{3}, \qquad b_2 = \frac{1}{3}$$

$$q_2(x) = xq_1(x) - \frac{1}{3}q_0(x)$$

$$= x^2 - \frac{1}{3} \qquad \langle q_2, q_2\rangle = \frac{8}{45}, \qquad b_3 = \frac{4}{15}$$

$$q_3(x) = xq_2(x) - \frac{4}{15}q_1(x)$$

$$= x^3 - \frac{3}{5}x \qquad \langle q_3, q_3\rangle = \frac{8}{175}, \qquad b_4 = \frac{9}{35}$$

$$\vdots$$

Clearly these calculations can be programmed so as to be carried out on the computer since we can program the calculation of $\int_{-1}^{1} p(x)\,dx$ for any polynomial, $p(x)$. Note that the existence of the three-term recurrence relation was crucial to our development of an efficient and systematic approach for finding the orthogonal polynomials. The simplification given by knowing $a_i = 0, i = 1, 2, \ldots$ is not peculiar to Legendre polynomials, but holds for a large class of inner-products (see Problem 10).

Finally, let us use Theorem 5.10 to find the best least-squares approximation of degree three or less to $f(x) = \cos(\pi x)$. Since the orthonormal polynomials $p_i(x)$ are given by $p_i(x) = q_i(x)/\|q_i\|$, we have

$$p_3^*(x) = \sum_{i=0}^{3} \langle f, p_i \rangle p_i(x) = \sum_{i=0}^{3} \left\langle f, \frac{q_i}{\|q_i\|} \right\rangle \frac{q_i(x)}{\|q_i(x)\|},$$

or

$$p_3^*(x) = \sum_{i=0}^{3} \frac{\langle f, q_i \rangle}{\langle q_i, q_i \rangle} q_i(x).$$

Since we calculated $\langle q_i, q_i \rangle$ in the process of finding the orthogonal polynomials $q_0(x), \ldots, q_3(x)$, we only need the numbers

$$\langle f, q_i \rangle = \int_{-1}^{1} \cos(\pi x) q_i(x)\, dx, \qquad i = 0, 1, 2, 3.$$

Since $\cos(\pi x)$ is an even function and $q_i(x)$ is an odd function for $i = 1$ and $i = 3$, we see immediately that $\langle f, q_1 \rangle = \langle f, q_3 \rangle = 0$. Next, we find $\langle f, q_0 \rangle = 0$ and integration by parts shows that

$$\langle f, q_2 \rangle = \int_{-1}^{1} \cos(\pi x) \left( x^2 - \frac{1}{3} \right) dx = \frac{-4}{\pi^2}.$$

Thus

$$p_3^*(x) = \frac{-45}{2\pi^2} \left( x^2 - \frac{1}{3} \right).$$

## PROBLEMS

1. Write a computer program to solve Problem A, Section 3, where the input is: $m$; weights $w_0, \ldots, w_m$; data points $(x_0, y_0), \ldots, (x_m, y_m)$; and the desired degree of approximation, $n$. Generate the orthogonal polynomials $q_i(x)$ according to (5.65) and find $p_n^*(x)$ by

$$p_n^*(x) = \sum_{i=0}^{n} \frac{\langle f, q_j \rangle}{\langle q_j, q_j \rangle} q_j(x).$$

   Evaluate $p_n^*(x)$ according to the recurrence (5.66). Test your program with the data in Example 5.13. (Also see Problem 10 for some programming simplifications.)

2. Reference to Problem $A$ shows that for $n = m$, $p_n^*(x)$ is the interpolating polynomial. Thus Theorem 5.10 provides another way to generate interpolating polynomials. Verify this directly for the discrete inner-product of Example 5.14, generating the orthogonal polynomials via (5.65) and finding $p_4^*(x)$ that interpolates $f(x) = 1/(2 + x)$.

3. Verify (5.57), (b), (c), and (d) for both inner-products $\langle f, g \rangle_1$ and $\langle f, g \rangle_2$.

4. Let $\langle f, g \rangle$ be either of the inner-products of (5.56). Show that $S = \{1, x, x^2, \ldots, x^n\}$, $n \geq 2$, cannot be an orthogonal set. (Hint: Consider $\langle 1, x^k \rangle$ for $k \geq 2$.)

5. For $p(x) \in \mathscr{P}_n$, we can write $p(x) = \sum_{j=0}^{n} \alpha_j p_j(x)$ by (5.60). Show the coefficients $\alpha_j$ are unique by supposing that $p(x) = \sum_{j=0}^{n} \beta_j p_j(x)$ is another expansion. (Hint: What is $\langle \theta, p_i \rangle$?)
6. Consider the inner-product $\langle f, g \rangle = \int_0^\infty e^{-x} f(x)g(x)\,dx$. Generate the first four monic polynomials that are orthogonal with respect to this inner-product (these are called the *Laguerre polynomials*).
7. For $f(x) = \sqrt{1 - x^2}$, find $p_n^*(x)$ when

$$\langle f, g \rangle = \int_{-1}^{1} \frac{f(x)g(x)}{\sqrt{1 - x^2}}\,dx$$

(recall Example 5.15(b)).
8. Show that $b_k$ in (5.65) is given by $b_k = \langle q_{k-1}, q_{k-1} \rangle / \langle q_{k-2}, q_{k-2} \rangle$. (Hint: Show by (5.60) that

$$x q_{k-2}(x) = \sum_{i=0}^{k-1} \alpha_i q_i(x).)$$

9. Write a computer program to do an $n$th degree discrete least-squares fit for $f(x) = 1/(2 + x)$ at $x_0, x_1, \ldots, x_m$; using the inner-product of Example 5.15(c) and (5.62). For $m = 20$ and $n = 10$, evaluate the approximation and print the errors for $x = -1 + ih$, $h = 0.01, 0 \leq i \leq 200$.

*10. Suppose $w(x)$ is an even function on $[-a, a]$ and $\langle f, g \rangle = \int_{-a}^{a} w(x)f(x)g(x)\,dx$. Show that the three-term recurrence of (5.65) becomes $q_i(x) = x q_{i-1}(x) - b_i q_{i-2}(x)$. Next, let $\langle f, g \rangle = \int_a^b w(x)f(x)g(x)\,dx$, $\alpha = (a + b)/2$ and suppose $w(\alpha - h) = w(\alpha + h)$ for all $h$ such that $0 \leq h \leq (b - a)/2$. Use the change of variable $x = y + \alpha$ to show that $q_i(x) = (x - \alpha)q_{i-1}(x) - b_i q_{i-2}(x)$. Show this simplification is also valid for the discrete inner-product $\langle f, g \rangle = \sum_{i=0}^{m} w_i f(x_i)g(x_i)$ whenever the $x_i$ are symmetrically spaced about $\alpha$ and the weights are symmetric (that is, $(x_i + x_{m-i})/2 = \alpha$ and $w_i = w_{m-i}$). Use this simplification in the program in Problem 1, with the data of Example 5.13.

*11. Do an operations count to get an estimate of the relative efficiency of solving Problem A (Section 5.3) using orthogonal polynomials (as in Problem 1) as against solving Problem A by the matrix method of Chapter 2.

### *5.3.2 Error Estimates for Least-Squares Approximations

Suppose $\{p_0(x), p_1(x), \ldots, p_k(x), \ldots\}$ are the orthonormal polynomials generated by the Gram-Schmidt theorem with respect to the inner-product $\langle f, g \rangle = \int_a^b f(x)g(x)w(x)\,dx$, and for each $k$ let

$$p_k^*(x) = \sum_{j=0}^{k} \langle f, p_j \rangle p_j(x),$$

as in (5.63). Then, by Theorem 5.10, if $m > k$, $\|f - p_m^*\| \leq \|f - p_k^*\|$ since $\mathscr{P}_k \subseteq \mathscr{P}_m$. By the Weierstrass theorem stated in Section 5.1, for any $\varepsilon > 0$, there exists a polynomial of degree $N$, $q_N(x)$, (where $N$ depends on $\varepsilon$) such that

$$\max_{a \leq x \leq b} |f(x) - q_N(x)| \equiv \|f - q_N\|_\infty < \varepsilon.$$

By the best least-squares approximation property of Theorem 5.10,

$$\begin{aligned}\|f - p_N^*\|^2 &= \int_a^b (f(x) - p_N^*(x))^2 w(x)\,dx \le \|f - q_N\|^2 \\ &= \int_a^b (f(x) - q_N(x))^2 w(x)\,dx \le \max_{a \le x \le b} |f(x) - q_N(x)|^2 \int_a^b w(x)\,dx \\ &< \varepsilon^2 \int_a^b w(x)\,dx \equiv \varepsilon',\end{aligned}$$

(since $\int_a^b w(x)\,dx$ is a constant). Thus we have $\lim_{k\to\infty} \|f - p_k^*\| = 0$, or the polynomials, $p_k^*(x)$, converge to $f(x)$ in the least-squares norm. By placing relatively mild restrictions on $f(x)$, we can even show uniform convergence, that is,

$$\lim_{k\to\infty} \|f - p_k^*\|_\infty = 0$$

(see Problem 3).

From Bessel's inequality we have that

$$\|f\|^2 = \int_a^b (f(x))^2 w(x)\,dx \ge \sum_{j=0}^{n} \langle f, p_j\rangle^2$$

holds for all $n$. Since $\|f\|^2$ is a finite number for any $f(x) \in C[a, b]$, the partial sums of the infinite series, $\sum_{j=0}^{\infty} \langle f, p_j\rangle^2$, form a nondecreasing sequence which is bounded from above, and hence the series is convergent. Thus we have $\lim_{j\to\infty} \langle f, p_j\rangle^2 = 0$ and so $\lim_{j\to\infty} |\langle f, p_j\rangle| = 0$. From Corollary 1, Theorem 5.10, we see that $\langle f - p_n^*, p_n^*\rangle = 0$ for each $n$. Using formula (5.63), we see that $\langle f, p_n^*\rangle = \langle f(x), \sum_{j=0}^{n} \langle f, p_j\rangle p_j(x)\rangle = \sum_{j=0}^{n} \langle f, p_j\rangle^2$. From this information we can calculate the least-squares error, $\|f - p_n^*\|$, since

$$\begin{aligned}\|f - p_n^*\|^2 &= \langle f - p_n^*, f - p_n^*\rangle = \langle f - p_n^*, f\rangle - \langle f - p_n^*, p_n^*\rangle \\ &= \langle f - p_n^*, f\rangle = \langle f, f\rangle - \langle f, p_n^*\rangle = \|f\|^2 - \sum_{j=0}^{n} \langle f, p_j\rangle^2. \qquad \textbf{(5.67)}\end{aligned}$$

In addition, since $\lim_{n\to\infty} \|f - p_n^*\| = 0$, we have by (5.67) that

$$\|f\|^2 = \sum_{j=0}^{\infty} \langle f, p_j\rangle^2 \quad \text{or} \quad \|f - p_n^*\|^2 = \sum_{j=n+1}^{\infty} \langle f, p_j\rangle^2. \qquad \textbf{(5.68)}$$

This formula is known as Parseval's Equality, and can be used in a very practical way to estimate the error of the $n$th degree least-squares approximation. We note that the error is explicitly dependent on the size of the Fourier coefficients, $\langle f, p_j\rangle$, $j = n + 1, n + 2, \ldots$.

We now examine one way to estimate the error in (5.68) without having to calculate each $\langle f, p_j\rangle$ for $j \ge n + 1$. We call on Theorem 5.2 which tells us that, for each $k$, there exists a unique polynomial, $\tilde{p}_k(x)$, such that $\max_{a \le x \le b} |f(x) - \tilde{p}_k(x)| \equiv \|f - \tilde{p}_k\|_\infty \le \|f - p\|_\infty$ for all $p(x) \in \mathscr{P}_k$, and again we let $E_k(f) \equiv \|f - \tilde{p}_k\|_\infty$. There are many results in the literature which give upper bounds for

$E_k(f)$. Any such result is usually called a Jackson Theorem in honor of Dunham Jackson who first presented broad significant results of this form. For instance, we already have a simple example of a Jackson Theorem via formula (5.30). That is, for $f(x) \in C^{k+1}[a, b]$ let $q_k(x) \in \mathscr{P}_k$ be the polynomial which interpolates $f(x)$ at the zeros of the shifted Chebyshev polynomial, $\tilde{T}_{k+1}(x)$. Then (5.36) yields

$$E_k(f) \equiv \|f - \tilde{p}_k\|_\infty \leq \|f - q_k\|_\infty \leq \frac{\max\limits_{a \leq x \leq b} |f^{(k+1)}(x)|}{2^k(k+1)!} \left(\frac{b-a}{2}\right)^{k+1}. \tag{5.69a}$$

A stronger Jackson theorem, which is beyond the scope of this text to prove, is (5.69b) (see Cheney 1966, p. 147). In (5.69b), we let $[a, b] = [-1, 1]$ and suppose that $f^{(m)}(x) \in C[-1, 1]$ with $k \geq m$. Then

$$E_k(f) \leq (\pi/2)^m \|f^{(m)}\|_\infty / [(k+1)(k) \cdots (k-m+2)]. \tag{5.69b}$$

We now use these bounds for $E_k(f)$ in estimating the error in (5.68). Since $\langle p_k, \tilde{p}_{k-1} \rangle = 0$ by Corollary 1, Theorem 5.10, then for $k \geq n+1$,

$$\langle f, p_k \rangle = \langle f - \tilde{p}_{k-1}, p_k \rangle = \int_a^b (f(x) - \tilde{p}_{k-1}(x)) p_k(x) w(x)\, dx. \tag{5.70}$$

Two ways to bound (5.70) immediately come to mind. In the first we use the familiar results for the absolute values of integrals and obtain

$$\begin{aligned} |\langle f, p_k \rangle| &\leq \int_a^b |f(x) - \tilde{p}_{k-1}(x)|\, |p_k(x)| w(x)\, dx \\ &\leq \max_{a \leq x \leq b} |f(x) - \tilde{p}_{k-1}(x)| \int_a^b |p_k(x)| w(x)\, dx \\ &= E_{k-1}(f) \int_a^b |p_k(x)| w(x)\, dx. \end{aligned} \tag{5.70a}$$

In the second we use Cauchy-Schwarz, and the orthonormality of the $p_k(x)$'s and obtain

$$\begin{aligned} |\langle f, p_k \rangle| &\leq \left(\int_a^b (f(x) - \tilde{p}_{k-1}(x))^2 w(x)\, dx\right)^{1/2} \left(\int_a^b p_k(x)^2 w(x)\, dx\right)^{1/2} \\ &\leq \max_{a \leq x \leq b} |f(x) - \tilde{p}_{k-1}(x)| \left(\int_a^b w(x)\, dx\right)^{1/2} (\|p_k\|^2) \\ &= E_{k-1}(f) \left(\int_a^b w(x)\, dx\right)^{1/2}. \end{aligned} \tag{5.70b}$$

We now give an example of how to use the bounds on the Fourier coefficients and the Jackson theorems to bound the least-squares error, $\|f - p_n^*\|$, in (5.68).

**Example 5.17.** Let $[a, b] = [-1, 1]$ and $\langle f, g \rangle = \int_{-1}^1 f(x)g(x)(1 - x^2)^{-1/2}\, dx$. Then, as in Example 5.15(b), the orthornormal polynomials are $p_k(x) = \sqrt{2/\pi}\, T_k(x)$ for $k \geq 1$. For (5.70a) we have

$$\int_a^b |p_k(x)| w(x)\, dx = \sqrt{\frac{2}{\pi}} \int_{-1}^1 |T_k(x)|(1 - x^2)^{-1/2}\, dx = \sqrt{\frac{2}{\pi}} \int_0^\pi |\cos(k\theta)|\, d\theta,$$

using $x = \cos(\theta)$. We leave to the reader to verify that $\int_0^\pi |\cos(k\theta)|\, d\theta = 2$, and so (5.70a) yields $|\langle f, p_k\rangle| \le (2\sqrt{2}/\sqrt{\pi})E_{k-1}(f)$. In (5.70b), $\int_a^b w(x)\,dx = \int_{-1}^1 (1 - x^2)^{-1/2}\,dx = \pi$, and so (5.70b) yields $|\langle f, p_k\rangle| \le \sqrt{\pi}E_{k-1}(f)$. In this example $\sqrt{\pi} \ge 2\sqrt{2}/\sqrt{\pi}$, thus (5.70a) yields a better (smaller) bound on $|\langle f, p_k\rangle|$. (This is usually true for other examples as well. However, $\int_a^b |p_k(x)|w(x)\,dx$ in (5.70a) can be a formidable task to compute, especially when the zeros of $p_k(x)$ are not known precisely, or when they vary irregularly for different values of $k$.) If we assume that $f(x) \in C^2[-1, 1]$, then by (5.69b) and (5.70a)

$$\langle f, p_k\rangle^2 \le \frac{8}{\pi} E_{k-1}^2(f) \le \pi^3\|f''\|_\infty^2/2k^2(k-1)^2.$$

Thus, by (5.68),

$$\|f - p_n^*\|^2 \le \frac{\pi^3}{2}\|f''\|_\infty^2 \sum_{k=n+1}^{\infty} (1/k^2(k-1)^2) \le \frac{\pi^2\|f''\|_\infty^2}{6(n-1)^3},$$

where the integral test for infinite series from calculus was used to provide a crude bound for the above series. We note that we would obtain a smaller error bound if $f(x) \in C^m[-1, 1]$ for $m > 2$ by virtue of (5.69b) if we assume that $\|f^{(m)}\|_\infty$ is not "too much larger" than $\|f''\|_\infty$.

## PROBLEMS

1. Using (5.67) and the expansion found in Problem 7 in the last section, determine $\|f - p_n^*\|^2$ for $n = 2, 4, 6$.
2. Find the Chebyshev series expansion for $f(x) = |x|$ and determine $\|f - p_n^*\|^2$ and $|f(1) - p_n^*(1)|$ for $n = 2, 4, 6$.

*3. a) Suppose $f \in C^2[-1, 1]$ and suppose the Chebyshev series expansion for $f(x)$ is

$$\sum_{i=0}^{\infty} \alpha_i p_i(x)$$

where $p_i(x) = \sqrt{2/\pi}T_i(x)$, $i \ge 1$; $p_0(x) = \sqrt{1/\pi}$; and

$$\langle f, g\rangle = \int_{-1}^{1} \frac{f(x)g(x)}{\sqrt{1 - x^2}}\,dx.$$

Show that the Chebyshev series converges uniformly to $f(x)$ for $-1 \le x \le 1$. (Hint: Show, by integrating by parts twice, that each $\alpha_i$ satisfies an inequality of the form $|\alpha_i| \le M/i^2$. Thus conclude the series $\sum_{i=0}^\infty \alpha_i p_i(x)$ converges uniformly to a continuous function $g(x)$. Show next that $\|f - g\| = 0$ by considering $\|f - g\| \le \|f - p_n^*\| + \|p_n^* - g\|$. Thus since $f$ and $g$ are continuous, $f(x) \equiv g(x)$.)

b) For any positive integer $n$, let $Q_{n-1}(x) \in \mathscr{P}_{n-1}$ interpolate $f(x)$ at $x_j = \cos[(2j - 1)\pi/2n]$, $1 \le j \le n$ (the zeros of $T_n(x)$). For any $x \in [-1, 1]$ define $R_{n-1}(f, x) \equiv f(x) - Q_{n-1}(x)$. Let $k$ be any positive integer where $k \ge n$ and write $k$ as $k = 2rn + \alpha$, $r \ge 1$ and $\beta \equiv |\alpha| \le n$. Show that $T_k(x_j) = (-1)^r T_\beta(x_j)$, $1 \le j \le n$, and therefore that $(-1)^r T_\beta(x)$ for $\beta < n$ or $\theta(x)$ for $\beta = n$ is the interpolating polynomial in $\mathscr{P}_{n-1}$ for $T_k(x)$. Show that $R_{n-1}(T_k; x) = 0$ for $k \le n - 1$ and $|R_{n-1}(T_k; x)| \le 2$ for $k \ge n$. From Part (a) we see that $f(x)$ can be expressed in a uniformly convergent Fourier expansion of the form,

$$f(x) = \sum_{k=0}^{\infty} a_k T_k(x).$$

Because of the uniform convergence we may write

$$R_{n-1}(f; x) = \sum_{k=0}^{\infty} a_k R_{n-1}(T_k; x) = \sum_{k=n}^{\infty} a_k R_{n-1}(T_k; x).$$

Use Part (a) to argue that $\lim_{n\to\infty} \|f - Q_{n-1}\|_\infty = 0$, and use a Jackson theorem to bound $\|R_{n-1}(f; x)\|_\infty$ for any $n$. (Note: If we drop the hypothesis that $f \in C^2[-1, 1]$, we may still argue that $Q_{n-1}(x)$ is uniformly convergent to $f(x)$ whenever $\sum_{k=0}^{\infty} |a_k|$ is convergent.)

4. Let $f(x) = \cos(x)$ and let

$$\sum_{i=0}^{\infty} \alpha_i p_i(x)$$

be the Chebyshev series for $\cos(x)$. By Problem 3, this series converges uniformly and absolutely to $\cos(x)$ for $-1 \le x \le 1$. Use the error bound for Chebyshev interpolation to show that $|\langle f, p_k\rangle| \le 4\sqrt{2}/(k!2^k\sqrt{\pi})$ as in Example 5.17. Since the series converges uniformly to $f(x)$, we have

$$f(x) - p_n^*(x) = \sum_{k=n+1}^{\infty} \alpha_k p_k(x).$$

Use this to show that $|f(x) - p_n^*(x)| \le 8/\pi \sum_{k=n+1}^{\infty} |T_k(x)|/2^k k!$. Recalling that $\|T_k\|_\infty = 1$ and using the Maclaurin's series expansion for $e^x$, with $x = \frac{1}{2}$, find an upper bound for $\|f - p_n^*\|_\infty$.

5. Verify that $\int_0^\pi |\cos(k\theta)|\, d\theta = 2$.

6. Determine the Chebyshev expansion for $f(x) = \arccos(x)$. Give both a lower and upper bound for $|f(1) - p_n^*(1)|$, using the integral test on

$$\sum_{i=n+1}^{\infty} \langle f, p_i\rangle p_i(x)$$

for $x = 1$. How many terms of the series must be taken in order that $\|f - p_n^*\|_\infty \le 10^{-8}$?

## *5.4 APPROXIMATION BY RATIONAL FUNCTIONS

In this section, we will briefly describe some methods for approximating functions $f(x)$* by *rational functions.* Recall that a rational function $r(x)$ is defined to be a quotient of polynomials. Thus

$$r(x) = \frac{p(x)}{q(x)} = \frac{p_0 + p_1 x + \cdots + p_n x^n}{q_0 + q_1 x + \cdots + q_m x^m} \tag{5.71}$$

where we assume that $p(x)$ and $q(x)$ are reduced to lowest terms (that is, $p(x)$ and $q(x)$ have no common factors). Further, if we wish to approximate $f(x)$ by $r(x)$, where $f(x)$ is in $C[a, b]$, then we will naturally insist that $q(x)$ does not vanish in $[a, b]$.

One reason for introducing rational functions as a means to approximate a function $f(x)$ is that ordinary polynomial approximation may not be practical. For example, in order to approximate $f(x) = \arccos(x)$ for $x$ in $[-1, 1]$ with an accuracy of $10^{-8}$, it can be shown that a polynomial of degree $10^4$ would be necessary. Clearly, since graphs of rational functions can assume shapes that the graph of a polynomial cannot, rational functions should provide a useful class of approximations. The theory of rational function approximation is still under active investigation and what has been developed is fairly complex, so we will not present it in detail here. Instead, we will illustrate some of the techniques and give some examples. (We emphasize that polynomial approximation is a special case of rational approximation with $m = 0$. So in theory, rational approximation must always yield better results, although the computation may become more cumbersome.)

The approach most widely used to choose a rational function approximation is that of *Padé approximation.* Suppose $f(x)$ has a Maclaurin's series expansion $f(x) = \sum_{i=0}^{\infty} a_i x^i$. In Padé approximation, we seek a rational function, $r(x)$, such that the Maclaurin's series expansion for $r(x)$ agrees with that of $f(x)$ to as many terms as possible. Thus, we seek $p(x)$ and $q(x)$ such that

$$f(x) - \frac{p(x)}{q(x)} = \sum_{i=s}^{\infty} c_i x^i \tag{5.72}$$

where $s$ is as large as possible. If $r(x)$ is to have a Maclaurin's series expansion, then clearly $q_0 \neq 0$ in (5.71). By dividing $p(x)$ and $q(x)$ by $q_0$, we may as well assume that $q(x)$ has the form $q(x) = 1 + q_1 x + \cdots + q_m x^m$. Hence, in (5.71), $r(x)$ has $(n + 1) + m$ parameters and we might hope that the series for $r(x)$ can be made to match that of $f(x)$ for at least the first $(n + 1) + m$ terms (that is, $s = n + m + 1$). As we shall see (Problem 1), it is not always possible to obtain $s \geq n + m + 1$. However, by choosing $m = 0$ and $p(x)$ to be the first $n + 1$ terms of the series for $f(x)$, we see we can always obtain $s \geq n + 1$.

To determine the coefficients of $r(x)$, we write (5.72) as

$$f(x)q(x) - p(x) = q(x) \sum_{i=s}^{\infty} c_i x^i.$$

Now, $q(x) \sum_{i=s}^{\infty} c_i x^i$ has the form $q(x) \sum_{i=s}^{\infty} c_i x^i = \sum_{i=s}^{\infty} d_i x^i$. Therefore, we want $p(x)$ and $q(x)$ so that the expression $f(x)q(x) - p(x)$ has the form

$$f(x)q(x) - p(x) = \sum_{i=s}^{\infty} d_i x^i,$$

where $s$ is as large as possible. That means we want to collect like powers of $x$ in the expression $f(x)q(x) - p(x)$ and choose the coefficients $p_j$ and $q_j$ so that as many powers of $x$ vanish as possible.

This program leads us to consider

$$f(x)q(x) - p(x) = \left(\sum_{i=0}^{\infty} a_i x^i\right)\left(\sum_{j=0}^{m} q_j x^j\right) - \sum_{j=0}^{n} p_j x^j$$
$$= (a_0 q_0 - p_0) + (a_0 q_1 + a_1 q_0 - p_1)x$$
$$+ (a_0 q_2 + a_1 q_1 + a_2 q_0 - p_2)x^2 + \cdots$$

In general, to get as many powers of $x$ to vanish as possible, we want

$$\sum_{i=0}^{k} a_i q_{k-i} = p_k, \qquad k = 0, 1, \ldots, n$$

$$\sum_{i=0}^{k} a_i q_{k-i} = 0, \qquad k = n+1, n+2, \ldots, s$$

where $q_{k-i} = 0$ when $k - i > m$.

**Example 5.18.** Let

$$f(x) = \cos(x) = 1 - \frac{x^2}{2!} + \frac{x^4}{4!} - \frac{x^6}{6!} + \frac{x^8}{8!} - \cdots.$$

Since $f(x)$ is even, we consider an approximation $r(x)$ of the form

$$r(x) = \frac{p_0 + p_2 x^2 + p_4 x^4 + \cdots + p_{2n} x^{2n}}{1 + q_2 x^2 + q_4 x^4 + \cdots + q_{2m} x^{2m}}.$$

Since $\cos(x)$ looks roughly like a quadratic polynomial on $[-1, 1]$, it is plausible to try $n = m + 1$. So, for example, we might try

$$r(x) = \frac{p_0 + p_2 x^2 + p_4 x^4}{1 + q_2 x^2}.$$

Since $r(x)$ has four parameters, we try to choose them so that $f(x) - r(x) = c_8 x^8 + c_{10} x_{10} + \cdots$. As above, we find that we want to match coefficients in as many powers of $x$ as possible in

$$(1 + q_2 x^2)\left(1 - \frac{x^2}{2!} + \frac{x^4}{4!} - \frac{x^6}{6!} + \cdots\right) = p_0 + p_2 x^2 + p_4 x^4.$$

Multiplying this out and equating like powers of $x$ leads to

$$1 = p_0$$

$$q_2 - \frac{1}{2!} = p_2$$

$$\frac{-q_2}{2!} + \frac{1}{4!} = p_4.$$

$$\frac{q_2}{4!} - \frac{1}{6!} = 0$$

The last equation gives us $q_2 = \frac{1}{30}$ and hence we find $p_0 = 1, p_2 = -\frac{7}{15}$ and $p_4 = \frac{1}{40}$. Therefore,

$$r(x) = \frac{1 - \dfrac{7}{15}x^2 + \dfrac{1}{40}x^4}{1 + \dfrac{1}{30}x^2}.$$

It is not hard to show (Problem 3) that $r(x)$ is a better approximation to $\cos(x)$ than is

$$h(x) = 1 - \frac{x^2}{2!} + \frac{x^4}{4!} - \frac{x^6}{6!},$$

which also has four parameters. (See Table 5.4)

**Table 5.4**

| $x$ | $\cos(x)$ | $r(x)$ |
|---|---|---|
| 0 | 1.0 | 1.0 |
| 0.25 | 0.96891242 | 0.96891242 |
| 0.5 | 0.87758256 | 0.87758264 |
| 0.75 | 0.73168887 | 0.73169095 |
| 1.0 | 0.54030231 | 0.54032258 |

Other frequently used methods for determining the coefficients of a rational function approximation are to select $r(x)$ so that

$$r(x_i) = f(x_i), \qquad i = 0, 1, \ldots, s \tag{5.73a}$$

or

$$\sum_{i=0}^{N} [f(x_i) - r(x_i)]^2 = \text{minimum}. \tag{5.73b}$$

A least-squares criterion similar to (5.73b) is to write $p(x) = \sum_{i=0}^{n} p_i\phi_i(x)$ and $q(x) = \sum_{i=0}^{m} q_i\phi_i(x)$, where $\phi_0(x), \phi_1(x), \ldots$ is an orthonormal set of polynomials. By analogy with Padé approximation, we ask that $p(x)$ and $q(x)$ be chosen so that

$$f(x)q(x) - p(x) = \sum_{i=s}^{\infty} \gamma_i\phi_i(x) \tag{5.73c}$$

where $s$ is as large as possible. There is empirical evidence to suggest that choosing the Chebyshev polynomials $T_i(x)$ as the orthonormal set is a good choice. Using Chebyshev polynomials in (5.73c) is particularly nice from a computational point of view because of the identity $2T_i(x)T_j(x) = T_{i+j}(x) + T_{i-j}(x)$ when $i \geq j$ (Problem 4).

## PROBLEMS

1. To see that $s = n + m + 1$ in (5.72) is not always possible, let

$$r(x) = \frac{p_0 + p_1 x}{1 + q_1 x} \qquad \text{and} \qquad f(x) = a_0 + a_1 x + a_2 x^2 + \cdots.$$

If $a_1 = 0$ and $a_2 \neq 0$, show that $s = 2$ is the best that can be done. If $a_1 \neq 0$, show that $s = 3$ is possible.

2. To see another limitation of Padé approximation, suppose we wish to approximate $f(x) = e^{3x}$ on $[-1, 1]$ by a quotient of linear polynomials as in Problem 1. By Problem 1, a Padé approximation exists with $s = 3$. Find the approximation and determine why it is not suitable for the interval $[-1, 1]$.

*3. Using the standard error estimates for alternating series, show in Example 5.18 that

$$|\cos(x) - r(x)| \leq \left(\frac{26}{x^2 + 30} \frac{x^8}{8!}\right).$$

Compare this estimate with the table in Example 5.18.

4. Using the change of variable $x = \cos(\theta)$, establish the identity $2T_i(x)T_j(x) = T_{i+j}(x) + T_{i-j}(x)$, for $i \geq j$.

5. The most straight-forward way to determine $p(x)$ and $q(x)$ in (5.73c) is by taking inner-products and obtaining

$$\langle f(x)q(x), \phi_k(x)\rangle = \langle p(x), \phi_k(x)\rangle, \qquad k = 0, 1, \ldots, n$$
$$\langle f(x)g(x), \phi_k(x)\rangle = 0, \qquad k = n + 1, \ldots, s.$$

For $\phi_i(x) = T_i(x)$ and $m = n = 3$, write out the resulting equations, using the identity of Problem 4. Using this and Problem 7 of Section 5.3.1, find a rational function approximation to $f(x) = \sqrt{1 - x^2}$. Use the inner product of Example 5.17.

# 6

# NUMERICAL INTEGRATION AND DIFFERENTIATION

## 6.1 INTRODUCTION

In Chapter 5 we developed numerous ways for efficiently approximating a function, $f(x) \in C[a, b]$. If $g(x)$ is any approximation whatsoever to $f(x)$, and we wish to approximate either the integral or derivative of $f(x)$, then our first inclination is to use either the integral or derivative of $g(x)$, respectively, as our approximation. In the case of approximating derivatives, we have already pointed out that the derivative of the interpolatory cubic spline is fairly reliable, whereas the derivative of the $n$th degree interpolating polynomial usually is not, due to its oscillatory behavior. One other fairly reliable means of derivative approximation is to use the derivative of the discrete least-squares approximation of Section 5.3.1. (This is especially true when the values of $f(x)$ are given in tabular form since this approximation tends to "smooth" the data when the degree of the approximation is fairly small.) In general, however, it remains true that numerical differentiation is a particularly unstable process and quite difficult to analyze carefully. We shall present in detail only one more method for numerical differentiation (based on Richardson extrapolation—Section 6.4). The rest of this chapter will be primarily devoted to numerical integration procedures.

## 6.2 INTERPOLATORY NUMERICAL INTEGRATION

For any function $f(x)$ which is integrable on the interval $[a, b]$, we define

$$I(f) \equiv \int_a^b f(x)w(x)\,dx \tag{6.1}$$

where $w(x)$ is a fixed nonnegative weight function as defined in (5.1a) or (5.1b). (Often we shall be using $w(x) \equiv 1$, but we shall see some important cases later where we shall desire more flexibility in choosing $w(x)$.) Any formula which approximates $I(f)$ is called a numerical integration or quadrature formula. As mentioned above, if $g(x)$ approximates $f(x)$ on $[a, b]$, then we expect $I(g) \approx I(f)$. For example, if $g(x)$ is a cubic interpolating spline, then it is usually true that $I(g)$ approximates $I(f)$ quite well. However, due to the complex formula for the cubic spline we shall not pursue this further here. We shall instead concentrate on the use of interpolating polynomials, and we shall find that they yield quite favorable results and lead to efficient, yet simple, formulas. Formulas of this type are called *interpolatory quadratures.*

Interpolatory quadratures all have one standard form. From (5.4), the interpolating polynomial in $\mathscr{P}_n$ at the points $\{x_j\}_{j=0}^n$ can be written as

$$p_n(x) = \sum_{j=0}^{n} f(x_j)\ell_j(x),$$

and we define $Q_n(f) \equiv I(p_n)$ as the quadrature formula to approximate $I(f)$. Thus

$$Q_n(f) \equiv I(p_n) = \int_a^b p_n(x)w(x)\,dx = \int_a^b \left( \sum_{j=0}^{n} f(x_j)\ell_j(x) \right) w(x)\,dx$$

$$= \sum_{j=0}^{n} f(x_j) \int_a^b \ell_j(x)w(x)\,dx \equiv \sum_{j=0}^{n} A_j f(x_j) \tag{6.2}$$

where

$$A_j \equiv \int_a^b \ell_j(x)w(x)\,dx, \qquad 0 \le j \le n.$$

The values $\{A_j\}_{j=0}^n$ are called the *weights* of the quadrature and the points $\{x_j\}_{j=0}^n$ are called the *nodes*. Thus an interpolatory quadrature formula,

$$Q_n(f) = \sum_{j=0}^{n} A_j f(x_j),$$

is nothing more than a *weighted* sum of the function values $f(x_0), f(x_1), \ldots, f(x_n)$. The weights, $A_j$, are computed once and for all and depend only on the nodes $x_0, x_1, \ldots, x_n$, the weight function $w(x)$ and the interval $[a, b]$. That is, the same weights are used no matter what function $f(x)$ appears in the integral $\int_a^b w(x)f(x)\,dx$ that we are trying to estimate. This simple form makes interpolatory quadrature formulas both easy to use and to analyze.

We note again that if $f(x)$ is a polynomial of degree $n$ or less (that is, $f(x) \in \mathscr{P}_n$) then $f(x) = p_n(x)$ by Theorem 5.3 ($f(x)$ is its own interpolating polynomial). Hence for any $f(x) \in \mathscr{P}_n$, $Q_n(f) = I(f)$, and so the quadrature of (6.2) gives the exact value for the integral. If for some integer $m$, $Q_n(f) = I(f)$ for all $f(x) \in \mathscr{P}_m$, then we say that the quadrature has *precision m*. From our remarks above, we see that any $(n + 1)$-point interpolatory quadrature, $Q_n(f)$, always has precision at least $n$. Later we shall develop quadratures with precision strictly greater than $n$.

Since $Q_n(f)$ from (6.2) has precision $n$, it must give the exact values of the integrals of $1, x, x^2, \ldots, x^n$. This gives us an alternative way to determine the quadrature weights, $\{A_j\}_{j=0}^n$, once the nodes, $\{x_j\}_{j=0}^n$, are given and fixed. This procedure is similar to the method of undetermined coefficients in that we have a linear system of $(n + 1)$ equations in $(n + 1)$ unknowns (the weights). To be specific, for $f_k(x) = x^k$, $0 \le k \le n$, we get the $(n + 1)$ equations

$$I(f_k) = \int_a^b x^k w(x)\,dx = Q_n(f_k) = \sum_{j=0}^{n} A_j x_j^k, \qquad 0 \le k \le n. \tag{6.3}$$

In (6.3) the nodes, $\{x_j\}_{j=0}^n$, were fixed beforehand and the weights are determined by solving the linear system. If, however, we treat the nodes, as well as the weights, as unknowns, then we have $2(n + 1)$ unknowns. Considering Eq. (6.3) we might reason, as Gauss did in 1814, that it could be possible to let $k$ range from 0 to $2n + 1$ so that (6.3) represents a system (nonlinear this time) of $(2n + 2)$ equations in $(2n + 2)$ unknowns. If (6.3) has a solution it will yield a quadrature of precision $(2n + 1)$. This is indeed possible and we will study Gaussian quadrature in Section 6.5. For now we will be content to illustrate the construction of a quadrature with the following simple example.

**Example 6.1.** Let $[a, b] = [-h, h]$ be the interval of integration, where $h$ is a positive constant, and let $w(x) \equiv 1$. We shall derive two simple quadrature formulas, the first by direct integration of the interpolating polynomial and the second by undetermined coefficients as in (6.3).

1. Let $x_0 = -h$ and $x_1 = h$, then the first-degree interpolating polynomial is

$$p_1(x) = f(-h)\frac{(h - x)}{2h} + f(h)\frac{(h + x)}{2h}.$$

Therefore

$$Q_1(f) = I(p_1) = \int_{-h}^{h} p_1(x)\,dx$$

$$= \frac{-f(-h)}{2h}\frac{(h - x)^2}{2}\bigg|_{-h}^{h} + \frac{f(h)}{2h}\frac{(h + x)^2}{2}\bigg|_{-h}^{h}.$$

So, $Q_1(f) = hf(-h) + hf(h)$, which is the familiar *trapezoidal rule* estimate for $\int_{-h}^{h} f(x)\,dx$.

2. Let $x_0 = -h, x_1 = 0$ and $x_2 = h$. By forcing the quadrature formula $Q_2(f) = A_0 f(x_0) + A_1 f(x_1) + A_2 f(x_2)$ to equal $\int_{-h}^{h} f(x)\,dx$ for $f(x) = 1, x, x^2$, we obtain the system:

$$I(1) = \int_{-h}^{h} 1\,dx = 2h = A_0 + A_1 + A_2$$

$$I(x) = \int_{-h}^{h} x\,dx = 0 = -A_0 h + A_2 h$$

$$I(x^2) = \int_{-h}^{h} x^2\,dx = \frac{2h^3}{3} = A_0 h^2 + A_2 h^2.$$

Solving these equations yields

$$A_0 = A_2 = \frac{h}{3} \qquad \text{and} \qquad A_1 = \frac{4h}{3}.$$

Thus

$$Q_2(f) = \frac{h}{3}[f(-h) + 4f(0) + f(h)]$$

which is the well-known *Simpson's rule* for estimating $\int_{-h}^{h} f(x)\,dx$. Note in this case that $I(x^3) = 0 = Q_2(x^3)$ but $I(x^4) \neq Q_2(x^4)$. Thus, Simpson's rule has precision 3 and integrates all cubic polynomials exactly. For small values of $h$, these formulas often provide quite good estimates. For example, for $h = 0.2$ and $f(x) = \cos(x)$, we have

$I(f) = 0.397339$, $Q_2(f) = 0.397342$, $Q_1(f) = 0.392027$, and for $f(x) = e^x$ and $h = 0.2$, we obtain

$$I(f) = 0.402672, \qquad Q_2(f) = 0.402676, \qquad Q_1(f) = 0.408027.$$

It is natural at this point to ask if the interpolatory quadratures of formula (6.2) will converge to $\int_a^b w(x)f(x)\,dx$ as we increase the number of interpolating points, that is, as $n \to \infty$. The following theorem provides a condition for which this result is true.

**Theorem 6.1.** For any positive integer $n$ and any $f \in C[a, b]$, let $Q_n(f) = \sum_{j=0}^{n} A_j^{(n)} f(x_j^{(n)})$ be the interpolatory quadrature given by (6.2). If there exists a constant $K > 0$ such that $\sum_{j=0}^{n} |A_j^{(n)}| \leq K$ for all $n$, then $\lim_{n\to\infty} Q_n(f) = I(f)$ for all $f \in C[a, b]$.

*Proof*: By the Weierstrass theorem, for any $\varepsilon > 0$ there exists a polynomial $q_N(x) \in \mathscr{P}_N$ (where $N$ depends on $\varepsilon$), such that

$$\max_{a \leq x \leq b} |f(x) - q_N(x)| = \|f - q_N\|_\infty \leq \varepsilon.$$

We note, since the quadrature formula is interpolatory, that $Q_n(q_N) = I(q_N)$, when $n \geq N$. Choosing $n \geq N$ and letting $\int_a^b w(x)\,dx = c$, we obtain

$$\begin{aligned} |I(f) - Q_n(f)| &= |I(f) - I(q_N) + Q_n(q_N) - Q_n(f)| \\ &\leq |I(f) - I(q_N)| + |Q_n(q_N) - Q_n(f)|. \end{aligned}$$

Now,

$$\begin{aligned} |I(f) - I(q_N)| &= \left| \int_a^b w(x)[f(x) - q_N(x)]\,dx \right| \\ &\leq \|f - q_N\|_\infty \int_a^b w(x)\,dx \leq c\varepsilon \end{aligned}$$

and

$$|Q_n(q_N) - Q_n(f)| = \left| \sum_{j=0}^{n} A_j^{(n)}[q_N(x_j^{(n)}) - f(x_j^{(n)})] \right| \leq \|f - q_N\|_\infty \sum_{j=0}^{n} |A_j^{(n)}| \leq K\varepsilon.$$

Thus, for any $\varepsilon > 0$, there is an integer $N$ such that $|I(f) - Q_n(f)| \leq (K + c)\varepsilon$ whenever $n \geq N$. Therefore, $\lim_{n\to\infty} Q_n(f) = I(f)$ for every $f$ in $C[a, b]$. ■

It can also be shown that the converse of this theorem is true. That is, if $\lim_{n\to\infty} Q_n(f) = I(f)$ for each $f \in C[a, b]$ (where $Q_n(f)$ is an interpolatory quadrature), then there must be a number $K$ such that $\sum_{j=0}^{n} |A_j^{(n)}| \leq K$ for all $n$. However, it goes beyond the scope of this text to prove the converse. Theorem 6.1

(and its converse) do have some practical applications in terms of selecting quadrature formulas. We note first that any interpolatory formula has the property that $Q_n(1) = \int_a^b w(x)\,dx$ since the constant polynomial 1 is integrated exactly. Since $w(x)$ is a nonnegative weight function, we must have

$$0 < I(1) = \int_a^b w(x)\,dx = Q_n(1) = \sum_{j=0}^{n} A_j^{(n)}.$$

Therefore, if the weights $A_j^{(n)}$ are all positive for $0 \le j \le n$, $\int_a^b w(x)\,dx = \sum_{j=0}^{n} |A_j^{(n)}|$. If for each $n$ we can choose the nodes $x_0^{(n)}, x_1^{(n)}, \ldots, x_n^{(n)}$ so that the corresponding weights $A_0^{(n)}, A_1^{(n)}, \ldots, A_n^{(n)}$ are positive, then the hypotheses of Theorem 6.1 are satisfied and convergence is guaranteed. A further advantage of positive-weight quadrature formulas is that they have good rounding error properties. For example, if $Q_N(f) = \sum_{j=0}^{N} A_j f(x_j)$, we can usually expect that the round-off error for the evaluation of the $f(x_j)$'s will be on the high side approximately as often as it is on the low side. Thus, if the $A_j$'s are all positive, the errors to the high side will tend to cancel the errors to the low side when forming the sum, $Q_N(f)$. Furthermore, the expected value of the total rounding error will be minimized if the $A_j$'s are nearly equal. (We shall see two examples of equal weight formulas when we investigate the composite trapezoidal rule and Gauss-Chebyshev quadrature.)

### 6.2.1 Transforming Quadrature Formulas to Other Intervals

Frequently we have a quadrature formula, $Q_n(g) = \sum_{j=0}^{n} A_j g(t_j)$ which is derived for a specific interval, say $[-1, 1]$, and is designed to approximate $\int_{-1}^{1} g(t)\,dt$. If the problem is to estimate $\int_a^b f(x)\,dx$, we can use the results of Section 5.2.4 to transform the formula from $[-1, 1]$ to $[a, b]$. To be specific, suppose $Q_n(g)$ is a given quadrature formula for $[-1, 1]$ and suppose $f(x)$ is defined on $[a, b]$. With the change of variable $x = \alpha t + \beta$, where $\alpha = (b - a)/2$ and $\beta = (a + b)/2$, we have

$$\int_a^b f(x)\,dx = \int_{-1}^{1} f(\alpha t + \beta)\alpha\,dt.$$

Letting $g(t) = \alpha f(\alpha t + \beta)$, we find $Q_n(g) = \sum_{j=0}^{n} \alpha A_j f(\alpha t_j + \beta)$ as an approximation to $\int_{-1}^{1} g(t)\,dt$. Thus we can define the corresponding transformed quadrature formula, $Q_n^*(f)$, for the interval $[a, b]$ by

$$Q_n^*(f) = \sum_{j=0}^{n} A_j^* f(x_j) = \frac{b-a}{2} \sum_{j=0}^{n} A_j f(x_j)$$
$$x_j = \frac{b-a}{2} t_j + \frac{a+b}{2}. \tag{6.4}$$

We leave as a problem for the reader to show that if $Q_n$ has precision $m$ on $[-1, 1]$, then $Q_n^*$ has precision $m$ on $[a, b]$.

### 6.2.2 Newton-Cotes Formulas

Given $\int_a^b f(x)\,dx$ to approximate, probably the most natural choice of nodes $x_i$ to use in a quadrature formula are nodes that are equally spaced in $[a, b]$. Let $h = (b - a)/n$ and let $x_i = x_0 + ih$, where $x_0 = a$. An interpolatory quadrature formula $Q_n(f) = \sum_{j=0}^{n} A_j f(x_j)$, constructed using these equally spaced nodes, is called a *closed* Newton-Cotes formula. The interpolatory quadrature formula constructed using the equally spaced nodes $y_i = a + ih$, $i = 1, 2, \ldots, n + 1$; $h = (b - a)/(n + 2)$ is called an *open* Newton-Cotes formula. (The closed formula uses the end-points $a$ and $b$, while the open formula has $a < y_1$, and $y_{n+1} < b$.)

From Example 6.1, with $h = 1$, we see that

$$Q_1(f) = f(-1) + f(1)$$

$$Q_2(f) = \frac{1}{3}[f(-1) + 4f(0) + f(1)]$$

are the two- and three-point closed Newton-Cotes rules for $[-1, 1]$. Using the results of Section 6.2.1 on these formulas, we have for an arbitrary interval $[a, b]$, that

$$T(f) = \frac{(b - a)}{2}[f(a) + f(b)] \tag{6.5}$$

$$S(f) = \frac{(b - a)}{6}\left[f(a) + 4f\left(\frac{a + b}{2}\right) + f(b)\right] \tag{6.6}$$

are respectively the closed two-point and three-point Newton-Cotes formulas. (Formula (6.5) is the trapezoidal rule for $[a, b]$ and (6.6) is Simpson's rule for $[a, b]$.) As two more examples of Newton-Cotes formulas, we have the three-point open formula for $[-1, 1]$ and the four-point closed formula for $[-1, 1]$ (see Problem 1):

$$\int_{-1}^{1} f(x)\,dx \approx \frac{2}{3}\left[2f\left(-\frac{1}{2}\right) - f(0) + 2f\left(\frac{1}{2}\right)\right]$$

$$\int_{-1}^{1} f(x)\,dx \approx \frac{1}{4}\left[f(-1) + 3f\left(-\frac{1}{3}\right) + 3f\left(\frac{1}{3}\right) + f(1)\right].$$

A table for the first few Newton-Cotes formulas is given in Chapter 7, where quadrature formulas are used as a tool for solving differential equations numerically.

Notice in the three-point open formula above, that not all the weights are positive. In fact, it is known for $n \geq 10$ that the weights of any Newton-Cotes formula are of mixed sign. As we have noted, this is a bad feature in terms of rounding error. Moreover, as $n$ increases, the weights themselves grow without bound, meaning that there are continuous functions for which the quadratures do not converge to the integral. For these reasons, higher order Newton-Cotes formulas are rarely used in practice. However, lower order Newton-Cotes formulas such as the trapezoidal rule and Simpson's rule are *extremely* useful, and the time invested in analyzing their properties in the next section will be well spent.

We now give one further type of quadrature which involves derivative evaluations of $f(x)$ as well as values of $f(x)$ itself. The cubic polynomial, $p_3(x)$, which satisfies $p_3(a) = f(a)$, $p_3'(a) = f'(a)$, $p_3(b) = f(b)$, and $p_3'(b) = f'(b)$ is given by formula (5.37) and we can approximate $I(f)$ by $\tilde{Q}_3(f) \equiv I(p_3)$. We could laboriously compute $I(p_3) \equiv \int_a^b p_3(x)\,dx$ by direct integration, but we can simplify this task by noting, since Simpson's rule has precision 3, that $S(p_3) = I(p_3)$. From Eq. (5.37) we can easily verify that $p_3(a) = f(a)$, $p_3((a+b)/2) = 1/2\,(f(a) + f(b)) + ((b-a)/8)(f'(a) - f'(b))$, and $p_3(b) = f(b)$. Then, using Eq. (6.6), we have

$$\tilde{Q}_3(f) = S(p_3) = \frac{(b-a)}{2}[f(a) + f(b)] + \frac{(b-a)^2}{12}[f'(a) - f'(b)],$$

or

$$\tilde{Q}_3(f) = T(f) + \frac{(b-a)^2}{12}[f'(a) - f'(b)]. \tag{6.7}$$

Because of the presence of $T(f)$ in this formula, $\tilde{Q}_3(f)$ is called the *corrected trapezoidal rule*, and is often denoted by $CT(f) \equiv \tilde{Q}_3(f)$. By the uniqueness of the Hermite interpolating polynomial in Theorem 5.7, if $f(x)$ is any polynomial of degree 3 or less, then $f(x) = p_3(x)$ and so $I(f) = CT(f)$. We can easily check that $I(x^4) \neq CT(x^4)$, and therefore $CT(f)$ has precision 3.

**Example 6.2.** Given $\int_{-1}^{1} \cos(x)\,dx$ to estimate, we find for $f(x) = \cos(x)$ that

$$I(f) = 1.683$$
$$T(f) = 1.081$$
$$CT(f) = T(f) + 0.561 = 1.642$$

This example indicates that when derivative data is readily available, the corrected trapezoidal rule may be expected to provide better answers. The theoretical analysis in the next section bears this out.

### 6.2.3 Errors of Quadrature Formulas

In this section, we consider ways of estimating the quadrature error $I(f) - Q_n(f)$ and derive specific estimates for the important cases of the trapezoidal rule, Simpson's rule, and the corrected trapezoidal rule. By formula (5.27), the error of regular polynomial interpolation, for $f(x) \in C^{n+1}[a, b]$, is given by

$$e_n(x) \equiv f(x) - p_n(x) = \frac{f^{(n+1)}(\xi)}{(n+1)!} W(x).$$

We can therefore express the error of interpolatory quadrature as

$$e_n = \int_a^b e_n(x) w(x)\,dx,$$

or

$$e_n \equiv \int_a^b f(x)w(x)\,dx - \int_a^b p_n(x)w(x)\,dx = \int_a^b \frac{f^{(n+1)}(\xi)}{(n+1)!} W(x)w(x)\,dx. \tag{6.8a}$$

Recalling that $f^{(n+1)}(\xi)$ is actually an unknown function of $x$, we see the right-hand integral in (6.8a) cannot usually be evaluated explicitly. We can, however, obtain a computable bound on the error, $e_n$, from (6.8a) by using

$$|e_n| \leq \max_{a \leq x \leq b} |f^{(n+1)}(x)| \int_a^b \frac{|W(x)|}{(n+1)!} w(x)\, dx. \tag{6.8b}$$

We will use an error bound similar to this in a later section, but for now we consider two other ways of simplifying (6.8a). Recall that a simple form of the Second Mean-Value Theorem for integrals states that if $g(x)$ and $h(x)$ are continuous and if $g(x)$ does not change sign in the interval $(a, b)$, then there exists a mean value point, $\eta \in (a, b)$, such that $\int_a^b g(x)h(x)\, dx = h(\eta) \int_a^b g(x)\, dx$. We can use this theorem immediately in (6.5) and (6.7). For (6.5), (6.8a) gives the error for the trapezoidal rule

$$I(f) - T(f) \equiv e^T = \frac{1}{2} \int_a^b f''(\xi)(x-a)(x-b)\, dx.$$

Now $[(x-a)(x-b)]$ is of constant sign on $(a, b)$, so the above mean-value theorem yields

$$e^T = \frac{f''(\eta)}{2} \int_a^b (x-a)(x-b)\, dx = -\frac{f''(\eta)}{12}(b-a)^3. \tag{6.9}$$

Since (6.7) was the result of Hermite interpolation, we must use (5.39) instead of (5.27) to obtain the error. From (5.39), where $p_3(x)$ is given as in (6.7), we have

$$f(x) - p_3(x) = \frac{f^{(iv)}(\xi)}{4!}(x-a)^2(x-b)^2.$$

Now $[(x-a)^2(x-b)^2] > 0$ for $x \in (a, b)$, so the mean-value theorem yields

$$I(f) - CT(f) \equiv e^{CT} = \int_a^b \frac{f^{(iv)}(\xi)}{4!}(x-a)^2(x-b)^2\, dx$$

or

$$e^{CT} = \frac{f^{(iv)}(\eta)}{4!} \int_a^b (x-a)^2(x-b)^2\, dx = \frac{f^{(iv)}(\eta)}{720}(b-a)^5. \tag{6.10}$$

The error for Simpson's rule is a bit more difficult to analyze, since in (6.8a) the function $W(x)$ changes sign in $(a, b)$. First, let $x_0 = a$, $x_1 = (a+b)/2$ and $x_2 = b$. We can get at the error in Simpson's rule by defining an auxiliary cubic polynomial, $p_3(x)$, such that $p_3(x_0) = f(x_0)$, $p_3(x_1) = f(x_1)$, $p_3(x_2) = f(x_2)$ and $p_3'(x_1) = f'(x_1)$ (we leave as Problem 4 to show that such a polynomial exists). Next, note that $S(f) = S(p_3)$ since $p_3(x)$ interpolates $f(x)$ at $x_0$, $x_1$, and $x_2$. Further, since Simpson's rule has precision 3, $S(p_3) = I(p_3)$. Therefore we obtain an alternative expression for the error by:

$$I(f) - S(f) = I(f) - S(p_3) = I(f) - I(p_3).$$

In order to use this expression, we need an error formula for $f(x) - p_3(x)$. This is easily obtained, as below, using a method similar to that in the proof of Theorem 5.5.

For a *fixed* $x$, different from $x_0$, $x_1$, and $x_2$, define $\phi(t)$ for $t \in [a, b]$ by

$$\phi(t) = [f(t) - p_3(t)] - [f(x) - p_3(x)]\frac{(t - x_0)(t - x_1)^2(t - x_2)}{(x - x_0)(x - x_1)^2(x - x_2)}.$$

Then $\phi(t)$ has at least 4 zeros in $[a, b]$, namely $x_0$, $x_1$, $x_2$, and $x$. By Rolle's theorem, $\phi'(t)$ has at least 3 zeros in $(a, b)$ which are between the 4 zeros of $\phi(t)$. By construction, we also see that $\phi'(x_1) = 0$ and therefore, $\phi'(t)$ has at least 4 *distinct* zeros in $(a, b)$. If $f \in C^4[a, b]$, we find (as in the proof of Theorem 5.5) that there is a point $\xi$ in $(a, b)$ (which depends on $x$) such that $\phi^{(iv)}(\xi) = 0$. Therefore,

$$f(x) - p_3(x) = \frac{f^{(iv)}(\xi)}{4!}(x - x_0)(x - x_1)^2(x - x_2).$$

Since the function $[(x - x_0)(x - x_1)^2(x - x_2)]$ does not change sign on $(a, b)$, we can now use the Second Mean-Value Theorem to deduce

$$I(f) - I(p_3) = \int_a^b \frac{f^{(iv)}(\xi)}{4!}(x - x_0)(x - x_1)^2(x - x_2)\,dx,$$

or

$$I(f) - S(f) = \frac{f^{(iv)}(\eta)}{4!}\int_a^b (x - x_0)(x - x_1)^2(x - x_2)\,dx. \qquad \textbf{(6.11a)}$$

We can obviously evaluate the above integral directly to obtain the error. To illustrate another idea, however, we use a standard trick to evaluate

$$\int_a^b (x - x_0)(x - x_1)^2(x - x_2)\,dx,$$

by choosing $\tilde{f}(x) = (x - x_1)^4$. From (6.11a), we have

$$I(\tilde{f}) - S(\tilde{f}) = \int_a^b (x - x_0)(x - x_1)^2(x - x_2)\,dx,$$

since the fourth derivative of $\tilde{f}(x) = (x - x_1)^4$ is the constant $4!$. Letting $h = (b - a)/2 = x_1 - x_0 = x_2 - x_1$, we find

$$\int_{x_0}^{x_2} (x - x_1)^4\,dx = \frac{(x - x_1)^5}{5}\bigg|_{x_0}^{x_2} = \frac{2h^5}{5}$$

and

$$S((x - x_1)^4) = \frac{2h}{6}[h^4 + h^4] = \frac{2h^5}{3}.$$

Thus for $\tilde{f}(x) = (x - x_1)^4$,

$$\int_a^b (x - x_0)(x - x_1)^2(x - x_2)\,dx = I(\tilde{f}) - S(\tilde{f}) = \frac{-4h^5}{15}.$$

Using this in (6.11a), with $e^S = I(f) - S(f)$, we have

$$e^S = \frac{-f^{(iv)}(\eta)}{90}\left(\frac{b-a}{2}\right)^5. \tag{6.11b}$$

**Example 6.3.** The sine-integral, $Si(x)$, is defined by

$$Si(x) = \int_0^x \frac{\sin(t)}{t}\,dt.$$

To illustrate the error analysis of this section, we estimate $Si(1)$ by the trapezoidal rule, Simpson's rule, and the corrected trapezoidal rule. In order to use the error bounds derived above on the estimates for $Si(1)$, we need in (6.9) to bound $|f''(\eta)|$ for $\eta \in [0, 1]$, and in (6.10) and (6.11b) we need to bound $|f^{(iv)}(\eta)|$ for $\eta \in [0, 1]$.

To obtain these bounds, we use the expansion

$$\sin(t) = t - \frac{t^3}{3!} + \frac{t^5}{5!} - \frac{t^7}{7!} + \cdots$$

to obtain for $f(t) = (\sin(t)/t)$:

$$f(t) = 1 - \frac{1}{3}\frac{t^2}{2!} + \frac{1}{5}\frac{t^4}{4!} - \frac{1}{7}\frac{t^6}{6!} + \cdots$$

$$f''(t) = -\frac{1}{3} + \frac{1}{5}\frac{t^2}{2!} - \frac{1}{7}\frac{t^4}{4!} + \cdots$$

$$f^{(iv)}(t) = \frac{1}{5} - \frac{1}{7}\frac{t^2}{2!} + \cdots.$$

Since all these series are alternating series, we see that $|f''(\eta)| \le 1/3$ and $|f^{(iv)}(\eta)| \le 1/5$, for $\eta \in [0, 1]$. Hence, the error bounds are: $|e^T| \le 1/36 = 0.027778$, $|e^{CT}| \le 1/3600 = 0.000278$ and $|e^S| \le 1/14400 = 0.000069$. From a table, we obtain $Si(1) = 0.946083$. Noting that $f(0) = 1$ and $f'(0) = 0$, we have

$$\begin{aligned} T(f) &= 0.920735; & e^T &= 0.025348 \\ CT(f) &= 0.945832; & e^{CT} &= 0.000251 \\ S(f) &= 0.946146; & e^S &= -0.000063. \end{aligned}$$

### 6.2.4 Composite Rules for Numerical Integration and the Euler-Maclaurin Formula

If we wish to derive a quadrature formula for the interval $[a, b]$ of high accuracy, we immediately see two obvious choices:

1. Take the number of nodes, $(n + 1)$, to be large so that the quadrature,

$$Q_n(f) = \sum_{j=0}^{n} A_j f(x_j),$$

is the integral of a high-degree interpolating polynomial.

2. Divide $[a, b]$ into several subintervals of small length and use a simple quadrature of fairly low precision on each, adding the results to obtain an approximation for $I(f)$.

The methods generated by (2) above are called *composite* quadratures, and we shall study them in this section. First we consider $N + 1$ points $\{x_j\}_{j=0}^{N}$, such that $a = x_0 < x_1 < \cdots < x_N = b$. From calculus we know that

$$I(f) = \int_a^b f(x)w(x)\,dx = \sum_{j=0}^{N-1} \int_{x_j}^{x_{j+1}} f(x)w(x)\,dx.$$

Thus approximating $I(f)$ by composite quadrature amounts to approximating each $\int_{x_j}^{x_{j+1}} f(x)w(x)\,dx$ by a low order quadrature, $Q_k^{(j)}(f)$, and adding the resulting approximations. The three rules specifically studied in Section 6.2.2 (trapezoidal, Simpson, and corrected trapezoidal) are very well suited for our purposes here, and we shall concentrate on their use. We shall see that their form is especially simple if we take the points $\{x_j\}_{j=0}^{N}$ to be equally spaced, that is, $x_j = a + jh$ for $h = (b - a)/N$ and $0 \leq j \leq N$. (Using these rules, of course, we take $w(x) \equiv 1$ in $I(f)$.)

First we consider the composite trapezoidal rule. From (6.5) and the error formula (6.9), we see that for any $j$

$$\int_{x_j}^{x_{j+1}} f(x)\,dx = \frac{h}{2}(f(x_j) + f(x_{j+1})) - \frac{f''(\eta_j)}{12}h^3, \qquad x_j < \eta_j < x_{j+1}.$$

Thus we define the composite trapezoidal rule, $T_N(f)$, as

$$T_N(f) = \sum_{j=0}^{N-1} \frac{h}{2}(f(x_j) + f(x_{j+1})) = h\sum_{j=1}^{N-1} f(x_j) + \frac{h}{2}(f(x_0) + f(x_N)). \tag{6.12a}$$

The error, $I(f) - T_N(f) \equiv e_N^T$, is the sum of the errors on each interval and so

$$e_N^T = \sum_{j=0}^{N-1} f''(\eta_j)(-h^3/12).$$

We pause here to give a lemma that will significantly simplify $e_N^T$ above, and can also be used when investigating the errors of other composite rules.

**Lemma 6.1.** Let $g(x) \in C[a, b]$ and let $\{a_j\}_{j=0}^{N-1}$ be any set of constants, all of which have the same sign. If $t_j \in [a, b]$ for $0 \leq j \leq N - 1$, then for some $\eta \in [a, b]$:

$$\sum_{j=0}^{N-1} a_j g(t_j) = g(\eta)\sum_{j=0}^{N-1} a_j.$$

*Proof*. Let $m = \min_{a \leq x \leq b}(g(x)) = g(y_m)$ and $M = \max_{a \leq x \leq b}(g(x)) = g(y_M)$. Then, if the $a_j$'s are nonnegative,

$$m\sum_{j=0}^{N-1} a_j \leq \sum_{j=0}^{N-1} a_j g(t_j) \leq M\sum_{j=0}^{N-1} a_j.$$

Define $r$ by $r = \sum_{j=0}^{N-1} a_j g(t_j)$ and let $G(x) \equiv g(x)\sum_{j=0}^{N-1} a_j$. Then $G(x) \in C[a, b]$, and by the above inequalities, $G(y_m) \leq r \leq G(y_M)$. By the intermediate-value theorem, there exists an $\eta \in [a, b]$ such that $G(\eta) = r$, or

$$g(\eta) \sum_{j=0}^{N-1} a_j = \sum_{j=0}^{N-1} a_j g(t_j).$$

The case where the $a_j$'s are all negative is similar and left to the reader. ■

Using $f''(x) = g(x)$ and $a_j = -h^3/12$ in Lemma 6.1, the expression for $e_N^T$ becomes

$$e_N^T = f''(\eta) \sum_{j=0}^{N-1} \left(\frac{-h^3}{12}\right) = -f''(\eta)\frac{Nh^3}{12} = -\frac{f''(\eta)h^2(b-a)}{12}. \quad \textbf{(6.12b)}$$

Continuing in the same fashion, from Simpson's rule, (6.6), and its error, we see that for any $j$

$$\int_{x_j}^{x_{j+1}} f(x)\,dx = \frac{h}{6}(f(x_j) + 4f((x_j + x_{j+1})/2) + f(x_{j+1})) - \frac{f^{(iv)}(\eta_j)}{90}\left(\frac{h}{2}\right)^5,$$

$$x_j < \eta_j < x_{j+1}.$$

Summing the approximations for each interval we have the composite Simpson's rule

$$S_N(f) = \frac{h}{6}\left\{f(x_0) + f(x_N) + 2\sum_{j=1}^{N-1} f(x_j) + 4\sum_{j=0}^{N-1} f((x_j + x_{j+1})/2)\right\}. \quad \textbf{(6.13a)}$$

Summing the individual errors for $0 \le j \le N-1$, and using Lemma 6.1 we obtain the error, $e_N^S \equiv I(f) - S_N(f)$, as

$$e_N^S = -\sum_{j=0}^{N-1} \frac{f^{(iv)}(\eta_j)}{90}\left(\frac{h}{2}\right)^5 = -\frac{f^{(iv)}(\eta)}{90} N\left(\frac{h}{2}\right)^5 = -\frac{f^{(iv)}(\eta)}{180}\left(\frac{h}{2}\right)^4 (b-a). \quad \textbf{(6.13b)}$$

We note that Simpson's rule in the form (6.13a) actually requires the evaluation of $f(x)$ at $(2N + 1)$ equally spaced points, $\{x_j\}_{j=0}^N$ and $\{(x_j + x_{j+1})/2\}_{j=0}^{N-1}$. For a fair comparison with the accuracy of the composite trapezoidal rule, we should actually display a form of the composite Simpson's rule using only $(N + 1)$ points, $\{x_j\}_{j=0}^N$. From above we see immediately that we must restrict $N$ to be even and investigate the formulas

$$\int_{x_j}^{x_{j+2}} f(x)\,dx = \frac{2h}{6}(f(x_j) + 4f(x_{j+1}) + f(x_{j+2})) - \frac{f^{(iv)}(\eta_j)}{90}h^5,\ x_j < \eta_j < x_{j+2},$$

for $j = 0, 2, 4, \ldots, N-2$. Doing this and summing we obtain

$$S'_N(f) = \frac{h}{3}\left(f(x_0) + f(x_N) + 2\sum_{j=1}^{(N-2)/2} f(x_{2j}) + 4\sum_{j=0}^{(N-2)/2} f(x_{2j+1})\right), \quad \textbf{(6.14a)}$$

and

$$e_N^{S'} = -\frac{f^{(iv)}(\eta)}{180} h^4(b-a). \tag{6.14b}$$

We leave to the reader to verify in the above manner that from the corrected trapezoidal rule, (6.7), and its error, (6.10), the composite corrected trapezoidal rule and its error are given by

$$\begin{aligned} CT_N(f) &= h\sum_{j=1}^{N-1} f(x_j) + \frac{h}{2}(f(x_0) + f(x_N)) + \frac{h^2}{12}(f'(a) - f'(b)) \\ &= T_N(f) + \frac{h^2}{12}(f'(a) - f'(b)) \end{aligned} \tag{6.15a}$$

$$e_N^{CT} = \frac{f^{(iv)}(\eta)}{720} h^4(b-a). \tag{6.15b}$$

The reader will note in his derivation of (6.15a) that the derivative evaluations of $f(x)$ at the interior nodes cancel out, and the only two required derivative evaluations are $f'(a)$ and $f'(b)$. This, plus the fact that it has precision 3 and has an $h^4$ term in the error, makes this rule computationally attractive.

Since the precision of a composite quadrature formula is not increased by taking $N$ larger and larger, we cannot use Theorem 6.1 to guarantee convergence. However, it is fairly easy to show that $\lim_{N\to\infty} T_N(f) = \lim_{N\to\infty} S_N(f) = I(f)$, for any $f \in C[a, b]$. To see this, we merely recall that $T_N(f) = \sum_{j=0}^{N-1} [f(x_j) + f(x_{j+1})](h/2)$. Since $[f(x_j) + f(x_{j+1})]/2$ is just an average, we have by the Intermediate-Value Theorem that there is a point $z_j \in [x_j, x_{j+1}]$ such that $f(z_j) = [f(x_j) + f(x_{j+1})]/2$. Thus $T_N(f) = \sum_{j=0}^{N-1} f(z_j)h$ is a Riemann sum, so from the definition of the definite integral it follows that $T_N(f) \to I(f)$ as $N \to \infty$. A similar analysis can be carried out for Simpson's rule. (See Problem 9 for a general result on the convergence of composite quadratures.)

Because Simpson's rule is derived from quadratic interpolation on the subintervals whereas the trapezoidal rule comes from linear interpolation we might jump to the conclusion that Simpson's rule is always preferable. This is often true, but in certain instances the trapezoidal rule yields surprisingly good results. For example, if $f(x)$ has the property that $f'(a) = f'(b)$, then the trapezoidal rule is equivalent to the corrected trapezoidal rule. (See (6.15a) and compare the error (6.15b) to the Simpson error (6.14b) for application of both rules to $(N + 1)$ points.) Davis and Rabinowitz (1967) is rich with comparative numerical examples from the literature using both rules.

**Example 6.4.** As an example showing how the error bounds of this section can be used, we consider again (as in Example 6.3), the problem of computing $Si(1)$. Suppose we want accuracy on the order of $10^{-8}$; how large must $N$ be in order that $|I(f) - T_N(f)| \le 10^{-8}$ and how large must $N$ be in order that $|I(f) - S'_N(f)| \le 10^{-8}$, where

$$I(f) = Si(1) = \int_0^1 \frac{\sin(t)}{t}\,dt?$$

Using the information derived in Example 6.3, we know that $|f''(x)| \le \frac{1}{3}$ and $|f^{(iv)}(x)| \le \frac{1}{5}$ for $0 \le x \le 1$. From (6.12b) and (6.14b), using $b - a = 1$ and $h = 1/N$, we have

$$|e_N^T| \le \frac{1}{36N^2}, \qquad |e_N^{S'}| \le \frac{1}{900N^4}.$$

A quick computation then shows that $N \ge 1667$ is required to make $|e_N^T| \le 10^{-8}$ and $N \ge 19$ is necessary to insure that $|e_N^{S'}| \le 10^{-8}$.

We conclude this section by showing how the composite trapezoidal rule may be used for the problem of finding an approximate value for a series, $S = \sum_{j=0}^{\infty} a_j$. We attack this problem by observing that often we can find a function $f(x)$ such that $f(j) = a_j$ for all $j$, and where $f(x)$ can be integrated directly. Then the $n$th partial sum, $S_n = \sum_{j=0}^{n} a_j$, is precisely equal to $(T_n(f) + 1/2(a_0 + a_n))$ by (6.12a) (where $[a, b] = [0, n]$ and $h = 1$). Thus we have the approximation

$$S_n - \frac{1}{2}(a_0 + a_n) = T_n(f) \approx \int_0^n f(x)\,dx \tag{6.16}$$

where, of course, $S \approx S_n$.

Formula (6.16) is called the Euler-Maclaurin summation formula, and essentially reverses the problem of numerical integration since we are now approximating an unknown sum by a known integral. Since $b - a = n$ in this case, the error formula (6.12b) is of little practical use in analyzing this approximation. Instead we define the function $P_1(x) = x - [x] - 1/2$, where $[x] \equiv$ greatest integer in $x$. We can easily verify the following identity (by using integration by parts on the second integral)

$$\frac{1}{2}(f(j) + f(j+1)) = \int_j^{j+1} f(x)\,dx + \int_j^{j+1} P_1(x) f'(x)\,dx. \tag{6.17}$$

Now summing (6.17) from $j = 0$ to $j = n$ we obtain

$$T_n(f) = \int_0^n f(x)\,dx + \int_0^n P_1(x) f'(x)\,dx. \tag{6.18}$$

Through somewhat cumbersome, but straight-forward manipulations, $P_1(x)$ can be expanded in a trigonometric Fourier sine series and (6.18) can be successively integrated by parts to yield

$$\begin{aligned}
T_n(f) &= (1/2)f(0) + f(1) + \cdots + f(n-1) + (1/2)f(n) \\
&= \int_0^n f(x)\,dx + \frac{B_2}{2!}(f'(n) - f'(0)) + \frac{B_4}{4!}(f'''(n) - f'''(0)) \\
&\quad + \cdots + \frac{B_{2j}}{(2j)!}(f^{(2j-1)}(n) - f^{(2j-1)}(0)) \\
&\quad + \int_0^n P_{2j+1}(x) f^{(2j+1)}(x)\,dx,
\end{aligned} \tag{6.19}$$

where

$$P_2'(x) = P_1(x), \qquad P_3'(x) = P_2(x), \text{ etc.,}$$

and where

$$B_{2j}/(2j)! \equiv (-1)^{j-1} \sum_{k=1}^{\infty} 2/(2k\pi)^{2j}, \qquad j > 1.$$

(Again see Davis and Rabinowitz (1967) for details.) The constants $B_{2j}$ are called the *Bernoulli numbers*, and their values can be found in most mathematical tables. The first few Bernoulli numbers of even order are

$$B_0 = 1, \qquad B_2 = \frac{1}{6}, \qquad B_4 = -\frac{1}{30}, \qquad B_6 = \frac{1}{42}.$$

One important result which follows from formula (6.19) is the following alternative bound for the error of the composite trapezoidal rule:

**Theorem 6.2.** Let $g(x) \in C^{2j+1}[a, b]$ and let $h = (b - a)/N$, then

$$-e_N^T \equiv T_N(g) - \int_a^b g(x)\,dx = \frac{B_2}{2!} h^2(g'(b) - g'(a))$$

$$+ \frac{B_4}{4!} h^4(g'''(b) - g'''(a)) + \cdots + \frac{B_{2j}}{(2j)!} h^{2j}(g^{(2j-1)}(b) - g^{(2j-1)}(a))$$

$$+ h^{2j+1} \int_a^b P_{2j+1}\left(N\frac{x-a}{b-a}\right) g^{(2j+1)}(x)\,dx. \qquad \textbf{(6.20a)}$$

*Proof*. In formula (6.19) define $f(x) \equiv g(a + hx)$. ■

From Theorem 6.2 we immediately have the result:

**Corollary.** Let $g(x) \in C^{(2j+1)}[a, b]$ with $g^{(2i-1)}(a) = g^{(2i-1)}(b)$ for $i = 1, 2, \ldots, j$. Then there exists a constant $C$ such that

$$|e_N^T| = \left|\int_a^b g(x)\,dx - T_N(g)\right| \le \frac{C}{N^{2j+1}}. \qquad \textbf{(6.20b)}$$

Formula (6.20b) emphasizes the fact that the more periodic character $g(x)$ has, then the stronger the composite trapezoidal rule becomes as a candidate for numerical integration.

Finally, the theory of Bernoulli polynomials can be used in deriving another form of the Euler-Maclaurin summation formula, although it is beyond our means in this brief introduction to summation to develop the properties of Bernoulli polynomials (the interested reader is referred to Ralston 1965). Using the notation of Theorem 6.2, it can be shown for $g \in C^{2j+2}[a, b]$, that

$$T_N(g) - \int_a^b g(x)\,dx = \sum_{i=1}^{j} \frac{B_{2i}}{(2i)!} h^{2i}[g^{(2i-1)}(b) - g^{(2i-1)}(a)]$$

$$+ (b-a)h^{2j+2} \frac{B_{2j+2}}{(2j+2)!} g^{(2j+2)}(\eta). \qquad \textbf{(6.20c)}$$

Note that (6.20c) is just formula (6.20a) rewritten, with a derivative evaluation replacing the integral that appears on the right-hand side of (6.20a).

**Example 6.5.** There is a variety of information that can be obtained from the Euler-Maclaurin summation formula. For example, consider the problem of evaluating the sum:

$$S = 1 + \frac{1}{2} + \frac{1}{3} + \cdots + \frac{1}{20} = \sum_{k=0}^{19} \frac{1}{k+1}.$$

To estimate $S$, let us choose $h = 1$ and $[a, b] = [0, 19]$ and let $g(x) = 1/(x + 1)$, so that $-e_{19}^T = T_{19}(g) - \int_0^{19} g(x)\,dx$. Since we have $S = T_{19}(g) + \frac{1}{2}[g(0) + g(19)] = T_{19}(g) + \frac{21}{40}$, we find

$$S = \int_0^{19} g(x)\,dx + \frac{21}{40} - e_{19}^T$$

where $-e_{19}^T$ is given by (6.20c). Now, $\int_0^{19} g(x)\,dx = \ln(x + 1)|_0^{19} = \ln(20)$, so $S = \ln(20) + \frac{21}{40} - e_{19}^T$. By evaluating a few of the terms in the expression for $e_{19}^T$, we can obtain estimates for $S$ (where the actual value of $S$ is 3.597740). For example,

$$(j = 0) \qquad S \approx S_0 = \ln(20) + \frac{21}{40} = 3.520732$$

$$(j = 1) \qquad S \approx S_1 = S_0 + \frac{B_2}{2!}[g'(19) - g'(0)] = S_0 + \left(\frac{1}{12}\right)\left(\frac{399}{400}\right) = 3.603857$$

$$(j = 2) \qquad S \approx S_2 = S_1 + \frac{B_4}{4!}[g'''(19) - g'''(0)] = S_1 - \frac{1}{720}[g'''(19) - g'''(0)] = 3.595524.$$

For $j = 2$, we find from (6.20c) that $S - S_2 = 19(B_6/6!)g^{(vi)}(\eta)$, which gives a bound of $|S - S_2| \le 19/42 = 0.45$. The crudeness of this bound brings home the fact that caution must be exercised in using the Euler-Maclaurin summation formula. Part of the problem is the growth of the Bernoulli numbers, for it can be shown that $|B_{2i}| \to \infty$ since $|B_{2i}| \approx 2(2i)!/(2\pi)^{2i}$.

Another interesting case of (6.20c) is given when $g(x)$ is a polynomial. For $g \in \mathscr{P}_{2j+1}$, we see that $g^{(2j+2)}(\eta) = 0$. To see how this can be used, consider the sum $S = 1 + 2 + \cdots + n$. Again, choose $h = 1$ and $[a, b] = [0, n - 1]$ and let $g(x) = x + 1$. Then $S = T_{n-1}(g) + \frac{1}{2}[g(0) + g(n - 1)] = T_{n-1}(g) + (n + 1)/2$. As before,

$$S = \int_0^{n-1} g(x)\,dx + \frac{n+1}{2} - e_{n-1}^T$$

but in this case, since $g'''(x) \equiv 0$, we have $-e_{n-1}^T = (B_2/2)[g'(n - 1) - g'(0)] = 0$. Therefore

$$S = \int_0^{n-1} g(x)\,dx + \frac{n+1}{2} = \frac{n^2 - 1}{2} + \frac{n+1}{2} = \frac{n(n+1)}{2},$$

which is the familiar formula for the sum of the first $n$ integers. This technique will clearly work to evaluate $S = 1^k + 2^k + \cdots + n^k$, for any positive integer $k$.

Finally, choosing $j = 1$ in formula (6.20c), we obtain the corrected trapezoidal rule $CT(g)$ and the error of the corrected trapezoidal rule, $e_N^{CT}$. Choosing successively larger values of $j$, we can derive higher-order corrections to the trapezoidal rule. For example, with $j = 2$, we obtain the quadrature formula

$$Q(g) = T_N(g) - \frac{h^2}{12}[g'(b) - g'(a)] + \frac{h^4}{720}[g'''(b) - g'''(a)]$$

and the error term

$$\int_a^b g(x) - Q(g) = -(b - a)h^6 \frac{B_6}{6!} g^{(vi)}(\eta).$$

## PROBLEMS

1. Using (6.3), construct the open Newton-Cotes formula for $\int_{-1}^{1} f(x)\,dx$ with nodes $-1/2, 0, 1/2$ and the closed Newton-Cotes formula with nodes $-1, -1/3, 1/3, 1$.
2. Construct the interpolatory quadrature for $\int_{-1}^{1} f(x)\,dx$, with nodes $-1, -1/2, 1/2, 1$.
3. Prove that the quadrature formula $Q_n^*$ given in (6.4) has precision $m$ on $[a, b]$ whenever $Q_n$ has precision $m$ on $[-1, 1]$. (Hint: If $f(x)$ is a polynomial in $x$ of degree $m$ or less, what is $g(t)$?)
4. Given $x_0$ and $h > 0$, let $x_1 = x_0 + h$, $x_2 = x_0 + 2h$. Suppose $f \in C^1[x_0, x_2]$. Using the method of undetermined coefficients, show there is a cubic polynomial $p(x)$ such that $p(x_i) = f(x_i)$, $i = 0, 1, 2$ and $p'(x_1) = f'(x_1)$.
5. Verify the error formula (6.15b) for the corrected trapezoidal rule.
6. How small must $h$ be in order that $|I(f) - S_N(f)| \leq 10^{-6}$, for $f(x) = \sin(10x)$ and $[a, b] = [0, \pi]$? How small must $h$ be in order that $|I(f) - T_N(f)| \leq 10^{-6}$?
7. Following Example 6.5, find a formula for $1^4 + 2^4 + \cdots + n^4$.
8. Show that $\lim_{N \to \infty} S_N(f) = I(f)$ for any $f \in C[a, b]$, where $S_N(f)$ denotes the composite Simpson's rule for $[a, b]$.

*9. Suppose $Q(f) = \sum_{j=0}^{n} A_j f(x_j)$ is a quadrature formula to approximate $I(f) = \int_{-1}^{1} f(x)\,dx$, where $\sum_{i=0}^{n} A_j = 2$. Let $Q_N$ be the composite formula corresponding to $Q$ applied to $[a, b]$ and suppose $g \in C[a, b]$. Using upper and lower Riemann sums, show that $Q_N(g) \to \int_a^b g(z)\,dz$.

10. Discuss how the composite Simpson's rule given in (6.13a) can be modified slightly so as to integrate cubic splines exactly (the only thing that must be considered is the possibility of unequal knot spacings).

## *6.3 ADAPTIVE QUADRATURE

Adaptive quadrature is a process designed to use a given set of quadrature rules to approximate $I(f)$ to within a given error tolerance, $\varepsilon > 0$, where the computer is programmed to make the decisions which produce the approximation with optimal efficiency, thus relieving the user (as much as possible) of the necessity of analyzing the accuracy of the results. We shall concentrate here on describing an adaptive quadrature routine based on composite Simpson rules, although extensions to other rules should become evident to the reader.

For the sake of intuition, let us suppose that $f(x)$ is "badly behaved" only on some small segment, $[\alpha, \beta]$, of the entire interval of integration $[a, b]$. Then composite Simpson's rules with relatively few points will produce fairly accurate approximations for $\int_a^\alpha f(x)\,dx$ and $\int_\beta^b f(x)\,dx$, although the entire estimate for $I(f) = \int_a^b f(x)\,dx$ may be badly in error. If we were to take the seemingly natural course of halving the step-size over $[a, b]$ we would not appreciably increase the accuracy of our estimates for $\int_a^\alpha f(x)\,dx$ and $\int_\beta^b f(x)\,dx$ and we still might not have an acceptable estimate for $I(f)$. Thus the new work done in $[a, \alpha]$ and $[\beta, b]$ is essentially wasted whereas more refinement may still be necessary on $[\alpha, \beta]$. Intuitively we see that what we must do for optimal efficiency is to improve our accuracy on $[\alpha, \beta]$ without any further refinement on the intervals where our approximations are acceptable. We must further keep in mind that we are asking the computer to internally decide when an approximation is acceptable on a given subinterval.

The following is one way to attack this problem. First compute Simpson's rule, (6.6) or (6.13a) with $N = 1$, to get $S_1(f)$, and then halve the step-size and compute $S_2(f)$ from (6.13a) with $N = 2$. If $|S_2(f) - S_1(f)| \le \varepsilon$, then we accept $S_2(f)$ as an approximation for $I(f)$. If $|S_2(f) - S_1(f)| > \varepsilon$, then further refinement is necessary. We first note that $S_2(f)$ is actually the sum of $S_1(f)$ on $[a, (a + b)/2]$ plus $S_1(f)$ on $[(a + b)/2, b]$. For the sake of notation we write these as $S_{1,1}(f)$ and $S_{1,2}(f)$, respectively, that is, $S_2(f) = S_{1,1}(f) + S_{1,2}(f)$. For the moment we concentrate on the interval $[a, (a + b)/2]$ and examine the accuracy of our approximation there to $\int_a^{(a+b)/2} f(x)\,dx$. We repeat the above process and compute $S_2(f)$ on $[a, (a + b)/2]$ which we denote by $S_{2,1}(f)$. If $|S_{2,1}(f) - S_{1,1}(f)| \le \varepsilon/2$, we accept $S_{2,1}(f)$ as an approximation for $\int_a^{(a+b)/2} f(x)\,dx$. If $|S_{2,1}(f) - S_{1,1}(f)| > \varepsilon/2$, we must continue this same process. For example, we note that $S_{2,1}(f)$ is the sum of $S_1(f)$ on $[a, (3a + b)/4]$ and $S_1(f)$ on $[(3a + b)/4, (a + b)/2]$, which we write as $S_{2,1}(f) = S_{1,1,1}(f) + S_{1,1,2}(f)$. As above, we apply $S_2(f)$ to $[a, (3a + b)/4]$ (denoted by $S_{2,1,1}(f)$) and ask if $|S_{2,1,1}(f) - S_{1,1,1}(f)| \le \varepsilon/4$. The first stage of the process is completed either when we find an integer $m_1$ such that $|S_{2,1,\ldots,1}(f) - S_{1,1,\ldots,1}(f)| < \varepsilon/2^{m_1}$ (we have our desired accuracy on the interval $[a, a + (a + b)/2^{m_1}]$), or when we decide that the bisection has been repeated so many times that it is not fruitful to continue (due to considerations of rounding error contamination, time and storage limitations, etc.) The latter case is handled by *a priori* limiting each stage to some fixed number of bisections. We then accept the last computation as our approximation on this subinterval and have the computer notify the user that the desired accuracy could not be achieved there.

Once each stage is completed we procede to the next stage until we work our way left to right across the entire interval $[a, b]$. Thus the procedure as a whole refines often where $f(x)$ is ill-behaved and less often where the integral can be more easily approximated. The process has the further advantage that no work is wasted and that the function need only be evaluated once at each point (if carefully programmed). At each level of each stage of the process, the subinterval under investigation is bisected and thus it seems reasonable that the error tolerance be halved. However, this in actuality is assuming that the errors are "worst-possible" and that they all are of the same sign, i.e., there is no error cancellation effect in going from stage to stage. In practice, of course, this is not usually true, and the error estimate tends to be too conservative. Some authors advocate using a factor of $r$, $1/2 < r < 1$, rather than $1/2$ at each level, that is, using $\{\varepsilon, \varepsilon r, \varepsilon r^2, \ldots\}$ instead of $\{\varepsilon, \varepsilon/2, \varepsilon/4, \ldots\}$ for error tolerances (see Shampine and Allen 1973).

The reader should note that the *exit criterion* or decision at what point to stop in each stage is based on comparing $S_2(f)$ to $S_1(f)$ on a subinterval of length $(b - a)/2^q$. If $I_q(f)$ denotes the integral of $f(x)$ over this subinterval, then what we really desire is that $|S_2(f) - I_q(f)| \leq \varepsilon_q$. On this subinterval, if $|S_2(f) - S_1(f)| \geq |S_2(f) - I_q(f)|$, then requiring that $|S_2(f) - S_1(f)| \leq \varepsilon_q$ is a valid stopping inequality. Clenshaw and Curtis (1960, pp. 197–205) have given an example where $|S_2(f) - S_1(f)| < \varepsilon_q$ but $|S_2(f) - I_q(f)|$ is much larger than $\varepsilon_q$. On the other hand, Rowland and Varol (1972, pp. 699–703) have shown that if $|f^{(iv)}(t_1) - f^{(iv)}(t_2)| \leq K|t_1 - t_2|$ for all $t_1$ and $t_2$ in the $q$th subinterval and $f^{(iv)}(x)$ is of constant sign there, then $|S_2(f) - S_1(f)| \geq |S_2(f) - I_q(f)|$. Hence the stopping inequality is valid in this case, although it is often too conservative since $|S_2(f) - S_1(f)|$ is usually much larger than $|S_2(f) - I_q(f)|$. The method of choosing exit criteria for adaptive routines is still the subject of extensive research and we shall not pursue it further here, except to say that as the lengths of the intervals become smaller the concept of relative error becomes increasingly more important in choosing exit criteria at each stage, and most modern adaptive routines take this into consideration. (See references to the important work of Lyness in Rowland and Varol (1972).)

## 6.4 RICHARDSON EXTRAPOLATION AND NUMERICAL DIFFERENTIATION

Richardson extrapolation is a procedure similar to Aitken's $\Delta^2$-process. That is, we compute several estimates of some quantity and then combine these estimates in such a way as to (hopefully) provide a better estimate. To be more precise, let us assume that an *unknown* quantity, $a_0$, is approximated by a quantity $A(h)$, where $A(h)$ can be explicitly computed for each value of $h$, $h \neq 0$. Further, we assume that $a_0 = \lim_{h \to 0} A(h)$ (typically, we only consider values of $h$ that are positive). Finally, we assume that

$$a_0 = A(h) + \sum_{i=k}^{m} a_i h^i + C_m(h)h^{m+1} \tag{6.21}$$

where $C_m(h)$ is a function of $h$, the constants $a_i$ are independent of $h$ and $a_k \neq 0$. We emphasize that to use the techniques of this section, we do not need to know explicitly what the coefficients $a_i$ are, nor do we need to know the function $C_m(h)$.

**Example 6.6.** As a simple example, suppose we wish to approximate $f'(\alpha)$, where $f(x)$ is differentiable at $x = \alpha$. From the definition of the derivative, we have that

$$f'(\alpha) = \lim_{h \to 0} \frac{f(\alpha + h) - f(\alpha)}{h}.$$

Although we will see this is not optimal, we might think of approximating $a_0 = f'(\alpha)$ by $A(h) = [f(\alpha + h) - f(\alpha)]/h$, for small $h$. Doing this, we certainly have that $\lim_{h \to 0} A(h) = a_0$. Using a Taylor's series expansion for $f(x)$ about $x = \alpha$, we easily see that $a_0$ and $A(h)$ are related by an expression of the form (6.21):

$$a_0 = A(h) - \left[\frac{f''(\alpha)}{2!} h + \cdots + \frac{f^{(m+1)}(\alpha)}{(m+1)!} h^m\right] - \frac{f^{(m+2)}(\eta_h)}{(m+2)!} h^{m+1}.$$

We have written the Taylor's series error term as

$$\frac{f^{(m+2)}(\eta_h)}{(m+2)!} h^{m+1}$$

to emphasize the dependence of the mean-value point, $\eta_h$, on the step size $h$. In the context of (6.21), we have that

$$C_m(h) = \frac{-f^{(m+2)}(\eta_h)}{(m+2)!}$$

and that

$$a_i = \frac{-f^{(i+1)}(\alpha)}{(i+1)!} \quad \text{for} \quad i = 1, 2, \ldots, m.$$

Note that the coefficients $a_i$ are independent of $h$ and that we probably do not know the values of $a_i$ (since we do not even know $f'(\alpha)$).

Another example of the situation described in (6.21) is provided by the composite trapezoidal rule (see in particular (6.20c)) where we have $a_0 = \int_a^b g(x)\,dx$, $A(h) = T_N(g)$ (here, $h = (b - a)/N$, so $T_N(g)$ is a function of $h$). The constants $a_i$ and the function $C_m(h)$ can be found from (6.20c).

Formula (6.21) provides the basis for a simple technique known as *Richardson extrapolation* or *extrapolation to the limit*. The basic idea is to compute two different approximations, $A(h_1)$ and $A(h_2)$, to $a_0$. We can then form a combination of $A(h_1)$ and $A(h_2)$ in such a way as to eliminate the $h^k$ term in (6.21), thus obtaining (for small $h$), a better approximation to $a_0$. This is usually done by choosing $h_1 = h$ and $h_2 = rh$, where $0 < r < 1$ (normally, $r = \frac{1}{2}$). Carrying this idea out, we find from (6.21):

$$\begin{aligned} a_0 &= A(h) + \sum_{i=k}^{m} a_i h^i + C_m(h) h^{m+1} \\ a_0 &= A(rh) + \sum_{i=k}^{m} a_i (rh)^i + C_m(rh)(rh)^{m+1}. \end{aligned} \tag{6.22}$$

Multiplying the top equation by $r^k$ and subtracting from the bottom equation in (6.22), we find:

$$a_0 - r^k a_0 = A(rh) - r^k A(h) + \sum_{i=k+1}^{m} a_i[r^i - r^k]h^i + [C_m(rh)r^{m+1} - C_m(h)r^k]h^{m+1}. \quad \textbf{(6.23)}$$

Thus, with $b_i = a_i[r^i - r^k]/[1 - r^k]$ and $\tilde{C}_m(h) = [C_m(rh)r^{m+1} - C_m(h)r^k]/[1 - r^k]$, we see from (6.23) that

$$a_0 = \frac{A(rh) - r^k A(h)}{1 - r^k} + \sum_{i=k+1}^{m} b_i h^i + \tilde{C}_m(h)h^{m+1}. \quad \textbf{(6.24)}$$

That is, $B(h) = [A(rh) - r^k A(h)]/[1 - r^k]$ is an approximation to $a_0$ which satisfies (6.21), except that the error begins with $h^{k+1}$ instead of $h^k$. Therefore we would tend to feel that $B(h)$ is probably more accurate (for small $h$) than $A(h)$.

Note that in order to form a new approximation from $A(h)$, all we needed to know was the integer $k$. That is, it is not necessary to know the coefficients $a_i$ in (6.21) or the function $C_m(h)$, but it is necessary to know $k$, where $a_0 - A(h) = a_k h^k + \cdots$. If $b_{k+1}$ in (6.24) is different from zero, we can carry out the extrapolation process again, finding an approximation whose error term starts with $h^{k+2}$. This idea can be mechanized most naturally as follows. For $m = 0, 1, 2, \ldots$ define $A_{0,m} = A(r^m h)$. By (6.24), extrapolation with $A_{0,m}$ and $A_{0,m+1}$ gives

$$A_{1,m} = \frac{A_{0,m+1} - r^k A_{0,m}}{1 - r^k} \quad \textbf{(6.25)}$$

and in general, we have

$$A_{i+1,m} = \frac{A_{i,m+1} - r^{k+i} A_{i,m}}{1 - r^{k+i}}, \qquad m = 0, 1, \ldots. \quad \textbf{(6.26)}$$

Using (6.26), we have an easily programmed algorithm to construct Table 6.1 one column at a time from left to right:

**Table 6.1.** Richardson Extrapolation of $n$th Order.

| | | | | | |
|---|---|---|---|---|---|
| $A_{0,0}$ | | | | | |
| $A_{0,1}$ | $A_{1,0}$ | | | | |
| $A_{0,2}$ | $A_{1,1}$ | $A_{2,0}$ | | | |
| $A_{0,3}$ | $A_{1,2}$ | $A_{2,1}$ | $A_{3,0}$ | | |
| $\vdots$ | $\vdots$ | $\vdots$ | $\vdots$ | $\ddots$ | |
| $A_{0,n}$ | $A_{1,n-1}$ | $A_{2,n-2}$ | $A_{3,n-3}$ | $\cdots$ | $A_{n,0}$ |

As with all extrapolation procedures, we must be somewhat careful. For example, if in (6.24) it turns out that $b_{k+1} = 0$, then it is possible that a further extrapolation using (6.25) might even cause a loss of accuracy in our approximation.

**Example 6.7.** An approximation that is sometimes used to estimate $f'(\alpha)$ is

$$A(h) = \frac{f(\alpha + h) - f(\alpha - h)}{2h}, \qquad h > 0. \tag{6.27}$$

Clearly, $\lim_{h \to 0} A(h) = f'(\alpha)$; moreover, using a Taylor's series expansion about $x = \alpha$, we see (Problem 3)

$$A(h) = f'(\alpha) + \frac{f'''(\alpha)}{3!} h^2 + \frac{f^{(v)}(\alpha)}{5!} h^4 + \cdots. \tag{6.28}$$

As frequently happens, only even powers of $h$ occur in (6.28). A moment's reflection shows that we need to use $r^2$, $r^4$, $r^6$, etc. in (6.26) in order to be successful in extrapolation. That is, since (Problem 4)

$$B(h) = \frac{A(rh) - r^2 A(h)}{1 - r^2} = f'(\alpha) + b_4 h^4 + \cdots \tag{6.29}$$

we see that the next extrapolation must be of the form $[B(rh) - r^4 B(h)]/[1 - r^4]$. In general, we obtain as in (6.26):

$$A_{i+1,m} = \frac{A_{i,m+1} - r^{k+2i} A_{i,m}}{1 - r^{k+2i}}, \qquad i = 0, 1, \ldots. \tag{6.30}$$

Formula (6.30) should be used whenever we have

$$a_0 = A(h) + a_k h^k + a_{k+2} h^{k+2} + a_{k+4} h^{k+4} + \cdots.$$

To demonstrate this on a specific function, consider the Gamma function which is defined by

$$\Gamma(x) = \int_0^\infty t^{x-1} e^{-t} \, dt, \qquad \text{for } x > 0.$$

Suppose we wish to estimate $\Gamma'(1.5)$, using (6.27) and Richardson extrapolation. Choosing $h = 0.4$ and $r = \frac{1}{2}$, we set up a table similar to Table 6.1, where the necessary values of $\Gamma(x)$ are obtained from a mathematical handbook. We find first

$$A_{0,0} = A(h) = \frac{\Gamma(1.9) - \Gamma(1.1)}{0.8} = 0.013019$$

$$A_{0,1} = A\left(\frac{h}{2}\right) = \frac{\Gamma(1.7) - \Gamma(1.3)}{0.4} = 0.027920$$

$$A_{0,2} = A\left(\frac{h}{4}\right) = \frac{\Gamma(1.6) - \Gamma(1.4)}{0.2} = 0.031258$$

$$A_{0,3} = A\left(\frac{h}{8}\right) = \frac{\Gamma(1.55) - \Gamma(1.45)}{0.1} = 0.032069.$$

Using (6.30) we then find:

$$\begin{array}{llll}
0.013019 & & & \\
0.027920 & 0.032887 & & \\
0.031258 & 0.032371 & 0.032337 & \\
0.032069 & 0.032339 & 0.032337 & 0.032337.
\end{array}$$

If fact, $\Gamma'(1.5) = 0.032338$ to the places shown.

### 6.4.1 Romberg Integration

Classical Romberg integration is a numerical integration procedure that amounts to using Richardson extrapolation on the composite trapezoidal rule $T_N(g)$. It is customary to use $h = b - a$ and $r = \frac{1}{2}$ when using Romberg integration to approximate $I(g) = \int_a^b g(x)\,dx$. Thus we define

$$T_{0,m} = \frac{b-a}{2^m}\left[\frac{1}{2}g_0 + g_1 + \cdots + g_{s-1} + \frac{1}{2}g_s\right] \tag{6.31}$$

where $g_i = g(x_i)$, $x_i = a + (i(b-a)/2^m)$ and $s = 2^m$. This choice of $h$ and $r$ means that we can use $T_{0,m}$ in the computation of $T_{0,m+1}$, saving us half the work in evaluating (or looking up) $g(x)$, when forming $T_{0,m+1}$.

By (6.20c), we see that error term contains only even powers of $h$, so (6.30) is the appropriate form of Richardson extrapolation to use. Thus we have

$$T_{1,m} = \frac{T_{0,m+1} - \frac{1}{4}T_{0,m}}{1 - \frac{1}{4}}$$

or, in general,

$$T_{i,m} = \frac{T_{i-1,m+1} - (\frac{1}{4})^i T_{i-1,m}}{1 - (\frac{1}{4})^i}, \qquad i = 1, 2, \ldots.$$

This can be simplified to the more convenient form

$$T_{i,m} = T_{i-1,m+1} + \frac{T_{i-1,m+1} - T_{i-1,m}}{4^i - 1}, \qquad i = 1, 2, \ldots. \tag{6.32}$$

Since Romberg integration is an extrapolation procedure, some care must be exercised in the application of Romberg integration. As a check (when forming a table similar to Table 6.1), we can calculate the ratios

$$R_{i,m} = \frac{T_{i,m} - T_{i,m-1}}{T_{i,m+1} - T_{i,m}}. \tag{6.33}$$

As is shown in Problem 6, we expect the ratios $R_{i,m}$ to have approximately the value $R_{i,m} \approx 4^{m+1}$. If this is not the case, we should suspect that roundoff error is contaminating our estimates and should stop the extrapolation at this point.

**Example 6.8.** As in Example 6.4, we consider the problem of calculating

$$Si(1) = \int_0^1 \frac{\sin(t)}{t}\,dt.$$

The exact value of $Si(1)$ to 8 places is $Si(1) = 0.94608307$. The result of Romberg integration is given in Table 6.2. Note that only nine evaluations of the integrand $\sin(t)/t$ are required to construct this table and that $T_{3,0}$ is correct to eight places.

**Table 6.2**

| $T_{0,i}$ | $T_{1,i}$ | $T_{2,i}$ | $T_{3,i}$ |
|---|---|---|---|
| 0.92073549 | | | |
| 0.93979328 | 0.94614588 | | |
| 0.94451352 | 0.94608693 | 0.94608300 | |
| 0.94569086 | 0.94608331 | 0.94608307 | 0.94608307 |

## PROBLEMS

1. Write a computer program to carry out an adaptive Simpson's rule quadrature. Test your program on the Fresnel integral

$$C(x) = \int_0^x \cos\left(\frac{\pi}{2}t^2\right)dt$$

estimating $C(5) = 0.5636312$ *to* 6 places.

2. Use Richardson extrapolation to estimate $f'(x)$ at $x = 1$, where $f(x) = \ln(x)$. In Table 6.1, use $n = 4$ and $h = 0.4$ and $r = \frac{1}{2}$. Try both forms of $A(h)$, that given in Example 6.6 and that given in Example 6.7.

3. Show that formula (6.28) is valid by writing down the Taylor's series expansion for $f(x + h)$ and $f(x - h)$ and expanding about $x = \alpha$ in both cases.

4. Verify the expansion in (6.29).

5. Write a computer program to carry out Romberg integration and test your program on the Fresnel integral given in Problem 1. To determine (in a rough fashion) the accuracy of your answers, print out the ratios defined in (6.33).

6. The fundamental assumption of Richardson extrapolation can be seen from (6.21). We are assuming that $a_0 - A(h) = a_k h^k + a_{k+1}h^{k+1} + \cdots$ and that eliminating $a_k h^k$ will lead to a better approximation. That is, we are assuming that $a_k h^k$ is the dominant term in the error $a_0 - A(h)$.

   a) Show that the assumption above means that the ratio

   $$[A_{0,m} - A_{0,m-1}]/[A_{0,m+1} - A_{0,m}]$$

   should be approximately $r^{-k}$.

   b) Use part (a) to establish that the ratios $R_{i,m}$ defined in (6.33) should be approximately $4^{m+1}$ if Romberg integration is proceeding without a substantial error.

7. Calculate the appropriate ratios, as defined in Problem 6, for the tables in Examples 6.7 and 6.8.
8. Show that $T_{1,m}$ defined in (6.32) corresponds to the composite Simpson's rule. (However, there is no relation between $T_{k,m}$ and Newton-Cotes rules for $k > 2$.)

## 6.5 GAUSSIAN QUADRATURE

As we previously remarked in Section 6.2, if we let the quadrature weights, $\{A_j\}_{j=0}^n$, *and* the quadrature nodes, $\{x_j\}_{j=0}^n$, be treated as unknown variables, then the equations of (6.3),

$$\int_a^b x^k w(x)\,dx = \sum_{j=0}^{n} A_j x_j^k, \qquad 0 \le k \le 2n+1, \tag{6.34}$$

represent a nonlinear system of $(2n + 2)$ equations in $(2n + 2)$ unknowns. In 1814, Gauss was able to show that this system of equations has a unique solution for the unknowns, $\{A_j\}_{j=0}^n$ and $\{x_j\}_{j=0}^n$ when $w(x) \equiv 1$ and $[a, b] = [-1, 1]$.

Once again we return to the notation of Section 5.3, where we let $\langle f, g\rangle \equiv \int_a^b f(x)g(x)w(x)\,dx$, and $\|f\| \equiv \langle f, f\rangle^{1/2}$. We let $\{q_j(x)\}_{j=0}^{\infty}$ denote the *monic* orthogonal polynomials and $\{p_j(x)\}_{j=0}^{\infty}$ the orthonormal polynomials with respect to this inner product, where degree $(q_j(x))$ = degree $(p_j(x)) = j$, for each $j$. We first prove the following theorem which localizes the zeros of each $p_j(x)$. (Recall that for each $j$, $p_j(x)$ is a constant multiple of $q_j(x)$, and thus they have the same zeros.)

**Theorem 6.3.** Let $\{p_j(x)\}_{j=0}^{\infty}$ be given as above. Then for each $n \ge 1$, the zeros of $p_n(x)$ are real and distinct and lie in the interval $(a, b)$.

*Proof.* Let $n \ge 1$ be fixed and suppose that none of the zeros of $p_n(x)$ are in $(a, b)$, so that $p_n(x)$ is of constant sign on $(a, b)$ (say $p(x) > 0$ for $x \in (a, b)$). By the orthogonality of $p_n(x)$ and $q_0(x)$, ($q_0(x) \equiv 1$), we would then have:

$$0 = \langle 1, p_n\rangle = \int_a^b p_n(x)w(x)\,dx > 0,$$

since we assume $w(x)$ is nonnegative and strictly positive for some subinterval of $(a, b)$. Thus our initial assumption is contradicted, and $p_n(x)$ must have at least one zero, $x_0$, in $(a, b)$.

If any zero, say $x_0$, is a multiple zero of $p_n(x)$, then $(x - x_0)^2$ factors $p_n(x)$ and so $r(x) \equiv p_n(x)/(x - x_0)^2$ is in $\mathscr{P}_{n-2}$. By Corollary 1 of Theorem 5.9, $\langle p_n, r\rangle = 0$, and so we can write

$$\begin{aligned} 0 = \langle p_n, r\rangle &= \int_a^b p_n(x)[p_n(x)/(x - x_0)^2]w(x)\,dx \\ &= \int_a^b \frac{p_n(x)^2}{(x - x_0)^2}\,w(x)\,dx > 0, \end{aligned}$$

which is again a contradiction. Hence we can infer that any zero of $p_n(x)$ lying in $(a, b)$ is simple.

Now let $\{x_0, x_1, \ldots, x_j\}$ be the zeros of $p_n(x)$ lying in $(a, b)$ and suppose that $j < n - 1$, that is, $p_n(x)$ has other zeros elsewhere. Since the zeros $\{x_i\}_{i=0}^{j}$ are simple, we can form the polynomial $p_n(x)[(x - x_0)(x - x_1)\cdots(x - x_j)]$. We write this polynomial as $r(x) \cdot [(x - x_0)^2(x - x_1)^2 \cdots (x - x_j^2)]$, where $r(x) \in \mathscr{P}_{n-j-1}$ and we note that $r(x)$ is of constant sign (say $>0$) on $(a, b)$. Again

$$\langle p_n(x), [(x - x_0)(x - x_1)\cdots(x - x_j)]\rangle = 0$$

by the above mentioned corollary, so

$$\begin{aligned} 0 &= \int_a^b p_n(x)[(x - x_0)(x - x_1)\cdots(x - x_j)]w(x)\,dx \\ &= \int_a^b r(x)[(x - x_0)^2(x - x_1)^2\cdots(x - x_j)^2]w(x)\,dx > 0, \end{aligned}$$

contradicting the assumption that $j < n - 1$. Thus, all the zeros of $p_n(x)$ are in $(a, b)$ and are simple. ∎

Recall that the Eqs. (6.34) are satisfied for $0 \le k \le 2n + 1$ if and only if $Q_n(f) = \sum_{j=0}^{n} A_j f(x_j)$ has precision $2n + 1$, that is, if $f(x)$ is any polynomial of degree $(2n + 1)$ or less, then $I(f) = Q_n(f)$. (Any interpolatory quadrature formula with this property is called *Gaussian.*) With this in mind we are able to prove the basic theorem of Gaussian quadrature.

**Theorem 6.4.** The formula $\int_a^b p(x)w(x)\,dx = \sum_{j=0}^{n} A_j p(x_j)$ holds for all $p(x)$ in $\mathscr{P}_{2n+1}$ if and only if $\{x_j\}_{j=0}^{n}$ are the zeros of $p_{n+1}(x)$ (as given in Theorem 6.3) and $\{A_j\}_{j=0}^{n}$ are given in (6.2).

*Proof*: 1. Let $p(x) \in \mathscr{P}_{2n+1}$ and let $p_{n+1}(x_j) = 0$, $0 \le j \le n$. By the Euclidean division algorithm, $p(x)$ can be written as $p(x) = p_{n+1}(x)S(x) + R(x)$, where $S(x)$ and $R(x)$ are in $\mathscr{P}_n$. (Note that since $p_{n+1}(x_j) = 0$, $0 \le j \le n$, then $p(x_j) = R(x_j)$, $0 \le j \le n$.) In addition, since the weights are given by (6.2), $I(R) = Q_n(R) = \sum_{j=0}^{n} A_j R(x_j)$ (since the quadrature formula has precision at least $n$). Using Corollary 1 of Theorem 5.9, $\langle p_{n+1}, S\rangle = 0$ so

$$\begin{aligned} I(p) \equiv \int_a^b p(x)w(x)\,dx &= \int_a^b p_{n+1}(x)S(x)w(x)\,dx + \int_a^b R(x)w(x)\,dx \\ &= \langle p_{n+1}, S\rangle + I(R) = 0 + \sum_{j=0}^{n} A_j R(x_j) \\ &= \sum_{j=0}^{n} A_j p(x_j) = Q_n(p). \end{aligned}$$

2. Now we assume that $\{x_j\}_{j=0}^{n}$ is any distinct set of points and $\int_a^b p(x)w(x)\,dx = \sum_{j=0}^{n} A_j p(x_j)$, for all $p(x) \in \mathscr{P}_{2n+1}$. Given any integer $k$, $0 \le k \le n$, let $r_k(x)$ be any polynomial of degree $k$ or less. Let $W(x) = \prod_{j=0}^{n} (x - x_j)$ and define $p(x) \equiv$

$r_k(x)W(x)$. Then $p(x) \in \mathscr{P}_{2n+1}$, and by our hypothesis, $I(p) = Q_n(p)$. Using this, we have

$$\langle r_k, W \rangle = \int_a^b r_k(x)W(x)w(x)\,dx = \int_a^b p(x)w(x)\,dx$$
$$= \sum_{j=0}^{n} A_j p(x_j) = \sum_{j=0}^{n} A_j r_k(x_j)W(x_j) = 0$$

since $W(x_j) = 0$, $0 \le j \le n$. Hence we have just shown that $W(x)$, a monic polynomial of degree $(n + 1)$, is orthogonal to any polynomial of degree $n$ or less. By Corollary 2, Theorem 5.9, $W(x) \equiv q_{n+1}(x)$, and thus $x_j$'s are the zeros of $p_{n+1}(x)$. Now that the nodes $x_j$ are known, we are nearly done. The fact that the weights $A_j$ are as in (6.2) follows immediately, since the first $n + 1$ equations on the right-hand side of (6.34) constitutes a linear system where the coefficient matrix is Vandermonde. Hence, there is one and only one choice for the weights and (6.2) gives the solution explicitly. ■

Gaussian quadratures are powerful numerical integration methods as the following corollary illustrates.

**Corollary.** Let $Q_n(f) = \sum_{j=0}^{n} A_j f(x_j)$ be Gaussian, then $\lim_{n\to\infty} Q_n(f) = I(f)$, for all $f(x) \in C[a, b]$.

*Proof*. By the definition of $\ell_k(x)$ ((5.2) in Chapter 5), for each $k$, $(\ell_k(x))^2 \in \mathscr{P}_{2n}$ and so $I(\ell_k^2) = Q_n(\ell_k^2)$. Thus

$$0 < \int_a^b (\ell_k(x))^2 w(x)\,dx = \sum_{j=0}^{n} A_j(\ell_k(x_j))^2 = A_k$$

since $(\ell_k(x_j))^2 = \delta_{jk}$. Thus the Gaussian quadrature weights are positive for each $n$, so by the remarks following Theorem 6.1 this corollary is proved. ■

We should also note that since this corollary shows that the weights are all positive, the formulas have nice rounding properties. There exist extensive tables for the weights and nodes of many common Gaussian quadratures in the literature. (For example, see Stroud and Secrest, 1966.) We shall now consider efficient methods for calculating the weights and nodes of any Gaussian formula.

To compute the nodes we take the obvious course of generating $q_{n+1}(x)$ by the three-term recurrence formula, (5.65). Since the zeros of $q_{n+1}(x)$ are simple and lie in $(a, b)$, Newton's method is ideally suited for computing these nodes.

Efficient computation of the weights is not so straightforward. We could use the computed values for the nodes and solve for the weights by formula (6.3) or we could use the formula $A_j = \int_a^b \ell_j(x)w(x)\,dx$, $0 \le j \le n$. There is a more efficient procedure, however, which uses the same recurrence relation as for $q_{n+1}(x)$, but has *different starting values* (otherwise we would be just generating the $q_j(x)$'s again).

We first define a new sequence of polynomials, $\{\phi_0, \phi_1, \phi_2, \ldots, \phi_{n+1}\}$, by $\phi_0(x) \equiv 0$, $\phi_1(x) \equiv \int_a^b w(x)\,dx$, and

$$\phi_k(x) = (x - a_k)\phi_{k-1}(x) - b_k\phi_{k-2}(x), \qquad \text{for } k \geq 2. \tag{6.35}$$

The constants $\{a_k\}_{k=2}^{n+1}$ and $\{b_k\}_{k=2}^{n+1}$ are the same as in (5.65), the three-term recurrence which we have just used to find $q_{n+1}(x)$ in order to compute the nodes. (The reader should note that $\phi_0(x)$ and $\phi_1(x)$ differ from $q_0(x)$ and $q_1(x)$, respectively. One should also note that the degree of $\phi_j(x)$ is $j - 1$.) We have introduced (6.35) so that we can calculate each $A_j$ by

$$A_j = \phi_{n+1}(x_j)/q'_{n+1}(x_j), \qquad 0 \leq j \leq n. \tag{6.36}$$

The validity of (6.36) is an immediate consequence of the following theorem.

**Theorem 6.5.** Let $\phi_{n+1}(x)$ and $q_{n+1}(x)$ be generated by (6.35) and (5.65), respectively. Then $\phi_{n+1}(x)$ can alternatively be written as

$$\phi_{n+1}(x) = \int_a^b \frac{q_{n+1}(t) - q_{n+1}(x)}{(t - x)}\, w(t)\,dt. \tag{6.37}$$

*Proof.* The proof is by induction and we leave to the reader to verify (6.37) for $n = -1$ and $n = 0$. Then, for $n \geq 1$, assume (6.37) is valid for all positive integers up to $n$. By the three-term recurrence, (5.65),

$$\int_a^b \frac{q_{n+1}(t) - q_{n+1}(x)}{(t - x)}\, w(t)\,dt$$

$$= \int_a^b \frac{[(t - a_{n+1})q_n(t) - b_{n+1}q_{n-1}(t) - (x - a_{n+1})q_n(x) + b_{n+1}q_{n-1}(x)]w(t)\,dt}{(t - x)}.$$

Adding and subtracting $xq_n(t)$ to the numerator of the integrand and recalling that $\int_a^b q_n(t)w(t)\,dt = 0$ reduces this expression to:

$$(x - a_{n+1}) \int_a^b \frac{q_n(t) - q_n(x)}{t - x}\, w(t)\,dt - b_{n+1} \int_a^b \frac{q_{n-1}(t) - q_{n-1}(x)}{t - x}\, w(t)\,dt$$

$$+ \int_a^b q_n(t)w(t)\,dt = (x - a_{n+1})\phi_n(x) - b_{n+1}\phi_{n-1}(x) = \phi_{n+1}(x). \quad \blacksquare$$

To see that (6.36) follows from this theorem, recall that $\ell_j(x)$ can be alternatively written as (Problem 5, Section 5.2.1): $\ell_j(x) = q_{n+1}(x)/[(x - x_j)q'_{n+1}(x_j)]$, $0 \leq j \leq n$. Thus, since $q_{n+1}(x_j) = 0$ and $q'_{n+1}(x_j) \neq 0$ (by Theorem 6.3), (6.37) yields

$$\frac{\phi_{n+1}(x_j)}{q'_{n+1}(x_j)} = \int_a^b \frac{q_{n+1}(t) - 0}{(t - x_j)q'_{n+1}(x_j)}\, w(t)\,dt = \int_a^b \ell_j(t)w(t)\,dt \equiv A_j.$$

**Example 6.9.** We pause here to present a particularly nice type of Gaussian quadrature, since we can derive closed-form formulas for its weights and nodes for any $n$. If we let $\langle f, g\rangle = \int_{-1}^1 f(x)g(x)(1 - x^2)^{-1/2}\,dx$, then the orthogonal polynomials are the Chebyshev polynomials of the first kind, $T_k(x) = \cos[k \cos^{-1}(x)]$. Thus we immediately have the nodes since the zeros of $T_{n+1}(x)$ are $x_j = \cos((2j + 1)\pi/(2n + 2))$, $0 \le j \le n$. To find the weights we must introduce the Chebyshev polynomials of the second kind, $U_k(x) \equiv \sin[(k + 1) \cos^{-1}(x)]/\sin[\cos^{-1}(x)]$, or $U_k(\cos(\theta)) = \sin[(k + 1)\theta]/\sin(\theta)$, for $x \equiv \cos(\theta)$. Now, obviously, $U_0(x) \equiv 1$ and $U_1(x) = U_1(\cos(\theta)) = \sin(2\theta)/\sin(\theta) = 2\cos(\theta) = 2x$. By elementary trigonometric identities we can show

$$\sin[(k + 2)\theta] = 2\cos(\theta)\sin[(k + 1)\theta] - \sin(k\theta),$$

and so, for $k \ge 1$,

$$\begin{aligned} U_{k+1}(x) \equiv U_{k+1}(\cos(\theta)) &\equiv \sin[(k + 2)\theta]/\sin(\theta) \\ &= 2\cos(\theta)\sin[(k + 1)\theta]/\sin(\theta) - \sin(k\theta)/\sin(\theta) \\ &= 2\cos(\theta)U_k(\cos(\theta)) - U_{k-1}(\cos(\theta)) = 2xU_k(x) - U_{k-1}(x). \end{aligned} \quad \textbf{(6.38a)}$$

Formula (6.38a) shows that $U_k(x)$ is a polynomial of degree $k$ with leading coefficient $2^k$ (since $U_0(x) \equiv 1$ and $U_1(x) = 2x$). Thus $V_k(x) \equiv 2^{-k}U_k(x)$ is monic in $\mathscr{P}_k$ and by (6.38a) satisfies the recurrence formula: $V_0(x) = 1$, $V_1(x) = x$,

$$V_k(x) = xV_{k-1}(x) - \frac{1}{4}V_{k-2}(x), \qquad k \ge 2. \quad \textbf{(6.38b)}$$

Similarly the monic Chebyshev polynomial of the first kind, $q_k(x) \equiv T_k(x)/2^{k-1}$, $k \ge 1$ satisfies (Problem 4)

$$q_k(x) = xq_{k-1}(x) - b_k q_{k-2}(x), \qquad k \ge 2,$$

where $b_2 = 1/2$ and $b_k = 1/4$ for $k \ge 3$. (This follows from $T_0(x) = 1$, $T_1(x) = x$, and $T_k(x) = 2xT_{k-1}(x) - T_{k-2}(x)$, $k \ge 2$.)

Now to use (6.36) we must find $\phi_{n+1}(x)$ from (6.35). We have $\phi_0(x) \equiv 0$, $\phi_1(x) = \int_{-1}^{1} (1 - x^2)^{-1/2}\, dx = \pi$, and $\phi_k(x) = x\phi_{k-1}(x) - b_k\phi_{k-2}(x)$, $k \ge 2$. If $k = 2$, $\phi_2(x) = x(\pi) - b_2(0) = \pi x$. If $k = 3$, $\phi_3(x) = \pi x^2 - \pi/4 = \pi V_2(x)$. Since $\phi_2(x) = \pi V_1(x)$ and for $k > 3$ (6.35) is $\phi_k(x) = x\phi_{k-1}(x) - \phi_{k-2}(x)/4$, we see by (6.38b) that $\phi_{k+1}(x) = \pi V_k(x)$, $k \ge 0$. Hence, $\phi_{n+1}(x) = \pi V_n(x) = \pi 2^{-n}U_n(x)$. Now

$$\frac{d}{dx}(T_{n+1}(x)) = \frac{d(\cos[(n + 1)\theta])}{d\theta}\frac{d\theta}{dx} = (n + 1)\frac{\sin[(n + 1)\theta]}{\sin(\theta)} = (n + 1)U_n(x).$$

Thus (6.36) becomes

$$A_j = \frac{\phi_{n+1}(x_j)}{q'_{n+1}(x_j)} = \frac{\pi 2^{-n}U_n(x_j)}{(n + 1)2^{-n}U_n(x_j)} = \frac{\pi}{(n + 1)}, \qquad 0 \le j \le n.$$

This yields the Gaussian formula

$$\int_{-1}^{1} f(x)(1 - x^2)^{-1/2}\, dx \approx \frac{\pi}{n + 1}\sum_{j=0}^{n} f(x_j). \quad \textbf{(6.39)}$$

This particular Gaussian quadrature is called a *Gauss-Chebyshev* quadrature and has especially nice rounding characteristics since the weights are all equal.

Perhaps the most commonly used Gaussian quadrature is obtained from $w(x) \equiv 1$ and $\langle f, g\rangle = \int_{-1}^{1} f(x)g(x)\, dx$. The orthogonal polynomials for this case are the Legendre polynomials. There are no nice closed-form formulas for the nodes and weights in this case. However, with the use of extensive previously tabulated results and our ability to easily translate the integral of integration from $[-1, 1]$ to $[a, b]$, this quadrature (called a *Gauss-Legendre* quadrature or sometimes even simply a *Gauss quadrature*) is a very practical tool.

If $f(x) \in C^{(2n+2)}[a, b]$, then it is possible to derive an error formula for Gaussian quadrature of the form (see Ralston, 1965):

$$R_n(f) \equiv I(f) - Q_n(f) = \frac{f^{(2n+2)}(\eta)}{(2n + 2)!} \int_a^b (q_{n+1}(x))^2 w(x)\, dx. \qquad \textbf{(6.40a)}$$

We shall take a slightly different approach which emphasizes the benefits of having precision $(2n + 1)$. Once again, we define the uniform degree of approximation as

$$E_m(f) = \min_{p(x) \in \mathscr{P}_m} \left\{ \max_{a \le x \le b} |f(x) - p(x)| \right\} = \min_{p(x) \in \mathscr{P}_m} \{\|f - p\|_\infty\} \equiv \|f - p_m^*\|_\infty.$$

Since $I(1) = Q_n(1)$ for any Gaussian formula, we have $\int_a^b w(x)\, dx = \sum_{j=0}^n A_j$. Since the formula is Gaussian, $A_j \ge 0$, $0 \le j \le n$, and $I(p) = Q_n(p)$ for all $p(x) \in \mathscr{P}_{2n+1}$. Thus, if $p_{2n+1}^*(x) \in \mathscr{P}_{2n+1}$ is the best uniform polynomial approximation to $f(x)$,

$$R_n(f) \equiv I(f) - Q_n(f) = (I(f) - I(p_{2n+1}^*)) - (Q_n(f) - Q_n(p_{2n+1}^*))$$

$$= \int_a^b (f(x) - p_{2n+1}^*(x))w(x)\, dx - \sum_{j=0}^n A_j(f(x_j) - p_{2n+1}^*(x_j)).$$

Therefore

$$|R_n(f)| \le E_{2n+1}(f) \left( \int_a^b w(x)\, dx + \sum_{j=0}^n A_j \right) = 2E_{2n+1}(f) \int_a^b w(x)\, dx. \qquad \textbf{(6.40b)}$$

Obviously, if any interpolatory quadrature, $Q_n(f) = \sum_{j=0}^n B_j f(x_j)$, has precision $m$ and $B_j \ge 0$, $0 \le j \le n$, the above argument could be repeated to yield $|R_n(f)| \le 2E_m(f) \int_a^b w(x)\, dx$. Since the precision is maximized when the quadrature is Gaussian, that is, $m = 2n + 1$, and since $E_{2n+1}(f) \le E_m(f)$ for $m < 2n + 1$, then (6.40b) represents an optimal error bound of this type. For practical use of (6.40b) we can call on any Jackson theorem (for instance, (5.69a) or (5.69b)) to yield a fairly simple bound on $|R_n(f)|$. We also note (Problem 6) that no formula of the type (6.34) can have precision equal to $2n + 2$. Thus Gaussian formulas are the oustide limits in increasing precision.

**Example 6.10.** As an illustration of the power of Gauss-Legendre quadrature, we consider again

$$Si(1) = \int_0^1 \frac{\sin(t)}{t}\, dt.$$

From a table, we obtain the weights and nodes for the fives point Gauss-Legendre formula for $[-1, 1]$:

$$
\begin{aligned}
x_0 &= -0.9061798459 & A_0 &= 0.2369268851\\
x_1 &= -0.5384693101 & A_1 &= 0.4786286705\\
x_2 &= 0.0 & A_2 &= 0.5688888889\\
x_3 &= 0.5384693101 & A_3 &= 0.4786286705\\
x_4 &= 0.9061798459 & A_4 &= 0.2369268851
\end{aligned}
$$

(The symmetric character of the weights and nodes should be expected since the $n$th degree Legendre polynomial is an even function when $n$ is even and an odd function when $n$ is odd). Since our problem is to estimate an integral on $[0, 1]$ rather than on $[-1, 1]$, we must use (6.4) where $a = 0$ and $b = 1$. With this change, the Gauss-Legendre 5-point formula provides the estimate of 0.94608307, which is correct to 8 places.

### 6.5.1 Interpolation at the Zeros of Orthogonal Polynomials

Gaussian quadrature provides a valuable link between the integral inner product, $\langle f, g\rangle \equiv \int_a^b f(x)g(x)w(x)\,dx$, and the discrete inner product,

$$\langle f, g\rangle_d = \sum_{j=0}^{n} A_j f(x_j)g(x_j),$$

where $A_j$ and $x_j$, $0 \le j \le n$, are the weights and the nodes respectively of the Gaussian quadrature $Q_n(f) \approx \int_a^b f(x)w(x)\,dx$. (Note that since the quadrature is Gaussian, then $A_j > 0$ and $x_j \in (a, b)$, $0 \le j \le n$, so that $\langle f, g\rangle_d$ is a well-defined discrete inner product.)

**Theorem 6.6.** Given $\langle f, g\rangle$ and $\langle f, g\rangle_d$ as above, let $\{p_0(x), p_1(x), p_2(x), \ldots\}$ be orthonormal polynomials (with degree $(p_j(x)) = j$) with respect to $\langle f, g\rangle$. Then $\{p_0(x), p_1(x), \ldots, p_n(x)\}$ is an orthonormal set with respect to $\langle f, g\rangle_d$.

*Proof*: Let $p(x) \equiv p_k(x)p_m(x)$, $k + m \le 2n$. Then, since $I(p) = Q_n(p)$, we have

$$
\begin{aligned}
\delta_{km} = \langle p_k, p_m\rangle &= \int_a^b p_k(x)p_m(x)w(x)\,dx = I(p) = Q_n(p)\\
&= \sum_{j=0}^{n} A_j p_k(x_j)p_m(x_j) = \langle p_k, p_m\rangle_d. \quad \blacksquare
\end{aligned}
$$

We can use Theorem 6.6 to give an easily computable formula for the polynomial, $P(x) \in \mathscr{P}_n$, that interpolates a function, $f(x)$, at the zeros of $p_{n+1}(x)$.

**Theorem 6.7.** Let $\{x_j\}_{j=0}^n$ be the zeros of $p_{n+1}(x)$ (as in Theorem 6.6), and let $P(x) = \sum_{k=0}^n \alpha_k p_k(x)$. If $\alpha_k = \sum_{j=0}^n A_j p_k(x_j)f(x_j)$, then $P(x_j) = f(x_j)$ for $0 \le j \le n$.

*Proof*: From Chapter 5 we know that the interpolating polynomial exists and is unique and thus can be written in the form $P(x) = \sum_{k=0}^n \alpha_k p_k(x)$. If $P(x_j) = f(x_j)$, then $\sum_{k=0}^n \alpha_k p_k(x_j) = f(x_j)$, $0 \le j \le n$. Now for $m$ fixed, multiplying both sides by $A_j p_m(x_j)$ and summing from $j = 0$ to $j = n$ yields

$$\sum_{j=0}^{n} A_j p_m(x_j) f(x_j) = \sum_{j=0}^{n} \left( A_j p_m(x_j) \sum_{k=0}^{n} \alpha_k p_k(x_j) \right)$$

$$= \sum_{k=0}^{n} \alpha_k \left( \sum_{j=0}^{n} A_j p_m(x_j) p_k(x_j) \right) = \sum_{k=0}^{n} \alpha_k \langle p_m, p_k \rangle_d = \alpha_m.$$

This holds for each $m$, $0 \le m \le n$, and so the proof is complete. ■

We next turn our attention to an estimate of the error that is made when interpolating at the zeros of orthogonal polynomials. Suppose $p(x) \in \mathscr{P}_n$ interpolates $f(x)$, $f^{(n+1)}(x) \in C[a, b]$, at *any* set of $(n + 1)$ distinct points, $\{z_j\}_{j=0}^{n}$, in $[a, b]$. Then we recall from formula (5.27), that for any $x \in [a, b]$,

$$f(x) - p(x) = \frac{f^{(n+1)}(\xi)}{(n + 1)!} W(x) \tag{6.41a}$$

where $\xi \in \operatorname{Spr}\{x, z_0, z_1, \ldots, z_n\}$ and where $W(x) = \prod_{j=0}^{n} (x - z_j)$. We have already seen the merits of interpolating at the zeros of the shifted Chebyshev polynomials, $\tilde{T}_{n+1}(x)$, in that this minimizes $\|W\|_\infty \equiv \max_{a \le x \le b} |W(x)|$ (see formula (5.35)). There are similar advantages in the interpolation given in Theorem 6.7. Recall, as in (5.31), if we square both sides of (6.41a), multiply by $w(x)$, integrate from $a$ to $b$, and take the square root we have

$$\|f - p\| = \left( \int_a^b (f(x) - p(x))^2 w(x)\, dx \right)^{1/2}$$

$$\le \frac{\max\limits_{a \le x \le b} |f^{(n+1)}(x)|}{(n + 1)!} \left( \int_a^b (W(x))^2 w(x)\, dx \right)^{1/2}$$

$$= \frac{\|f^{(n+1)}\|_\infty}{(n + 1)!} \|W\|. \tag{6.41b}$$

Now let $W(x)$ be any monic polynomial of degree $(n + 1)$, as above, and let $q_{n+1}(x)$ be the monic orthogonal polynomial, as before. Then $r(x) \equiv (W(x) - q_{n+1}(x)) \in \mathscr{P}_n$, so $\langle r, q_{n+1} \rangle = 0$. Now, we have

$$\|W\|^2 = \int_a^b (W(x))^2 w(x)\, dx = \int_a^b (q_{n+1}(x) + r(x))^2 w(x)\, dx$$

$$= \int_a^b (q_{n+1}(x))^2 w(x)\, dx + 2 \int_a^b q_{n+1}(x) r(x) w(x)\, dx + \int_a^b (r(x))^2 w(x)\, dx$$

$$= \|q_{n+1}\|^2 + 2\langle q_{n+1}, r \rangle + \|r\|^2 = \|q_{n+1}\|^2 + \|r\|^2.$$

Therefore $\|q_{n+1}\| \le \|W\|$ for all such $W(x)$, and so the error bound, (6.41b), is minimized for $W(x) = q_{n+1}(x)$, that is, for interpolation at the zeros of the orthogonal polynomial. In the special case where $[a, b] = [-1, 1]$ and $w(x) = 1$, the minimum bound is achieved by interpolating at the zeros of the $(n + 1)$st-degree Legendre polynomial.

Another indication of the power of interpolating at the zeros of orthogonal polynomials is provided by the following theorem, which guarantees least-squares convergence of the interpolating polynomials.

**Theorem 6.8. Erdös-Turán.** Let $P_n(x) \in \mathscr{P}_n$ interpolate $f(x)$ at the zeros of $p_{n+1}(x)$ (as in Theorem 6.7). Then,

$$\|f - P_n\| \equiv \left(\int_a^b (f(x) - P_n(x))^2 w(x)\, dx\right)^{1/2} \to 0, \qquad \text{as } n \to \infty.$$

*Proof*: Again, for each $n$, let $p_n^*(x) \in \mathscr{P}_n$ be the best uniform approximation to $f(x)$, and let

$$E_n(f) \equiv \max_{a \le x \le b} |f(x) - p_n^*(x)| \equiv \|f - p_n^*\|_\infty.$$

Then, by the triangle inequality, $\|f - P_n\| \le \|f - p_n^*\| + \|p_n^* - P_n\|$. Now,

$$\|f - p_n^*\|^2 = \int_a^b (f(x) - p_n^*(x))^2 w(x)\, dx \le (E_n(f))^2 \int_a^b w(x)\, dx.$$

Also, since $(p_n^*(x) - P_n(x))^2 \in \mathscr{P}_{2n}$ and the $n$th Gaussian quadrature is exact in $\mathscr{P}_{2n+1}$ and has positive weights with $\int_a^b w(x)\, dx = \sum_{j=0}^n A_j$, we see that

$$\begin{aligned}
\|p_n^* - P_n\|^2 &= \int_a^b (p_n^*(x) - P_n(x))^2 w(x)\, dx \\
&= \sum_{j=0}^n A_j (p_n^*(x_j) - P_n(x_j))^2 = \sum_{j=0}^n A_j (p_n^*(x_j) - f(x_j))^2 \\
&\le (E_n(f))^2 \sum_{j=0}^n A_j = (E_n(f))^2 \int_a^b w(x)\, dx.
\end{aligned}$$

Since $\int_a^b w(x)\, dx$ is a constant and $E_n(f) \to 0$ as $n \to \infty$, then both $\|f - p_n^*\|$ and $\|p_n^* - P_n\| \to 0$ as $n \to \infty$. ■

Gaussian quadratures are particularly effective in approximating Fourier coefficients, $\langle f, p_k \rangle \equiv \int_a^b f(x) p_k(x) w(x)\, dx$, but provide the following result which may be somewhat surprising at first reading. Let $f(x)$ be approximated by the truncated Fourier expansion, $f(x) \approx F_n(x) \equiv \sum_{k=0}^n \langle f, p_k \rangle p_k(x)$. Using the Gaussian quadrature to approximate $\langle f, p_k \rangle$ we get $\langle f, p_k \rangle \approx \alpha_k \equiv \sum_{j=0}^n A_j f(x_j) p_k(x_j)$. Then, $F_n(x) \approx \sum_{k=0}^n \alpha_k p_k(x)$, but we notice this is precisely the interpolating polynomial, $P_n(x)$, as in Theorem 6.7.

### 6.5.2 Interpolation Using Chebyshev Polynomials

In this section, we will consider some of the practical aspects of interpolation at the zeros of $T_{n+1}(x)$. We have already seen from (5.36) that there are advantages in using these interpolation points. The example given below establishes a useful discrete orthogonality property determined by the zeros of $T_{n+1}(x)$. This relation (and a companion result given in (6.43)) can be used in a variety of ways.

**Example 6.11.** Let $\langle f, g\rangle = \int_{-1}^{1} f(x)g(x)(1 - x^2)^{-1/2}\,dx$ and $x_j = \cos[(2j + 1)\pi/2(n + 1)]$, $0 \le j \le n$. For this inner product, the orthonormal polynomials are $p_0(x) = 1/\sqrt{\pi}$ and $p_k(x) = \sqrt{2/\pi}\,T_k(x)$, for $k \ge 1$ (see Example 5.15b, Chapter 5). By Example 6.9, $A_j = \pi/(n + 1)$, $0 \le j \le n$. Thus, by Theorem 6.6, the following relationship holds for $k + m \le 2n$:

$$\frac{2}{n+1}\sum_{j=0}^{n} T_k(x_j)T_m(x_j) = \begin{cases} 0, & k \ne m \\ 1, & k = m > 0 \\ 2, & k = m = 0. \end{cases} \tag{6.42}$$

Formula (6.42) very closely resembles another discrete orthogonality relation for the Chebyshev polynomials (displayed in Example 5.15c) which we can express for $k + m \le 2n$ as:

$$\frac{2}{n}\sum_{j=0}^{n}{}'' \; T_k(t_j)T_m(t_j) = \begin{cases} 0, \; k \ne m \\ 1, \; k = m, & 1 \le k \le n - 1 \\ 2, \; k = m, & k = 0 \quad \text{or} \quad k = n. \end{cases} \tag{6.43}$$

In (6.43), the points $t_j$ are given by $t_j = \cos[j\pi/n]$, $0 \le j \le n$, and the double prime denotes halving the first and last terms. Since the validity of (6.42) rests upon the fact that (6.39) is a Gaussian quadrature for $I(f) = \int_{-1}^{1} f(x)(1 - x^2)^{-1/2}\,dx$, the similarity of (6.42) and (6.43) leads us to suspect that (6.43) also represents a quadrature for $I(f) = \int_{-1}^{1} f(x)(1 - x^2)^{-1/2}\,dx$, of the form

$$Q_n'(f) = \frac{\pi}{n}\sum_{j=0}^{n}{}'' f(t_j). \tag{6.44}$$

In investigating the quadrature $Q_n'(f)$, we note first that $I(1) = Q_n'(1) = \pi$ and $I(T_k) = Q_n'(T_k) = 0$, for $1 \le k \le 2n - 1$. However, $I(T_{2n}) = 0$ but $Q_n'(T_{2n}) = \pi$, and so the precision of this quadrature is $(2n - 1)$. To verify these results, we note that $I(T_k) = \langle 1, T_k\rangle = 0$, for $k \ge 1$. To determine $Q_n'(T_k)$, we first suppose $1 \le k \le n$ and use (6.43) with $m = 0$ to find $Q_n'(T_k) = 0$. For $n < k \le 2n$, we let $k = n + m$ and observe that $T_k(t_j) = T_n(t_j)T_m(t_j)$. Thus, using (6.43), we see that $Q_n'(T_k) = 0$ for $n < k < 2n$ and that $Q_n'(T_{2n}) = \pi$. We note that $Q_n'(f)$ is not Gaussian since its precision is $(2n - 1)$ instead of $(2n + 1)$, but it is a powerful formula as we can see from the remarks following (6.40b). The nodes $t_j$ are often called the "practical" Chebyshev nodes. Also observe that if we make the transformation $x = \cos\theta$, then $I(f) = \int_0^{\pi} f(\cos\theta)\,d\theta$. From (6.44), we see that $Q_n'(f)$ is the composite trapezoidal rule for approximating $\int_0^{\pi} f(\cos\theta)\,d\theta$.

From Theorem 6.7 and formula (6.39) we easily find that the polynomial of degree $n$ or less which interpolates $f(x)$ at $x_j = \cos[(2j + 1)\pi/2(n + 1)]$,

$0 \le j \le n$, is given by

$$P(x) = \beta_0/2 + \sum_{k=1}^{n} \beta_k T_k(x), \qquad \beta_k = \frac{2}{n+1} \sum_{j=0}^{n} T_k(x_j) f(x_j). \tag{6.45a}$$

A similar formula can be developed for interpolation at the points $t_j = \cos(j\pi/n)$. However, since (6.44) is not a Gaussian formula, we cannot directly apply Theorem 6.7. However, we can use (6.43) and mimic the proof of Theorem 6.7 to show that the polynomial of degree $n$ or less which interpolates $f(x)$ at $t_j = \cos(j\pi/n)$, $0 \le j \le n$, is (see Problem 8):

$$\tilde{P}(x) = \sum_{k=0}^{n}{}'' \gamma_k T_k(x), \qquad \text{where} \qquad \gamma_k = \frac{2}{n} \sum_{j=0}^{n}{}'' T_k(t_j) f(t_j). \tag{6.45b}$$

We pause to mention that the interpolating polynomials constructed via (6.45a) and (6.45b) can be easily evaluated at a point $x = \alpha$, since they have the form $p(x) = \sum_{k=0}^{n} b_k r_k(x)$, where $\{b_k\}_{k=0}^{n}$ are known constants and each $r_k(x)$ is a constant multiple of $p_k(x)$. We leave to the reader to verify that the efficient algorithm of formula (5.66) can be modified to accomplish this task with $(2n - 1)$ multiplications, once the three-term recursion is known for the $r_k(x)$'s. For example (see Problem 9), if $r_k(x) = T_k(x)$, then we have $T_0(x) = 1$, $T_1(x) = x$, and $T_k(x) = 2xT_{k-1}(x) - T_{k-2}(x)$, for $k \ge 2$. The modification of (5.66) yields, in this case,

$$\sum_{k=0}^{n} b_k T_k(\alpha) = s_0(\alpha) - \alpha s_1(\alpha), \tag{6.45c}$$

where $s_k(\alpha)$ is defined by $s_{n+1}(\alpha) = s_{n+2}(\alpha) = 0$ and for $k = n, n-1, \ldots, 0$, by

$$s_k(\alpha) - 2\alpha s_{k+1}(\alpha) + s_{k+2}(\alpha) = b_k.$$

(Note in this particular example, since the coefficient of $s_{k+2}(\alpha)$ equals 1, only $(n + 1)$ multiplications are required.)

We turn now to the infinite Chebyshev expansion for $f(x)$. Recall that $p_0(x) = 1/\sqrt{\pi}T_0(x)$ and $p_k(x) = \sqrt{2/\pi}T_k(x)$ for $k \ge 1$, so $\langle f, p_0 \rangle = 1/\sqrt{\pi}\langle f, T_0 \rangle$ and $\langle f, p_k \rangle = \sqrt{2/\pi}\langle f, T_k \rangle$, $k \ge 1$. Thus the Fourier-Chebyshev expansion for $f(x)$, for $-1 \le x \le 1$ is:

$$f(x) \sim \langle f, p_0 \rangle p_0(x) + \sum_{k=1}^{\infty} \langle f, p_k \rangle p_k(x). \tag{6.46a}$$

Often, for the sake of convenience, the normalizing constants of the $p_k(x)$'s are multiplied together and (6.46a) is written in the equivalent form

$$f(x) \sim \sum_{k=0}^{\infty}{}' a_k T_k(x), \qquad a_k = \frac{2}{\pi} \int_{-1}^{1} f(x) T_k(x)(1 - x^2)^{-1/2}\, dx \tag{6.46b}$$

where the prime denotes that the first term is halved.

Using the Gauss-Chebyshev quadrature (6.39) to approximate $a_k$ for $0 \le k \le n$, we obtain the interpolating polynomial $P(x)$ of formula (6.45a) as an approximation for $f(x)$, where each $\beta_k = (2/(n+1)) \sum_{j=0}^{n} T_k(x_j) f(x_j)$ is our approximation to $a_k$. An alternative approach is to use $\tilde{P}(x)$ of formula (6.45b) as our approximation to $f(x)$, where $\gamma_k = (2/n) \sum_{j=0}^{\prime\prime\, n} T_k(t_j) f(t_j)$ is our approximation to $a_k$, $0 \le k \le n$.

To further assess the accuracy of these interpolations, let us assume that the expansion (6.46b) is absolutely and uniformly convergent to $f(x)$ (which is the case, for example, if $f \in C^2[-1, 1]$). We first consider $\tilde{P}(x)$ in (6.45b), and for a fixed value of $m$, $0 \le m \le n$, we consider the approximation of $\gamma_m$ for $a_m$. We can easily verify by trigonometric identities that if $k = 2rn + \alpha$; $r = 0, 1, 2, \ldots, |\alpha| \equiv \beta \le n$; then $T_k(t_j) = T_\beta(t_j)$, $0 \le j \le n$. Making use of (6.43), we obtain for any $m$,

$$
\begin{aligned}
\gamma_m &= \frac{2}{n} \sum_{j=0}^{n}{}'' \left( \sum_{k=0}^{\infty}{}' a_k T_k(t_j) \right) T_m(t_j) \\
&= \sum_{k=0}^{\infty}{}' a_k \left( \frac{2}{n} \sum_{j=0}^{n}{}'' T_k(t_j) T_m(t_j) \right) \\
&= a_m + (a_{2n-m} + a_{2n+m}) + (a_{4n-m} + a_{4n+m}) + \cdots
\end{aligned}
\quad \textbf{(6.47a)}
$$

and the resulting approximation of (6.45b),

$$
f(x) \approx \sum_{m=0}^{n}{}'' \gamma_m T_m(x). \quad \textbf{(6.47b)}
$$

Similarly, if we write $k = 2r(n+1) + \alpha$, we can easily establish the identity $T_\alpha(x_j) = (-1)^r T_{2r(n+1) \pm \alpha}(x_j)$, $0 \le j \le n$. Using this and (6.42) in the same manner as above, we obtain, for any $m$,

$$
\beta_m = a_m - (a_{2n+2-m} + a_{2n+2+m}) + (a_{4n+4-m} + a_{4n+4+m}) - \cdots \quad \textbf{(6.48a)}
$$

which yields the approximation

$$
f(x) \approx \sum_{m=0}^{n}{}' \beta_m T_m(x). \quad \textbf{(6.48b)}
$$

We leave (Problem 10) for the reader to verify that $\left| f(x) - \sum_{m=0}^{\prime\prime\, n} \gamma_m T_m(x) \right|$ and $\left| f(x) - \sum_{m=0}^{\prime\, n} \beta_m T_m(x) \right|$ are both bounded by $2 \sum_{m=n+1}^{\infty} |a_m|$. Thus, by Section 5.3.2, the error of (6.47b) and (6.48b) can never exceed twice the error of the truncated series, $\left| f(x) - \sum_{m=0}^{\prime\, n} a_m T_m(x) \right|$. We thus note that, if the magnitudes of the Fourier coefficients are rapidly decreasing, then both (6.47b) and (6.48b) are good approximations. We also note from (6.47a) and (6.48a) that both $\gamma_m$ and $\beta_m$ are most likely to agree closely with $a_m$ when $m$ is small. We expect the most discrepency in (6.47a) when $m = n - 1$ and in (6.48a) when $m = n$, where we see that $\gamma_{n-1} \approx a_{n-1} + a_{n+1}$ and $\beta_n \approx a_n - a_{n+2}$, respectively. However, if $n$ is sufficiently large, then the coefficients $a_{n-1}$, $a_n$, $a_{n+1}$, and $a_{n+2}$ should be relatively small and should not significantly affect the accuracy.

### 6.5.3 Clenshaw-Curtis Quadrature

We note that any time we have an approximation for $f(x)$ of the form, $f(x) \approx p(x) \equiv \sum_{k=0}^{n} b_k T_k(x)$, then we can easily construct a numerical integration formula from it. This follows from the simple observation that indefinite integration of $T_k(x)$ yields

$$\begin{aligned} \int T_k(x)\,dx &= -\int \cos(k\theta)\sin(\theta)\,d\theta \\ &= -\frac{1}{2}\int (\sin[(k+1)\theta] - \sin[(k-1)\theta])\,d\theta \\ &= \frac{1}{2}\left(\frac{T_{k+1}(x)}{k+1} - \frac{T_{k-1}(x)}{k-1}\right), \qquad k \geq 2, \end{aligned} \tag{6.49a}$$

and

$$\int T_0(x)\,dx = T_1(x), \qquad \int T_1(x)\,dx = \frac{1}{4}(T_0(x) + T_2(x)). \tag{6.49b}$$

Given $p(x)$ as above, then the result of the indefinite integration of $p(t)$ is an expression of the form

$$\int_{-1}^{x} p(t)\,dt = \sum_{k=0}^{n} b_k \int_{-1}^{x} T_k(t)\,dt \equiv \sum_{k=0}^{n+1} A_k T_k(x) \equiv P(x). \tag{6.50}$$

Using the integral formulas (6.49a) and (6.49b) and equating like coefficients of each $T_k(x)$, we have the following equations for each $A_k$:

$$A_{n+1} = \frac{b_n}{2(n+1)}, \qquad A_n = \frac{b_{n-1}}{2n}, \qquad A_1 = b_0 - \frac{b_2}{2}, \tag{6.51a}$$

and

$$A_k = \frac{1}{2k}(b_{k-1} - b_{k+1}), \qquad \text{for } 2 \leq k \leq n-1.$$

To obtain $A_0$, we notice that $P(-1) = 0$ and since $T_k(-1) = (-1)^k$, we find

$$A_0 = A_1 - A_2 + A_3 + \cdots + (-1)^n A_{n+1}. \tag{6.51b}$$

This technique is valid for all $x$ in $[-1, 1]$, and, moreover, $P(x)$ can be efficiently evaluated at any $x$ by (6.45c). If $x = 1$ and $p(x)$ in (6.50) is given by either the interpolating polynomials (6.45a) or (6.45b), then $P(1)$ is an interpolatory quadrature for $I(f) \equiv \int_{-1}^{1} f(x)\,dx$. Thus, (6.51) can be written in the form $Q_n(f) = \sum_{j=0}^{n} A_j f(z_j)$. Although it is not computationally efficient to do this, it can be shown in either case that the weights are positive so the convergence result of Theorem 6.1 is applicable. Formulas of this type are called *Clenshaw-Curtis* quadratures.

Finally, we remark that we seem to have placed a great deal of emphasis on Chebyshev-type approximations in these sections. However, both from a theoretical background and from practical experience, Chebyshev methods have proven to yield excellent procedures in terms of truncation errors and round-off propagation.

**Example 6.12.** To illustrate some of these ideas, let $f(x) = \sin(x)/x$ for $-1 \le x \le 1$. First, using (6.45b), we construct the interpolating polynomial for $f(x)$, with $n = 4$. Since $f(x)$ is an even function on $[-1, 1]$ and $T_k(x)$ is odd when $k$ is odd, we see that $\gamma_1$ and $\gamma_3$ in (6.45b) are zero. For even $k$, since $f(x)$ and $T_k(x)$ are even, we have (by symmetry between $t_1$ and $t_3$ and between $t_0$ and $t_4$).

$$\gamma_k = \frac{2}{4}[f(t_0)T_k(t_0) + 2f(t_1)T_k(t_1) + f(t_2)T_k(t_2)].$$

Now $T_k(t_0) = 1$ for $k = 0, 2, 4$; $T_k(t_1)$ has the value 1 for $k = 0$, 0 for $k = 2$ and $-1$ for $k = 4$; $T_k(t_2)$ has the value 1 for $k = 0$, $-1$ for $k = 2$ and 1 for $k = 4$. Thus, using $f(0) = 1$,

$$\gamma_0 = \frac{1}{2}\left[f(1) + 2f\left(\cos\frac{\pi}{4}\right) + f(0)\right] = 1.839461$$

$$\gamma_2 = \frac{1}{2}[f(1) - f(0)] = -0.079265$$

$$\gamma_4 = \frac{1}{2}\left[f(1) - 2f\left(\cos\frac{\pi}{4}\right) + f(0)\right] = 0.002010.$$

Thus $p(x) = 0.919731 - 0.079265T_2(x) + 0.001005T_4(x)$, where the rapid decrease of the coefficients is typical of a well-behaved function $f(x)$.

To demonstrate Clenshaw-Curtis quadrature, we derive a polynomial $P(x)$ that approximates

$$Si(x) = \int_0^x \frac{\sin(t)}{t}\,dt.$$

Writing the interpolating polynomial $P(x)$ above in the form of (6.50), we have $b_0 = 0.919731$, $b_1 = 0$, $b_2 = -0.079265$, $b_3 = 0$, and $b_4 = 0.001005$. Thus, from (6.51a) and (6.51b), we obtain $A_5 = 0.000101$, $A_4 = 0$, $A_3 = -0.013378$, $A_2 = 0$, $A_1 = 0.959364$, and $A_0 = 0.946087$. Thus

$$P(x) = 0.946087 + 0.959364T_1(x) - 0.013378T_3(x) + 0.000101T_5(x)$$

is an approximation for $\int_{-1}^{x} (\sin(t)/t)\,dt$. Since $\sin(t)/t$ is even, we have

$$\int_{-1}^{0} \frac{\sin(t)}{t}\,dt = \int_0^1 \frac{\sin(t)}{t}\,dt = Si(1).$$

Thus $P(0) = 0.946807$ is an estimate to $Si(1) = 0.946083$. Moreover, we can use $P(x)$ to provide an estimate $Si(x)$, $0 < x < 1$, by observing (again since the integrand is even) that

$$\int_{-1}^{-1+x} \frac{\sin(t)}{t}\,dt = \int_{1-x}^{1} \frac{\sin(t)}{t}\,dt = Si(1) - Si(1-x).$$

Therefore, for $0 < x < 1$, we have the approximation

$$Si(1-x) \approx P(0) - P(-1+x)$$

which is an easily computed and accurate approximation. For example, with $x = 0.5$, $Si(0.5) = 0.493107$ while $P(0) - P(-0.5) = 0.493060$, which is in error by 0.000047.

## PROBLEMS

1. Using the result of Theorem 6.4, find the weights and nodes of the two- and three-point Gauss-Legendre quadrature formulas. Do this by finding the weights by undetermined coefficients. (The second- and third-degree monic Legendre polynomials are respectively: $P_2(x) = x^2 - (1/3)$, $P_3(x) = x^3 - (3/5)x$. The weights can be verified by checking precision in (6.34).)
2. Use the three-point Gauss-Legendre formula of Problem 1 and the five-point formula given in Example 6.10 to estimate
   a) $\int_{-1}^{1} \sin(3x)\,dx$
   b) $\int_{1}^{3} \ln(x)\,dx$
   c) $\int_{1}^{2} e^{x^2}\,dx$.
3. Write a computer program to generate the $n$th Legendre polynomial from the three-term recurrence relation and find the zeros by Newton's method. Next, use (6.36) to find the Gauss-Legendre quadrature weights. Check your results for various values of $n$ against tabulated formulas.
4. Let $q_k(x) = (1/2^{k-1})T_k(x)$ denote the monic Chebyshev polynomial of the first kind. Verify that $q_k(x) = xq_{k-1}(x) - b_k q_{k-2}(x)$, $k \geq 2$, where $b_2 = 1/2$ and $b_k = 1/4$, $k \geq 3$.
5. Use (6.40a) and (6.40b) to bound the error made in estimating the integral in Problem 2(a) by the three-point Gauss-Legendre formula (that is, $n = 2$ in (6.40a) and (6.40b)). Use (6.40b) and an appropriate Jackson theorem to bound the error for five-point Gauss-Legendre formula applied to the integral in 2(a).
6. Show that no matter how the nodes and weights of a quadrature formula

$$Q_n(f) = \sum_{j=0}^{n} A_j f(x_j)$$

   are chosen, the formula cannot have precision greater than $2n + 1$. (Hint: Assume $Q_n(f)$ is designed to approximate $\int_a^b f(x)w(x)\,dx$. Use Theorem 6.4 and find a polynomial $p(x) \in \mathscr{P}_{2n+2}$ for which $Q_n(p) \neq \int_a^b p(x)w(x)\,dx$.)
7. The complete elliptic integral

$$K = \int_0^1 \frac{dx}{\sqrt{(1-x^2)(1-k^2x^2)}}$$

is usually written as

$$K = \int_0^{\pi/2} \frac{d\theta}{\sqrt{1 - k^2 \sin^2 \theta}},$$

where $0 \le k < 1$. For $k = 1/2$, the value of $K$ is 1.6858. Using the fact that the integrand is even, estimate $K$ by Gauss-Chebyshev quadrature.

8. Verify in (6.45b) that $\tilde{P}(x)$ interpolates $f(x)$ at $t_j$, $j = 0, 1, \ldots, n$.
9. Verify the recursion (6.45c) for evaluating a finite Chebyshev series.

*10. Suppose that $f(x) = \sum_{m=0}^{\infty}{}' a_m T_m(x)$, where this Chebyshev expansion converges uniformly and absolutely to $f(x)$ for $-1 \le x \le 1$. For $\gamma_m$ defined in (6.47a) and $\beta_m$ defined in (6.48a), show that $|f(x) - \sum_{m=0}^{n}{}'' \gamma_m T_m(x)|$ and $|f(x) - \sum_{n=0}^{m}{}' \beta_m T_m(x)|$ are each bounded by $2 \sum_{m=n+1}^{\infty} |a_m|$. (Note that in conjunction with (5.70a) or (5.70b), this bound gives us a way to estimate interpolation errors.)

*11. This rather long problem gives a sharper convergence result for the uniform convergence of best least-squares approximations. Suppose $\langle f, g \rangle = \int_a^b f(x)g(x)w(x)\,dx$ and suppose $p_i(x)$, $i = 0, 1, \ldots$ are the orthonormal polynomials associated with this inner-product. Let $P_m(x) = \sum_{j=0}^{m} \langle f, p_j \rangle p_j(x)$ be the best least-squares approximation to $f(x)$.

a) Show: $P_m(x) = \int_a^b f(t)G_m(x, t)w(t)\,dt$, where $G_m(x, t) = \sum_{j=0}^{m} p_j(x)p_j(t)$. (The function $G_m(x, t)$ is usually called a "kernel.")

b) Let $q_i(x)$, $i = 0, 1, \ldots$ denote the monic orthogonal polynomials generated by the three-term recurrence relation (5.65). Then $p_i(x) = \lambda_i q_i(x)$, where

$$\lambda_i = \frac{1}{\|q_i\|}, \qquad \|q_i\| = \sqrt{\langle q_i, q_i \rangle}.$$

Thus (5.65) can be written so as to generate the $p_i$, where

$$p_k(x) = \lambda_k^{-1}(x - a_k)\lambda_{k-1}p_{k-1}(x) - \lambda_k^{-1}b_k\lambda_{k-2}p_{k-2}(x), \qquad k \ge 2.$$

Rewrite this as: $p_k(x) = (A_k x + B_k)p_{k-1}(x) - C_k p_{k-2}(x)$ and prove the *Christoffel-Darboux Identity*:

$$\frac{\lambda_m}{\lambda_{m+1}}[p_{m+1}(x)p_m(t) - p_{m+1}(t)p_m(x)] = (x - t)\sum_{j=0}^{m} p_j(x)p_j(t).$$

(Hint: Multiply the rewritten recurrence relation for $p_{j+1}(x)$ by $p_j(t)$ and the relation for $p_{j+1}(t)$ by $p_j(x)$. Thus obtain an expression for $(x - t)p_j(x)p_j(t)$ and sum these expressions.)

c) Show that $\int_a^b w(t)G_m(x, t)\,dt = 1$ and hence deduce that $f(x) = \int_a^b f(x)G_m(x, t)w(t)\,dt$.

d) Using the above results, show that $f(x) - P_m(x) = \int_a^b [f(x) - f(t)]G_m(x, t)w(t)\,dt$.

e) From the Christoffel-Darboux identity, we have a useful expression for the kernel. For example, suppose $|\lambda_m/\lambda_{m+1}| \le C$ for $m = 0, 1, \ldots$ and suppose $f(x)$ is differentiable at $x = x_0$, $a < x_0 < b$. Further suppose that $|f(x) - f(x_0)| \le K|x - x_0|$ for all $x \in [a, b]$ and that $|p_i(x_0)| \le M$ for $i = 0, 1, \ldots$. Show that $\lim_{m\to\infty} P_m(x_0) = f(x_0)$. (Hint: Let $g(t) = [f(x_0) - f(t)]/[x_0 - t]$ and use (b) and the fact that $\langle g, p_i \rangle \to 0$ as $i \to \infty$ (see Section 5.3.2).)

f) To see that the conditions of (e) are not too hard to satisfy, what are the constants $C$ and $M$ in (e) when the $p_i(x)$ are the Chebyshev polynomials? For the Chebyshev polynomials, estimate $f(x_0) - P_m(x_0)$ when $f \in C^2[-1, 1]$.

## 6.6 TRIGONOMETRIC POLYNOMIALS AND THE FAST FOURIER TRANSFORM

As we have seen in Chapter 5, interpolating data or functions given at equally spaced points by algebraic polynomials $p(x) = a_n x^n + \cdots + a_1 x + a_0$ may lead to poor approximations. However, particularly in engineering or scientific problems, we may have no other practical alternative than to use information given at equally spaced points. In this case, we shall see that trigonometric polynomials (as opposed to algebraic polynomials) provide a good tool for approximation. The set of all $n$th-degree trigonometric polynomials, $\mathscr{T}_n$, is all functions which can be written in the form:

$$p(\theta) = a_0 + \sum_{j=1}^{n} a_j \cos(j\theta) + \sum_{j=1}^{n} b_j \sin(j\theta) \qquad \textbf{(6.52)}$$

where $\theta$ is given in radians and $a_j$ and $b_j$ are constants. The most natural interval to consider for an approximation problem involving trigonometric polynomials is the interval $[0, 2\pi]$, and trigonometric polynomials are well suited to approximate functions which are continuous on $[0, 2\pi]$ and which satisfy $f(0) = f(2\pi)$. This latter condition allows us to extend each such $f(\theta)$ to the entire real axis such that this extension is continuous and periodic with period $2\pi$. For notation, we define this set of functions as

$$C_{2\pi} = \{f(\theta) | f(\theta) \text{ continuous for } 0 \le \theta \le 2\pi, f(0) = f(2\pi)\}.$$

These approximations can be easily modified to suit any interval $[a, b]$ and functions which are not $2\pi$-periodic, but we omit the details in the interest of brevity.

The power of trigonometric polynomial approximation is based on classical Fourier series (see Example 5.15(a)). Classical Fourier series uses the inner-product

$$\langle f, g \rangle = \int_0^{2\pi} f(\theta) g(\theta)\, d\theta \qquad \textbf{(6.53)}$$

for which we have (see Problem 1)

$$\int_0^{2\pi} \cos(k\theta)\cos(j\theta)\, d\theta = \begin{cases} 0, & k \ne j \\ \pi, & k = j, \quad k \ne 0 \\ 2\pi, & k = j = 0 \end{cases}$$

$$\int_0^{2\pi} \cos(k\theta)\sin(j\theta)\, d\theta = 0, \qquad \text{for all } k \text{ and } j \qquad \textbf{(6.54)}$$

$$\int_0^{2\pi} \sin(k\theta)\sin(j\theta)\, d\theta = \begin{cases} 0, & k \ne j \\ \pi, & k = j > 0. \end{cases}$$

Thus, for each $n$, the set of functions

$$S_n = \left\{ \frac{1}{\sqrt{2\pi}}, \frac{1}{\sqrt{\pi}} \cos(j\theta), \frac{1}{\sqrt{\pi}} \sin(j\theta) \right\}_{j=1}^{n} \tag{6.55}$$

is an orthonormal set in $C_{2\pi}$. Furthermore each trigonometric polynomial, $p(\theta)$ in $\mathscr{T}_n$, can be written as a linear combination of elements of $S_n$.

Hence, given any $f(\theta) \in C_{2\pi}$, we can mimic the proof of Theorem 5.10 and see that $P_n^*(\theta)$ given by

$$\begin{aligned} P_n^*(\theta) = \left\langle f, \frac{1}{\sqrt{2\pi}} \right\rangle \frac{1}{\sqrt{2\pi}} + \sum_{j=1}^{n} \left\langle f, \frac{\cos(j\theta)}{\sqrt{\pi}} \right\rangle \frac{\cos(j\theta)}{\sqrt{\pi}} \\ + \sum_{j=1}^{n} \left\langle f, \frac{\sin(j\theta)}{\sqrt{\pi}} \right\rangle \frac{\sin(j\theta)}{\sqrt{\pi}} \end{aligned} \tag{6.56a}$$

satisfies $\|f - P_n^*\| \leq \|f - p\|$, for all $p(\theta) \in \mathscr{T}_n$, that is,

$$\int_0^{2\pi} (f(\theta) - P_n^*(\theta))^2 \, d\theta \leq \int_0^{2\pi} (f(\theta) - p(\theta))^2 \, d\theta, \qquad p(\theta) \in \mathscr{T}_n.$$

(The reader should compare (6.56a) with (5.63) and see that the result is equally valid with trigonometric polynomials replacing the algebraic polynomials in Chapter 5.) We note that (6.56a) may be written in the equivalent, but slightly more simplified form:

$$\begin{aligned} P_n^*(\theta) &= \frac{A_0}{2} + \sum_{j=1}^{n} A_j \cos(j\theta) + \sum_{j=1}^{n} B_j \sin(j\theta), \\ A_j &= \frac{1}{\pi} \langle f, \cos(j\theta) \rangle = \frac{1}{\pi} \int_0^{2\pi} f(\theta) \cos(j\theta) \, d\theta, \\ B_j &= \frac{1}{\pi} \langle f, \sin(j\theta) \rangle = \frac{1}{\pi} \int_0^{2\pi} f(\theta) \sin(j\theta) \, d\theta. \end{aligned} \tag{6.56b}$$

As in the case of the closely related Chebyshev series expansions, we can guarantee that $P_n^*(\theta)$ converges uniformly to $f(\theta)$ for $0 \leq \theta \leq 2\pi$ by imposing mild smoothness conditions on $f(\theta)$ (for instance, $f''(\theta) \in C_{2\pi}$). Thus trigonometric polynomials will give good approximations, but in order to provide a practical computational tool, it is necessary to develop algorithms for approximating the Fourier coefficients $A_j$ and $B_j$ in (6.56b). It turns out that the composite trapezoidal rule is a *Gaussian* quadrature formula with respect to trigonometric polynomials and hence will give good estimates of the integrals in (6.56b). Since the composite trapezoidal rule is based on equally spaced points, we see that trigonometric polynomials provide a practical and useful form of approximation that is well-suited for functions or data given at equally spaced points.

To be specific, let

$$\theta_j = j\frac{2\pi}{2m} = \frac{j\pi}{m}, \qquad j = 0, 1, \ldots, 2m \tag{6.57}$$

be equally spaced in $[0, 2\pi]$ and let $T_{2m}(g)$ denote the composite trapezoidal rule applied to $g(\theta)$ on $[0, 2\pi]$. Thus

$$T_{2m}(g) = \frac{\pi}{m}\sum_{j=0}^{2m}{}'' g(\theta_j) \approx \int_0^{2\pi} g(\theta)\, d\theta.$$

Since $g \in C_{2\pi}$, we know that $g(0) = g(2\pi)$ (or $g(\theta_0) = g(\theta_{2m})$). When we wish, then, we can write $T_{2m}(g)$ as

$$T_{2m}(g) = \frac{\pi}{m}\sum_{j=0}^{2m-1} g(\theta_j).$$

To show that this quadrature is Gaussian with respect to trigonometric polymials, we note that

$$T_{2m}(\cos(k\theta)) = \frac{\pi}{m}\sum_{j=0}^{2m}{}'' \cos\frac{jk\pi}{m}$$

$$T_{2m}(\sin(k\theta)) = \frac{\pi}{m}\sum_{j=0}^{2m}{}'' \sin\frac{jk\pi}{m}.$$

Using Lemma 5.1 and (6.54), we see (Problem 3) that $T_{2m}(\cos(k\theta)) = 0 = \int_0^{2\pi} \cos(k\theta)\, d\theta$, for $1 \le k \le 2m - 1$ and $T_{2m}(1) = 2\pi = \int_0^{2\pi} d\theta$. Since $\sin(k\theta)$ is an odd function about the center of the interval of integration, $\theta = \pi$, it is easy to show (Problem 4) that $T_{2m}(\sin(k\theta)) = 0 = \int_0^{2\pi} \sin(k\theta)\, d\theta$ for $1 \le k \le 2m - 1$. Thus, we have that $T_{2m}(p(\theta)) = \int_0^{2\pi} p(\theta)\, d\theta$ for any trigonometric polynomial of degree $2m - 1$ or less, and this is the Gaussian feature of the composite trapezoidal rule for $\mathcal{T}_{2m-1}$.

### 6.6.1 Least-squares Fits and Interpolation at Equally Spaced Points

Since the composite trapezoidal rule is Gaussian with respect to trigonometric polynomials, it is not surprising (as in Section 6.5.1) that there is a discrete innerproduct, using the composite trapezoidal rule weights and nodes, for which the first $2m$ functions in (6.55) are orthonormal. More precisely, the $2m$ functions

$$\frac{1}{\sqrt{2\pi}}, \quad \frac{1}{\sqrt{\pi}}\cos(\theta), \ldots, \quad \frac{1}{\sqrt{\pi}}\cos((m-1)\theta), \quad \frac{1}{\sqrt{2\pi}}\cos(m\theta),$$

$$\frac{1}{\sqrt{\pi}}\sin(\theta), \ldots, \quad \frac{1}{\sqrt{\pi}}\sin((m-1)\theta) \tag{6.58}$$

are orthonormal with respect to the discrete inner-product

$$\langle f, g \rangle_d = \frac{\pi}{m} \sum_{j=0}^{2m-1} f\left(\frac{j\pi}{m}\right) g\left(\frac{j\pi}{m}\right) \tag{6.59}$$

(see Problem 5). Thus, given the problem of finding a $k$th-degree trigonometric polynomial to minimize the expression

$$\sum_{j=0}^{2m-1} [y_j - p(\theta_j)]^2$$

where $\{y_j\}_{j=0}^{2m-1}$ is a given set of values, we find (Problem 6) that the solution for $k \leq m - 1$ is given by

$$p_k(\theta) = \frac{a_0}{2} + \sum_{j=1}^{k} a_j \cos(j\theta) + \sum_{j=1}^{k} b_j \sin(j\theta)$$

$$a_j = \frac{1}{m} \langle f, \cos(j\theta) \rangle_d = \frac{1}{m} \sum_{r=0}^{2m-1} y_r \cos \frac{rj\pi}{m}, \tag{6.60}$$

$$b_j = \frac{1}{m} \langle f, \sin(j\theta) \rangle_d = \frac{1}{m} \sum_{r=0}^{2m-1} y_r \sin \frac{rj\pi}{m},$$

where $f(\theta)$ is any function in $C_{2\pi}$ such that $f(\theta_r) = y_r$, $0 \leq r \leq 2m - 1$.

**Example 6.13.** In order to demonstrate the calculations in (6.60), we let the data $y_j$ be given by

$$y_j = \frac{\sin \theta_j}{\theta_j}, \qquad \theta_j = j\frac{2\pi}{10} = \frac{j\pi}{5}, \qquad j = 0, 1, \ldots, 9;$$

where we define $y_0 = 1$. In this case, $m = 5$ and we calculate as in (6.60)

$$a_0 = \frac{1}{5} \sum_{j=0}^{9} \frac{\sin(\theta_j)}{\theta_j} = 0.7061$$

$$a_1 = \frac{1}{5} \sum_{j=0}^{9} \frac{\sin(\theta_j)}{\theta_j} \cos(\theta_j) = 0.0823$$

$$b_1 = \frac{1}{5} \sum_{j=0}^{9} \frac{\sin(\theta_j)}{\theta_j} \sin(\theta_j) = 0.1501$$

Similarly, we find $a_2 = 0.0918$ and $b_2 = 0.0759$, so that $p_2(\theta) = 0.3531 + 0.0823 \cos(\theta) + 0.1501 \sin(\theta) + 0.0918 \cos(2\theta) + 0.0759 \sin(2\theta)$ is the best least-squares fit of degree 2 to this data on the 10 points $\theta_j$.

When $k = m$ in (6.60), the situation is somewhat altered. As was the case in Section 6.5.1, we find that the trigonometric polynomial $p_m(\theta)$ given below in

(6.61) is an *interpolating* polynomial. (Note in (6.61), that $p_m(\theta)$ is still the best least-squares approximation to the data $y_0, y_1, \ldots, y_{2m-1}$ with respect to the orthonormal set (6.58).) We have

$$p_m(\theta) = \sum_{j=0}^{m}{}'' a_j \cos(j\theta) + \sum_{j=1}^{m-1} b_j \sin(j\theta) \tag{6.61}$$

where the coefficients $a_j$ and $b_j$ are as given in (6.60) and the double prime means that the first and last terms are halved. To see that $p_m(\theta_r) = y_r$ for $r = 0, 1, \ldots, 2m - 1$, we write out $p_m(\theta_r)$ obtaining

$$\begin{aligned} p_m(\theta_r) &= \sum_{s=0}^{m}{}'' \cos(s\theta_r) \frac{1}{m} \sum_{j=0}^{2m-1} y_j \cos(s\theta_j) \\ &\quad + \sum_{s=1}^{m-1} \sin(s\theta_r) \frac{1}{m} \sum_{j=0}^{2m-1} y_j \sin(s\theta_j) \\ &= \sum_{j=0}^{2m-1} y_j \frac{1}{m} \sum_{s=0}^{m}{}'' \cos(s\theta_r) \cos(s\theta_j) \\ &\quad + \sum_{j=0}^{2m-1} y_j \frac{1}{m} \sum_{s=1}^{m-1} \sin(s\theta_r) \sin(s\theta_j) \\ &= \sum_{j=0}^{2m-1} y_j \frac{1}{m} \sum_{s=0}^{m}{}'' \cos\left(s(r-j)\frac{\pi}{m}\right). \end{aligned}$$

To obtain the last equality, we use the fact that $\sin(s\theta_r)\sin(s\theta_j) = 0$ for $s = 0$ and $m$, and then using the identity for $\cos(A - B)$. Using Lemma 5.1 (Problem 7), we see that the inner sum is zero unless $r = j$, in which case the inner sum is 1. Thus $p_m(\theta_r) = y_r$ for $r = 0, 1, \ldots, 2m - 1$.

In considering (6.60) or (6.61), we see that the bulk of the computation involves calculating the coefficients $a_j$ and $b_j$. In many engineering problems, the *power spectrum*, which is a plot of the numbers $a_j^2 + b_j^2$ as a function of $j$, has significance. Generally, we are interested in the behavior of the function $f(x)$ defined by the inverse Fourier transform

$$f(x) = \int_{-\infty}^{\infty} F(t)e^{2\pi ixt}\,dt = \int_{-\infty}^{\infty} F(t)[\cos(2\pi xt) + i\sin(2\pi xt)]\,dt \tag{6.62a}$$

where $i = \sqrt{-1}$. Typically, $F(t)$ decays rapidly for $|t|$ large, so that

$$\int_{-a}^{a} F(t)\cos(2\pi xt)\,dt + i\int_{-a}^{a} F(t)\sin(2\pi xt)\,dt \tag{6.62b}$$

provides a good estimate for $f(x)$. The composite trapezoidal rule gives good approximations to the integrals in (6.62b), which again leads to sums of the form

$$\sum_{j=0}^{2m-1} F(t_j)\cos(\lambda t_j) \qquad \text{and} \qquad \sum_{j=0}^{2m-1} F(t_j)\sin(\lambda t_j)$$

where the points $t_j$ are equally spaced in $[-a, a]$. Thus sums of this form occur frequently in important applications other than interpolation and efficient techniques such as the *fast Fourier transform* have been developed to compute them.

### *6.6.2 The Fast Fourier Transform

The fast Fourier transform (FFT) is an efficient procedure for organizing the computation of the $a_j$'s and $b_j$'s (and thus $p_k(\theta)$) in (6.60) or (6.61). As an example, suppose we wish to compute the $2m$ coefficients $a_0, a_1, \ldots, a_m, b_1, \ldots, b_{m-1}$ that are needed to calculate the interpolating polynomial in (6.61). We can verify that the direct calculation of the coefficients takes approximately $(2m)^2$ multiplications and $(2m)^2$ additions. By efficient programming, the fast Fourier transform can reduce the number of operations to approximately $2m \ln(2m)$ operations. For small values of $m$ this change is perhaps not significant, but for large values of $m$ (say, $m = 10^4$ or more) this amount is tremendously significant (see Problem 8), and reduces formerly impractical computations to ones that are fairly easily handled.

It is most convenient to define $C_j \equiv a_j + ib_j$, $(i \equiv \sqrt{-1})$. Using Euler's formula, $e^{ij\theta} = \cos(j\theta) + i\sin(j\theta)$, we have

$$C_j = a_j + ib_j = \frac{1}{m}\sum_{r=0}^{2m-1} f(\theta_r)e^{ij\theta_r}. \tag{6.63}$$

Now let $M$ and $K$ be any two integers such that $2m = MK$ where $K$ is even. Next, set

$$j = j_1 M + j_0, \qquad 0 \le j_0 < M,$$

and

$$r = r_1 K + r_0, \qquad 0 \le r_0 < K.$$

We leave to the reader (Problem 9) to verify that any sum of the form, $\sum_{r=0}^{2m-1} v_r$ can be expressed as the double sum $\sum_{r_0=0}^{K-1}\sum_{r_1=0}^{M-1} v_{r_1K+r_0}$. Using this fact we can write (with $\exp(x) \equiv e^x$),

$$C_j = \frac{1}{m}\sum_{r_0=0}^{K-1}\sum_{r_1=0}^{M-1} f(\theta_{r_1K+r_0})\exp[i(j_1M + j_0)(r_1K + r_0)2\pi/MK]$$

$$= \frac{1}{m}\sum_{r_0=0}^{K-1}\sum_{r_1=0}^{M-1} f(\theta_{r_1K+r_0})\exp[2\pi i(j_1r_1 + j_1r_0/K + j_0r_1/M + j_0r_0/MK)]$$

$$= \frac{1}{m}\sum_{r_0=0}^{K-1}\exp\left[2\pi i\left(j_1 + \frac{j_0}{M}\right)\frac{r_0}{K}\right]\sum_{r_1=0}^{M-1} f(\theta_{r_1K+r_0})\exp\left[\frac{2\pi i j_0 r_1}{M}\right]. \tag{6.64}$$

In the inner sum of the last expression we used the reduction

$$\exp\left[2\pi i\left(j_1 r_1 + \frac{j_0 r_1}{M}\right)\right] = \exp[2\pi i j_1 r_1]\exp\left[\frac{2\pi i j_0 r_1}{M}\right]$$
$$= \exp\left[\frac{2\pi i j_0 r_1}{M}\right],$$

which is valid since $j_1$ and $r_1$ are integers (and so $\exp[2\pi i j_1 r_1] \equiv 1$). Thus the inner sum is independent of $j_1$. This is what makes the FFT "fast," that is to say, the same inner sum is used to compute each $C_j$ of the form $C_j = C_{j_1 M + j_0}$ for any *fixed value* of $j_0$. For example, if $j_0 = 3$, then the same inner sum is used to compute $C_3, C_{M+3}, C_{2M+3}, \ldots$.

We now define

$$C(j_0, r_0) \equiv \frac{1}{m}\sum_{r_1=0}^{M-1} f(\theta_{r_1 K + r_0})\exp\left[\frac{2\pi i j_0 r_1}{M}\right], \tag{6.65}$$

and so (6.64) becomes (for $j = j_1 M + j_0$)

$$C_j = \sum_{r_0=0}^{K-1} C(j_0, r_0)\exp\left[2\pi i\left(j_1 + \frac{j_0}{M}\right)\frac{r_0}{K}\right]. \tag{6.66}$$

Thus far we see that we must: (1) Form a table of $C(j_0, r_0)$ for $0 \le j_0 < M$, $0 \le r_0 < K$. (2) Compute each $C_j$ according to formula (6.66). To be more precise, let us now investigate the number of multiplications required in the above scheme:

a) Each entry $C(j_0, r_0)$ of the table requires $M$ multiplications. The table has $MK$ entries, so step (1) requires $M(MK)$ multiplications.
b) With the table constructed, it takes $K$ multiplications to compute each $C_j$. There are $m = MK/2$ coefficients $C_j$, so step (2) requires less than $K(MK/2)$ multiplications. (In fact, calculating $C_0$ and $C_m$ requires no multiplications, so we need only $K(MK/2 - 1)$ multiplications. However, $K(MK/2)$ is an easier bound to deal with, as we shall see. Additionally, the multiplications in (6.65) and (6.66) are "complex" multiplications due to the presence of the complex exponential.)

The total number of multiplications in (a) and (b) is then less than $M^2K + MK^2/2 = MK(M + K/2)$. The straightforward evaluation of $C_j$ from (6.63) requires $2m$ multiplications for each $j$, or a total of $2m(m) = MK(MK/2) = M^2K^2/2$. For example, if $m = 64$ and we select $M = 8$ and $K = 16$ ($2m = 128 = MK = 8(16)$), then FFT requires $MK(M + K/2) = 8(16)(8 + 16/2) = 128(16)$ multiplications, and straightforward computation requires $2m(m) = 128(64)$ multiplications, which is larger than FFT by a factor of 4.

Thus far, however, we have not utilized the FFT scheme to its full potential. For example we note that formula (6.65) for each $C(j_0, r_0)$ looks very much like formula (6.63) for each $C_j$, where in (6.65), $M$ and $j_0$ play the roles of $2m$ and $j$, respectively, in (6.63). We may then correctly reason that we can use the FFT scheme to construct the required table of $C(j_0, r_0)$'s. More precisely, fix $r_0$ and consider construction of the $r_0$ column of the table—$C(0, r_0)$, $C(1, r_0), \ldots,$ $C(M - 1, r_0)$. Thus if $M = 2s$, then for $0 \leq k \leq M - 1$,

$$\begin{aligned} C(k, r_0) &= \frac{1}{m} \sum_{r_1=0}^{M-1} f(\theta_{r_1K+r_0}) \exp\left[\frac{2\pi i k r_1}{M}\right] \\ &= \frac{1}{s} \sum_{r_1=0}^{2s-1} \frac{1}{K} f(\theta_{r_1K+r_0}) \exp\left[\frac{\pi i k r_1}{s}\right] \end{aligned} \tag{6.67}$$

but this is precisely the form of (6.63) with the role of $f(\theta_r)$ being played by $g(\theta_{r_1}) \equiv (1/K)f(\theta_{r_1K+r_0})$.

Once again we let $u$ and $v$ be two integers such that $2s = u \cdot v$ and repeat the same process on each $C(k, r_0)$, $0 \leq k \leq M - 1$, instead of on each $C_j$, $1 \leq j \leq m$. We shall not write this out again in detail except to note that with the above modification of indices the same program may be used. Counting multiplications as outlined in (a) and (b) yields $u(uv) + v(uv) = u^2v + uv^2$ multiplications to compute $C(0, r_0)$, $C(1, r_0), \ldots, C(M - 1, r_0)$. (Note that we have a factor in (b) of $v(uv)$ instead of $K(MK/2)$ in the previous calculation. This is because there are $2s = uv$ quantities in the latter versus $m = (MK/2)$ quantities in the former.) Thus each column of the table requires $u^2v + uv^2 = uv(u + v)$ multiplications or a total of $Kuv(u + v)$ for the entire table since there are $K$ rows. Now having constructed the table, (b) still tells us we need about $K(MK/2)$ more multiplications or a total of $Kuv(u + v) + K(MK/2) = Kuv(u + v) + K^2uv/2 = Kuv(u + v + K/2)$.

If, in the previous step, $u$ can be written in the factored form $u = 2\sigma = \mu v$, we can again use the FFT scheme to economize our computation. In general we find that if $2m = h_1h_2 \cdots h_k$; then the multiplications necessary are $h_1h_2 \cdots h_k \times (h_1 + h_2 + \cdots + h_{k/2})$. A typical choice is $2m = 2^k$, which yields $2^k(2 + 2 + \cdots + 2 + 1) = 2^k(2k - 1)$ multiplications. Straightforward calculation requires $(2m)(m) = (2^k)(2^{k-1})$ multiplications. For example, let $m = 1024 = 2^{10}$ (which is not an exceptionally large number of data points). Then $k = 11$, $2^k(2k - 1) = 43{,}008$ and $2^k(2^{k-1}) = 2{,}097{,}152$. The FFT has a factor of approximately 49 times fewer multiplications.

## PROBLEMS

1. Verify the identities in (6.54).
2. Using (6.56b), find $P_n^*(\theta)$ for $f(\theta) = \theta(\theta - 2\pi)$, for $n = 2, 3$, and 4. What is $\|f - P_n^*\|$?
3. Verify that $T_{2m}(\cos(k\theta)) = \int_0^{2\pi} \cos(k\theta)\, d\theta$, $0 \leq k \leq 2m - 1$. (See page 277.)

4. Verify that $T_{2m}(\sin(k\theta)) = \int_0^{2\pi} \sin(k\theta)\, d\theta$, $1 \le k \le 2m - 1$. (See page 277.)
5. Verify that the functions in (6.58) are orthonormal with respect to the inner-product in (6.59).
6. Show that $p_k(\theta)$ defined in (6.60) minimizes $\sum_{j=0}^{2m-1} [y_j - p(\theta_j)]^2$ among all $p(\theta) \in \mathscr{T}_n$.
7. Show that $1/m \sum_{s=0}^{\prime\prime\, m} \cos(s(r - j)\pi/m) = \delta_{rj}$.
8. Compare the numbers $10^4 10^4$ and $10^4 \ln(10^4)$.
9. Show that a sum of the form $\sum_{r=0}^{2m-1} v_r$ can be rewritten as $\sum_{r_0=0}^{K-1} \sum_{r_1=0}^{M-1} v_{r_1 K + r_0}$, when $2m = MK$ and $r = r_1 K + r_0$, $0 \le r_0 < K$.

*10. As an example, let $2m = 128 = MK = (8)(16)$ and sketch out the details of how the tables can be constructed for the FFT. Write a program to implement the FFT, testing your routine by calculating numerically the coefficients $A_j$ and $B_j$ for $f(\theta) = \theta(\theta - 2\pi)$, $0 \le j \le 64$.

# 7

# NUMERICAL SOLUTION OF ORDINARY DIFFERENTIAL EQUATIONS

## 7.1 INTRODUCTION

As the reader is probably aware, differential equations serve as *mathematical descriptions* for many physical problems and phenomena. A few examples of commonly occurring differential equations are the following:

$$\text{a)}\ ay''(x) + by'(x) + cy(x) = F(x),$$

$$\text{b)}\ EIy^{(iv)}(x) = w(x),$$

$$\text{c)}\ y''(x) = \frac{w}{H}\sqrt{1 + (y'(x))^2},$$

and

$$\text{d)}\ \frac{d}{dx}\left(p(x)\frac{dy}{dx}\right) + q(x)y = r(x).$$

Equation (a) occurs in the study of vibrating or oscillating mechanical systems or in electrical circuits while (b) arises in the study of beam deflections, (c) deals with problems in cable suspension, and (d) arises when a separation of variables technique can be applied to partial differential equations describing important problems in mathematical physics and engineering. Finding accurate and efficient procedures for solving differential equations has long been a problem of importance. However, in many practical situations, an analytical solution (that is, a closed-form mathematical solution) is either impossible to find or extremely difficult to evaluate. In recent years, numerical procedures for approximating solutions have become increasingly popular, thanks in large part to the power of modern, high-speed computers. The basic ideas involved in deriving numerical methods for differential equations are relatively simple and do not require an extensive theoretical background (although a detailed analysis of the convergence of a method may be quite difficult). In this section, we will present most of the fundamentals and terminology necessary for an understanding of many of the more popular methods now in use.

The first problem we consider is the *initial-value problem*:

$$y'(x) = f(x, y(x)), \qquad a \le x \le b \tag{7.1a}$$

$$y(a) = y_0. \tag{7.1b}$$

In this problem, we are given a finite interval $[a, b]$ and a function $f(x, y)$. We seek a function $y(x)$ defined on $[a, b]$, such that $y'(x) \equiv f(x, y(x))$ for $a \le x \le b$ *and* such that $y(x)$ satisfies the initial condition $y(a) = y_0$, where $y_0$ is some prescribed value. Thus an initial-value problem consists of two parts: (7.1a) the differential equation $y' = f(x, y)$ which gives the desired relationship between $y(x)$ and $y'(x)$, and (7.1b) the initial condition $y(a) = y_0$. The equation $y' = f(x, y)$ is called a *first-order* differential equation since the highest derivative that appears is the first derivative. In a later section, we will show how the methods developed for solving the first-order equation (7.1) can be applied to simultaneously solve *systems* of first-order differential equations. Further, we will show how to convert an $n$th-order differential equation into a system of first-order equations. Thus methods for handling the apparently simple case considered in (7.1) are, in fact, applicable to systems of higher-order equations. Therefore we will concentrate almost exclusively on numerical procedures for the initial-value problem (7.1).

**Example 7.1.** Some examples of initial-value problems are:

$$\begin{aligned} y' &= y, \qquad 0 \le x \le 1 \\ y(0) &= 1 \end{aligned} \tag{7.2a}$$

$$\begin{aligned} y' &= 2xy, \qquad 0 \le x \le 1 \\ y(0) &= 1 \end{aligned} \tag{7.2b}$$

$$\begin{aligned} y' &= y^2(x - 6)/x^3, \qquad 1 \le x \le 2 \\ y(1) &= -\tfrac{1}{2} \end{aligned} \tag{7.2c}$$

$$\begin{aligned} y'' &= (z^2 - 1)/y, \qquad 0 \le x \le 1 \\ z'' &= -z \end{aligned} \tag{7.2d}$$

$$y(0) = 1, \qquad y'(0) = 0, \qquad z(0) = 0, \qquad z'(0) = 1.$$

The differential equations in (7.2a) and (7.2b) are first-order *linear* equations (since $y$ and $y'$ occur linearly). The differential equation in (7.2c) is a first-order *nonlinear* equation because of the presence of $y^2$. Finally, (7.2d) provides an example of a second-order system of nonlinear differential equations. The solutions are given in Problem 1. A more realistic example of a nonlinear second-order system is provided by (4.0a), the equations of motion for a ballistic reentry trajectory. This equation cannot be solved in closed form and must be solved numerically.

As examples in this chapter, we will frequently use differential equations like (7.2a) and (7.2b). These equations are quite simple and not at all representative of the sorts of differential equations that occur in practice. Our purpose in using these equations as examples is merely to clarify the presentation. We return now to the simple initial-value problem $y' = y$ with $y(0) = 1$, as in Example 7.1. The differential equation $y' = y$ has many solutions, all of the form $y(x) = ce^x$ where $c$ can be any constant (it is easy to verify that the relationship $y' = y$ holds when $y = ce^x$). So the set of functions $y = ce^x$ forms a one-parameter family of solutions for $y' = y$, with each member of the family determined by the value chosen for $c$. Although $y = ce^x$ solves $y' = y$ regardless of the value of $c$, there is only one choice of $c$ for which $y(0) = 1$, namely $c = 1$. Thus the solution of the initial-value problem is $y = e^x$.

For the general case, $y' = f(x, y)$ with $y(a) = y_0$, we call each function $y(x)$ that satisfies $y' = f(x, y)$, an *integral curve*. Thus solving an initial-value problem is a two-stage process: (i) determine the integral curves for $y' = f(x, y)$; (ii) find the member of this family of integral curves that satisfies $y(a) = y_0$. In the above example, we note that $y' = y$, $y(0) = y_0$ can always be solved no matter what value $y_0$ has (in fact, the solution is $y = y_0e^x$). Geometrically, this says that given any point in the $xy$-plane of the form $(0, y_0)$, there is a solution of $y' = y$ passing through this point (see Fig. 7.1).

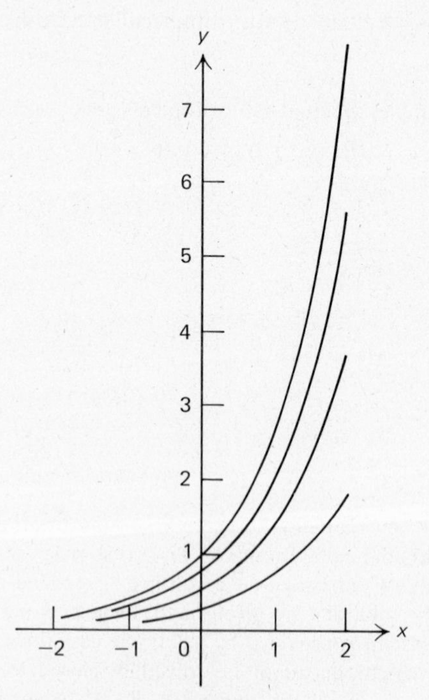


**Fig. 7.1** Integral curves for $y' = y$

The idea of a *flow field* or *direction field* is related to the geometry of Fig. 7.1. Given the differential equation $y' = f(x, y)$, we observe that the function $f(x, y)$ gives the slope of the integral curve at the point $(x, y)$. We can represent this information pictorially, as in Fig. 7.2, which shows the flow field for $y' = y$. A little imagination shows that we can use the flow field to get a rough idea of the shapes of the integral curves by drawing curves through the flow field, where the slopes of these curves are dictated by Fig. 7.2. If two integral curves $y_1(x)$ and $y_2(x)$ are near to each other at $x = x_1$, their slopes $y_1'(x_1)$ and $y_2'(x_1)$ will determine whether they remain near each other as $x$ moves away from $x_1$. Thus given nearby points $(x_1, y_1)$ and $(x_1, y_2)$ in the $xy$-plane, if $f(x_1, y_1)$ is substantially different from $f(x_1, y_2)$, then the integral curves through $(x_1, y_1)$ and $(x_1, y_2)$ are diverging from each other (at least for $x$ near $x_1$). The simple geometric observations provided in Figs. 7.1 and 7.2 will be of value to us in understanding the numerical methods presented later.

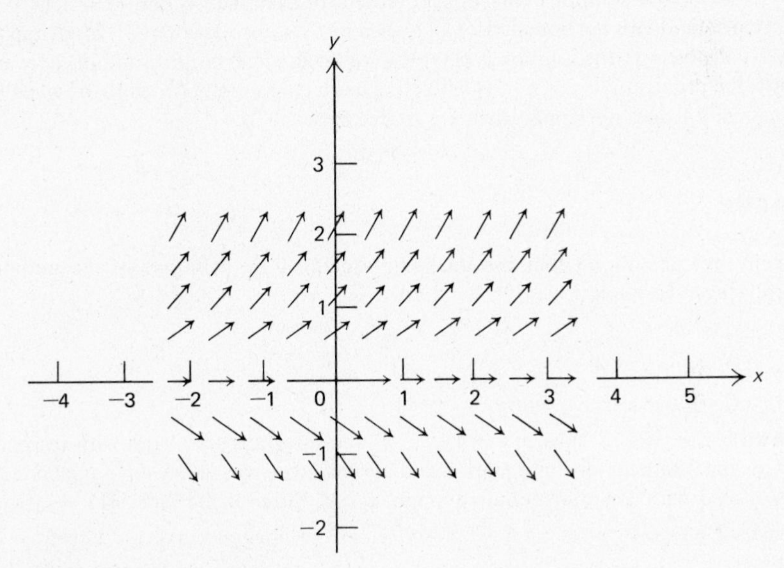


**Fig. 7.2** The flow field for $y' = y$

We conclude this section with a well-known existence and uniqueness theorem for the initial-value problem (7.1). For a proof, we refer the reader to Coddington and Levinson (1955).

**Theorem 7.1.** Let $f(x, y)$ be continuous for $a \le x \le b$ and for all $y$. Suppose further that $f(x, y)$ satisfies the Lipschitz condition

$$|f(x, u) - f(x, v)| \le L|u - v| \qquad \textbf{(7.3)}$$

for $a \leq x \leq b$ and all $u$ and $v$, where $L$ is a constant. Then, for any number $y_0$, there exists a unique differentiable function $y(x)$ defined on $[a, b]$, such that $y' = f(x, y)$ and $y(a) = y_0$.

From the mean-value theorem, it is easy to see that $f(x, y)$ satisfies the Lipschitz condition (7.3) if the partial derivative, $f_y(x, y)$, is bounded in absolute value for $x$ in $[a, b]$ and all $y$. To see this, suppose $|f_y(x, y)| \leq L$, then

$$f(x, u) - f(x, v) = f_y(x, \theta)(u - v)$$

where $\theta$ is between $u$ and $v$. Hence if $|f_y(x, y)| \leq L$ for $a \leq x \leq b$, then $|f(x, u) - f(x, v)| \leq L|u - v|$.

**Example 7.2.** For the problem $y' = y$, $y(a) = y_0$, we have $f(x, y) = y$ and hence $|f(x, u) - f(x, v)| \leq |u - v|$. Therefore solutions exist uniquely on any interval $[a, b]$. For the problem $y' = 2xy$, $y(0) = y_0$, we have $f_y(x, y) = 2x$. Thus on any interval $[0, b]$, $|f(x, u) - f(x, v)| \leq 2b|u - v|$, so a unique solution exists. For problem (c) of Example 7.1, $f_y(x, y) = 2y(x - 6)/x^3$. This partial derivative is not bounded for all $y$ when $1 \leq x \leq b$, so Theorem 7.1 is not applicable. In fact (see Problem 1), the solution is $y(x) = x^2/(x - 3)$, which becomes infinite as $x$ nears 3. Similarly, the problem $y' = y/(x - 1)$, $y(0) = y_0$ has a unique solution on $[0, b]$ when $b < 1$, but Theorem 7.1 does not apply when $1 \leq b$ (see Problem 2).

## PROBLEMS

1. Verify by substitution that the following functions are solutions of the initial-value problems in Example 7.1:
   b) $y(x) = e^{x^2}$
   c) $y(x) = x^2/(x - 3)$
   d) $y(x) = \cos(x)$, $z(x) = \sin(x)$
2. Rewrite $y' = y/(x - 1)$ as $y'/y = 1/(x - 1)$ and integrate both sides with respect to $x$. Draw the flow field for this equation and conclude you can satisfy any initial conditions associated with this differential equation, except those of the type $y(1) = y_0$, $y_0 \neq 0$.
3. Find all integral curves for $y' = x + 1$. Find the integral curve for $y' = x + 1$ that satisfies
   a) $y(0) = 0$,
   b) $y(0) = 1$,
   c) $y(1) = 5$.
4. Determine the Lipschitz constant for
   a) $f(x, y) = x^2 y, \quad -1 \leq x \leq 1;$
   b) $f(x, y) = xe^{-y^2}, \quad -1 \leq x \leq 1;$
   c) $f(x, y) = \dfrac{(x^4 - x^2 - 2)^{33}}{(x^2 - 5)^6}, \quad -1 \leq x \leq 1.$

5. Derive the closed form solution to the Allen and Eggers reentry model, (4.0b), as follows.
   a) Suppose the position vector $\mathbf{R}(t)$ is given by $\mathbf{R}(t) = [x_1(t), x_2(t), x_3(t)]^T$. Write (4.0b) out componentwise, obtaining the three equations

$$\ddot{x}_i(t) = -Ke^{-\lambda H}|\dot{\mathbf{R}}|\dot{x}_i(t), \qquad i = 1, 2, 3.$$

   b) Since $\mathbf{R}(t)$ is a vector from the center of the earth to the point-mass, we see that $\mathbf{R}(t)$ is perpendicular to the local horizon. Show that the dot product $\mathbf{R}(t) \cdot \dot{\mathbf{R}}(t)$ is given by $\mathbf{R}(t) \cdot \dot{\mathbf{R}}(t) = -|\mathbf{R}(t)|\,|\dot{\mathbf{R}}(t)| \sin(\alpha)$. Note that $\alpha$ is a constant since the motion is in a straight line.
   c) If $R_e$ is the radius of the earth and if $H$ denotes the altitude of the point-mass above the earth's surface, then $|\mathbf{R}(t)| = R_e + H(t)$. From this, show that

$$\frac{dH}{dt} = \frac{\mathbf{R}(t) \cdot \dot{\mathbf{R}}(t)}{|\mathbf{R}(t)|}$$

   d) Using (a), (b), and (c), derive the closed form expression for $\dot{\mathbf{R}}(t)$ given in Example 4.0.
6. One of the early iterative methods designed for solving $y' = f(x, y)$, $y(a) = y_0$, was Picard's method. In this method we let $y^{(0)}(x) \equiv y_0$, and since $\int_a^x y'(t)\,dt = y(x) - y(a) = \int_a^x f(t, y(t))\,dt$, we define a sequence of functions, $\{y^{(k)}(x)\}_{k=1}^{\infty}$, by

$$y^{(k+1)}(x) = y(a) + \int_a^x f(t, y^{(k)}(t))\,dt, \qquad k \geq 0.$$

   a) Apply Picard's method to the first two parts of Example 7.1 and to the problem, $y' = \sin(x + y)$, $y(0) = 0$. (After this work, the reader should quickly see the drawbacks of this method—repeated indefinite integrations, slowness of convergence, etc. Hence Picard's method is seldom used for practical computation, but it is used extensively in developing the theory of differential equations.)
   *b) The reader should note that Picard's method is actually a special case of *fixed-point* iteration as in Section 2.4, and Section 4.3. From this background the reader should be able to find sufficient conditions to insure the convergence of Picard's method. (Hint: Consider Theorem 7.1.)

## 7.2 TAYLOR'S SERIES METHODS

Taylor's series methods will serve to demonstrate some of the basic ideas used in developing numerical methods for ordinary differential equations. Given the initial-value problem (7.1), we seek an approximation to the solution $y(x)$. This approximation might take the form of a polynomial, as in the method of collocation (see Section 7.7). Normally, however, the approximation to $y(x)$ will be a discrete set of values $y_1, y_2, \ldots, y_n$ which approximate $y(x_1), y(x_2), \ldots, y(x_n)$ with $a < x_1 < x_2 < \cdots < x_n \leq b$. Then, since $y_i \approx y(x_i)$, the set of points $(x_i, y_i)$ could be plotted to provide an approximation to $y(x)$. In some applications, only an approximation to $y(b)$ is wanted. In this case, with $x_n = b$, the numbers $y_1, y_2, \ldots, y_{n-1}$ become just intermediate steps.

Taylor's series methods are a special case of a more general class of methods called *one-step* methods. The general one-step method is an iteration that takes the form

$$y_{i+1} = y_i + h\Phi(x_i, y_i; h), \qquad i = 0, 1, \ldots, n - 1.$$

The function $\Phi(x_i, y_i; h)$ is called the *increment function* and tells us how to proceed from an estimate $y_i$ for $y(x_i)$ to an estimate $y_{i+1}$ for $y(x_{i+1})$. Essentially, the only one-step methods we will consider are Taylor's series methods and the closely related Runge-Kutta methods. The interested reader is referred to Henrici (1962) for a more detailed treatment of one-step methods. One notable advantage of any one-step method is our ability to easily change the size of the step-size $h$ as we proceed from $a$ to $b$. For example, if an error analysis (see Section 7.2.3) indicates that $y_{i+1}$ may not be a good approximation for $y(x_{i+1})$, then we can easily replace $h$ in the above formula by a smaller value, say $h/2$, and repeat the calculation and the estimation of its accuracy.

### 7.2.1 Euler's Method

The simplest Taylor's series method is Euler's method, which is suggested by Figs. 7.1 and 7.2. Euler's method is based on the fact that if we know the value of the solution $y(x)$ at a point $x = x_0$, then we also know the slope of the integral curve there, since $y'(x_0) = f(x_0, y(x_0))$. It seems reasonable that we might proceed on a line of slope $y'(x_0)$ from the point $(x_0, y(x_0))$ and still be close to $y(x)$, at least for $x$ near $x_0$. For small $h$, this expectation leads us to approximate $y(x_0 + h)$ by $y(x_0 + h) \approx y(x_0) + hy'(x_0)$. To formalize these ideas, we note from (7.1) that we do know one point on the solution curve, since we are given $y(a) = y_0$. As a next step, we choose a value $n$, let $h = (b - a)/n$ and set $x_i = x_0 + ih$, where $x_0 = a$. Since $y(a) = y(x_0) = y_0$, we have an approximation $y_1 = y_0 + hf(x_0, y_0)$ to $y(x_1)$. Unfortunately, the point $(x_1, y_1)$ is probably not on the solution curve $y(x)$ (see Fig. 7.3). If $f(x, y)$ is well behaved and if $y_1$ is near $y(x_1)$, we can hope that $f(x_1, y_1) \approx f(x_1, y(x_1))$. Thus we next proceed along a line from $(x_1, y_1)$ of slope $f(x_1, y_1)$, finding $y_2 = y_1 + hf(x_1, y_1)$ as an approximation to $y(x_2)$. Another way of looking at this is to realize that the point $(x_1, y_1)$ is probably on an integral curve, $y_1(x)$, that is near the solution curve $y(x)$. In other words, we have a point $(x_1, y_1)$ on the curve $y_1(x)$, where $y_1(x)$ is the solution of the initial-value problem $u' = f(x, u)$, $u(x_1) = y_1$. We have no information as to how to return from this integral curve to the solution curve $y(x)$ (the integral curve that also satisfies $y(a) = y_0$). Therefore, the best we can hope to do is to try to remain on this new integral curve, $y_1(x)$. We thus repeat the same process and obtain $y_2 = y_1 + hf(x_1, y_1)$ as our approximation for $y(x_2)$ at $x = x_2$. We note that $y_2$ is an approximation for $y_1(x_2)$ which in turn is an approximation for $y(x_2)$, or $y_2$ is a "two-stage" approximation for $y(x_2)$. Continuing this process, we have Euler's method which is defined by the iteration

$$y_{i+1} = y_i + hf(x_i, y_i), \qquad i = 0, 1, \ldots, n - 1. \tag{7.4}$$

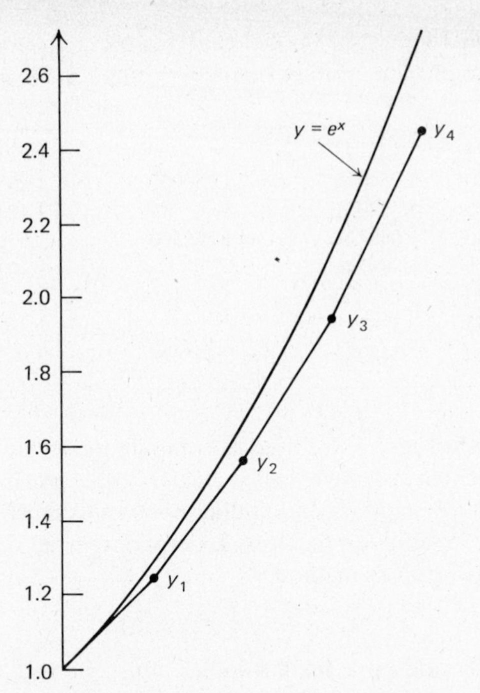


**Fig. 7.3** Euler's method with $h = 0.25$ for $y' = y$, $y(0) = y_0$

By the same reasoning as above, we see that $y_i$, $1 \leq i \leq n$, is an $i$th stage approximation for $y(x_i)$. This analysis is also true for any one-step method and it emphasizes how the error can propagate when $i$ is very large. We should note that any mathematical error analysis will have to include the extreme case where the integral curves get successively further and further from the solution curve. (See Fig. 7.3.) The rather unsophisticated mathematical error bounds we will derive are usually overly pessimistic, but a realistic appraisal of the error at each step is difficult to obtain. Figure 7.3 shows the first four steps of Euler's method for $h = 0.25$. From either Fig. 7.1 or Fig. 7.2, we note for this particular differential equation, that as $h$ decreases we expect our estimates of the solution to increase but remain below the true solution (see Problem 2). Also, for this particular differential equation, it is easy to show directly that the numerical estimates converge to the solution $y(x) = e^x$ as $h \to 0$, provided there are *no* round-off errors (see Problem 3).

**Example 7.3.** Consider the initial-value problem $y' = y$, $y(0) = 1$, which has $y(x) = e^x$ as a solution. In Table 7.1 we have tabulated the results of using Euler's method on $[0, 1]$, with $h = \frac{1}{4} = 0.25$ and $h = \frac{1}{8} = 0.125$. For this differential equation, the iteration, (7.4), reduces to simply $y_{i+1} = (1 + h)y_i$, with $0 \leq i \leq 3$ for $h = 0.25$ and $0 \leq i \leq 7$ for $h = 0.125$ (we start the iteration with $y_0 = 1$ in both cases).

**Table 7.1**

| $x$ | $e^x$ | $h = 0.25$ | $h = 0.125$ |
|---|---|---|---|
| 0.125 | 1.133148 | | $y_1 = 1.125000$ |
| 0.250 | 1.284025 | $y_1 = 1.250000$ | $y_2 = 1.265625$ |
| 0.375 | 1.454991 | | $y_3 = 1.423828$ |
| 0.500 | 1.648721 | $y_2 = 1.562500$ | $y_4 = 1.601807$ |
| 0.625 | 1.868246 | | $y_5 = 1.802032$ |
| 0.750 | 2.117000 | $y_3 = 1.953125$ | $y_6 = 2.027287$ |
| 0.875 | 2.398875 | | $y_7 = 2.280697$ |
| 1.000 | 2.718282 | $y_4 = 2.441406$ | $y_8 = 2.565784$ |

The subroutine in Fig. 7.4 was used to compute the results given in Table 7.1. The subroutine executes in a mode called *partial double precision*, which provides more accuracy than single precision but does not cost as much in terms of storage and execution time as double precision. To explain partial double precision, we note from the form of Euler's method

$$y_{i+1} = y_i + hf(x_i, y_i)$$

that we are typically adding a small number, $hf(x_i, y_i)$, to a number, $y_i$, that is usually large relative to $hf(x_i, y_i)$. In order to preserve accuracy, we make this *one*

```
      SUBROUTINE EULER(Y,X,H,N)
      DOUBLE PRECISION Y
C
C  THIS SUBROUTINE USES EULER'S METHOD WITH PARTIAL DOUBLE PRECISION
C  TO SOLVE THE INITIAL-VALUE PROBLEM Y'(X)=F(X,Y), WITH INITIAL
C  CONDITION Y(X0)=Y0.  A SINGLE PRECISION FUNCTION SUBPROGRAM F(X,Y)
C  MUST BE PROVIDED.  THE MAIN PROGRAM MUST SUPPLY A DOUBLE PRECISION
C  INITIAL-VALUE Y=Y0, A STARTING VALUE X=X0 AND STEP-SIZE H IN
C  SINGLE PRECISION, AND AN INTEGER N=NUMBER OF STEPS TO EXECUTE.
C
      ISTEP=0
C
C  PRINT INITIAL CONDITIONS
C
      PRINT 100,ISTEP,X,Y
C
C  EXECUTE EULER'S METHOD
C
      DO 1 ISTEP=1,N
      SY=Y
      Y=Y+H*F(X,SY)
      X=X+H
    1 PRINT 100,ISTEP,X,Y
  100 FORMAT(1H ,I3,2E20.7)
      RETURN
      END
```

**Figure 7.4**

addition, $y_i + hf(x_i, y_i)$, in double precision and store the result, $y_{i+1}$, as a double precision number ($y_i$ is also a double precision number of course). The function $f(x_i, y_i)$ is calculated in single precision and in this subroutine, $y_i$ is passed to $f$ in single precision (by the statement $SY = Y$). Thus, since the bulk of the computation is normally done in evaluating $f(x, y)$, we have not appreciably increased storage or execution, but we have lessened the effects of a potential source of rounding error. By way of example, we ran the subroutine with $N = 101$ and $H = 0.01$ both in single precision and partial double precision. In single precision, we have $y_{100} = 2.704736$ and in partial double precision we had $y_{100} = 2.704813$, where $y_{100} \approx y(1)$. Using a smaller step-size $h$ would result in even more of a difference. Partial double precision is usually recommended for all the one-step methods.

### 7.2.2 Taylor's Series Methods of Order *k*

In the last section, we saw that Euler's method takes the form $y_{i+1} = y_i + hf(x_i, y_i)$ when applied to the initial-value problem $y' = f(x, y)$, $y(x_0) = y_0$. For $i = 0$, we have $y_1 = y_0 + hf(x_0, y_0) = y(x_0) + y'(x_0)h$, since $y'(x_0) = f(x_0, y(x_0))$. Thus another way to view Euler's method is to think of $y_1$ as being the result of neglecting the higher-order terms in a Taylor's series expansion for $y(x_1)$. That is, since $x_1 = x_0 + h$, we have the expansion

$$y(x_1) = y(x_0) + y'(x_0)h + y''(x_0)\frac{h^2}{2!} + \cdots$$

$$+ y^{(k)}(x_0)\frac{h^k}{k!} + y^{(k+1)}(\theta)\frac{h^{k+1}}{(k+1)!}, \tag{7.5}$$

when the solution $y(x)$ is sufficiently differentiable. Setting $k = 1$ in (7.5) and neglecting higher-order terms, we have $y_1 = y(x_0) + hy'(x_0)$, which is the Euler's method estimate to $y(x_1)$. If we knew some of the higher derivatives of $y(x)$ evaluated at $x = x_0$, then we suspect we could use (7.5) to find a better estimate to $y(x_1)$ by carrying more terms (i.e., by choosing $k > 1$).

In fact, the differential equation $y' = f(x, y)$ does provide us with as many derivatives, $y''(x_0), y'''(x_0), \ldots$ as we care to calculate. To see this, we note that if $y'(x) = f(x, y(x))$, then by taking the total derivative, we have

$$y''(x) = \frac{d}{dx}[f(x, y(x))] = f_x(x, y) + f_y(x, y)\frac{dy}{dx}. \tag{7.6}$$

Simplifying (7.6) by using $y' = f(x, y)$, we have $y''(x) = f_x(x, y) + f_y(x, y)f(x, y)$. Thus, $y''(x_0) = f_x(x_0, y_0) + f_y(x_0, y_0)f(x_0, y_0)$ and we have the (presumably) more accurate estimate $y_1$ to $y(x_1)$

$$y_1 = y_0 + hf(x_0, y_0) + \frac{h^2}{2}[f_x(x_0, y_0) + f_y(x_0, y_0)f(x_0, y_0)].$$

Continuing, we next find that

$$y'''(x) = f_{xx}(x, y) + 2f_{xy}(x, y)f(x, y) + f_{yy}(x, y)f^2(x, y) + f_y(x, y)f_x(x, y) + f_y^2(x, y)f(x, y).$$

Thus we see that in principle we can compute $y^{(i)}(x)$ (if it exists) for as many values $i$ as we choose, which means that (7.5) can be used to estimate $y(x_1)$. In general, if we knew $y(x_i)$, we could use (7.5) with $x_i$ replacing $x_0$ to estimate $y(x_{i+1})$.

In fact, unless $y(x)$ is a polynomial, we can never hope that a truncated Taylor's series will provide more than just a good estimate, $y_{i+1}$, to $y(x_{i+1})$. Nevertheless, if $y_i$ is a good approximation to $y(x_i)$, then we would expect

$$\begin{aligned} y'(x_i) &\approx f(x_i, y_i) \\ y''(x_i) &\approx f_x(x_i, y_i) + f_y(x_i, y_i)f(x_i, y_i) \\ y'''(x_i) &\approx f_{xx}(x_i, y_i) + 2f_{xy}(x_i, y_i)f(x_i, y_i) + f_{yy}(x_i, y_i)f^2(x_i, y_i) \\ &\quad + f_y(x_i, y_i)f_x(x_i, y_i) + f_y^2(x_i, y_i)f(x_i, y_i). \end{aligned}$$

To exploit these ideas, let $f^{(j)}(x, y)$ denote the $j$th total derivative of $f(x, y)$ with respect to $x$ (that is, $f^{(j)}(x, y) = y^{(j+1)}(x)$), so that we have

$$y(x_{i+1}) = y(x_i) + hT_k(x_i, y(x_i); h) + y^{(k+1)}(\theta)\frac{h^{k+1}}{(k+1)!} \tag{7.7}$$

where

$$T_k(x, y; h) = f(x, y) + f'(x, y)\frac{h}{2!} + \cdots + f^{(k-1)}(x, y)\frac{h^{k-1}}{k!}.$$

We are then led to the iteration

$$y_{i+1} = y_i + hT_k(x_i, y_i; h), \qquad i = 0, 1, \ldots, n-1. \tag{7.8}$$

The method described by (7.8) is called a *kth-order Taylor's series method.* Note that (7.8) has the form of a one-step method as defined in Section 7.2. For $k = 1$, formula (7.8) reduces to Euler's method.

For values of $k$ larger than 3, it is clear that a major effort is required to compute the necessary derivatives of $f(x, y)$. Moreover, in many practical problems, it may be impossible or extremely costly to find and evaluate the necessary partial derivatives of $f(x, y)$. This is one reason that higher-order Taylor's series methods are not frequently used. As we shall see shortly, Runge-Kutta methods provide a good compromise, replacing suitably chosen combinations of evaluations of $f(x, y)$ for derivatives of $f(x, y)$. However, as demonstrated in the next example, the problem of obtaining higher derivatives is not always as formidable as we might first think.

**Example 7.4.** Consider the initial-value problem (b) in Example 7.1 for $0 \le x \le 1$:

$$y' = 2xy, \qquad y(0) = 1.$$

In this case, we quickly find

$$y'' = 2y + 2xy' = 2y(1 + 2x^2)$$
$$y''' = 2y'(1 + 2x^2) + 8xy = 4xy(3 + 2x^2).$$

Hence, a third-order Taylor's series method is provided by the iteration

$$y_{i+1} = y_i + h\left[2x_iy_i + y_i(1 + 2x_i^2)h + 2x_iy_i(3 + 2x_i^2)\frac{h^2}{3}\right].$$

Most computer installations have symbol manipulation packages such as FORMAC which can be used to automate the computation of higher derivatives for algebraic expressions such as $y' = 2xy$ or even for more cumbersome expressions such as

$$y' = \frac{x^2 \sin(y)}{1 + y^2}.$$

For certain types of differential equations, a Taylor's series might be appropriate if a symbol manipulation package is available.

Before continuing, we wish to observe that the implementation of most numerical methods for solving $y' = f(x, y)$ will require the evaluation of $f(x, y)$ at various points $(x_i, y_i)$. In fact, unlike the simple examples we have considered so far, the evaluation of $f(x, y)$ may be quite time consuming. To give an indication of what can be involved in these evaluations, let us return to the equations of motion for a ballistic reentry trajectory, (4.0a). For a given position vector $\mathbf{R}(t)$ and velocity vector $\dot{\mathbf{R}}(t)$, we see from (4.0a) that we need to evaluate the gravity vector $\mathbf{G}(\mathbf{R})$ and the scalar drag force $-\overline{q}C_DA/M$. In simple textbook problems, gravity is often assumed to be constant; however, to obtain an accurate solution of (4.0a), we cannot assume constant gravity. What is required for an accurate numerical solution is a good mathematical model for the earth's gravity field. Thus, given a point $\mathbf{R}$ in space, we need to be able to determine a good approximation to the gravity vector, $\mathbf{G}(\mathbf{R})$. This is frequently done in practice by converting $\mathbf{R}$ to spherical coordinates $(r, \theta, \lambda)$ and then evaluating various partial derivatives of expressions that have the form

$$\sum_{n=0}^{N} \sum_{m=0}^{n} \frac{P_{nm}(\theta)}{r^{n+1}} [a_{nm} \cos(m\lambda) + b_{nm} \sin(m\lambda)].$$

These expressions arise from the truncation of the theoretical spherical harmonic expansion for the earth's gravity potential. The coefficients $a_{nm}$ and $b_{nm}$ are usually determined from satellite data, using least-squares fits. The functions $P_{nm}(\theta)$ can be evaluated through a series of recurrence formulas. Even for a modest value of $N$, say $N = 10$ or $N = 20$, the evaluation of these expressions is quite time consuming.

When drag forces are present, as in (4.0a), the term $-\bar{q}C_D A/M$ needs to be evaluated also. Recall that the function $\bar{q}$ is given by

$$\bar{q} = \frac{\rho(H)|\dot{\mathbf{R}}|^2}{2},$$

where $\rho(H)$ is atmospheric density at an altitude $H$. The drag coefficient, $C_D$, is usually an experimentally determined function of Mach and altitude where the Mach number, $m$, is given from $a_s = a_1\sqrt{T(H)}$ and $m = |\dot{\mathbf{R}}|/a_s$, with $a_1$ a constant and $T(H)$ denoting an atmospheric temperature function of altitude. The basic steps in evaluating the drag forces on a body moving with velocity $\dot{\mathbf{R}}$ at a point $\mathbf{R}$ are:

a) determine the altitude $H$;
b) given $H$, determine $T(H)$—this is usually done by interpolation in a table of values for $T(H_0)$, $T(H_1)$, $\cdots$;
c) determine the density, $\rho(H)$—$\rho(H)$ is usually given as a set of empirical formulas depending on $T(H)$;
d) determine the Mach number, $m$;
e) evaluate $C_D$ as a function of $H$ and $m$.

In the scheme above, step (e) is normally the most costly. Determining $C_D$ may depend on a long sequence of table look-ups, interpolation, and the evaluation of a series of empirical formulas, where the tables and empirical formulas are derived from theoretical aerodynamic considerations, wind-tunnel data, and flight-test data.

In the discussion above, we have tried to illustrate with a specific case, that evaluating $f(x, y)$ in the equation $y' = f(x, y)$ can be costly in terms of storage and execution time. The example above also serves to illustrate the fact that many realistic differential equations cannot be solved except by numerical means. A further observation that can be drawn from this example is that the function $f(x, y)$ may not be very smooth. The roughness of $f(x, y)$ can have a drastic effect on the size of the steps that can be taken in a numerical method, requiring a large number of evaluations of $f(x, y)$ to maintain any accuracy in the solution.

### 7.2.3 Error Analysis for One-step Methods

It is fairly easy to derive a simple error bound for the general one-step method

$$y_{i+1} = y_i + h\Phi(x_i, y_i; h), \qquad i = 0, 1, \ldots, n-1; \tag{7.9}$$

where (7.9) is used to estimate the solution of the initial-value problem $y' = f(x, y)$, $y(x_0) = y_0$, for $a \leq x \leq b$. This error bound for the general method in (7.9) can then be interpreted for special cases like Euler's method, $k$th order Taylor's series methods, and Runge-Kutta methods.

The first step in obtaining the error bound is to define the *local truncation error*, $\tau_i$, by the equation

$$y(x_{i+1}) = y(x_i) + h\Phi(x_i, y(x_i); h) + h\tau_i \tag{7.10}$$

where $y(x)$ is the solution to the intitial-value problem. Recalling that $y_{i+1} = y_i + h\Phi(x_i, y_i)$, we see that $\tau_i$ defines by how much $y_{i+1}$ would miss $y(x_{i+1})$ if we were starting at $x_i$ with the exact answer, that is, if $y_i = y(x_i)$. For example, in the case of Euler's method, $\Phi(x, y; h) = f(x, y)$, so

$$y(x_{i+1}) = y(x_i) + hf(x_i, y(x_i)) + h\tau_i.$$

For this case, we can determine $\tau_i$, since

$$y(x_{i+1}) = y(x_i) + hy'(x_i) + \frac{h^2}{2!}y''(\theta_i), \qquad x_i < \theta_i < x_{i+1}.$$

Thus, recalling that $y'(x_i) = f(x_i, y(x_i))$, we have for Euler's method that

$$\tau_i = \frac{h}{2}y''(\theta_i) \tag{7.11}$$

for some $\theta_i$ in $(x_i, x_{i+1})$. The total truncation error, $y_{i+1} - y(x_{i+1})$, consists of two parts:

a) the local truncation error $h\tau_i$ that we would make at $x_i$ in passing from $y_i$ to $y_{i+1}$ if $y_i$ were equal to $y(x_i)$,

b) the accumulated error that we have made in going from $x_0$ to $x_i$ and using $y_i$ instead of $y(x_i)$ in (7.9).

Setting $e_j = y_j - y(x_j)$ for $j = 0, 1, \ldots, n$ we have from (7.9) and (7.10) that $e_{i+1} = y_i - y(x_i) + h[\Phi(x_i, y_i; h) - \Phi(x_i, y(x_i); h)] - h\tau_i$, or

$$e_{i+1} = e_i + h[\Phi(x_i, y_i; h) - \Phi(x_i, y(x_i); h)] - h\tau_i. \tag{7.12}$$

In order to make (7.12) useful as a computational tool, we assume that the local truncation errors are bounded and that the increment function $\Phi(x, u; h)$ satisfies a Lipschitz condition in $u$. In particular, let

$$\tau = \max_{0 \le i \le n-1} |\tau_i|$$

and suppose there is a constant $L$ such that

$$|\Phi(x, u; h) - \Phi(x, v; h)| \le L|u - v|$$

for all $h > 0$ and all $x$, $a \le x \le b$. With these preliminaries, we can prove:

**Theorem 7.2.** For $0 < j \le n$, $\quad |e_j| \le \tau(e^{L(x_j - x_0)} - 1)/L.$

*Proof*: Using the Lipschitz condition in (7.12), we have $|e_{i+1}| \leq |e_i|(1 + hL) + h|\tau_i| \leq |e_i|(1 + hL) + h\tau$ for any $i$, $0 \leq i \leq n - 1$. Throwing the $i$th error back on the $(i - 1)$st error, we see $|e_{i+1}| \leq |e_{i-1}|(1 + hL)^2 + h\tau[1 + (1 + hL)]$. Continuing, we finally obtain

$$|e_{i+1}| \leq |e_0|(1 + hL)^{i+1} + h\tau[1 + (1 + hL) + \cdots + (1 + hL)^i].$$

Since $e_0 = y_0 - y(x_0)$, it follows that $e_0 = 0$. Using the familiar formula for a finite geometric summation, we then have

$$|e_{i+1}| \leq h\tau \frac{(1 + hL)^{i+1} - 1}{(1 + hL) - 1} \leq \frac{\tau}{L}[(1 + hL)^{i+1} - 1].$$

We next estimate $(1 + hL)^{i+1}$ by observing (for $\alpha > 0$), that $(1 + \alpha)^{i+1} \leq e^{(i+1)\alpha}$ (Problem 5). Therefore, $(1 + hL)^{i+1} \leq e^{(i+1)hL} = e^{(x_{i+1} - x_0)L}$. ■

This bound is overly pessimistic, due to a number of fairly crude approximations. However, it does illustrate the essential features of the error in a one-step method. (See Problem 6 for a more careful approach and Problem 7 for an approximate error bound.) In the special case of Euler's method, we see that Theorem 7.2 takes the form

$$|e_j| \leq \frac{hM_2}{2L}(e^{L(x_j - x_0)} - 1)$$

where $M_2 = \max_{a \leq x \leq b} |y''(x)|$ and where $f$ satisfies $|f(x, u) - f(x, v)| \leq L|u - v|$ for all $u$ and $v$ and for all $x$ in $[a, b]$. For a $k$th order Taylor's series method, we see (Problem 8) that the local truncation errors are bounded by $\tau = M_{k+1}h^k/(k + 1)!$, where $M_{k+1} = \max_{a \leq x \leq b} |y^{(k+1)}(x)|$. Thus Theorem 7.2 leads us to

$$|e_j| \leq \frac{h^k M_{k+1}}{L(k + 1)!}(e^{L(x_j - x_0)} - 1)$$

as an error bound for the $k$th order Taylor's series method.

We note that the error bound in Theorem 7.2 generally provides qualitative rather than quantitative information. For example, to use the bound in the special case of Euler's method requires that we have an upper bound for $|y''(x)|$, while in reality we do not even know $y(x)$, let alone $y''(x)$. What we do know from the bound is that the error for a $k$th order Taylor's series method satisfies an inequality of the form $|e_j| \leq Mh^k e^{L(x_j - x_0)}$ where $M$ and $L$ are some constants (probably not known). However, this information does allow us to establish convergence (see Problem 9) and also tells us something about the rate of convergence. Finally, we note that Theorem 7.2 does not take into account the possibility of round-off error, and we treat this complication later.

**Example 7.5.** In Example 7.3, we used Euler's method to estimate the solution of $y' = y$, $y(0) = 1$ for $0 \leq x \leq 1$. To see what sort of information is provided by Theorem 7.2, we apply the theorem to this problem with $h = 0.125$. In this case, $f(x, u) = u$, so that $|f(x, u) - f(x, v)| = |u - v|$ meaning that the Lipschitz constant $L$ is 1. Since $x_0 = 0$, we find

$$|e_j| \leq \frac{hM_2}{2}(e^{x_j} - 1).$$

Since we also know the solution, $y(x) = e^x$, we see that $M_2 = e$ and hence

$$|e_j| \leq \frac{he}{2}(e^{x_j} - 1) = \frac{h}{2}(e^{1+x_j} - e).$$

For $h = 0.125$ and $j = 4$, we have $|y_4 - y(0.5)| \leq (e^{1.5} - e)/16 \approx 0.11$ and for $j = 8$, we find $|y_8 - y(1)| \leq (e^2 - e)/16 \approx 0.29$. From Example 7.3, we see the actual errors at $x_4$ and $x_8$ are 0.05 and 0.15 respectively. A somewhat tighter bound of $|y_4 - y(0.5)| \leq 0.07$ can be established (Problem 6).

### 7.2.4 Runge-Kutta Methods

The general one-step method takes the form

$$y_{i+1} = y_i + h\Phi(x_i, y_i; h)$$

so we can think of the number $h\Phi(x_i, y_i; h)$ as representing an approximation to the amount by which the solution $y(x)$ increases (or decreases) as we move from $x_i$ to $x_{i+1}$. Since the nature of the flow field in a neighborhood of $(x_i, y(x_i))$ governs the change in $y(x)$ between $x_i$ and $x_{i+1}$, it seems plausible that we might "sample" the flow field at various points near $(x_i, y(x_i))$ in order to get some indication of changes to be expected in $y(x)$ for $x_i \leq x \leq x_{i+1}$. In the real situation, of course, we do not have $y(x_i)$, so we must think of sampling at points $(\theta_1, \gamma_1), \ldots, (\theta_k, \gamma_k)$ which are near $(x_i, y_i)$. In particular, we could specify the increment function by

$$\Phi(x_i, y_i; h) = A_1 f(\theta_1, \gamma_1) + A_2 f(\theta_2, \gamma_2) + \cdots + A_k f(\theta_k, \gamma_k) \qquad \textbf{(7.13)}$$

so that $\Phi(x_i, y_i; h)$ is a weighted average of the slopes near $(x_i, y_i)$. For example, Euler's method uses just the one sample point $(x_i, y_i)$ with $A_1 = 1$.

The problem with (7.13) is in choosing the weights $A_j$ and the sample points $(\theta_j, \gamma_j)$. The Runge-Kutta methods that are developed in this section provide one logical means for choosing the weights and sample points. Runge-Kutta methods (named for the German mathematicians C. Runge and M. W. Kutta, who made important contributions around the turn of the century) are quite similar to Taylor's series methods but do not require the evaluation of derivatives of $f(x, y)$ (which, as we recall, is one major deficiency of a Taylor's series method). In order to demonstrate the essential ideas, we will develop a second-order Runge-Kutta method which uses the form of (7.13) with two sample points. As one sample point, let us choose $(x_i, y_i)$; and for a sample point nearby to $(x_i, y_i)$, let us choose $(x_i + \beta h, y_i + \alpha h f(x_i, y_i))$, where $\alpha$ and $\beta$ are as yet undetermined. This choice of a nearby sample point is not as unnatural as it might seem at first glance, since we want to sample near the integral curve passing through $(x_i, y_i)$. (For example, with $\alpha = \beta = 1$, we would be using the sample point $(x_{i+1}, \tilde{y}_{i+1})$, where $\tilde{y}_{i+1}$ is the estimate of $y(x_{i+1})$ given by Euler's method.) In general, for two sample points we have $y_{i+1} = y_i + h\Phi(x_i, y_i; h)$, where $\Phi(x_i, y_i; h)$ is defined by

$$y_{i+1} = y_i + h[A_1 f(x_i, y_i) + A_2 f(x_i + \beta h, y_i + \alpha h f(x_i, y_i))]. \tag{7.14}$$

The basis for a Runge-Kutta method is to choose the parameters $A_1$, $A_2$, $\alpha$, and $\beta$ so that $\Phi(x_i, y_i; h)$, given by (7.14), agrees as well as is possible with the increment function $T_k(x_i, y_i; h)$ for a Taylor's series method.

To see how this is done, we first expand $f(x_i + \beta h, y_i + \alpha h f(x_i, y_i))$ in a Taylor's series to find

$$f(x_i + \beta h, y_i + \alpha h f(x_i, y_i)) = f(x_i, y_i) + f_x(x_i, y_i)\beta h + f_y(x_i, y_i)\alpha h f(x_i, y_i) + E$$

where $E$ denotes the remainder in the expansion. If we let $K_1 = \alpha h f(x_i, y_i)$, we see that $E$ has the form $E = Ch^2$ since

$$2E = f_{xx}(\eta, \delta)(\beta h)^2 + 2f_{xy}(\eta, \delta)\beta h K_1 + f_{yy}(\eta, \delta)(K_1)^2$$

where $(\eta, \delta)$ is the mean-value point for the Taylor's series expansion in two variables. Collecting like terms, we find that the increment function in (7.14) reduces to

$$\begin{aligned}\Phi(x_i, y_i; h) = (A_1 + A_2)f(x_i, y_i) + A_2 h[\beta f_x(x_i, y_i)\\ + \alpha f_y(x_i, y_i)f(x_i, y_i) + Ch].\end{aligned} \tag{7.15a}$$

Recalling that $T_2(x_i, y_i; h)$ is given by

$$T_2(x_i, y_i; h) = f(x_i, y_i) + \frac{h}{2}[f_x(x_i, y_i) + f_y(x_i, y_i)f(x_i, y_i)], \tag{7.15b}$$

we see that (7.15a) can be made to agree fairly well with (7.15b) if we equate like coefficients in (7.15a) and (7.15b) and thereby choose the parameters so that

$$\begin{aligned} A_1 + A_2 &= 1 \\ A_2\beta &= \tfrac{1}{2} \\ A_2\alpha &= \tfrac{1}{2}. \end{aligned} \tag{7.16}$$

With these choices, we will have

$$\Phi(x_i, y_i; h) = T_2(x_i, y_i; h) + A_2 Ch^2.$$

We observe that (7.16) reduces to $\beta = \alpha$, $A_2 = 1/2\alpha$, and $A_1 = 1 - (1/2\alpha)$. The fact that there are infinitely many solutions for the parameters is a peculiarity of Runge-Kutta methods. One natural choice is $\alpha = \frac{1}{2}$, giving the sequence

$$y_{i+1} = y_i + hf\left(x_i + \frac{h}{2}, y_i + \frac{h}{2}f(x_i, y_i)\right)$$

which is usually called the *modified Euler's method.* Another choice is $\alpha = 1$, which leads to the *improved Euler's method* (often called *Heun's method*)

$$y_{i+1} = y_i + \frac{h}{2}[f(x_i, y_i) + f(x_{i+1}, y_i + hf(x_i, y_i))].$$

We see that these Runge-Kutta methods do not require the evaluation of the derivative of $f(x, y)$, as does the second-order Taylor's series method. By Problem 10, we see that the truncation error, as defined in (7.10), is of the form $\tau_i = C_i h^2$. From Theorem 7.2, the errors $|y_i - y(x_i)|$ are bounded by an expression of the form $Mh^2$ (hence the term *second-order*).

```
      SUBROUTINE HEUN(Y,X,H,N)
      DOUBLE PRECISION Y
C
C   THIS SUBROUTINE USES HEUN'S METHOD WITH PARTIAL DOUBLE PRECISION
C   TO SOLVE THE INITIAL-VALUE PROBLEM Y'(X)=F(X,Y), WITH INITIAL
C   CONDITION Y(XO)=YO.  A SINGLE PRECISION FUNCTION SUBPROGRAM F(X,Y)
C   MUST BE PROVIDED.  THE MAIN PROGRAM MUST SUPPLY A DOUBLE PRECISION
C   INITIAL-VALUE Y=YO, A STARTING VALUE X=XO AND STEP-SIZE H IN
C   SINGLE PRECISION, AND AN INTEGER N=NUMBER OF STEPS TO EXECUTE.
C
      ISTEP=0
C
C   PRINT INITIAL CONDITIONS
C
      PRINT 100,ISTEP,X,Y
C
C   EXECUTE HEUN'S METHOD
C
      DO 1 ISTEP=1,N
      YS=Y
      YP1=F(X,YS)
      X=X+H
      YS=Y+H*YP1
      YP2=F(X,YS)
      Y=Y+.5*H*(YP1+YP2)
    1 PRINT 100,ISTEP,X,Y
  100 FORMAT(1H ,I3,2E20.7)
      RETURN
      END
```

**Figure 7.5**

**Example 7.6.** The subroutine given in Fig. 7.5 uses Heun's method (the improved Euler's method). As an illustration of the sort of results that Heun's method might give, we ran the problem $y' = 2xy$, $y(0) = 1$ (with $h = 0.125$). The results are shown in Table 7.2. By contrast, Euler's method with $h = 0.125$ gives $y_8 = 2.820158$, where $y_8 \approx y(1)$. For $h = 0.01$, Heun's method provides $y_{100} = 2.718186$ where $y(1) = 2.718282$ (an error of 0.000096).

There are many different Runge-Kutta methods of order 2, 3, . . . , all of which can be developed in much the same way as we did for the second-order Runge-Kutta method. The determination of these methods rapidly becomes tedious (Problem 11)

**Table 7.2**

| $x$ | $y_i$ | $y(x_i)$ |
|---|---|---|
| 0.125 | 1.015625 | 1.015748 |
| 0.250 | 1.064224 | 1.064494 |
| 0.375 | 1.150485 | 1.150993 |
| 0.500 | 1.283060 | 1.284025 |
| 0.625 | 1.476020 | 1.477904 |
| 0.750 | 1.751332 | 1.755055 |
| 0.875 | 2.142987 | 2.150338 |
| 1.000 | 2.703847 | 2.718282 |

and we close this section by giving a fourth-order method that is frequently used in practice. For a given step-size $h$, we define

$$y_{i+1} = y_i + \frac{1}{6}(K_1 + 2K_2 + 2K_3 + K_4) \tag{7.17}$$

where

$$K_1 = hf(x_i, y_i)$$

$$K_2 = hf\left(x_i + \frac{h}{2}, y_i + \frac{1}{2}K_1\right)$$

$$K_3 = hf\left(x_i + \frac{h}{2}, y_i + \frac{1}{2}K_2\right)$$

$$K_4 = hf(x_i + h, y_i + K_3).$$

The method in (7.17) has a truncation error $\tau$ bounded by a term of the form $Mh^4$ and requires four evaluations of $f(x, y)$ per step. While higher-order Runge-Kutta methods can be found, it is known that a method of order $k$ requires at least $k + 1$ evaluations of $f(x, y)$ per step when $k \geq 5$. Consequently, a loss of computational efficiency may result when a Runge-Kutta method of order 5 or greater is used. This, in part, explains the popularity of the classical fourth-order method, particularly in realistic problems where the evaluation of $f(x, y)$ is complicated. Finally, the Runge-Kutta methods given in (7.15), (7.17), and Problem 11 all have a particularly simple form. The general form for a Runge-Kutta method is

$$y_{i+1} = y_i + h\sum_{r=1}^{m} A_r K_r$$

where the $A_r$ are constant weights,

$$K_r = f(x_i + \alpha_r h, y_i + \beta_{r1}K_1 + \beta_{r2}K_2 + \cdots + \beta_{r,r-1}K_{r-1})$$

and $\alpha_1 = 0$.

## PROBLEMS

1. Test Euler's method on the differential equations of Example 7.1, (7.2a), (7.2b), and (7.2c). Use step-sizes $h = 1/n$, for $n = 4, 8, 16$, and 32.

2. Given the differential equation $y' = y$, $y(0) = 1$ as in Example 7.3, suppose that Euler's method is employed first with a step-size $h$ and then with a step-size $h_1 = h/2$. Suppose that $y_i$ is the $i$th estimate of Euler's method using $h$, and $u_i$ is the $i$th estimate using $h_1$. Show that $y_{i+1} = (1 + h)y_i$ and $u_{i+1} = (1 + h_1)u_i$. For a point $x_i = x_0 + ih = ih$, note that $y_i \approx y(x_i)$ and $u_{2i} \approx y(x_i)$. Show that $y_i < u_{2i} < y(x_i)$.

*3. Again, suppose that Euler's method is employed to estimate the solution of $y' = y$, $y(0) = 1$ and suppose that $\bar{x}$ is fixed, $\bar{x} > 0$. For $h_n = \bar{x}/n$, let $y_n$ denote the estimate of $y(\bar{x})$ given by Euler's method with step-size $h_n$. Show from a definition of $e^x$, that

$$\lim_{n\to\infty} y_n = y(x).$$

(Hint: By Problem 2, there is a closed form expression for $y_n$.)

4. Program the 2nd- and 3rd-order Taylor's series methods and test your program as in Problem 1.

5. Using the Taylor's series expansion for $e^x$, expanded about $x = 0$, show that $(1 + \alpha)^k \le e^{\alpha k}$, where $\alpha \ge 0$.

6. In the proof of Theorem 7.2, let $\tilde{\tau}_j = \max_{0 \le i \le j-1} |\tau_i|$ and let $L_j$ be the relevant Lipschitz constant for $[x_0, x_j]$. Show that

$$|e_j| \le \frac{\tilde{\tau}_j}{L_j}(e^{L_j(x_j - x_0)} - 1).$$

Show that this simple observation leads to the improved bound mentioned in Example 7.5.

7. For Euler's method, we know from (7.11) that $\tau_i = (h/2)y''(\theta_i)$, $x_i < \theta_i < x_{i+1}$. Since we do not know $y(x)$, we cannot hope to know $y''(\theta_i)$, but we can give an approximation for $|y''(x)|$, $x_i \le x \le x_{i+1}$, by observing that $y'' = f_x(x, y) + f_y(x, y)f(x, y)$. For the equations in Problem 1, compute $y''$ and determine for a fixed $y$ whether the maximum of $|y''|$ occurs at $x_i$ or $x_{i+1}$. With this expression for $y''$, use $|f_x(x, y_i) + f_y(x, y_i)f(x, y_i)|$ with $x$ being either $x_i$ or $x_{i+1}$ to approximate $|\tau_i|$. Modify the program in Problem 1, using $|e_{i+1}| \le |e_i|(1 + hL) + h|\tau_i|$ to compute and print a running estimate of $|e_i|$, $1 \le i \le n$. Note that these estimates are not rigorous.

8. Show that a bound for $\tau$ in a $k$th order Taylor's series method is $M_{k+1}h^k/(k + 1)!$

9. As in Problem 3, suppose $\bar{x}$ is fixed, $a < \bar{x} \le b$, and suppose $h_n = (\bar{x} - a)/n$. Let $y_n$ denote the estimate of $y(\bar{x})$ given by a $k$th order Taylor's series method using step-size $h_n$. If $y(x)$ has a $(k + 1)$st derivative that is continuous on $[a, b]$, use Problem 8 and Theorem 7.2 to show that $\lim_{n\to\infty} y_n = y(\bar{x})$.

*10. Show that the truncation errors, $\tau_i$, for a 2nd-order Runge-Kutta method are of the form $\tau_i = C_i h^2$. (Hint: Write $y(x_{i+1}) = y(x_i) + h\Phi(x_i, y(x_i); h) + h\tau_i$ and $y(x_{i+1}) = y(x_i) + hT_2(x_i, y(x_i); h) + (h^3/3!)y'''(\theta)$ and use the fact that

$$\Phi(x_i, y(x_i); h) = T_2(x_i, y(x_i); h) + Q_i h^2.)$$

11. Let $\Phi(x_i, y_i; h) = \frac{1}{9}(2K_1 + 3K_2 + 4K_3)$ where

$$K_1 = f(x_i, y_i), \qquad K_2 = f\left(x_i + \frac{h}{2}, y_i + \frac{1}{2}K_1\right), \qquad K_3 = f\left(x_i + \frac{3h}{4}, y_i + \frac{3}{4}K_2\right).$$

Show that $y_{i+1} = y_i + h\Phi(x_i, y_i; h)$ is a 3rd-order Runge-Kutta method. (3rd-order Runge-Kutta methods usually have two free parameters determined in a similar manner as was the parameter $\alpha$ in (7.16). The method above is one of many possible 3rd-order Runge-Kutta methods.)

12. Program and test the 3rd- and 4th-order Runge-Kutta methods given in this section on the equations in Problem 1.

## 7.3 PREDICTOR-CORRECTOR METHODS

In this section, we consider *predictor-corrector* methods for numerically solving initial-value problems, which are not usually of the form of one-step methods. The idea behind the particular predictor-corrector methods that we investigate here comes from the observation that we can integrate both sides of $y' = f(x, y)$, and by using $y(x_0) = y_0$, we can represent the solution, $y(x)$, mathematically by

$$y(x) - y(x_0) = \int_{x_0}^{x} y'(t)\,dt = \int_{x_0}^{x} f(t, y(t))\,dt. \tag{7.18}$$

Therefore, from (7.18), we have for $a \leq x \leq b$ that $y(x) = y_0 + \int_{x_0}^{x} f(t, y(t))\,dt$. In fact, if we know the solution at any point $x_q$, then as in (7.18), we have

$$y(x) = y(x_q) + \int_{x_q}^{x} f(t, y(t))\,dt. \tag{7.19}$$

Equation (7.19) is a formal identity and does not appear at first to have any computational utility, since we do not know $y(t)$ and hence cannot integrate $f(t, y(t))$. However, if we replace $\int_{x_q}^{x} f(t, y(t))\,dt$ by a quadrature formula, we have the approximation

$$y(x) \approx y(x_q) + \sum_{j=0}^{k} A_j f(t_j, y(t_j)), \qquad x_q \leq t_0 < t_1 < \cdots < t_k \leq x.$$

To put this in a more familiar context, if we knew $y(x_0), y(x_1), \ldots, y(x_i)$, then we could estimate $y(x_{i+1})$ from a quadrature formula that used nodes $x_i, x_{i-1}, \ldots, x_{i-p}$. For example, using the $(p + 1)$-point *open* Newton-Cotes formula

$$\sum_{j=0}^{p} A_j g(x_{i-j}) \approx \int_{x_{i-p-1}}^{x_{i+1}} g(x)\,dx,$$

(see Section 6.2.2), we have

$$y(x_{i+1}) \approx y(x_{i-p-1}) + \sum_{j=0}^{p} A_j f(x_{i-j}, y(x_{i-j})). \tag{7.20}$$

But, given that we have an initial-value problem, the only place we know $y(x)$ *a priori* is at $x = x_0$. Thus, in order to use (7.20), we must assume that we have approximations $y_k$ to $y(x_k)$, for $i - p - 1 \leq k \leq i$.

With this motivation, we are led from (7.20) to the procedure

$$y_{i+1} = y_{i-p-1} + \sum_{j=0}^{p} A_j f(x_{i-j}, y_{i-j}), \qquad i = p + 1, p + 2, \ldots. \qquad \textbf{(7.21)}$$

This is really a twofold approximation, where we first replace the integral in (7.19) by a quadrature formula and then replace the function evaluation $f(x_{i-j}, y(x_{i-j}))$ by $f(x_{i-j}, y_{i-j})$.

**Example 7.7.** The open three-point Newton-Cotes rule approximation for

$$\int_{x_{i-3}}^{x_{i+1}} f(x, y(x))\, dx \text{ is } \frac{4h}{3}\,[2f(x_{i-2}, y(x_{i-2})) - f(x_{i-1}, y(x_{i-1})) + 2f(x_i, y(x_i))].$$

So we are led by (7.21) to

$$y_{i+1} = y_{i-3} + \frac{4h}{3}\,[2f(x_{i-2}, y_{i-2}) - f(x_{i-1}, y_{i-1}) + 2f(x_i, y_i)] \qquad \textbf{(7.22)}$$

for $i = 3, 4, \ldots$. The first thing we notice is that in order to start this procedure, we need four starting values $y_0$, $y_1$, $y_2$, and $y_3$. We are given $y_0$, but we need to furnish approximations $y_1$, $y_2$, and $y_3$. (These starting values could, for example, be provided by a one-step method such as a Runge-Kutta method.)

As a more specific example, let us consider $y' = 2xy$, $y(0) = 1$ for $0 \leq x \leq 1$. For $h = 0.125$, the procedure above reduces to

$$y_{i+1} = y_{i-3} + \frac{1}{3}\,[2x_{i-2}y_{i-2} - x_{i-1}y_{i-1} + 2x_i y_i]$$

for $i = 3, 4, 5, 6, 7$. The solution of the problem is $y(x) = e^{x^2}$ and so that we may concentrate on the error of the method uninfluenced by any initial errors in the starting values, we choose starting values $y_0$, $y_1$, $y_2$, and $y_3$ which are correct to eight places, finding the following:

| $x_i$ | $y_i$ | $y(x_i)$ | $y(x_i) - y_i$ |
|---|---|---|---|
| 0.500 | 1.28369 | 1.28403 | 0.00034 |
| 0.625 | 1.47718 | 1.47790 | 0.00072 |
| 0.750 | 1.75379 | 1.75506 | 0.00127 |
| 0.875 | 2.14804 | 2.15034 | 0.00230 |
| 1.000 | 2.71375 | 2.71828 | 0.00453 |

The ideas embodied in (7.21) can clearly be extended in several different directions. First of all, we need not have restricted ourselves to an open Newton-Cotes formula. We could have chosen an interpolatory formula of the form

$$\sum_{j=0}^{p} B_j f(x_{i-j}, y_{i-j}),$$

where

$$\sum_{j=0}^{p} B_j g(x_{i-j}) \approx \int_{x_i}^{x_{i+1}} g(x)\, dx$$

(that is, we allow nodes $x_{i-j}$ that are to the left of the interval of integration, $(x_i, x_{i+1})$). For example, the quadrature formula

$$\frac{h}{2}[-g(x_{i-1}) + 3g(x_i)] \approx \int_{x_i}^{x_{i+1}} g(x)\, dx$$

leads us to consider a method

$$y_{i+1} = y_i + \frac{h}{2}[-f(x_{i-1}, y_{i-1}) + 3f(x_i, y_i)], \qquad i = 1, 2, \ldots.$$

This is a specific example of the general formula

$$y_{i+1} = y_i + \sum_{j=0}^{p} B_j f(x_{i-j}, y_{i-j}).$$

Any formula of this type is called an *Adams-Bashforth formula* and can be obtained by integrating the backwards version of the Newton form of the interpolating polynomial given in (5.25). Euler's method can be represented as a formula of this type (Problem 5).

Another possibility would be to use a composite quadrature formula. On the other hand, Gaussian quadrature formulas are not immediately applicable, since one essential ingredient of a procedure like (7.21) is that we use equally spaced points, estimating $y_{i+1}$ from $y_i, y_{i-1}, \ldots$. A more interesting and fruitful possibility is provided by the closed Newton-Cotes formulas. By analogy with (7.21), we consider

$$y_{i+1} = y_{i-p} + \sum_{j=0}^{p+1} B_j f(x_{i+1-j}, y_{i+1-j}), \qquad i = p, p+1, \ldots \tag{7.23}$$

where

$$\sum_{j=0}^{p+1} B_j g(x_{i+1-j}) \approx \int_{x_{i-p}}^{x_{i+1}} g(x)\, dx$$

is a closed Newton-Cotes formula for the interval $[x_{i-p}, x_{i+1}]$. For example, Simpson's rule would lead us to

$$y_{i+1} = y_{i-1} + \frac{h}{3}[f(x_{i-1}, y_{i-1}) + 4f(x_i, y_i) + f(x_{i+1}, y_{i+1})], \qquad i = 1, 2, \ldots. \tag{7.24}$$

The first thing we notice about (7.23) or the special case (7.24) is that the unknown

$y_{i+1}$ appears on both sides of the equation. That is, (7.23) has the form

$$y_{i+1} = y_{i-p} + \sum_{j=1}^{p+1} B_j f(x_{i+1-j}, y_{i+1-j}) + B_0 f(x_{i+1}, y_{i+1}) \qquad \textbf{(7.25)}$$

which we rewrite as $y_{i+1} = Q_i + B_0 f(x_{i+1}, y_{i+1})$. Since $y_{i+1}$ is an unknown quantity and $Q_i$ is known, this equation is suggestive of the fixed-point equation $t = Q_i + B_0 f(x_{i+1}, t)$ which in turn suggests that we might solve (7.25) for $y_{i+1}$ by a fixed-point iteration. For the case of (7.24) above, we would be using the iteration

$$t_{k+1} = Q_i + \frac{h}{3} f(x_{i+1}, t_k), \qquad k = 0, 1, \ldots$$

where $Q_i = y_{i-1} + (h/3)[f(x_{i-1}, y_{i-1}) + 4f(x_i, y_i)]$.

In order to efficiently use iteration, we need a good starting value $t_0$ (that is, a good estimate of the solution $y_{i+1}$ of (7.25)). An open Newton-Cotes formula or an Adams-Bashforth formula could be used to give a starting value. An example should serve to clarify this discussion. Suppose we were to use the open formula (7.22) to provide an initial estimate $y_{i+1}^{(0)}$ for $y(x_{i+1})$, and then use a closed formula like (7.24) to "refine" this estimate. In this case, we would be calculating:

$$y_{i+1}^{(0)} = y_{i-3} + \frac{4h}{3}[2f(x_{i-2}, y_{i-2}) - f(x_{i-1}, y_{i-1}) + 2f(x_i, y_i)] \qquad \textbf{(7.26a)}$$

$$y_{i+1}^{(k+1)} = y_{i-1} + \frac{h}{3}[f(x_{i-1}, y_{i-1}) + 4f(x_i, y_i)] + \frac{h}{3} f(x_{i+1}, y_{i+1}^{(k)}), \qquad k = 0, 1, 2, \ldots. \qquad \textbf{(7.26b)}$$

Since the main cost of any scheme for numerically solving a differential equation is usually in evaluating $f(x, y)$, it is clearly inefficient to iterate (7.26b) often. In fact, we normally do not iterate more than once or twice. If we iterate just once, we obtain $y_{i+1}$ from $y_i, y_{i-1}, \ldots$ by using a pair of formulas such as (7.22) and (7.24) in conjunction to yield:

$$\tilde{y}_{i+1} = y_{i-3} + \frac{4h}{3}[2f(x_{i-2}, y_{i-2}) - f(x_{i-1}, y_{i-1}) + 2f(x_i, y_i)] \qquad \textbf{(7.26c)}$$

$$y_{i+1} = y_{i-1} + \frac{h}{3}[f(x_{i-1}, y_{i-1}) + 4f(x_i, y_i) + f(x_{i+1}, \tilde{y}_{i+1})]. \qquad \textbf{(7.26d)}$$

(To avoid confusion and complication, we assume through the rest of this chapter that just one iteration is used.)

Formulas (7.26c) and (7.26d) together comprise an example of a predictor-corrector method (Milne's method). Formula (7.26c) is called the *predictor* and (7.26d) is the *corrector*. (It is also customary, for obvious reasons, to call the predictor an *explicit* formula and the corrector an *implicit* formula.) We list below three other predictor-corrector methods, where to shorten the description, we use the natural notation $y'_j = f(x_j, y_j)$ and $\tilde{y}'_j = f(x_j, \tilde{y}_j)$.

$$\text{(Heun's Method)} \quad \begin{cases} \tilde{y}_{i+1} = y_i + hy'_i \\ y_{i+1} = y_i + \dfrac{h}{2}[y'_i + \tilde{y}'_{i+1}] \end{cases}$$

$$\text{(Improved Heun's Method)} \quad \begin{cases} \tilde{y}_{i+1} = y_i + \dfrac{h}{2}[-y'_{i-1} + 3y'_i] \\ y_{i+1} = y_i + \dfrac{h}{2}[y'_i + \tilde{y}'_{i+1}] \end{cases}$$

$$\text{(Adams-Moulton Method)} \quad \begin{cases} \tilde{y}_{i+1} = y_i + \dfrac{h}{24}[-9y'_{i-3} + 37y'_{i-2} - 59y'_{i-1} + 55y'_i] \\ y_{i+1} = y_i + \dfrac{h}{24}[y'_{i-2} - 5y'_{i-1} + 19y'_i + 9\tilde{y}'_{i+1}] \end{cases}$$

There is some confusion of names in the last formula. Many authors refer to the corrector formula only as an Adams-Moulton formula, while others call it the Adams-Moulton predictor-corrector method. Note that Heun's method (often called the improved Euler method) is in reality a one-step method—in fact, if we combine the formulas and write them as

$$y_{i+1} = y_i + \frac{h}{2}[f(x_i, y_i) + f(x_{i+1}, y_i + hf(x_i, y_i))]$$

we find that it is a second-order Runge-Kutta method. (See Section 7.2.4.) The improved Heun's method uses an Adams-Bashforth predictor and a trapezoidal rule corrector, while the Adams-Moulton method uses a four-point Adams-Bashforth predictor and a corrector derived from a quadrature formula for the interval $[x_i, x_{i+1}]$, based on the nodes $x_{i-2}$, $x_{i-1}$, $x_i$ and $x_{i+1}$. This corrector is a formula of the type

$$y_{i+1} = y_i + \sum_{j=0}^{m} B_j f(x_{i+1-j}, y_{i+1-j})$$

which is called an *Adams-Moulton corrector*. These formulas can be obtained from integrating the backwards form of the Newton interpolating polynomial (5.25) at $x_{i+1}, x_i, \ldots, x_{i+1-m}$. Finally, although we do not wish at present to go into the details, we note that (see Section 7.6) some care must be exercised in choosing a predictor and a corrector that are compatible. We would not, for example, use a trapezoidal rule corrector with a four-point Adams-Bashforth predictor, since (roughly speaking) the predictor would be more accurate than the corrector.

**Example 7.8.** In this example, Milne's method was employed on two simple problems: $y' = 2xy$, $y(0) = 1$, and $y' = -y$, $y(0) = 1$. The results for the first problem illustrate the success that the idea of first predicting and then correcting can have. The results for the second problem demonstrate a potential computational difficulty that must be guarded against in any "multistep" method, the problem of stability (which we consider again in Section 7.6.2).

We used Milne's method (7.26c) and (7.26d), for $y' = 2xy$, $y(0) = 1$ with $h = 0.125$, with the following results

| $x_i$ | $\tilde{y}_i$ | $y_i$ | $y(x_i)$ |
|---|---|---|---|
| 0.500 | 1.28369 | 1.28403 | 1.28403 |
| 0.625 | 1.47730 | 1.47791 | 1.47790 |
| 0.750 | 1.75403 | 1.75506 | 1.75506 |
| 0.875 | 2.14863 | 2.15032 | 2.15034 |
| 1.000 | 2.71541 | 2.71821 | 2.71828 |

These results should be contrasted with the results of Example 7.7, which used only the predictor component of Milne's method for each step.

We ran the same problem with $h = 0.0625$ and found $y_{16} = 2.718277$ where $y_{16} \approx y(1) = 2.718282$. To demonstrate the effect of iterating the corrector formula, we modified the subroutine so that the corrector was iterated ten times at each step and obtained a value $y_{16} = 2.718296$ which is actually a slightly worse answer than the one we found with no iteration. We should not attach too much significance to this one particular experiment, since other test cases will show the iterated corrector to be slightly more accurate than a noniterated corrector. The point is that an excessive number (say 10) of iterations of the corrector at each step is usually wasteful in terms of computation. For example, in this problem, we had $y_{15} = 2.408274$. The predicted value $\tilde{y}_{16}$ was $y_{16}^{(0)} = \tilde{y}_{16} = 2.718139$, the first corrected value $y_{16}^{(1)} = 2.718289$, then $y_{16}^{(2)} = 2.718295$, $y_{16}^{(3)} = 2.718296$ and no change thereafter (i.e., $y_{16}^{(i)} = y_{16}^{(3)}$, $i = 4, 5, \ldots, 10$). Problem 6 shows the theoretical rate of convergence to be expected when the corrector is iterated. In the next section, we give some more qualitative results which bear on the proper matching of predictor-corrector pairs and step-size control.

While Milne's method was fairly successful in solving $y' = 2xy$, $y(0) = 1$, it is not recommended for general computational use, since it suffers from a form of instability. The Adams-Moulton method (which has a comparable truncation error) is usually preferred over Milne's method. To show the sort of errors that can be introduced by instability, we used Milne's method for $y' = -y$, $y(0) = 1$, with $h = 0.1$ and for $0 \le x \le 16$, obtaining:

| $x_i$ | $y_i$ | $y(x_i)$ | Relative Error |
|---|---|---|---|
| 5. | 0.673770E−02 | 0.673795E−02 | 0.000036 |
| 10. | 0.460999E−04 | 0.453999E−04 | 0.014122 |
| 11. | 0.157779E−04 | 0.167017E−04 | 0.055309 |
| 12. | 0.496962E−05 | 0.614421E−05 | 0.191171 |
| 13. | 0.766646E−06 | 0.226033E−05 | 0.660826 |
| 14. | −0.106799E−05 | 0.831529E−06 | 2.284367 |
| 15. | −0.210973E−05 | 0.305906E−06 | 7.896741 |
| 16. | −0.295945E−05 | 0.112535E−06 | 27.298027 |

The large relative errors produced by Milne's method in this problem are due to the method's instability. (We have left the topic of stability until Section 7.6.2, but the results above should be compared with those in Example 7.11.)

In the next section, we will briefly discuss certain aspects of error estimation and stability for predictor-corrector methods. Before doing this, some general observations about one-step and multistep methods are in order. (By a multistep method, we mean any method which explicitly uses more than one previous approximation (that is, $y_i, y_{i-1}, y_{i-2}, \ldots$) to calculate $y_{i+1}$. Methods such as (7.20) or the predictor-corrector methods discussed above are examples of multistep methods.) As we have mentioned, many differential equations arising from practical problems are not nearly so simple as the equations we have been using in our examples. In particular, evaluation of $f(x, y)$ may well involve table look-ups and the calculation of quite a few auxiliary quantities. Hence the number of evaluations of $f(x, y)$ required to pass from $y_i$ to $y_{i+1}$ may be an important criterion for the selection of a method. A method like (7.20) or an Adams-Bashforth method requires one new evaluation of $f(x, y)$ per step. A predictor-corrector method clearly requires (effectively) two new evaluations per step, since we need to evaluate $f(x_{i+1}, \tilde{y}_{i+1})$ in the corrector step and then need to evaluate $f(x_{i+1}, y_{i+1})$ to make the next predictor step. By contrast, a 4th-order Runge-Kutta method such as (7.17) requires four evaluations per step. For comparison of accuracy, we note (without proof) that the Adams-Moulton predictor-corrector above has a theoretical error $|y(x_j) - y_j|$ that is roughly comparable for a given step-size $h$ to the theoretical error for a 4th-order Runge-Kutta method. For many differential equations, Taylor's series methods cannot even be considered, owing to the impossibility or difficulty of computing the necessary partials of $f(x, y)$.

One especially good feature of predictor-corrector methods is that they provide us with the ability to continuously monitor the quantities $|y_{i+1} - \tilde{y}_{i+1}|$ and these quantities can be used to give an estimate of the per-step error. This feature can be used as a step-size control (much as in an adaptive quadrature routine) where the step-size is increased or decreased according to tests on $|y_{i+1} - \tilde{y}_{i+1}|$. Runge-Kutta methods can also be modified for this sort of step-size control, but at the expense of an additional evaluation per step. We note, however, that adjusting the step-size of a predictor-corrector method will usually require either new starting values or an interpolation between old values. For example, to decrease the step-size from $h$ to $h/2$ in the improved Heun's method means we need an estimate $y_i$ to $y(x_i)$ and an estimate $y_{i-1}$ to $y(x_i - h/2)$, in order to calculate an estimate $y_{i+1}$ to $y(x_i + h/2)$.

Positive features of Runge-Kutta methods include ease of programming, less need for concern about stability, and no need for special starting values. There are also geometric distinctions between one-step and multistep formulas. Multistep formulas rely heavily on past history (that is, values $y_i, y_{i-1}, \ldots$) to estimate $y_{i+1}$, whereas Runge-Kutta methods probe out ahead of $x_i$, sampling the flow field to estimate $y_{i+1}$, not explicitly caring about past history (except for the value $y_i$). Taylor's series methods are a middle ground, looking neither ahead nor back and basing an estimate of $y_{i+1}$ solely on the information at hand at $(x_i, y_i)$.

Finally, we mention that not all multistep methods need be derived from quadratures. For example, *finite-difference* schemes are based on approximating the derivative. To be more explicit, if $y' = f(x, y)$ is the equation to be solved, then we can construct a number of approximations to $y'$. For instance, using the approximation $y'(x_i) \approx [y(x_{i+1}) - y(x_i)]/h$ in $y'(x) = f(x, y(x))$ leads to

$$\frac{y_{i+1} - y_i}{h} = f(x_i, y_i), \quad i = 0, 1, \ldots$$

which is Euler's method in a slightly different guise. A frequently used approximation to the derivative is the centered-difference formula (see (6.27)), wherein $y'(x_i) \approx [y(x_{i+1}) - y(x_{i-1})]/2h$. This leads to the method

$$y_{i+1} = y_{i-1} + 2hf(x_i, y_i), \qquad i = 1, 2, \ldots$$

(which has poor stability characteristics and requires two starting values $y_0$ and $y_1$). Besides finite-difference schemes, there are integral and derivative approximations which use estimates of $y(x_j)$ and $y'(x_j)$ (Hermite methods), methods which utilize theoretical estimates of $y_{i+1} - \tilde{y}_{i+1}$ (for instance, Hamming's method). In addition, there are special methods for special types of equations: Störmer-Cowell methods for equations of the form $y'' = f(x, y)$, special methods for "stiff" systems, etc.

## 7.4 ROUND-OFF ERRORS

So far, in our discussion of numerical methods for ordinary differential equations, we have not considered the effects of rounding error, except to advocate the use of partial double precision. In fact, round-off error will limit (in some cases, severely) the accuracy that can be attained. Unfortunately, while round-off errors are important in the practical setting of solving differential equations, their effects are complicated and difficult to analyze. In this introductory section, we will illustrate rounding errors by showing their effects in Euler's method and, in Section 7.6.2, we will return to the topic when stability is discussed. Thus, in lieu of a detailed analysis of rounding error, we will analyze Euler's method in order to outline the dimensions of the problem.

Since Euler's method is given by

$$y_{i+1} = y_i + hf(x_i, y_i), \qquad i = 0, 1, 2, \ldots$$

there are two places for rounding error to occur, namely in the evaluation of $f(x_i, y_i)$ and in the formation of $y_i + hf(x_i, y_i)$. The use of partial double precision will help in controlling errors when $y_i + hf(x_i, y_i)$ is formed, but we cannot usually avoid making errors in evaluating $f(x_i, y_i)$. As a simple example, for the differential equation $y' = x \sin y$, calculating $f(x_i, y_i)$ would involve both round-off errors and the errors made by the computer approximation for $\sin(y_i)$. In a more realistic setting, evaluating $f(x_i, y_i)$ may involve table look-ups, interpolation, and other approximations which are dictated by the problem and for which we cannot control the accuracy. As a final observation, we note that the step-size $h$ should be chosen (if possible) with reference to how accurately the particular machine can represent the numbers $x_i = x_0 + ih$. For example, on hexadecimal machines the number $h = \frac{1}{1000}$ may not be stored internally as accurately as the number $h = \frac{1}{1024}$.

Theoretically, Euler's method produces the estimates $y_{i+1} = y_i + hf(x_i, y_i)$. In fact, we see from the discussion above that the computer actually calculates the numbers

$$u_{i+1} = u_i + hf(x_i, u_i) + \varepsilon_i$$

where we use $\varepsilon_i$ to denote the "local rounding error" in passing from the machine estimate $u_i$ to the next machine estimate $u_{i+1}$. To simplify our analysis, we next make the idealistic assumption that $y_0$ and the numbers $x_i$ are exactly representable in the machine. As in Theorem 7.2, we suppose that $f(x, y)$ satisfies the Lipschitz condition $|f(x, v) - f(x, w)| \leq K|v - w|$ for $x_0 \leq x \leq x_n$ and all $v$ and $w$. We next define $\delta_i = y_i - u_i$, so that $\delta_i$ represents the difference between what is calculated in the machine and what Euler's method would produce if no errors were made. With these preliminaries, we have

$$y_{i+1} - u_{i+1} = y_i + hf(x_i, y_i) - u_i - hf(x_i, u_i) - \varepsilon_i$$

or

$$\delta_{i+1} = \delta_i + h[f(x_i, y_i) - f(x_i, u_i)] - \varepsilon_i.$$

As in Theorem 7.2, with $e_i = y(x_i) - y_i$, we again have

$$e_{i+1} = e_i + h[f(x_i, y(x_i)) - f(x_i, y_i)] + h\tau_i.$$

Thus the error between the true solution at $x_i$ and the machine approximation $u_i$, which we denote as $E_i = y(x_i) - u_i$, is given by $E_i = e_i + \delta_i$. Hence, using the equations for $e_i$ and $\delta_i$ above,

$$E_{i+1} = E_i + h[f(x_i, y(x_i)) - f(x_i, u_i)] + h\tau_i - \varepsilon_i.$$

If we let $\tau = \max |\tau_k|$ and $\varepsilon = \max |\varepsilon_k|$, we can use the Lipschitz condition for $f(x, y)$ to find

$$|E_{i+1}| \leq |E_i|(1 + hK) + h\tau + \varepsilon.$$

Proceeding exactly as in the proof of Theorem 7.2, we then obtain

$$|E_{i+1}| \leq (1 + hK)^{i+1}|E_0| + (h\tau + \varepsilon)[1 + (1 + hK) + \cdots + (1 + hK)^i].$$

Since we are assuming that $y_0 = u_0$, we have $E_0 = 0$ and thus

$$|E_{i+1}| \leq (h\tau + \varepsilon)\frac{(1 + hK)^{i+1} - 1}{hK}$$

which, using the bound $(1 + hK)^{i+1} \leq e^{(i+1)hK}$, leads to

$$|y(x_{i+1}) - u_{i+1}| \leq \frac{e^{K(x_{i+1}-x_0)} - 1}{K}\left(\tau + \frac{\varepsilon}{h}\right).$$

The above inequality is both significant and disheartening because of the factor $(\tau + (\varepsilon/h))$. We recall from (7.11) that for Euler's method the local truncation error is $\tau_k = (h/2)y''(\theta_k)$. Hence, although $\tau$ decreases as $h$ decreases, the term $\varepsilon/h$ *increases* with decreasing $h$. This agrees with our intuition, for while we expect Euler's method to yield smaller truncation errors as $h$ decreases, we also take more steps as $h$ decreases and hence expect a larger influence from rounding. For a given machine, a given program, and a given differential equation, there is clearly an optimum value of $h$ for which the quantity $(\tau + (\varepsilon/h))$ is a minimum. Unfortunately, in all but the most simple cases, we cannot expect to do more than make an educated guess at what this optimum step-size might be. Although we have considered only Euler's method in describing the propagation of rounding error, similar results hold for other single-step and multistep methods.

**Example 7.9.** As a demonstration of the effects of a decreasing step-size, consider the equation $y' = 2xy$, $y(0) = 1$ which has $y(x) = e^{x^2}$ as the solution. In this example, we use Euler's method with step-size $h = 1/n$, finding $u_n$ as an estimate to $e = y(1) = 2.71828$ (to six places). Some representative results are tabulated below.

| $n$ | $u_n$ | $n$ | $u_n$ |
|---|---|---|---|
| 100 | 2.72706 | 700 | 2.71797 |
| 200 | 2.72238 | 1100 | 2.71666 |
| 300 | 2.72061 | 2200 | 2.71386 |
| 500 | 2.71894 | 3300 | 2.71140 |
| 600 | 2.71845 | 4400 | 2.70894 |

As the table shows (for this particular differential equation and this particular machine), the results improve as $h$ decreases to about $h = 1/600$. As $h$ continues to decrease, the answers become progressively worse until for $h = 1/4400$ the estimate at $x = 1$ is actually worse than that computed with $h = 1/100$.

## 7.5 *n*th-ORDER DIFFERENTIAL EQUATIONS AND SYSTEMS OF DIFFERENTIAL EQUATIONS

So far, we have only considered first-order differential equations, $y' = f(x, y)$, $y(x_0) = y_0$. In fact, most differential equations encountered in scientific and engineering problems are equations that are second order or higher. To be specific, the general $n$th-order initial-value problem takes the form

$$y^{(n)} = f(x, y, y', \ldots, y^{(n-1)}), \qquad y(x_0) = y_0, y'(x_0) = y'_0, \ldots, y^{(n-1)}(x_0) = y_0^{(n-1)}.$$

The $n$th-order equation above can be converted into an equivalent system of $n$ first-order equations by setting $u_1(x) = y(x)$, $u_2(x) = y'(x), \ldots, u_n(x) = y^{(n-1)}(x)$. When we do this, we obtain the system

$$\begin{aligned} u_1'(x) &= u_2(x) \\ u_2'(x) &= u_3(x) \\ &\vdots \\ u_n'(x) &= f(x, u_1(x), u_2(x), \ldots, u_n(x)). \end{aligned}$$

We next set

$$\mathbf{u}(x) = \begin{bmatrix} u_1(x) \\ u_2(x) \\ \vdots \\ u_n(x) \end{bmatrix}, \qquad \mathbf{F}(x, \mathbf{u}(x)) = \begin{bmatrix} u_2(x) \\ u_3(x) \\ \vdots \\ f(x, u_1(x), u_2(x), \ldots, u_n(x)) \end{bmatrix}$$

to obtain an equivalent first-order system of equations in vector form, namely

$$\mathbf{u}'(x) = \mathbf{F}(x, \mathbf{u}(x)), \mathbf{u}(x_0) = \mathbf{u}_0$$

where the vector $\mathbf{u}_0$ of initial-conditions is given by $\mathbf{u}_0 = [y_0, y_0', \ldots, y_0^{(n-1)}]^T$. In this reformulation, $\mathbf{u}'(x) = \mathbf{F}(x, \mathbf{u}(x))$, we are using differentiation and integration in a component-wise sense, so for example,

$$\int_{x_0}^{x} \mathbf{F}(t, \mathbf{u}(t))\, dt = \begin{bmatrix} \int_{x_0}^{x} u_2(t)\, dt \\ \vdots \\ \int_{x_0}^{x} u_{n-1}(t)\, dt \\ \int_{x_0}^{x} f(t, u_1(t), \ldots, u_n(t))\, dt \end{bmatrix}.$$

To solve the problem $\mathbf{u}'(x) = \mathbf{F}(x, \mathbf{u}(x))$, we could apply any of the preceding numerical schemes, where the only difference is that we calculate vector quantities instead of scalar quantities. For example, Euler's method becomes

$$\mathbf{u}_{i+1} = \mathbf{u}_i + h\mathbf{F}(x_i, \mathbf{u}_i); \qquad i = 0, 1, 2, \ldots$$

and the improved Heun's method is now

$$\tilde{\mathbf{u}}_{i+1} = \mathbf{u}_i + \frac{h}{2}[-\mathbf{F}(x_{i-1}, \mathbf{u}_{i-1}) + 3\mathbf{F}(x_i, \mathbf{u}_i)]$$

$$\mathbf{u}_{i+1} = \mathbf{u}_i + \frac{h}{2}[\mathbf{F}(x_i, \mathbf{u}_i) + \mathbf{F}(x_i, \tilde{\mathbf{u}}_{i+1})].$$

A second-order Runge-Kutta method, the original Heun's method, will illustrate more clearly the programming considerations that must be taken into account.

First, we consider the general first-order system of two equations:

$$v' = g(x, v, w)$$
$$w' = f(x, v, w)$$

with initial-conditions, $v(x_0) = v_0$, $w(x_0) = w_0$. For this system, Heun's method is

$$\mathbf{u}_{i+1} = \mathbf{u}_i + \frac{h}{2}[\mathbf{K}_1 + \mathbf{K}_2]$$

where $\mathbf{K}_1 = \mathbf{F}(x_i, \mathbf{u}_i)$ and $\mathbf{K}_2 = \mathbf{F}(x_{i+1}, \mathbf{u}_i + h\mathbf{F}(x_i, \mathbf{u}_i))$. Writing this out, we find

$$\mathbf{K}_1 = \begin{bmatrix} g(x_i, v_i, w_i) \\ f(x_i, v_i, w_i) \end{bmatrix}, \qquad \mathbf{K}_2 = \begin{bmatrix} g(x_{i+1}, v_i + hg(x_i, v_i, w_i), w_i + hf(x_i, v_i, w_i)) \\ f(x_{i+1}, v_i + hg(x_i, v_i, w_i), w_i + hf(x_i, v_i, w_i)) \end{bmatrix}.$$

The special case of a second-order initial-value problem is important, since it often occurs in problems of mechanics, electronics, etc. For this case, $y'' = f(x, y, y')$, $y(0) = y_0$, and $y'(0) = y'_0$, we set $v = y$ and $w = y'$. As before, we find

$$v' = w$$
$$w' = f(x, v, w).$$

Thus, using $g(x, v, w) = w$, we find $\mathbf{K}_1$ and $\mathbf{K}_2$ as above to be

$$\mathbf{K}_1 = \begin{bmatrix} w_i \\ f(x_i, v_i, w_i) \end{bmatrix}, \qquad \mathbf{K}_2 = \begin{bmatrix} w_i + hf(x_i, v_i, w_i) \\ f(x_{i+1}, v_i + hw_i, w_i + hf(x_i, v_i, w_i)) \end{bmatrix}.$$

Thus a program would proceed in a serial fashion, finding first $f(x_i, v_i, w_i)$, then $w_i + hf(x_i, v_i, w_i)$, then $f(x_{i+1}, v_i + hw_i, w_i + hf(x_i, v_i, w_i))$, and using them in the above fashion for each $i$.

## PROBLEMS

1. Verify that the nonlinear differential equation $y' = -y^2$, $y(0) = -1$ has the solution $y = 1/(x - 1)$. Program the explicit method given in (7.22) and the explicit two-point Adams-Bashforth method (given following Example 7.7). Test your program on the differential equation above, obtaining an estimate to $y(0.95)$. Use step-size $h = 0.05$, $h = 0.01$, and $h = 0.005$.
2. Add the appropriate correctors to your program in Problem 1 (i.e., program the improved Heun's method and Milne's method). Test your program as in Problem 1 and verify the improvement in accuracy.
3. Program the Adams-Moulton method and the 4th-order Runge-Kutta method of (7.17). Test these programs as in Problem 1 (use exact starting values for the Adams-Moulton method). Contrast your results. Also test the Adams-Moulton program on the problems in Example 7.8 and contrast the results with Milne's method.

4. As an experiment to show that predictors and correctors should be "balanced," write a program using the 1-point predictor from Heun's method and the 4-point corrector from the Adams-Moulton method. Test this program as in Problem 1.
5. Suppose a 1-point quadrature rule $A_1 g(x_i)$ is used to approximate $\int_{x_i}^{x_{i+1}} g(x)\,dx$. Show that if this predictor is used with no corrector, then Euler's method results.
6. Recall that the general corrector formula (7.25) can be written as

$$y_{i+1} = Q_i + B_0 f(x_{i+1}, y_{i+1})$$

where $Q_i$ and $x_{i+1}$ are fixed. Thus a corrector formula is, in reality, a fixed-point equation of the form $t = Q_i + B_0 f(x_{i+1}, t)$. From Section 6.2.1, we know that $B_0$ has the form $B_0 = hA_0$, where $A_0$ is some constant. Suppose $t = y^*_{i+1}$ is a solution to $t = Q_i + B_0 f(x_{i+1}, t)$. Using the results of Section 4.3, show that if $h$ is small enough and if the predicted value, $\tilde{y}_{i+1}$, is near enough to $y^*_{i+1}$, then a corrector iteration such as (7.26b) will converge. (Hint: Let $\gamma(t) = Q_i + B_0 f(x_{i+1}, t)$ and find $\gamma'(t)$.) What is the rate of convergence of (7.26b)?
7. Using the method of undetermined coefficients, find constants $a_0$, $a_1$, $a_2$, and $a_3$ so that

$$a_0 p(-2h) + a_1 p(-h) + a_2 p(0) + a_3 p(h) = p'(0),$$

for $p(x) = 1$, $p(x) = x$, $p(x) = x^2$, $p(x) = x^3$. Translate this "numerical differentiation" formula to the interval $[x_{i-2}, x_{i+1}]$, so that $y'(x_i) \approx a_0 y(x_{i-2}) + a_1 y(x_{i-1}) + a_2 y(x_i) + a_3 y(x_{i+1})$.
8. The results of Problem 7 provide an example of a *finite-difference* method

$$a_0 y_{i-2} + a_1 y_{i-1} + a_2 y_i + a_3 y_{i+1} = f(x_i, y_i), \qquad i = 2, 3, \ldots$$

for the problem $y'(x) = f(x, y(x))$. Test this method on the equation in Problem 1, using exact starting values.
9. Carry out an experiment similar to that in Example 7.9, applying Heun's method and the improved Heun's method to the equation of Problem 1. For $h = 0.95/n$, try to determine (roughly) the largest value of $n$ that can be used before the accuracy of the approximations to $y(0.95)$ begin to degrade.

*10. Program Heun's method and the 4th-order Runge-Kutta method in (7.17) for the system of two 2nd-order equations:

$$y'' = (z^2 - 1)/y$$
$$z'' = -z$$

where $0 \le x \le 1$, $y(0) = 1$, $y'(0) = 0$, $z(0) = 0$, $z'(0) = 1$ (see Problem 1, Section 7.1). Note that the corresponding first-order vector system will be four dimensional.

## 7.6 ERRORS IN PREDICTOR-CORRECTOR METHODS

In this section we will consider some aspects of errors in predictor-corrector methods and how we can make use of this information by monitoring the error estimates and then modifying the step-size $h$, when necessary, to preserve accuracy.

We first write a general predictor-corrector method as

$$\tilde{y}_{i+1} = y_{i-q} + h \sum_{j=0}^{m} A_j f(x_{i-j}, y_{i-j}) \tag{7.27a}$$

$$y_{i+1} = y_{i-p} + h \sum_{j=1}^{n} B_j f(x_{i+1-j}, y_{i+1-j}) + hB_0 f(x_{i+1}, \tilde{y}_{i+1}). \tag{7.27b}$$

In these formulas, we assume the predictor is derived from an $(m + 1)$-point quadrature formula for $[x_{i-q}, x_{i+1}]$ and the corrector from an $(n + 1)$-point quadrature formula for $[x_{i-p}, x_{i+1}]$. The presence of the step-size $h$ multiplying the quadrature weights comes from formula (6.4).

The first task in an error analysis is to determine the local truncation error; which is the error, $y(x_{i+1}) - y_{i+1}$, that would be made if we used the exact values of the solution for the quantities $y_i, y_{i-1}, \ldots$, in (7.27). To simplify the notation, we write the right-hand side of (7.27a) as $P$ and the right-hand side of (7.27b) as $Q + hB_0 f(x_{i+1}, \tilde{y}_{i+1})$, so that

$$\tilde{y}_{i+1} = P \tag{7.28a}$$

$$y_{i+1} = Q + hB_0 f(x_{i+1}, \tilde{y}_{i+1}). \tag{7.28b}$$

Since

$$y(x_{i+1}) = y(x_{i-q}) + \int_{x_{i-q}}^{x_{i+1}} y'(x)\,dx = y(x_{i-p}) + \int_{x_{i-p}}^{x_{i+1}} y'(x)\,dx,$$

we have

$$y(x_{i+1}) = P + E_P \tag{7.28c}$$

$$y(x_{i+1}) = Q + hB_0 f(x_{i+1}, y(x_{i+1})) + E_C \tag{7.28d}$$

where $E_P$ and $E_C$ denote the quadrature errors that would be made if exact values of $y'(x_k)$ were used in the quadrature sums (recall that we are assuming that $y_k = y(x_k)$, $k = 0, 1, \ldots, i$).

Using (7.28b) and (7.28d) and the mean-value theorem, we have

$$\begin{aligned} y(x_{i+1}) - y_{i+1} &= hB_0[f(x_{i+1}, y(x_{i+1})) - f(x_{i+1}, \tilde{y}_{i+1})] + E_C \\ &= hB_0 f_y(x_{i+1}, \eta)(y(x_{i+1}) - \tilde{y}_{i+1}) + E_C. \end{aligned}$$

The term $y(x_{i+1}) - \tilde{y}_{i+1}$ in the above equation can be determined from (7.28a) and (7.28c), giving

$$y(x_{i+1}) - \tilde{y}_{i+1} = E_P.$$

Thus we have the local truncation error, $h\tau_i \equiv y(x_{i+1}) - y_{i+1}$, given by

$$h\tau_i = y(x_{i+1}) - y_{i+1} = hB_0 f_y(x_{i+1}, \eta)E_P + E_C. \tag{7.29}$$

Figures 7.6 and 7.7 give the weights and error terms for some quadrature formulas. These error terms can be used in (7.29) to derive the local truncation error. For

Open Newton-Cotes formulas: $\int_{x_{i-q}}^{x_{i+1}} y'(x)\,dx = h \sum_{j=0}^{q-1} A_j y'(x_{i-j}) + E_P$

| $q$ | $A_0$ | $A_1$ | $A_2$ | $A_3$ | $E_P$ |
|---|---|---|---|---|---|
| 1 | 2 | | | | $\frac{h^3}{3} y'''(\eta)$ |
| 2 | $\frac{3}{2}$ | $\frac{3}{2}$ | | | $\frac{3h^3}{4} y'''(\eta)$ |
| 3 | $\frac{8}{3}$ | $-\frac{4}{3}$ | $\frac{8}{3}$ | | $\frac{14h^5}{45} y^{(v)}(\eta)$ |
| 4 | $\frac{55}{24}$ | $\frac{5}{24}$ | $\frac{5}{24}$ | $\frac{55}{24}$ | $\frac{95h^5}{144} y^{(v)}(\eta)$ |

Adams-Bashforth formulas: $\int_{x_i}^{x_{i+1}} y'(x)\,dx = h \sum_{j=0}^{r} A_j y'(x_{i-j}) + E_P$

| $r$ | $A_0$ | $A_1$ | $A_2$ | $A_3$ | $E_P$ |
|---|---|---|---|---|---|
| 0 | 1 | | | | $\frac{h^2}{2} y''(\eta)$ |
| 1 | $\frac{3}{2}$ | $-\frac{1}{2}$ | | | $\frac{5h^3}{12} y'''(\eta)$ |
| 2 | $\frac{23}{12}$ | $-\frac{16}{12}$ | $\frac{5}{12}$ | | $\frac{3h^4}{8} y''''(\eta)$ |
| 3 | $\frac{55}{24}$ | $-\frac{59}{24}$ | $\frac{37}{24}$ | $-\frac{9}{24}$ | $\frac{251h^5}{720} y^{(v)}(\eta)$ |

**Fig. 7.6** Explicit (predictor) formulas

example, from Fig. 7.7 (or from (6.11b), we know the error in using a Simpson's rule corrector is given by $E_C = (-h^5/90)y^{(v)}(\alpha)$. Using the open three-point Newton-Cotes predictor, we find $E_P = (14h^5/45)y^{(v)}(\gamma)$. See Fig. 7.6. Hence, by (7.29), the local truncation error for Milne's method is given by

$$h\tau_i = \frac{h}{3} f_y(x_{i+1}, \eta) \frac{14h^5}{45} y^{(v)}(\gamma) - \frac{h^5}{90} y^{(v)}(\alpha).$$

Equation (7.29) suggests strongly that predictor-corrector pairs should be

Closed Newton-Cotes formulas: $\int_{x_{i-p}}^{x_{i+1}} y'(x)\,dx = h\sum_{j=0}^{p} B_j y'(x_{i+1-j}) + E_C$

| $p$ | $B_0$ | $B_1$ | $B_2$ | $B_3$ | $B_4$ | $E_C$ |
|---|---|---|---|---|---|---|
| 1 | $\frac{1}{2}$ | $\frac{1}{2}$ | | | | $-\frac{h^3}{12}y'''(\eta)$ |
| 2 | $\frac{1}{3}$ | $\frac{4}{3}$ | $\frac{1}{3}$ | | | $-\frac{h^5}{90}y^{(v)}(\eta)$ |
| 3 | $\frac{3}{8}$ | $\frac{9}{8}$ | $\frac{9}{8}$ | $\frac{3}{8}$ | | $-\frac{3h^5}{80}y^{(v)}(\eta)$ |
| 4 | $\frac{14}{45}$ | $\frac{64}{45}$ | $\frac{24}{45}$ | $\frac{64}{45}$ | $\frac{14}{45}$ | $-\frac{8h^7}{945}y^{(vii)}(\eta)$ |

Adams-Moulton formulas: $\int_{x_i}^{x_{i+1}} y'(x)\,dx = h\sum_{j=0}^{r} B_j y'(x_{i+1-j}) + E_C$

| $r$ | $B_0$ | $B_1$ | $B_2$ | $B_3$ | $E_C$ |
|---|---|---|---|---|---|
| 0 | 1 | | | | $-\frac{h^2}{2}y''(\eta)$ |
| 1 | $\frac{1}{2}$ | $\frac{1}{2}$ | | | $-\frac{h^3}{12}y'''(\eta)$ |
| 2 | $\frac{5}{12}$ | $\frac{8}{12}$ | $-\frac{1}{12}$ | | $-\frac{h^4}{24}y''''(\eta)$ |
| 3 | $\frac{9}{24}$ | $\frac{19}{24}$ | $-\frac{5}{24}$ | $\frac{1}{24}$ | $-\frac{19h^5}{720}y^{(v)}(\eta)$ |

**Fig. 7.7** Implicit (corrector) formulas

"balanced" so that the order of $E_C$ is as comparable as possible with the order of $hE_P$. That is, we would not want to use a very accurate corrector and a crude predictor, for then $\tau_i$ would have the same order as the predictor (in effect, the corrector could not overcome the predictor error). For Milne's method, the predictor and corrector are not precisely balanced, since $h\tau_i$ is of the form $h\tau_i = C_1h^6 + C_2h^5$. However, we note that if the two-point open Newton-Cotes formula were used as the predictor, then $E_P = (3h^3/4)y'''(\mu)$, so that $h\tau_i$ would be of the form $h\tau_i = C_1h^4 + C_2h^5$.

### 7.6.1 Step-size Control for Predictor-Corrector Methods

By making some "reasonable" assumptions, we can use the difference between the predicted value, $\tilde{y}_{i+1}$, and the corrected value, $y_{i+1}$, to make running error estimates. To be specific, let $e_k$ denote the actual error $y(x_k) - y_k$ (rather than the local truncation error). As a typical example of the sort of result that can be obtained, we will show for the Adams-Moulton method that

$$e_{i+1} \approx e_i + \frac{19}{270}(\tilde{y}_{i+1} - y_{i+1}).$$

This sort of approximation means that we can monitor the errors in a predictor-corrector method in terms of a known quantity, $\tilde{y}_{i+1} - y_{i+1}$. Thus we can decrease the step-size, when necessary, as we progress with the calculation.

To see how these approximations are derived, we note from (7.27) and (7.28), that we have

$$y(x_{i+1}) - \tilde{y}_{i+1} = e_{i-q} + h \sum_{j=0}^{m} A_j[f(x_{i-j}, y(x_{i-j})) - f(x_{i-j}, y_{i-j})] + E_p \quad \textbf{(7.30a)}$$

$$\begin{aligned} y(x_{i+1}) - y_{i+1} = e_{i-p} &+ h \sum_{j=1}^{n} B_j[f(x_{i+1-j}, y(x_{i+1-j})) \\ &- f(x_{i+1-j}, y_{i+1-j})] + hB_0[f(x_{i+1}, y(x_{i+1})) \\ &- f(x_{i+1}, \tilde{y}_{i+1})] + E_C, \end{aligned} \quad \textbf{(7.30b)}$$

where $E_P$ is the error of the predictor in integrating $y'(x)$ on $[x_{i-q}, x_{i+1}]$ and $E_C$ is the error of the corrector in integrating $y'(x)$ on $[x_{i-p}, x_{i+1}]$. Replacing the terms $[f(x_k, y(x_k)) - f(x_k, y_k)]$ by $f_y(x_k, \eta_k)e_k$, we obtain an expression for (7.30b) of the form

$$e_{i+1} = e_{i-p} + h \sum_{j=1}^{n} \beta_j e_{i+1-j} + h\beta_0(y(x_{i+1}) - \tilde{y}_{i+1}) + E_C$$

which, in view of (7.30a), reduces to

$$e_{i+1} = e_{i-p} + h\left[\sum_{j=1}^{n} \beta_j e_{i+1-j} + \beta_0 e_{i-q} + h\beta_0 \sum_{j=0}^{m} \alpha_j e_{i-j}\right] + h\beta_0 E_p + E_C. \quad \textbf{(7.31)}$$

Thus the actual error is composed of two parts: (1) a linear combination of the previous errors, $e_i, e_{i-1}, \ldots$ and (2) the error $E_C + h\beta_0 E_p$ that comes from the approximations to the integral of $y'(x)$. We next make the assumption that $e_{i-p}$ is the dominant portion of the error $e_{i+1}$ that is due to the previous errors (since the other errors are all multiplied by $h$). We also assume that $h\beta_0 E_p$ is small with respect to $E_C$ (which is the case for most of the specific predictor-corrector methods we have considered, as the reader may verify from Figs. 7.6 and 7.7). Thus, we are led to the approximation

$$y(x_{i+1}) - y_{i+1} \approx e_{i-p} + E_C. \tag{7.32a}$$

Similar reasoning in (7.30a) gives us the approximation

$$y(x_{i+1}) - \tilde{y}_{i+1} \approx e_{i-q} + E_P. \tag{7.32b}$$

By making one more assumption, we will be able to use the quantity $y_{i+1} - \tilde{y}_{i+1}$ to estimate $E_C$ in (7.32a). First, let us suppose that $E_C$ and $E_p$ are given by

$$E_C = C_1 h^k y^{(r)}(\alpha), \qquad E_p = C_2 h^k y^{(r)}(\gamma).$$

That is, we suppose both the predictor and the corrector have error terms which essentially differ only by $C_1$ and $C_2$ (this supposition often holds true, as a check of Figs. 7.6 and 7.7 will show). Subtracting (7.32b) from (7.32a), we have

$$\tilde{y}_{i+1} - y_{i+1} \approx e_{i-p} + E_C - e_{i-q} - E_p.$$

For the moment, let us assume that $p = q$ (which is the case for all methods we have considered, except for Milne's method). In this case, we find

$$\tilde{y}_{i+1} - y_{i+1} \approx h^k(C_1 y^{(r)}(\alpha) - C_2 y^{(r)}(\gamma)).$$

If we further suppose that $y^{(r)}(x)$ is nearly constant in the interval of integration, $[x_{i-p}, x_{i+1}]$, we have

$$\tilde{y}_{i+1} - y_{i+1} \approx h^k y^{(r)}(\alpha)(C_1 - C_2)$$

which means that in (7.32a)

$$y(x_{i+1}) - y_{i+1} \approx e_{i-p} + \frac{C_1}{C_1 - C_2}(\tilde{y}_{i+1} - y_{i+1}) \tag{7.33}$$

since

$$h^k y^{(r)}(\alpha) \approx \frac{\tilde{y}_{i+1} - y_{i+1}}{C_1 - C_2}.$$

Thus, despite the fact that we do not know $y^{(r)}(x)$ and so do not know $E_C$, we can use $\tilde{y}_{i+1} - y_{i+1}$, to estimate $h^k y^{(r)}(\alpha)$ and, hence, $E_C$. To summarize our assumptions, we have supposed that $h$ is small enough so that $y^{(r)}(x)$ is nearly constant in $[x_{i-p}, x_{i+1}]$, terms $he_k$ are negligible with respect to $e_k$, and $hE_P$ is negligible with respect to $E_C$. Two specific cases are provided by the improved Heun's method where $C_1 = -\frac{1}{12}$ and $C_2 = \frac{5}{12}$, so that

$$e_{i+1} \approx e_i + \frac{1}{6}(\tilde{y}_{i+1} - y_{i+1})$$

and the Adams-Moulton method where $C_1 = -\frac{19}{720}$ and $C_2 = \frac{251}{720}$, so that

$$e_{i+1} \approx e_i + \frac{19}{270}(\tilde{y}_{i+1} - y_{i+1}).$$

We emphasize that these are only estimates, but they can be used as a guide to suggest when the step size $h$ should be decreased (or increased).

**Example 7.10.** In this example, the improved Heun's method with step-size $h = 0.0625$ was used on the problem $y' = 2xy$, $y(0) = 1$, for $0 \leq x \leq 1$. For this method we expect $e_{i+1} \approx e_i + [\tilde{y}_{i+1} - y_i]/6$, and we have tabulated some of the results in Table 7.3, where $d_i = [\tilde{y}_i - y_i]/6$. As we see from Table 7.3, the relationship $e_{i+1} \approx e_i + d_{i+1}$ indicates the trends of the local errors fairly well. If we assume that exact starting values are used so that $e_0 = e_1 = 0$, we would expect that $e_2 \approx d_2$, $e_3 \approx e_2 + d_3 \approx d_2 + d_3$ or, in general, that $e_{16} \approx d_2 + d_3 + \cdots + d_{16}$. Since there are so many approximations, such an estimate would have to be used with caution. For this particular differential equation, we find $d_2 + d_3 + \cdots + d_{16} = -0.00399$, whereas in fact, $e_{16} = -0.00541$ (our "estimated" error is less than 75% of the true error). From this we see that the quantities $d_i$ can serve as indicators of increasing local truncation errors, but cannot be used as guaranteed error bounds. As a point of interest, when the problem was run with $h_1 = h/2 = 0.03125$, the error was $y(1) - y_{32} = -0.00143$, which is about what is to be expected because the improved Heun's method is a 2nd-order method, so that the errors behave like $Kh^2$. (Since $h_1^2 = \frac{1}{4}h^2$, we expect decreasing the step-size by 2 will decrease the errors by about a factor of 4). Also, for this reduced step-size, we find $d_2 + d_3 + \cdots + d_{32} = -0.00107$, which for this problem, is a rough indication of the true error.

**Table 7.3**

| $i$ | $x_i$ | $y_i$ | $e_i$ | $d_i$ |
|---|---|---|---|---|
| 3 | 0.1875 | 1.03584 | −0.00006 | |
| 4 | 0.2500 | 1.06461 | −0.00012 | −0.00005 |
| 7 | 0.4375 | 1.21137 | −0.00042 | |
| 8 | 0.5000 | 1.28461 | −0.00058 | −0.00014 |
| 11 | 0.6875 | 1.60567 | −0.00142 | |
| 12 | 0.7500 | 1.75692 | −0.00187 | −0.00034 |
| 15 | 0.9375 | 2.41242 | −0.00416 | |
| 16 | 1.0000 | 2.72369 | −0.00541 | −0.00083 |

### 7.6.2 Stability of Predictor-Corrector Methods

We will not consider the problem of determining guaranteed error bounds for predictor-corrector formulas, since their derivation is somewhat more complicated than the error analysis made for one-step methods (as in Theorem 7.2). The interested reader may find these error bounds in most advanced numerical analysis texts. We wish to mention briefly the subject of *stability* for multistep methods. The topic of stability is fairly complex and we cannot do proper justice to the subject here. We hope, by considering two examples, to give the reader a feeling for what the problems of stability are. Essentially, the problem is that extraneous "solutions" may creep into the computation (due to rounding errors), and these extraneous solutions may damp out the true solution.

As an illustration, let us consider the two explicit formulas

$$y_{i+1} = y_{i-1} + 2hf(x_i, y_i) \tag{7.34a}$$

$$y_{i+1} = y_i + \frac{h}{2}[-f(x_{i-1}, y_{i-1}) + 3f(x_i, y_i)] \tag{7.34b}$$

where (7.34a) is the one-point open Newton-Cotes formula and (7.34b) is an Adams-Bashforth formula. To further simplify and clarify the discussion, let us suppose that we are using these formulas by themselves (as in Example 7.7), and not as one-half of a predictor-corrector pair. We note that both methods have comparable local truncation errors ($\tau_i = (h^2/3)y'''(\eta)$ for (7.34a) and $\tau_i = (5h^2/12)y'''(\eta)$ for (7.34b)). In fact, on the basis of these local truncation errors, we would judge (7.34a) as a more accurate formula. However, as we shall see, (7.34a) is more subject to problems of stability than is (7.34b).

We observe that both methods require two starting values $y_0$ and $y_1$. Moreover, once these starting values are prescribed, the respective sequences $\{y_k\}$ are completely determined for $k = 2, 3, \ldots$. In (7.34a), for example, given $y_0$ and $y_1$, we can uniquely find $y_k$ for any value of $k$, (say $k = 100$), by computing $y_j = y_{j-1} + 2hf(x_j, y_j)$ for *each* $j$, $2 \leq j \leq 100$. Both (7.34a) and (7.34b) are examples of *difference equations*, and if $f(x, y)$ is simple enough there are techniques for solving either equation for $y_k$ *without* having to calculate each $y_j$, $2 \leq j \leq k$ (i.e., $y_k$ can be given in closed form as a function of $k$). In general, $f(x, y)$ is not simple enough for us to do this, but, to illustrate the concept of stability, we shall consider one elementary example in which it can be done. Let us focus on the simple problem $y'(x) = \alpha y(x)$, $y(0) = y_0$ which has the solution $y(x) = y_0 e^{\alpha x}$. For this special problem, (7.34a) and (7.34b) reduce to (respectively):

$$y_{i+1} = y_{i-1} + 2h\alpha y_i, \qquad i = 1, 2, \ldots \tag{7.34c}$$

$$y_{i+1} = y_i - \frac{h}{2}\alpha y_{i-1} + \frac{3h}{2}\alpha y_i, \qquad i = 1, 2, \ldots. \tag{7.34d}$$

Before continuing, let us consider the second-order difference equation with constant coefficients

$$u_{i+1} + au_i + bu_{i-1} = 0, \qquad i = 1, 2, 3, \ldots \tag{7.35}$$

where starting values $u_0$ and $u_1$ are given, and where $a$ and $b$ are given constants. Note that (7.34c) and (7.34d) are special cases of (7.35). Once again, when $u_0$ and $u_1$ are given, the sequence $\{u_k\}$ defined by (7.35) is determined completely ($u_2 = -au_1 - bu_0$, $u_3 = -au_2 - bu_1$, etc.). This time, though, we can give a *formula* for $u_k$ in terms of $u_0$ and $u_1$. The formula, as established in Problem 2, is analogous to the technique used for solving a second-order differential equation:

1. Let $r_1$ and $r_2$ be roots of $p(x) = 0$, where $p(x) = x^2 + ax + b$, with $a$ and $b$ as in (7.35) and where we assume for simplicity that $r_1 \neq r_2$ and both are real.

2. Determine constants $c_1$ and $c_2$ such that $c_1 + c_2 = u_0$, $c_1 r_1 + c_2 r_2 = u_1$.

Once this is done, we have for all $k$, that

$$u_k = c_1(r_1)^k + c_2(r_2)^k. \tag{7.36}$$

Equation (7.36) is quite satisfactory as a closed-form mathematical expression which gives $u_k$, for any $k$, in terms of $u_0$ and $u_1$. Moreover, (7.36) will allow us to analyze the convergence of (7.34c) and (7.34d) to the solution $y = y_0 e^{\alpha x}$ of our simple problem.

In particular, for (7.34c), $p(x) = x^2 - 2\alpha h x - 1$, which has zeros $r_1$ and $r_2$ given by

$$\frac{2h\alpha \pm \sqrt{4\alpha^2 h^2 + 4}}{2} = h\alpha \pm \sqrt{1 + h^2\alpha^2}. \tag{7.37a}$$

Formula (7.34d) has

$$p(x) = x^2 - \left(1 + \frac{3h\alpha}{2}\right) + \frac{h\alpha}{2},$$

with zeros $\tilde{r}_1$ and $\tilde{r}_2$ given by

$$\frac{1 + \dfrac{3h\alpha}{2} \pm \sqrt{1 + h\alpha + \dfrac{9h^2\alpha^2}{4}}}{2}. \tag{7.37b}$$

In order to estimate these roots for small $h$, we use the Taylor's series expansion

$$\sqrt{1 + v} = 1 + \frac{1}{2} v - \frac{1}{8} v^2 + \cdots$$

in both (7.37a) and (7.37b). Neglecting the terms of the form $h^p$, for $p \geq 2$, we find the roots in (7.37a) are approximately $(1 + h\alpha)$ and $(-1 + h\alpha)$. Similarly the roots in (7.37b) are approximately $(1 + h\alpha)$ and $(h\alpha/2)$.

Thus, given starting values $y_0$ and $y_1$, there are constants $c_1$, $c_2$, $d_1$, and $d_2$ such that

$$y_i \approx c_1(1 + h\alpha)^i + c_2(-1 + h\alpha)^i \tag{7.38a}$$

$$y_i \approx d_1(1 + h\alpha)^i + d_2\left(\frac{h\alpha}{2}\right)^i \tag{7.38b}$$

where (7.38a) and (7.38b) are good approximations (for small $h$) to the values $y_i$ produced by (7.34c) and (7.34d), respectively. Let us fix a number $\tilde{x}$ and suppose that the step-size $h$ is chosen so that $\tilde{x} = ih$. We recall from calculus that

$$\lim_{i \to \infty} \left(1 + \frac{w}{i}\right)^i = e^w$$

and hence we have

$$\lim_{i\to\infty} (1 + h\alpha)^i = \lim_{i\to\infty} \left(1 + \frac{\alpha\bar{x}}{i}\right)^i = e^{\alpha\bar{x}}.$$

Thus, for large $i$, we see that the approximations in (7.38) become respectively (where $y_i \approx y(\bar{x})$):

$$y_i \approx c_1 e^{\alpha\bar{x}} + c_2(-1 + h\alpha)^i \tag{7.39a}$$

$$y_i \approx d_1 e^{\alpha\bar{x}} + d_2 \left(\frac{h\alpha}{2}\right)^i. \tag{7.39b}$$

Now, $(-1 + h\alpha)^i = (-1)^i(1 - h\alpha)^i$, and we have for large $i$ that $(1 - h\alpha)^i \approx e^{-\alpha\bar{x}}$. Finally, assuming that $h$ is small, $(h\alpha/2)^i$ tends to 0. Thus, we find for large $i$, that

$$y_i \approx c_1 e^{\alpha\bar{x}} + (-1)^i c_2 e^{-\alpha\bar{x}} \tag{7.40a}$$

$$y_i \approx d_1 e^{\alpha\bar{x}} \tag{7.40b}$$

where (7.40a) is an approximation to the estimates for $y(x)$ produced by (7.34c) and (7.40b) is an approximation to the estimates for $y(x)$ produced by (7.34d).

We note that the two approximations for the solutions of the difference equations in *both* (7.39a) and (7.39b) contain an approximation for the true solution, $y(\bar{x}) = y_0 e^{\alpha\bar{x}}$, and also contain an extraneous term, for example, $c_2(-1 + h\alpha)^i$ in (7.39a) and $d_2(h\alpha/2)^i$ in (7.39b). If we could perform our calculations in exact arithmetic, i.e., all values are represented exactly and no rounding errors are committed, then we could mathematically show that $c_1 = d_1 = y_0$ and $c_2 = d_2 = 0$. Exact arithmetic is obviously not possible in practice, however, and the extraneous parts of (7.39a) and (7.39b) will contribute to each $y_i$ generated by (7.34c) or (7.34d). Now we can see how the approximations in (7.40) illustrate what stability means. For if $\alpha < 0$ and $c_2 \neq 0$, the presence of the extraneous term $e^{-\alpha\bar{x}}$ in (7.40a) will overpower the term $e^{\alpha\bar{x}}$ representing the true solution, particularly if $\bar{x}$ is large.

On the other hand, rounding errors also introduce a component involving $(h\alpha/2)^i$ into (7.34d), but this component will tend to die out and not affect the accuracy severely. To summarize, both methods may produce extraneous solutions, but in one case the extraneous solution damps out while in the other (if $\alpha < 0$) it grows. Thus, if a method (such as (7.34a)) can be significantly affected by the extraneous part of the solution of its difference equation, it is said to be *unstable*. Our discussion has been limited to the simple problem $y' = \alpha y$, but should indicate what is meant by stability. For a more detailed discussion, and descriptions of other types of stability, there are a number of references (Lambert 1973).

**Example 7.11.** To illustrate the preceding discussion, we ran the problem $y' = -y$, $y(0) = 1$ with a step-size $h = 0.02$, for the methods of both (7.34a) and (7.34b). As expected, for "large" values of $x$, the approximations found by (7.34a) became progressively worse. In the table below, the column headed $A$ denotes the results found at $x$ by (7.34a) and those headed $B$ denote the results found at $x$ by (7.34b).

| $x$ | $A$ | $B$ | $e^{-x}$ |
|---|---|---|---|
| 4.00 | 0.18358E−01 | 0.18328E−01 | 0.18316E−01 |
| 8.00 | 0.23806E−02 | 0.33590E−03 | 0.33546E−03 |
| 16.00 | 0.60926E+01 | 0.11283E−06 | 0.11254E−06 |

We note that at $x = 16$ (after 800 steps), (7.34a) is producing nonsense, while (7.34b) is more nearly correct. Since $y(x) = e^{-x}$ is fairly small, relative errors are more meaningful than absolute errors. When these are listed, as below, the difference in the two methods becomes dramatically distinct, where at $x = 16$, the relative error in column $A$ is ten orders of magnitude larger than the relative error in column $B$. The columns headed $EA$ and $EB$ denote the relative errors:

| $x$ | $EA$ | $EB$ |
|---|---|---|
| 4.00 | 0.229E−02 | 0.655E−03 |
| 8.00 | 0.610E+01 | 0.131E−02 |
| 16.00 | 0.541E+08 | 0.258E−02 |

## 7.7 THE METHOD OF COLLOCATION

The method of collocation is a procedure unlike any of the single-step or multistep methods that we described in the previous sections. The output from a collocation method is a function (usually a polynomial) that approximates the solution of a differential equation throughout some range, $a \le x \le b$. This is in contrast to single-step and multistep methods which produce a discrete set of values $y_1, y_2, \ldots, y_n$. The basic ideas involved in collocation are fairly simple. As a starting point, let us consider the general initial-value problem

$$y'(x) = f(x, y(x)), \qquad y(0) = y_0, \qquad 0 \le x \le b. \tag{7.41}$$

We have chosen the initial condition to be specified at $x_0 = 0$ in order to simplify the development a bit, although this choice of $x_0$ is not a requirement.

Suppose next that there is a polynomial $p(x) = a_0 + a_1x + \cdots + a_nx^n$ that is a good approximation to the solution, $y(x)$, of (7.41). In order to determine $p(x)$, we consider the expression

$$e(x) = p'(x) - f(x, p(x)). \tag{7.42}$$

Our object is to find the coefficients $a_0, a_1, \ldots, a_n$ of $p(x)$ so as to make the error function, $e(x)$, as small as possible. This can be done in a variety of ways. For example, we might ask that $a_0, a_1, \ldots, a_n$ be chosen so as to minimize either of the expressions (7.43a) or (7.43b):

$$\int_0^b [e(x)]^2\,dx \tag{7.43a}$$

$$\sum_{j=0}^{m} [e(t_j)]^2, \qquad 0 \le t_0 < t_1 < \cdots < t_m \le b. \tag{7.43b}$$

Thus (7.43a) and (7.43b) represent least-squares problems which (we note without further discussion) are related to methods which are termed *Galerkin procedures.* If $f(x, y)$ should be nonlinear in $y$ (and if $f(x, y)$ is not too complicated), the parameters $a_0, a_1, \ldots, a_n$ in (7.43) can be found by solving the "normal equations" (see Problem 10).

Rather than using (7.43a) or (7.43b) as a criterion to determine the coefficients of $p(x)$, the method of collocation proceeds by asking that $p(x)$ be chosen so that $e(t_i) = 0$ at certain selected "collocation points," $t_0, t_1, \ldots, t_n$. (We note in passing that *collocation* is a term used interchangeably in the literature with *interpolation.*) Thus collocation leads to the system of $n + 1$ equations

$$e(t_i) = 0, \qquad i = 0, 1, \ldots, n.$$

In practice, we normally ask that $p(x)$ satisfy the initial condition, so that the approximate solution, $p(x)$, has the correct starting value. If $p(0) = y_0$, then clearly $a_0 = y_0$, so we are left with just $n$ coefficients $a_1, a_2, \ldots, a_n$ to find. We specify these by choosing $n$ collocation (or interpolation) points $t_1, t_2, \ldots, t_n$ and insisting that

$$\begin{aligned} e(t_1) &= 0 \\ e(t_2) &= 0 \\ &\vdots \\ e(t_n) &= 0. \end{aligned} \tag{7.44}$$

Writing out a representative equation, $e(t_i) = 0$, for the system (7.44), we find the equation has the form

$$a_1 + 2a_2t_i + \cdots + na_nt_i^{n-1} - f(t_i, y_0 + a_1t_i + \cdots + a_nt_i^n) = 0. \tag{7.45}$$

If $f(x, y)$ is a nonlinear function of $y$, we see from (7.45) that (7.44) is a nonlinear system of $n$ equations in $n$ unknowns. Newton's method for several variables as given in Chapter 4 would then provide a tool for solving this system.

**Example 7.12.** Consider the initial-value problem $y' = 2xy$, $y(0) = 1$, $0 \le x \le 1$. In this example, we will determine a cubic polynomial, $p(x)$, such that $p(0) = 1$ and

$$p'(t_i) - 2t_ip(t_i) = 0, \qquad i = 1, 2, 3.$$

As collocation points, we select the zeros of the Chebyshev polynomial $T_3(x)$, translated to the interval $[0, 1]$. Thus $t_1 = (2 - \sqrt{3})/4$, $t_2 = 0.5$, $t_3 = (2 + \sqrt{3})/4$. Since $p(x) = 1 + a_1x + a_2x^2 + a_3x^3$ and $p'(x) = a_1 + 2a_2x + 3a_3x^2$, we are led to the system:

$$(1 - 2t_1^2)a_1 + 2t_1(1 - t_1^2)a_2 + t_1^2(3 - 2t_1^2)a_3 = 2t_1$$
$$(1 - 2t_2^2)a_1 + 2t_2(1 - t_2^2)a_2 + t_2^2(3 - 2t_2^2)a_3 = 2t_2$$
$$(1 - 2t_3^2)a_1 + 2t_3(1 - t_3^2)a_2 + t_3^2(3 - 2t_3^2)a_3 = 2t_3$$

which, for this example, is a $(3 \times 3)$ *linear* system of equations for the unknowns $a_1$, $a_2$, and $a_3$. (This system is linear since $f(x, y) = 2xy$ is a linear function in $y$.) Solving this linear system, we find

$$p(x) = 1 + 0.15529x - 0.33882x^2 + 1.88235x^3$$

as an approximation to the true solution $y(x) = e^{x^2}$. The following table lists representative values of $p(x)$ and $y(x)$.

| $x$ | $p(x)$ | $y(x)$ |
|---|---|---|
| 0.25 | 1.04706 | 1.06449 |
| 0.50 | 1.22823 | 1.28403 |
| 0.75 | 1.72000 | 1.75505 |
| 1.00 | 2.69882 | 2.71828 |

There are a number of observations we should make about collocation. First, there are many variations which can be useful. For example, we need not have used algebraic polynomials as an approximating form but could have chosen $p(x) = \sum_{i=0}^{n} a_i\phi_i(x)$ as our basic approximating form, where the functions $\phi_i(x)$ might have been cubic splines, trigonometric functions, etc. Further, by differentiating the equation $y' = f(x, y)$ with respect to $x$, we could have insisted that $e(t_i) = 0$ and $e'(t_i) = 0$ at the collocation points (this would be a form of Hermite interpolation). In selecting the collocation points $t_i$ in the above example, we chose a natural set of points with regard to the approximating form, assuming that the zeros of the Chebyshev polynomial $T_n(x)$ are well suited for algebraic polynomial interpolation. If we were using trigonometric polynomials as an approximating form, we might consider using equally spaced collocation points. Finally, collocation methods can be used in boundary-value problems (see Section 7.8) as well as in initial-value problems. In fact, this is the most frequent application of collocation.

In the example above, we see that the estimates to $y(x)$ are off by about as much as 4%. By choosing a polynomial of higher degree, we could have gotten a better approximation, but at the expense of having to solve a larger system. However, even the collocation solution found in Example 7.12 is useful in that it provides an approximation to $y(x)$ *throughout* the interval. That is, collocation can be used to estimate, in a rough fashion, the behavior of the solution in the interval in question. This feature of collocation is useful in problems of engineering design where a solution is sought that is optimal with respect to various engineering parameters. Finally, we note that the collocation solution provides an approximation $p(x)$ that can be evaluated at any point $x$. The utility of having a "closed form" approximation to the solution, $y(x)$, is rarely commented on, and we wish to discuss this briefly.

Suppose $p(x)$ is an approximation to $y(x)$, where $y(x)$ is the solution of $y' = f(x, y)$, $y(x_0) = y_0$. It seems plausible, from an intuitive point of view, that we might set $y(x) = p(x) + c(x)$ and use the equation

$$p'(x) + c'(x) = f(x, p(x) + c(x)), \qquad p(x_0) + c(x_0) = y_0 \tag{7.46}$$

to solve for the "correction function," $c(x)$. If $p(x)$ is a known function, then (7.46) reduces to the initial-value problem

$$c'(x) = f(x, p(x) + c(x)) - p'(x), \qquad c(x_0) = y_0 - p(x_0). \tag{7.47}$$

It is plausible then that we might use any of the single-step or multistep methods developed previously to solve (7.47) for the correction, $c(x)$. Having $c(x)$, we could take $p(x) + c(x)$ as our approximation to $y(x)$. The hope in doing this is that the approximation, $p(x)$, would carry the bulk of the estimation to $y(x)$ and that $c(x)$ would be used to "fine-tune" the approximation. In this way, we might hope to use a larger step-size in solving (7.47) than we could use for the same accuracy if we tried to solve $y' = f(x, y)$ directly. This technique is widely used in the area of celestial mechanics (e.g., Encke's method) and we demonstrate this in Example 7.13.

**Example 7.13.** In this example, we return again to the equation $y' = 2xy$, $y(0) = 1$ used in Example 7.12. We use the collocation solution $p(x)$ found in Example 7.12 as an approximation, and then use Euler's method to solve

$$c'(x) = f(x, p(x) + c(x)) - p'(x), \qquad c(0) = 0. \tag{7.48}$$

In the table below, we tabulate the results of this experiment, where we used $h = 0.0625$. In the column headed $E$, we list the results gotten by solving $y' = 2xy$ with Euler's method. In the column headed $CE$, we list the estimates given by $c_i + p(x_i)$, where $c_i$ is the result of solving (7.48) with Euler's method.

| $x_i$ | $y(x_i)$ | $CE$ | $E$ |
|---|---|---|---|
| 0.25 | 1.06449 | 1.05239 | 1.04755 |
| 0.50 | 1.28403 | 1.27368 | 1.23936 |
| 0.75 | 1.75505 | 1.75039 | 1.65010 |
| 1.00 | 2.71828 | 2.69672 | 2.46401 |

The program was rerun with $h = 0.02$, yielding at $x = 0.5$, $CE = 1.28095$, $E = 1.26926$ and at $x = 1.$, $CE = 2.71109$, $E = 2.63082$.

Techniques such as the one illustrated above are useful when the evaluation of $f(x, y)$ is expensive. For example, problems in celestial mechanics typically involve solving the equations of motion for a body in vacuum, where the only force present is gravity. An accurate evaluation of gravity at a point in space usually requires the evaluation of a *spherical-harmonic* series expansion, which is time consuming. However, an approximate closed-form solution for the trajectory exists if we assume inverse-square gravity. This approximate trajectory, the Kepler solution, could play the same role in the equations of motion that $p(x)$ plays in (7.48). Thus we would be solving for the small perturbation that is necessary to correct the Kepler solution to the true trajectory.

## 7.8 BOUNDARY VALUE PROBLEMS FOR ORDINARY DIFFERENTIAL EQUATIONS

The type of problem we will consider in this section is called a *boundary value* problem, where an interval $[a, b]$ is given and we seek a function $y(x)$ such that

$$y'' = f(x, y, y'); \qquad y(a) = \alpha, \qquad y(b) = \beta. \tag{7.49}$$

Thus, in (7.49), instead of prescribing $y(a)$ and $y'(a)$ as in an initial-value problem, we prescribe values for $y(x)$ at $x = a$ and at $x = b$. Such problems are also called *two-point boundary value* problems and the prescribed conditions $y(a) = \alpha$ and $y(b) = \beta$ are called the *boundary conditions.* Equations of this sort arise quite frequently in scientific and engineering applications and often pose a difficult challenge to the numerical analyst. The equation (7.49) may have no solution, a unique solution, or infinitely many solutions (see Problem 7). Problems less specific than (7.49) often occur. For example, the boundary conditions in (7.49) need not be of the form $y(a) = \alpha$ and $y(b) = \beta$, but might take the form $k_1 y(a) + k_2 y'(a) = \alpha$ and $l_1 y(b) + l_2 y'(b) = \beta$ (or even more complicated forms). In addition, the function $f(x, y, y')$ may contain a parameter, $\lambda$, giving rise to an *eigenvalue problem,* where we also must determine those values $\lambda$ for which a solution exists and then find the solution. However, in order to keep our discussion to a reasonable length, we will restrict ourselves to the problem given in (7.49).

### 7.8.1 Finite Difference Methods

A class of methods called *finite difference methods* are frequently employed to solve (7.49) numerically. In a finite difference procedure, we choose equally spaced points $a = x_0 < x_1 < x_2 < \cdots < x_{n-1} < x_n = b$, where $x_i = x_0 + ih$. We next replace the derivatives in (7.49) by suitably chosen difference quotients defined in terms of the points $x_i$. A typical choice is to use the *centered difference* approximations (see Section 6.4)

$$\begin{aligned} y'(x_i) &\approx \frac{y(x_{i+1}) - y(x_{i-1})}{2h} \\ y''(x_i) &\approx \frac{y(x_{i+1}) - 2y(x_i) + y(x_{i-1})}{h^2}. \end{aligned} \tag{7.50}$$

When these approximations are used in (7.49), we obtain a system of equations of the form

$$\frac{y_{i+1} - 2y_i + y_{i-1}}{h^2} = f\left(x_i, y_i, \frac{y_{i+1} - y_{i-1}}{2h}\right), \qquad \begin{cases} i = 1, 2, \ldots, n-1 \\ y_0 = \alpha,\ y_n = \beta \end{cases} \tag{7.51}$$

In (7.51), we have designated $y_1, y_2, \ldots, y_{n-1}$ as the unknowns, since any solution $y(x)$ of (7.49) will probably not satisfy (7.51) exactly. We think of the system (7.51) as an approximation to the equation in (7.49) and hope that for a sufficiently small spacing, $h$, that the solutions $y_i$ of (7.51) are good approximations to $y(x_i)$.

If the function $f(x, y, y')$ is nonlinear in $y$ and $y'$, we are faced with solving a nonlinear system of $(n - 1)$ equations in $(n - 1)$ unknowns. This circumstance would require that we use a scheme like Newton's method for several variables to determine $y_1, y_2, \ldots, y_{n-1}$. Rather than considering this difficult problem, we consider instead a special case of (7.49) given by

$$y'' - p(x)y' - q(x)y = r(x); \qquad y(a) = \alpha, \qquad y(b) = \beta. \tag{7.52}$$

The special case given in (7.52) is an important special case and contains, for example, many "Sturm-Liouville" problems. If we now make the finite difference approximations given in (7.50), we are led to

$$\frac{y_{i+1} - 2y_i + y_{i-1}}{h^2} - p(x_i)\frac{y_{i+1} - y_{i-1}}{2h} - q(x_i)y_i = r(x_i), \quad 1 \le i \le n-1 \tag{7.53a}$$

which is a *linear* system of equations. Multiplying (7.53a) by $-h^2/2$ and collecting like terms, we find for $1 \le i \le n - 1$, that (7.53a) becomes

$$\left(-\frac{1}{2} - p(x_i)\frac{h}{4}\right) y_{i-1} + \left(1 + q(x_i)\frac{h^2}{2}\right) y_i + \left(-\frac{1}{2} + p(x_i)\frac{h}{4}\right) y_{i+1} = -\frac{h^2}{2} r(x_i). \tag{7.53b}$$

Thus, (7.53b) is a tridiagonal linear system of equations and can be solved quickly by the factorization method of Chapter 2. There is the question, of course, as to whether the system (7.53b) has a solution or not. It is easy to show (see Problem 8) that if $q(x)$ is continuous and positive for $a \le x \le b$ and if $h$ is chosen so that $(h/2)|p(x)| \le 1$, then the coefficient matrix for (7.53b) is diagonally dominant and hence non-singular. (See Theorem 2.3.)

### 7.8.2 Shooting Methods

A "shooting method" is an aptly named procedure for solving (7.49) which is applicable whether $f(x, y, y')$ is linear or nonlinear. The basic idea is to *guess* an initial-value $y'(a) = \alpha_1$ to go along with the left-hand boundary condition $y(a) = \alpha$. Then the *initial-value* problem

$$y'' = f(x, y, y'); \qquad y(a) = \alpha, \qquad y'(a) = \alpha_1 \tag{7.54}$$

is solved using any of the single-step or multistep methods of this chapter. Naturally, we do not expect that this solution will satisfy $y(b) = \beta$, but we can hope to adjust the condition on $y'(a)$ so that we converge to $\beta$ for $x = b$. To see how this is usually carried out, let $y_1(x)$ be the result of solving

$$y'' = f(x, y, y'); \qquad y(a) = \alpha, \qquad y'(a) = \alpha_1,$$

and let $y_2(x)$ be the result of solving

$$y'' = f(x, y, y'); \qquad y(a) = \alpha, \qquad y'(a) = \alpha_2.$$

For simplicity, suppose $y_1(b) < \beta < y_2(b)$. Then there is a number $\lambda$ such that $\lambda y_1(b) + (1 - \lambda)y_2(b) = \beta$. We next try $\alpha_3 = \lambda\alpha_1 + (1 - \lambda)\alpha_2$ and find $y_3(x)$ to solve

$$y'' = f(x, y, y'); \qquad y(a) = \alpha, \qquad y'(a) = \alpha_3.$$

In the above example, $y_1(x)$ undershot the desired ending condition $\beta$ and $y_2(x)$ overshot $\beta$. If $y_3(x)$ is still not near enough to $\beta$ at $x = b$, we choose a new value $\alpha_4$ for the initial conditions $y(a) = \alpha$, $y'(a) = \alpha_4$ and try again.

### 7.8.3 Collocation Methods

Collocation methods are also suitable for problems of the form (7.49), and we describe them briefly. Suppose $l(x)$ is a linear polynomial such that $l(a) = \alpha$, $l(b) = \beta$ and suppose for each $i$, that $q_i(x)$ is a polynomial of degree $i$ such that $q_i(a) = q_i(b) = 0$. If we define $p(x)$ by

$$p(x) = l(x) + \sum_{i=2}^{n} a_i q_i(x) \tag{7.55}$$

where the coefficients $a_i$ are as yet not specified, then we have $p(a) = \alpha$, $p(b) = \beta$. Choosing collocation points $t_2, t_3, \ldots, t_n$, we can determine $p(x)$ by insisting that

$$p''(t_i) = f(t_i, p(t_i), p'(t_i)), \qquad i = 2, 3, \ldots, n.$$

As in Section 7.7, we note that we need not restrict ourselves to polynomials in (7.55), but may choose any suitable approximating form.

## PROBLEMS

1. Program the Adams-Moulton method for $y' = 2xy$, $y(0) = 1$. As in Example 7.10, keep track of the estimates $d_i$ and compare the estimated error with the true error.
2. Consider the difference equation (7.35) given by $u_{i+1} + au_i + bu_{i-1} = 0$; $i = 1, 2, \ldots$. Let $p(x) = x^2 + ax + b$ and suppose $p(x) = (x - r_1)(x - r_2)$, where $r_1 \neq r_2$ and where $r_1$ and $r_2$ are real numbers. Show by direct substitution that the sequences $\{v_i\}_{i=0}^{\infty}$ and $\{w_i\}_{i=0}^{\infty}$ defined by $v_k = r_1^k$ and $w_k = r_2^k$ satisfy the relation (7.35). Next, show that the sequence $\{z_i\}_{i=0}^{\infty}$ defined by $z_k = \alpha v_k + \beta w_k$ also satisfies (7.35), where $\alpha$ and $\beta$ are any two constants. Finally, show that if $u_0$ and $u_1$ are any two prescribed initial-values, that

there exist constants $c_1$ and $c_2$ such that

$$c_1 v_0 + c_2 w_0 = u_0$$
$$c_1 v_1 + c_2 w_1 = u_1.$$

Thus conclude that the solution to $u_{i+1} + au_i + bu_{i-1} = 0$ which starts out $u_0, u_1, \ldots$ is given by $u_i = c_1 v_i + c_2 w_i$, for $i = 0, 1, 2, \ldots$.

3. Using the technique of Problem 2, solve the difference equation $u_{i+1} - 5u_i + 6u_{i-1} = 0$; $u_0 = 3$, $u_1 = 5$. Verify the closed-form solution for $i = 2, 3, 4$. Program this problem and also the problem that has the different initial conditions: $u_0 = 1/3$, $u_1 = 2/3$. Verify that the first terms of the solution found on the computer are correct in the first problem, but that rounding errors are contaminating the computer solution to the second problem. Analyze the difference in accuracy for these two problems, recalling our discussion of stability.
4. If, in Problem 2, $p(x) = (x - r_1)^2$, show that a solution is given by $u_i = c_1(r_1)^i + c_2 i(r_1)^{i-1}$. Use this result to solve the problem $u_{i+1} - 4u_i + 4u_{i-1} = 0$, $u_0 = 1$, $u_1 = 3$.
5. By neglecting terms of order $h^2$, show that the two 2nd-order Runge-Kutta methods defined in Section 7.2.4 are stable in the sense of Section 7.6.2.
6. Write a program to use the method of collocation to estimate a solution to $y'' + y = 0$; $y(0) = 1$, $y'(0) = 0$, for $0 \le x \le 2$. Use $p(x) = 1 + a_2 x^2 + \cdots + a_n x^n$ as an approximating form and insert this directly into the equation $y'' + y = 0$. (Note that $p(0) = y(0)$ and $p'(0) = y'(0)$, when $p(x)$ has the form given above.) Use the zeros of the shifted Chebyshev polynomials, $\tilde{T}_k(x)$, as collocation points, for $k = 1, 2, \ldots, 10$.
7. Consider the boundary value problem $y'' + y = 0$; $y(0) = 0$ and $y(b) = \beta$. Recall that the most general solution of the equation $y'' + y = 0$ is $y(x) = A\cos(x) + B\sin(x)$. Find choices for $b$ and $\beta$ for which the boundary value problem has:
   a) no solution,
   b) exactly one solution,
   c) infinitely many solutions.
8. Show that the coefficient matrix for (7.53b) is diagonally dominant when $q(x)$ is continuous and positive on $[a, b]$ and when $h$ is chosen so that $(h/2)|p(x)| \le 1$, for $a \le x \le b$.
9. Program a shooting method and the finite-difference method for the problem $y'' + y = 0$; $y(0) = 0$, $y(1) = 0.841471$ (Note: $\sin(1) = 0.841471$).

*10. Let $F(a_0, a_1, \ldots, a_n)$ equal either $\int_0^b [e(x)]^2\,dx$ or $\sum_{j=0}^{m} [e(t_j)]^2$ as in (7.43a) or (7.43b), respectively. As in the section on Newton's method in several variables in Chapter 4, the derivative of $F(\mathbf{a})$, $\mathbf{a} \equiv (a_0, a_1, \ldots, a_n)$, equals the $(1 \times (n + 1))$ Jacobian (or gradient),

$$\mathbf{F}'(\mathbf{a}) = \left(\frac{\partial F(\mathbf{a})}{\partial a_0}, \frac{\partial F(\mathbf{a})}{\partial a_1}, \ldots, \frac{\partial F(\mathbf{a})}{\partial a_n}\right).$$

Solving the $(n + 1)$ equations, $\mathbf{F}'(\mathbf{a}) = \mathbf{0}$, in the $(n + 1)$ unknowns, $a_0, a_1, \ldots, a_n$, is a condition for minimizing $F(\mathbf{a})$, and thus yielding a polynomial to minimize (7.43a) or (7.43b). (We will pursue this further in Chapter 8.) The set of equations $\mathbf{F}'(\mathbf{a}) = \mathbf{0}$ is usually nonlinear and thus calls for a technique such as Newton's method. Perform the above computations on the specific problem given in Example 7.12.

# 8

# OPTIMIZATION

## 8.1 INTRODUCTION

Optimization is a broad topic which treats some very modern and practical problems. These range from basic problems, such as a store manager's selection of products to maximize his profits or a farmer's choice of crops, fertilizers, and pesticides to maximize his returns, to modern control theory problems, such as determining a trajectory for a satellite rendezvous which uses a minimum amount of fuel. Optimization is thus a modern and on-going topic of practical research. As one might suspect, however, such a broad and powerful tool is quite deep theoretically and full of pitfalls for one who tries to use it without a sound knowledge of its basic principles. We must therefore limit ourselves in this text to a brief and somewhat superficial survey of certain optimization techniques, with a clear warning that caution must be exercised in most cases. Some of the material in this chapter depends on the use of vectors and vector norms (Chapter 2); the reader should also be familiar with Newton's method in several variables (Chapter 4). The material on linear programming in Section 8.5, however, is primarily self-contained and can be read without these prerequisites.

## 8.2 EXTENSIONS OF RESULTS FROM CALCULUS

We shall first consider the problem of finding the relative maxima or minima (relative extrema) of a real-valued function of $n$ variables, $f(x_1, x_2, \ldots, x_n) \equiv f(\mathbf{x})$: $R^n \to R$, where $\mathbf{x} \equiv (x_1, x_2, \ldots, x_n)^T$. A vector $\mathbf{x}^*$ is a *relative maximum* for $f$ on a domain $D \subseteq R^n$ if there exists an open sphere $S$ centered at $\mathbf{x}^*$ such that $f(\mathbf{x}^*) > f(\mathbf{x})$ for every $\mathbf{x}$ in $D \cap S$. (Note that $f(\mathbf{x}^*) > f(\mathbf{x})$ is well-defined since $f(\mathbf{x}^*)$ and $f(\mathbf{x})$ are real numbers.) A similar definition holds for a relative minimum. The problem of finding extrema is difficult both in terms of theory and computation. Letting $z = f(x_1, x_2, \ldots, x_n)$, our problem is to find the relative minima and maxima for $z$ as the vector $\mathbf{x} = (x_1, x_2, \ldots, x_n)^T$ ranges through some region $D$ in $R^n$. For $n = 2$, $z = f(x_1, x_2)$ can be graphed as a surface above the $x_1x_2$-plane (see Fig. 8.1). We often use this as a model for our intuition and think of $z = f(x_1, x_2, \ldots, x_n)$ as being a surface in $R^{n+1}$, where we realize that in the graph, the replacement of the $x_1x_2$-plane by $R^n$ is somewhat artificial (again see Fig. 8.1).

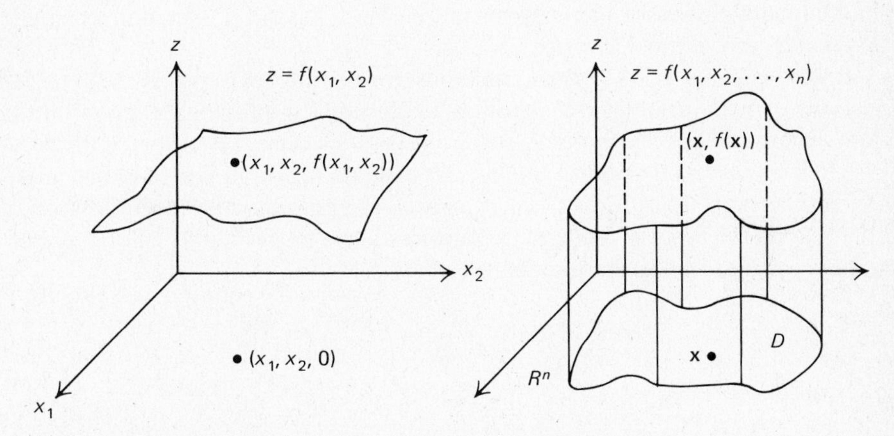


**Fig. 8.1** Visualizing $z = f(x_1, x_2, \ldots, x_n)$

Most often we like to visualize the surface $z = f(x_1, x_2, \ldots, x_n)$ in terms of a topographical map, where the mountain peaks represent the maxima and the valley bottoms represent the minima. We must be careful to keep in mind that the "surfaces" for $n > 2$ can be much more complicated than the surfaces in $n = 2$ with which we are familiar. We must also realize that we can probably only hope to develop techniques which locate relative extrema, and hope that the absolute extrema are among the relative extrema that we are able to locate. This situation is often likened to a person walking in dense fog in mountainous terrain. The bottom of a valley can be located by a process of continuous descent, but the path to the minima may be far from optimal, and furthermore the person has no idea of whether the surrounding valleys (if any exist) are deeper.

In order to gain more insight and a better intuitive feeling, let us consider the basic problem in just one variable, that is, $n = 1$, and see which results from calculus extend to higher dimensions. The reader should recall that if he wishes to minimize or maximize a real-valued function, $f(x)$, for $a \leq x \leq b$, then the typical procedure is to find the derivative, $f'(x)$, and then values $x^*$ such that $f'(x^*) = 0$. Even in this seemingly elementary case we can encounter many theoretical and practical difficulties. First of all, we must know that $f'(x)$ exists for $a < x < b$ and be able to calculate it (which we may or may not be able to do in closed form). Next we must find the zeros of $f'(x)$; this we have seen in Chapter 4 can be a formidable task. Given such a zero $x^*$ (a *critical point*), we must still determine whether it represents a minimum, a maximum, or an inflection point. All of this time we must be aware that an extremum can exist at the endpoint of the interval, and also we must check the points where the derivative does not exist. Thus the simple task of *setting the derivative equal to zero* may not be so "simple" after all. As an elementary example, consider the function $f(x) = |x|$ for $-1 \leq x \leq 1$. This function obviously has extrema at $x = -1$, 0, and 1; yet none of these represent a zero of the derivative.

Some calculus texts present methods for finding extrema of real-valued functions of two variables (see Thomas 1972), but few present the problem for more than two variables. In the two-variable case, suppose $z = f(x, y)$. If the point $(x^*, y^*)$ is an extremal point in the *interior* of some region in the $xy$-plane, and if tangent planes to the surface $z = f(x, y)$ exist at all points in a neighborhood of $(x^*, y^*)$, then the tangent plane to the surface at $(x^*, y^*, f(x^*, y^*))$ will be parallel to the $xy$-plane. Hence the gradient vector,

$$\nabla f(x^*, y^*) \equiv \left(\frac{\partial f(x^*, y^*)}{\partial x}, \frac{\partial f(x^*, y^*)}{\partial y}\right),$$

equals (0, 0). (We recall from calculus (see Thomas 1972, pp. 683–6 for example), that the normal line of the tangent plane at $(x^*, y^*, f(x^*, y^*))$ has direction numbers

$$\left(\frac{\partial f(x^*, y^*)}{\partial x}, \frac{\partial f(x^*, y^*)}{\partial y}, -1\right).$$

Hence, if $\nabla f(x^*, y^*) = (0, 0)$, the normal is parallel to the $z$-axis. Thus the tangent plane at this point is horizontal, since $(x^*, y^*)$ is extremal.) Once again we should be aware that $\nabla f(u, v) = \mathbf{0}$ does not imply that the point $(u, v)$ is extremal.

To see the analogy of this two-variable result to the familiar one-variable case, we recall from Chapter 4 that the derivative of the function $\mathbf{f}: R^n \to R^p$, that is,

$$\mathbf{f}(\mathbf{x}) \equiv \mathbf{f}(x_1, x_2, \ldots, x_n) = \begin{bmatrix} f_1(x_1, x_2, \ldots, x_n) \\ f_2(x_1, x_2, \ldots, x_n) \\ \vdots \\ f_p(x_1, x_2, \ldots, x_n) \end{bmatrix}$$

is the $(p \times n)$ Jacobian matrix,

$$J(\mathbf{x}) = \left(\frac{\partial f_i(x_1, x_2, \ldots, x_n)}{\partial x_j}\right).$$

In this chapter we are considering the problem of finding the extremal points of a *real-valued* function, $f: R^n \to R$, so $p = 1$. In this case the Jacobian matrix is actually a $(1 \times n)$ vector which we commonly call the *gradient of f*, that is, for $p = 1$,

$$\nabla f(\mathbf{x}) \equiv J(x_1, x_2, \ldots, x_n) = \left(\frac{\partial f(x_1, x_2, \ldots, x_n)}{\partial x_1}, \frac{\partial f(x_1, x_2, \ldots, x_n)}{\partial x_2}, \ldots, \frac{\partial f(x_1, x_2, \ldots, x_n)}{\partial x_n}\right). \tag{8.1}$$

The reader will recall that it has been our custom in this text to regard a vector $\mathbf{x} \in R^n$ as a column vector, that is, $\mathbf{x}$ is $(n \times 1)$. Since the gradient, $\nabla f(\mathbf{x})$, in (8.1) is a $(1 \times n)$ matrix, we must be a bit careful to avoid confusion with our usual notation. For this reason we shall hereafter in this chapter use the notation that $\mathbf{g}(\mathbf{x}) \equiv \mathbf{g}(x_1, x_2, \ldots, x_n) \equiv (\nabla f(\mathbf{x}))^T$, that is, $\mathbf{g}(\mathbf{x})$ is an $(n \times 1)$ representation of the gradient. We shall use $\nabla f(\mathbf{x})$ and $\mathbf{g}(\mathbf{x})$ interchangeably when there is no danger of confusion. Hence, from above, we see that finding $(x^*, y^*)$ such that $\nabla f(x^*, y^*) = (0, 0)$ is the direct vector analog of "setting the derivative equal to zero." Furthermore, we can extend the procedure to higher dimensions, as we shall see momentarily.

First, however, we note that we can think of the gradient $\mathbf{g}(\mathbf{x})$ in (8.1) as a function from $R^n \to R^n$, that is, for every $n$-tuple $\mathbf{x} = (x_1, x_2, \ldots, x_n)$, $\mathbf{g}(\mathbf{x})$ is also an $n$-tuple given by $\nabla f(\mathbf{x})^T$ in (8.1). Thus it is possible for $\mathbf{g}(\mathbf{x})$ to have a derivative. If so, since $\mathbf{g}: R^n \to R^n$, the derivative of $\mathbf{g}(\mathbf{x})$ will be the $(n \times n)$ Jacobian matrix

$$H(\mathbf{x}) \equiv g'(\mathbf{x}) \equiv g'(x_1, x_2, \ldots, x_n) \equiv \left(\frac{\partial g_i(x_1, x_2, \ldots, x_n)}{\partial x_j}\right), \tag{8.2a}$$

for $1 \leq i \leq n$ and $1 \leq j \leq n$. Since $H(\mathbf{x})$ is the derivative of $\mathbf{g}(\mathbf{x})$ it must also be the second derivative of $f(\mathbf{x})$ and is called the *Hessian* of $f(\mathbf{x})$. As we shall see, the Hessian will provide us with an analog of the second derivative test. Writing $H(\mathbf{x})$ in terms of $f(\mathbf{x})$ we have, since

$$g_i(\mathbf{x}) = \frac{\partial f(\mathbf{x})}{\partial x_i},$$

that

$$H(\mathbf{x}) \equiv f''(\mathbf{x}) = \left(\frac{\partial^2 f(x_1, x_2, \ldots, x_n)}{\partial x_i\, \partial x_j}\right). \tag{8.2b}$$

In the single-variable case, a differentiable function can have an extremum at the endpoint of an interval without having its derivative equal to zero there. Similarly, we must also be careful of boundary points of regions in higher-dimensional problems. Thus in this section we shall only consider derivative techniques for "nonboundary" points. Loosely speaking, we say that a vector **c** in a region $D$ of $R^n$ is an *interior point of D* if there exists a number $\varepsilon$ and a sphere of radius $\varepsilon > 0$, centered at **c**, such that the sphere is contained in $D$. For example, if $D$ is the unit disk in the $xy$-plane, $D = \{(x, y): x^2 + y^2 \leq 1\}$, then the point $(x, y)$ is interior to $D$ only if $x^2 + y^2 < 1$. Intuitively, if $(x, y)$ is anywhere inside the unit circle, $x^2 + y^2 = 1$, we can draw a sphere centered at $(x, y)$ with a radius small enough that all points within this sphere lie in $D$.

With these preliminaries, we are now able to state (without proof) multivariable analogs of the familiar derivative procedures of calculus for finding extrema of functions of a single variable.

**Theorem 8.1.** Let $f(\mathbf{x})$ map $D$ into the reals where $D \subseteq R^n$. If **c** is an interior point of $D$ at which $f$ is differentiable and at which $f$ has a relative extremum, then $\nabla f(\mathbf{c})^T \equiv \mathbf{g}(\mathbf{c}) = \mathbf{\theta}$.

Now if **c** is a relative minimum of $f(\mathbf{x})$, then **c** is a relative maximum of the function $(-f(\mathbf{x}))$. Thus a minimization problem can always be viewed as a maximization problem and vice-versa. We do, however, need to be concerned with inflection or saddle points, which seem to be more prevalent and harder to distinguish in higher-dimensional problems. (As in the familiar single-variable case, an *inflection* or *saddle* point is a vector where the derivative equals zero, but is not a relative extremum.) As a simple example, let $n = 3$, and let $f(x_1, x_2, x_3) = x_1x_2x_3$. Then $\nabla f(\mathbf{x}) = (x_2x_3, x_1x_3, x_1x_2)$ and $\mathbf{x} = \mathbf{\theta}$ is a critical point of $f(\mathbf{x})$. However, since $f(\mathbf{x})$ assumes both positive and negative values in every neighborhood of the origin, then $\mathbf{x} = \mathbf{\theta}$ must be a saddle point. Another relatively simple example of a saddle point is the origin for the familiar hyperbolic paraboloid $z = x^2/a^2 - y^2/b^2$.

To determine whether a critical point is an extremal point or a saddle point, we have the following multidimensional analog of the second derivative test:

**Theorem 8.2.** Let $f: D \to R$, $D \subseteq R^n$, have continuous second partial derivatives in the neighborhood of a critical point **c** in $R^n$. If the second derivative (the Hessian), $H(\mathbf{c})$ of (8.2b), satisfies

a) $\mathbf{x}^T H(\mathbf{c})\mathbf{x} > 0$, for all $\mathbf{x} \neq \mathbf{\theta}$ in $R^n$, then **c** is a relative minimum;

b) $\mathbf{x}^T H(\mathbf{c})\mathbf{x} < 0$, for all $\mathbf{x} \neq \mathbf{\theta}$ in $R^n$, then **c** is a relative maximum;

Finally, if $\mathbf{x}^T H(\mathbf{c})\mathbf{x}$ has both positive and negative values as **x** ranges through $R^n$, then **c** is a saddle point.

If $(\partial^2 f(\mathbf{c}))/(\partial x_i\, \partial x_j) = (\partial^2 f(\mathbf{c}))/(\partial x_j\, \partial x_i)$ for all $i$ and $j$ (as is implied by the hypotheses above), then $H(\mathbf{c})$ in (8.2b) is symmetric. If $\mathbf{x}^T H(\mathbf{c})\mathbf{x} > 0$ $(<0)$ for all $\mathbf{x} \neq \mathbf{\theta}$ in $R^n$, then we recall from Chapters 2 and 3 that $H(\mathbf{c})$ is positive (negative) definite, which is equivalent (see Problem 7) to the eigenvalues of $H(\mathbf{c})$ being all positive (negative). In these cases, $H(\mathbf{c})$ is non-singular.

## 8.3 DESCENT METHODS

Since maximization problems are easily transposed (as above) into minimization problems, we shall concentrate without loss of generality on the latter. We see by Theorem 8.1, that we should try to find the zeros of $f'(\mathbf{x}) \equiv \nabla f(\mathbf{x}) \equiv \mathbf{g}(\mathbf{x})^T$ and then use Theorem 8.2 to test them. Since $\nabla f(\mathbf{x})^T \equiv \mathbf{g}(\mathbf{x})\colon R^n \to R^n$, we immediately recall that Newton's method for several variables (Section 4.5) is specifically designed to solve a problem like $\nabla f(\mathbf{x})^T = \mathbf{0}$. Since $g'(\mathbf{x}) = H(\mathbf{x})$, as in 8.2b, then Newton's method (formula (4.51)) for this problem becomes:

$$\mathbf{x}_0, \text{ an initial guess for the minimum point,}$$
$$\mathbf{x}_{i+1} = \mathbf{x}_i - (H(\mathbf{x}_i))^{-1}\mathbf{g}(\mathbf{x}_i), \qquad i \geq 0, \tag{8.3a}$$

or, equivalently, solve

$$H(\mathbf{x}_i)\mathbf{z}_i = -\mathbf{g}(\mathbf{x}_i), \qquad \text{where} \qquad \mathbf{z}_{i+1} \equiv \mathbf{x}_{i+1} - \mathbf{x}_i, \qquad i \geq 0. \tag{8.3b}$$

We immediately see difficulties with this approach. The computation of $H(\mathbf{x}_i)$ can often be quite cumbersome or impossible to perform for practical purposes, or $H(\mathbf{x}_i)$ may be singular or not even exist. We also note that the sequence $\{\mathbf{x}_i\}$ in (8.3a) may not converge to a zero of $\mathbf{g}(\mathbf{x})$ unless the initial guess, $\mathbf{x}_0$, is sufficiently close to that zero. Thus we see that a good strategy for finding a zero of $\mathbf{g}(\mathbf{x})$ might be to develop a rather crude method for obtaining an initial approximation to that zero and then use Newton's method (or a modification of it to ease the computational difficulties) to refine that initial approximation. (This is exactly analogous to our suggestions for root-finding in Chapter 4, where we advocated first using crude techniques to isolate zeros and then using Newton's method to converge quickly (quadratically) to them.) Thus we shall now turn to the task of developing this package, consisting of a technique called *steepest descent* to give first approximations to local minima and a modification of Newton's method to improve its computational efficiency.

### 8.3.1 Steepest Descent

To intuitively describe the method of steepest descent, we return to our analogy of a person trying to descend to the bottom of a valley in heavy fog. At any particular point, the natural course would be to descend in the direction in which the slope of the land is steepest and to continue in that direction as long as descent continues. When a point is reached where the path would start to ascend if continued, the person stops, chooses a new direction of steepest descent, and repeats the process. If the valley were shaped like the bottom of a sphere, direction would never need to be changed and the bottom would be reached along the original path. However, if the valley were shaped like a long narrow kidney bean and the descent were started at the opposite end from the true bottom, it could take a considerable time along a very irregular path to reach the bottom.

This would clearly not be an optimal way of reaching the bottom, but the person would at least be assured of eventually getting arbitrarily close to the goal (making moderate assumptions about the walls of the valley).

We wish to develop our minimization technique to mathematically model the above process. In doing so, we must be able to answer two questions at each step of the process: (1) At any given point $\mathbf{x}$ in $R^n$, which direction represents the direction of steepest descent? (2) How far should one go in that direction before changing directions?

The first question is difficult to analyze for $n > 2$ because we cannot construct an actual graph of the surface that is not somewhat artificial. For $n = 2$, however, we can answer this question with basic calculus. Since we wish to proceed in the direction in which the slope is steepest, we must be able to examine the slope in any direction. This is precisely the concept of *directional derivative*, which we shall now briefly review. Given a surface $z = f(x, y)$, a fixed point $\mathbf{P}_0 \equiv (x_0, y_0, f(x_0, y_0))$ on the surface, and a vertical plane $\Pi$ containing $\mathbf{P}_0$ where $\Pi$ makes an angle $\phi$ with the $x$-axis, then the plane and surface have a curve of intersection which also contains $\mathbf{P}_0$. If the surface is sufficiently smooth, this curve will have a tangent at $\mathbf{P}_0$, which is called the *directional derivative of* $z = f(x, y)$ *in the direction* $\phi$ *at the point* $\mathbf{P}_0$ (see Fig. 8.2).

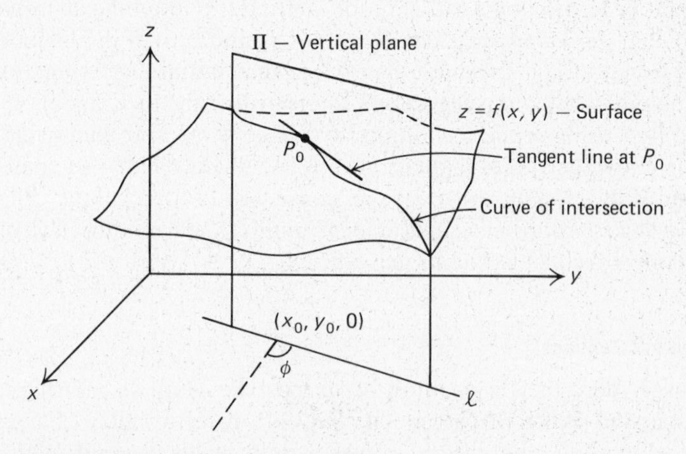


**Fig. 8.2** Directional derivative at $\mathbf{P}_0$ for $z = f(x, y)$

Since $\Pi$ is a vertical plane it has the same equation as its projection line $\ell$ in the $xy$-plane. If $s$ represents the distance along the line $\ell$ from the point $(x_0, y_0, 0)$, then $\ell$ can be represented by the parametric equations:

$$x = x_0 + s\cos(\phi), \qquad y = y_0 + s\cos\left(\frac{\pi}{2} - \phi\right) = y_0 + s\sin(\phi). \quad \textbf{(8.4)}$$

The slope of the tangent line (the directional derivative) is therefore given by

$$\frac{df}{ds} = \lim_{s \to 0} \frac{f(x_0 + s\cos(\phi), y_0 + s\sin(\phi)) - f(x_0, y_0)}{s}. \tag{8.5a}$$

By the chain rule and the equations (8.4) for $\ell$, (8.5a) becomes

$$\frac{df}{ds} = \frac{\partial f}{\partial x}\frac{dx}{ds} + \frac{\partial f}{\partial y}\frac{dy}{ds} \equiv \frac{\partial f(x_0, y_0)}{\partial x}\cos(\phi) + \frac{\partial f(x_0, y_0)}{\partial y}\sin(\phi). \tag{8.5b}$$

Letting $\mathbf{u} = (\cos(\phi), \sin(\phi))$, we can express (8.5b) symbolically as the inner-product

$$\frac{df}{ds} = \langle \nabla f(x_0, y_0), \mathbf{u} \rangle. \tag{8.5c}$$

(To avoid unnecessary theoretical technicalities, we shall adopt the following notation: If $\mathbf{x}$ and $\mathbf{y}$ are $m$-tuples with components $\{x_j\}_{j=1}^m$ and $\{y_j\}_{j=1}^m$ respectively, we shall regard the inner-product $\langle \mathbf{x}, \mathbf{y} \rangle$ as

$$\langle \mathbf{x}, \mathbf{y} \rangle = x_1 y_1 + x_2 y_2 + \cdots + x_m y_m$$

regardless of whether $x$ and/or $y$ are row vectors or column vectors.)

Now we wish to find the angle $\phi_0$ which yields the direction for which $|df/ds|$ is maximized, i.e., where the slope is steepest. We can do this by differentiating (8.5b) with respect to $\phi$, setting it equal to zero, and solving for $\phi_0$. The steps are as follows:

$$\frac{d}{d\phi}\left(\frac{df}{ds}\right) = \frac{\partial f(x_0, y_0)}{\partial x}(-\sin(\phi_0)) + \frac{\partial f(x_0, y_0)}{\partial y}(\cos(\phi_0)) = 0.$$

Thus with

$$\lambda_1 = \frac{\partial f(x_0, y_0)}{\partial x} \quad \text{and} \quad \lambda_2 = \frac{\partial f(x_0, y_0)}{\partial y},$$

we see $\tan(\phi_0) = \lambda_2/\lambda_1$. For $r = \sqrt{\lambda_1^2 + \lambda_2^2}$, it follows that $\cos(\phi_0) = \lambda_1/r$ and $\sin(\phi_0) = \lambda_2/r$, so the direction vector $\mathbf{u}$ in (8.5c) that is in the direction of maximum increase is $\mathbf{u} = (1/r)\,\nabla f(x_0, y_0)$. Thus $\mathbf{u}$ is in the direction of the gradient when the directional derivative is maximized. Since the gradient represents the direction of the outward normal, and we wish the direction of steepest *descent*, we thus see that this direction is given by the negative gradient, $-\nabla f(x_0, y_0)$.

We pause here to give an intuitive argument that this result is true in higher dimensions as well. Suppose we are at the vector $\tilde{\mathbf{x}}$ and wish to move to a vector of the form $\mathbf{x} = \tilde{\mathbf{x}} + \alpha\mathbf{e}$, where $\alpha > 0$ is small and $\mathbf{e}$ is a unit vector determining the direction in which we wish to move so that the object function $f$ decreases most rapidly. If $\alpha$ is sufficiently small and $\nabla f$ is continuous, then by the Mean Value Theorem in several variables (see Section 4.5) we have the approximate equality

$$f(\mathbf{x}) \approx f(\tilde{\mathbf{x}}) + \langle \mathbf{x} - \tilde{\mathbf{x}}, \nabla f(\tilde{\mathbf{x}}) \rangle.$$

(To have equality we should use $\nabla f(\mathbf{z})$ rather than $\nabla f(\tilde{\mathbf{x}})$ where $\mathbf{z}$ is on the line between $\mathbf{x}$ and $\tilde{\mathbf{x}}$.) Defining the unit vector $\mathbf{u}$ to be $\mathbf{u} \equiv \nabla f(\tilde{\mathbf{x}})/\|\nabla f(\tilde{\mathbf{x}})\|_2$ and noting that $(\mathbf{x} - \tilde{\mathbf{x}}) = \alpha\mathbf{e}$, the above approximation becomes

$$f(\mathbf{x}) - f(\tilde{\mathbf{x}}) \approx \alpha\|\nabla f(\tilde{\mathbf{x}})\|_2\langle \mathbf{e}, \mathbf{u}\rangle.$$

Thus the direction of maximum change to make the object function smaller forces $\langle \mathbf{e}, \mathbf{u}\rangle$ to be as large as possible negatively. From the Cauchy-Schwarz inequality (Chapter 2), $|\langle \mathbf{e}, \mathbf{u}\rangle| \leq \|\mathbf{e}\|_2\ \|\mathbf{u}\|_2$ with equality holding if and only if $\mathbf{u}$ is a multiple of $\mathbf{e}$. Thus to make $\langle \mathbf{e}, \mathbf{u}\rangle$ as large as possible negatively, we must choose $\mathbf{e} = -\mathbf{u}$, or $\mathbf{e}$ is in the direction of the negative gradient, $-\nabla f(\mathbf{x})$.

Thus we have the answer to our first question: If we are at a point $\mathbf{x}$ in $R^n$ and wish to move in the direction of steepest descent of the function $f(\mathbf{x})$, we proceed in the direction of the negative gradient,

$$-\nabla f(\mathbf{x})^T \equiv -\nabla f(x_1, x_2, \ldots, x_n)^T \equiv -\mathbf{g}(x_1, x_2, \ldots, x_n).$$

Thus our algorithm for steepest descent will generate a sequence of vectors $\{\mathbf{x}_k\}$, with $\mathbf{x}_0$ an initial guess for the minimum point, by

$$\mathbf{x}_{k+1} = \mathbf{x}_k - \alpha_k\mathbf{g}_k, \qquad k \geq 0, \tag{8.6}$$

where $\mathbf{g}_k \equiv \mathbf{g}(\mathbf{x}_k)$ and where $\alpha_k$ is a positive constant telling us how far to go in the direction $(-\mathbf{g}_k)$ in order to proceed from $\mathbf{x}_k$ to $\mathbf{x}_{k+1}$.

Thus, choosing a value for $\alpha_k$ for each $k$ is equivalent to answering our second question. To see how this might be done, suppose that $\mathbf{x}_k$ is not the minimal point. Then for sufficiently small $\alpha$, $f(\mathbf{x}_k - \alpha\mathbf{g}_k) < f(\mathbf{x}_k)$, since $-\alpha\mathbf{g}_k$ is in the direction of steepest descent from $\mathbf{x}_k$. Thus we wish to choose $\alpha = \alpha_k > 0$ to minimize the expression $f(\mathbf{x}_k - \alpha\mathbf{g}_k) \equiv h(\alpha)$. Since, for a given $k$, $\mathbf{x}_k$ and $\mathbf{g}_k$ are fixed, $h(\alpha)$ is actually a real-valued function of the real variable $\alpha$. Thus $h(\alpha)$ can be minimized by the familiar methods of single-variable calculus, i.e., set

$$h'(\alpha) \equiv \frac{d}{d\alpha} f(\mathbf{x}_k - \alpha\mathbf{g}_k) = 0 \tag{8.7}$$

and solve for $\alpha$, usually taking the smallest positive value of $\alpha$ that satisfies (8.7). The computation of the derivative in (8.7) and the solution for its zeros may be so cumbersome as to be impractical. In this case we can resort to cruder methods such as calculating $h(\alpha)$ for various values of $\alpha > 0$. For small $\alpha$, $h(\alpha)$ should initially decrease. (If it doesn't, $\alpha$ has not been chosen small enough initially.) The values $h(\alpha)$ will eventually start to increase as $\alpha$ gets larger, so after we locate two values of $\alpha$ (the first to the left of $\alpha_k$ and the second to the right of $\alpha_k$) we can close in on $\alpha_k$ by a bisection technique which proceeds exactly as in Chapter 4, except that now we are searching for a minimum of $h(\alpha)$ instead of for one of its zeros. As an alternative to the bisection approach, we can instead search until we find three values of $\alpha$, where $h(\alpha)$ is smaller at the middle $\alpha$ than at the other two. We can then fit a quadratic interpolating polynomial through these three values of $h(\alpha)$, minimizing the quadratic, and accepting the value of $\alpha$ where the minimum occurs as $\alpha_k$ in (8.6). This then completes our description of how to use (8.6), the steepest descent algorithm for finding the relative minimum of a function $f: R^n \to R$.

There are several variations of the method of (8.6), each called a *gradient method*, which we will not pursue here. We wish to reiterate that the steepest descent method is often used to obtain a first approximation for a faster, but more local, technique such as the Newton-type methods we will present in the next section. There are some problems, however, where it is advantageous to use steepest descent as the entire procedure.

**Example 8.1.** Let us now reconsider the problem of an over-determined linear system of equations (Chapter 2), in light of the material of this chapter. Let $m > n$, and let $A$ be an $(m \times n)$ matrix and $\mathbf{b}$ an $(m \times 1)$ vector. Our problem is to find an $(n \times 1)$ vector $\mathbf{x} \equiv (x_1, x_2, \ldots, x_n)^T$ such that $f(\mathbf{x}) \equiv \|A\mathbf{x} - \mathbf{b}\|_2^2 = \mathbf{x}^T A^T A\mathbf{x} - 2\mathbf{x}^T A^T\mathbf{b} + \mathbf{b}^T\mathbf{b}$ is minimized. We leave to the reader to verify that $\nabla f(\mathbf{x})^T = 2A^T A\mathbf{x} - 2A^T\mathbf{b}$, and so $\nabla f(\mathbf{x})^T = \mathbf{\theta}$ when $A^T A\mathbf{x} = A^T\mathbf{b}$. Thus we arrive at the "normal" equations, just as in Chapter 2. We also can easily see that $f''(\mathbf{x}) \equiv H(\mathbf{x}) = 2A^T A$, and so by Theorem 8.2, the zero of $\nabla f(\mathbf{x})$ is unique if $A^T A$ is positive definite. (From Chapter 3, $A^T A$ is positive definite if and only if $A$ has $n$ linearly independent columns. We note in the special problem of obtaining the polynomial of degree $(n - 1)$ which provides the best least-squares fit to $m$ data points, $A$ is the $(m \times n)$ Vandermonde-type matrix which has rank $n$, and so the polynomial is unique.)

If we choose to solve this problem by the method of steepest descent using (8.6), instead of directly solving the normal equations (such as by Gauss elimination), then for each $k$, we must minimize

$$f(\mathbf{x}_k - \alpha\mathbf{g}_k) = \mathbf{x}_k^T A^T A\mathbf{x}_k - 2\alpha\mathbf{g}_k^T A^T A\mathbf{x}_k + \alpha^2\mathbf{g}_k^T A^T A\mathbf{g}_k - 2\mathbf{x}_k^T\mathbf{b} + 2\alpha\mathbf{g}_k^T\mathbf{b} + \mathbf{b}^T\mathbf{b} \equiv h(\alpha),$$

as a function of $\alpha$. Now,

$$h'(\alpha) = -2\mathbf{g}_k^T A^T A\mathbf{x}_k + 2\alpha\mathbf{g}_k^T A^T A\mathbf{g}_k + 2\mathbf{g}_k^T\mathbf{b}.$$

Recalling that $\mathbf{g}_k = 2A^T A\mathbf{x}_k - 2A^T\mathbf{b}$, setting $h'(\alpha_k) = 0$ yields $\alpha_k = \mathbf{g}_k^T\mathbf{g}_k/2\mathbf{g}_k^T A^T A\mathbf{g}_k$ for use in the steepest descent algorithm (8.6).

Suppose the rows of the normal equations, $A^T A\mathbf{x} = A^T\mathbf{b}$, are divided by appropriate constants so that the diagonal entries of $A^T A$ are all equal to 1, that is, $A^T A\mathbf{x} = A^T\mathbf{b}$ has the equivalent form $Q\mathbf{x} = \mathbf{c}$ with $q_{ii} = 1$, $1 \le i \le n$. Suppose also that we modify the steepest descent procedure on this problem and choose $\alpha_k = 1/2$ for each $k$. Under this modification the gradient becomes $\mathbf{g}_k = 2Q\mathbf{x}_k - 2\mathbf{c}$, and the iteration of (8.6) becomes

$$\mathbf{x}_{k+1} = \mathbf{x}_k - \frac{1}{2}\mathbf{g}_k = \mathbf{x}_k - Q\mathbf{x}_k + \mathbf{c} = (I - Q)\mathbf{x}_k + \mathbf{c},$$

which is precisely the Jacobi iteration technique on $Q\mathbf{x} = \mathbf{c}$ (see Chapter 2).

### 8.3.2 Quasi-Newton Methods

We earlier pointed out some difficulties in using Newton's method (8.3a) or (8.3b) as the entire minimization process. One of these difficulties was our lack of anything beyond a local convergence theorem for the method. This problem is due mainly to the fact that it is possible to have $f(\mathbf{x}_{k+1}) > f(\mathbf{x}_k)$ (where $\mathbf{x}_{k+1}$ and $\mathbf{x}_k$ are generated by (8.3a)) if $\mathbf{x}_k$ is remote from the minimum point. Thus we are led to consider a modified Newton algorithm of the form:

$$\begin{aligned} &\mathbf{x}_0, \text{ an initial guess for the minimum point,} \\ &\mathbf{x}_{k+1} = \mathbf{x}_k - \alpha_k(H(\mathbf{x}_k))^{-1}\mathbf{g}(\mathbf{x}_k), \qquad k \geq 0, \end{aligned} \tag{8.8}$$

where $H(\mathbf{x}_k)$ and $\mathbf{g}(\mathbf{x}_k)$ are again the Hessian matrix and the transpose of gradient vector, respectively. The scalar $\alpha_k$ is again chosen in the same manner as in the method of steepest descent, (8.6), that is, it is chosen to minimize $k(\alpha) \equiv f(\mathbf{x}_k - \alpha(H(\mathbf{x}_k))^{-1}\mathbf{g}(\mathbf{x}_k))$. Once again, since $\mathbf{x}_k$ is known, $k(\alpha)$ is a real-valued function of the real variable $\alpha$. Hence $\alpha_k$ can be obtained from the solution of

$$k'(\alpha) \equiv \frac{d}{d\alpha} f(\mathbf{x}_k - \alpha H_k^{-1}\mathbf{g}_k) = 0, \tag{8.9}$$

and the remarks following (8.7) concerning the computation of $\alpha_k$ pertain to this equation as well.

We still have the usually formidable computational problems of calculating $H_k^{-1}$ at each step. We now turn to quasi-Newton methods which use an initial estimate and information from previous calculations to generate an estimate, $Q_k$, to the inverse Hessian, $H_k^{-1}$, at each step. The iteration for any such technique then becomes

$$\mathbf{x}_{k+1} = \mathbf{x}_k - \alpha_k Q_k \mathbf{g}_k, \tag{8.10}$$

where $\alpha_k$ is again chosen to minimize $f$ in the direction $-Q_k\mathbf{g}_k$ (see (8.7) and (8.9)).

The theory necessary to develop these methods rigorously is beyond the scope of this text, so we will use a more intuitive approach. In order to see how the approximations, $Q_k$, to the inverse Hessian, $H_k^{-1}$, might be formed, let us suppose that $f(\mathbf{x})$: $R^n \to R$ can be written as an $n$-dimensional quadratic form

$$f(\mathbf{x}) = \gamma + \mathbf{b}^T\mathbf{x} + \frac{1}{2}\mathbf{x}^T A\mathbf{x} \tag{8.11}$$

where $\gamma$ is a scalar constant, $\mathbf{b}$ is a constant vector in $R^n$, and $A$ is a symmetric positive definite $(n \times n)$ constant matrix. If $f(\mathbf{x})$ has this form, then the gradient, $\mathbf{g}(\mathbf{x})$, is given by $\mathbf{g}(\mathbf{x}) = \mathbf{b} + A\mathbf{x}$ and the Hessian is the constant matrix $H(\mathbf{x}) = A$. (By Theorems 8.1 and 8.2, the unique critical point, $\mathbf{x}^* = -A^{-1}\mathbf{b}$, is a minimum since $f''(\mathbf{x}^*) = A$ is positive definite. Thus, for the simple function $f(\mathbf{x})$ given in (8.11), we already have the minimum. However, we are only using (8.11) as a model to lead us to minimization techniques for more complicated functions.)

For the function in (8.11), we have by the mean-value theorem, that

$$\mathbf{g}(\mathbf{x}_{k+1}) = \mathbf{g}(\mathbf{x}_k) + H(\mathbf{x}_k)(\mathbf{x}_{k+1} - \mathbf{x}_k) \tag{8.12}$$

where (8.12) is an identity since $H(\mathbf{x}_k)$ is the constant matrix $A$. Thus we see that the inverse Hessian has the property that

$$H_k^{-1}(\mathbf{g}_{k+1} - \mathbf{g}_k) = \mathbf{x}_{k+1} - \mathbf{x}_k. \tag{8.13}$$

If we have an approximation, $Q_k$, to the inverse Hessian (as in the iteration defined by (8.10)), it seems plausible that we might choose an $(n \times n)$ matrix, $D_k$, to *correct* $Q_k$ in the sense that the equation

$$(Q_k + D_k)(\mathbf{g}_{k+1} - \mathbf{g}_k) = \mathbf{x}_{k+1} - \mathbf{x}_k \tag{8.14}$$

is satisfied when $\mathbf{x}_{k+1}$ is given by (8.10). Setting $Q_{k+1} = Q_k + D_k$, we then proceed with the next step of (8.10), going in the direction $-Q_{k+1}\mathbf{g}_{k+1}$.

This procedure condenses into the following algorithm for minimizing $f(\mathbf{x})$:

a) Choose an initial estimate $\mathbf{x}_0$ and an initial approximation, $Q_0$, to the inverse Hessian. The usual choice for $Q_0$ is the identity matrix, $I$, leaving us to proceed initially in the direction of the negative gradient
b) For any $k$, choose $\alpha_k > 0$ to minimize $f(\mathbf{x}_k - \alpha Q_k \mathbf{g}_k)$, as a function of $\alpha$
c) Set $\mathbf{x}_{k+1} = \mathbf{x}_k - \alpha_k Q_k \mathbf{g}_k$
d) Compute $\mathbf{g}_{k+1} \equiv \nabla f(\mathbf{x}_{k+1})$ and determine a matrix $D_k$ so that

$$(Q_k + D_k)(\mathbf{g}_{k+1} - \mathbf{g}_k) = \mathbf{x}_{k+1} - \mathbf{x}_k.$$

e) Let $Q_{k+1} = Q_k + D_k$ and return to step (b).

As a further motivation for this procedure, we note that when we are sufficiently near a minimum of $f(x)$, then $f(x)$ can be well approximated by a quadratic form. Furthermore, it can be shown that if the procedure outlined above is followed in the case that $f(\mathbf{x})$ *is* a quadratic form, then the minimum is *found* in $n$ steps or less. Thus, when we are near a minimum (so that $f(\mathbf{x})$ is very nearly a quadratic form), we expect that convergence will be rapid.

There are a number of possible choices for the correction matrix, $D_k$, which we cannot derive in this text. However, for a variety of substantial reasons, Shanno (1970) advocates the use of (8.15) below, where $\mathbf{r}_k = \mathbf{x}_{k+1} - \mathbf{x}_k$ and $\mathbf{y}_k = \mathbf{g}_{k+1} - \mathbf{g}_k$. (In the following formulas the reader should be alert to the fact that if $\mathbf{w}$ is any vector in $R^n$ (an $(n \times 1)$ column vector), then $(\mathbf{w}\mathbf{w}^T)$ is an $(n \times n)$ matrix, but $\mathbf{w}^T\mathbf{w}$ is a scalar.)

$$D_k = t\frac{\mathbf{r}_k\mathbf{r}_k^T}{\mathbf{r}_k^T\mathbf{y}_k} + \frac{((1-t)\mathbf{r}_k - Q_k\mathbf{y}_k)((1-t)\mathbf{r}_k - Q_k\mathbf{y}_k)^T}{((1-t)\mathbf{r}_k - Q_k\mathbf{y}_k)^T\mathbf{y}_k}. \tag{8.15}$$

In (8.15), $t$ is an arbitrary scalar parameter. If $t = 0$, (8.15) yields

$$D_k = \frac{(\mathbf{r}_k - Q_k\mathbf{y}_k)(\mathbf{r}_k - Q_k\mathbf{y}_k)^T}{(\mathbf{r}_k - Q_k\mathbf{y}_k)^T\mathbf{y}_k}, \tag{8.16a}$$

which is known as the Barnes-Rosen method. If $t = 1$, (8.15) yields

$$D_k = \frac{\mathbf{r}_k\mathbf{r}_k^T}{\mathbf{r}_k^T\mathbf{y}_k} - \frac{Q_k\mathbf{y}_k\mathbf{y}_k^T Q_k}{\mathbf{y}_k^T Q_k\mathbf{y}_k}, \tag{8.16b}$$

which is known as the Fletcher-Powell method. (We leave to the reader to verify the important fact that $D_k$ in (8.15) satisfies (8.14).) Shanno (1970) advocates other choices for $t$, which we cannot explore here. However, Fletcher-Powell is usually preferred over Barnes-Rosen, and has the theoretically important property that if $Q_0$ is positive definite, then the successive $Q_k$'s remain positive definite.

## PROBLEMS

1. Let $\mathbf{x} = (x_1, x_2, x_3)$ and $f(\mathbf{x}) = x_1x_2^2 + x_2x_3^2 + x_3^3$. Compute the gradient and Hessian of $f$ and find all values of $\mathbf{x}$ for which $\nabla f(\mathbf{x}) = (0, 0, 0)$. Is the Hessian singular or non-singular at these values?
2. Restate Theorems 8.1 and 8.2 in componentwise fasion for $n = 1$ and $n = 2$. Compare your results with the usual theorems from calculus in these cases.
3. Let $f_1(x_1, x_2) = x_1^4 + x_2^4$ and $f_2(x_1, x_2) = x_1^4 - x_2^4$. Determine if the origin (0, 0) is a relative extremum or saddle point for each of these two functions.
4. a) If $f(\mathbf{x}) = \|A\mathbf{x} - \mathbf{b}\|_2^2$ as in Example 8.1, verify that $\nabla f(\mathbf{x})^T = 2A^TA\mathbf{x} - 2A^T\mathbf{b}$ and $H(\mathbf{x}) \equiv f''(\mathbf{x}) = 2A^TA$.

   b) Write a computer program utilizing the steepest descent method for minimizing this $f(\mathbf{x})$.

   c) Use this program to find the best cubic least-squares polynomial fitting the data points: $(-2, 1)$, $(-1, 2)$, $(0, 1)$, $(1, 1)$, and $(2, 2)$.
5. Let $f_1(x_1, x_2) = x_1^2 - x_1 + x_2 - 1$ and $f_2(x_1, x_2) = x_1^3 - x_1^2 + 2x_1x_2 - x_1 - x_2 - 2$. Define $F(x_1, x_2) \equiv f_1(x_1, x_2)^2 + f_2(x_1, x_2)^2$. Use an initial guess of $(\xi_0, \zeta_0) \equiv (0, 0)$ and apply four iterations of the steepest descent method toward minimizing $F(x_1, x_2)$. Note that any absolute minimum of $F(x_1, x_2)$ is a simultaneous solution of $f_1(x_1, x_2) = 0$, $f_2(x_1, x_2) = 0$. Loosely compare the amount of necessary computation in using Newton's method in two variables for the simultaneous solution of $f_1 = f_2 = 0$ versus the Newton method of this chapter in minimizing $F$. (This analysis should indicate why steepest descent is advocated above.)
6. a) Given a vector $\mathbf{x} \in R^n$, a function $f\colon R^n \to R$, its gradient $\mathbf{g}(\mathbf{x}) \equiv \nabla f(\mathbf{x})^T$, and an $(n \times n)$ positive definite matrix $Q$, let $k(\alpha) \equiv f(\mathbf{x} - \alpha Q\mathbf{g}(\mathbf{x}))$, $\alpha > 0$. Write a computer program using a bisection technique to locate the smallest $\alpha$ minimizing $k(\alpha)$.

   b) Implementing the program in Part (a), write a steepest descent program for $n = 2$ and test it on the Rosenbrock function $f(x_1, x_2) = 100(x_2 - x_1^2) + (1 - x_1)^2$ with the initial estimate $(1, -1.2)$.

   c) Repeat the problem in Part (b) using a Fletcher-Powell program.
7. Show that a real, symmetric matrix is positive-definite if and only if all of its eigenvalues are positive.

## 8.4 LAGRANGE MULTIPLIERS

The minimization problems we have considered thus far are said to be *unconstrained*. This essentially means that we have an object function, $f\colon R^n \to R$, and we wish to minimize it without imposing any other conditions on it. There are many practical problems that arise, however, in which we want to minimize a function, $f: R^n \to R$ subject to certain restraints or limits on the vectors in the domain of $f$ that we wish to consider. Perhaps we can best illustrate this by example.

**Example 8.2.** Suppose we wish to minimize the object function $f(\alpha, \beta, \gamma) = \alpha^2 + (\beta - 2)^2 + \gamma^2$ subject to the constraints that we only wish to minimize $f(\alpha, \beta, \gamma)$ for vectors $(\alpha, \beta, \gamma)$ that satisfy $g_1(\alpha, \beta, \gamma) = \alpha^2 + \beta^2 - 1 = 0$ and $g_2(\alpha, \beta, \gamma) = \alpha^2 + \gamma^2 - 1 = 0$. Geometrically, this problem is equivalent to finding the minimum distance from the curve of intersection of $g_1(\alpha, \beta, \gamma) = g_2(\alpha, \beta, \gamma) = 0$ to the point $\mathbf{P} \equiv (0, 2, 0)$ (see Fig. 8.3). From the constraint equations, we have $\beta^2 = 1 - \alpha^2$ and $\gamma^2 = 1 - \alpha^2$, so the object function under these two constraints becomes

$$\begin{aligned} f(\alpha, \beta, \gamma) &= \alpha^2 + (1 - \alpha^2) - 4\sqrt{1 - \alpha^2} + 4 + (1 - \alpha^2) \\ &= -\alpha^2 - 4\sqrt{1 - \alpha^2} + 6 \equiv h(\alpha). \end{aligned}$$

Now, $h'(\alpha) = -2\alpha + 4\alpha(1 - \alpha^2)^{-1/2}$, so $h'(0) = 0$ and $\alpha = 0$ is a critical point. Furthermore, $h''(\alpha) = -2 + 4(1 - \alpha^2)^{-1/2} + 4\alpha^2(1 - \alpha^2)^{-3/2}$, and so $h''(0) = 2 > 0$, and thus $h(\alpha)$ is minimized for $\alpha = 0$ (and $h(0) = 2$). (Therefore, since $f(\alpha, \beta, \gamma)$ equals the square of the distance, the minimum distance is $\sqrt{h(0)} = \sqrt{2}$, as we can clearly see from Fig. 8.3.)

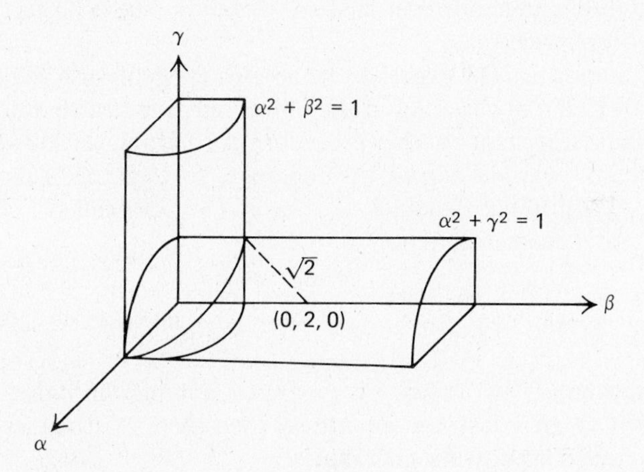


**Fig. 8.3** Geometry of Example 8.2

We chose the above example because of its simplicity, in that we were able to use the equations $g_1(\alpha, \beta, \gamma) = g_2(\alpha, \beta, \gamma) = 0$ to explicitly eliminate the variables $\beta$ and $\gamma$ from the object function $f(\alpha, \beta, \gamma)$ and thereby were able to minimize the object function in terms of the single variable $\alpha$. More often than not, however, the constraint equations will be too complex to explicitly eliminate variables as we were able to do in the above example. For instance, if we had had $g_1(\alpha, \beta, \gamma) = e^{\alpha\beta} - \alpha\beta^2\gamma + \cos(\gamma) = 0$ and $g_2(\alpha, \beta, \gamma) = \cos(\alpha\beta) - \sin(\beta\gamma) = 0$, we would probably not have been able to explicitly eliminate $\beta$ and $\gamma$ from the object function. However, the two simultaneous equations in three variables, $g_1(\alpha, \beta, \gamma) = 0$ and $g_2(\alpha, \beta, \gamma) = 0$, can still be envisioned as two restrictions (the equations) in three free parameters (the three variables). Using two of the free parameters to fulfill the two restrictions, we are still left with one free variable with which we can theoretically minimize the object function. Generalizing this notion to higher dimensions, suppose we have an object function of $n$ variables, $f(x_1, x_2, \ldots, x_n)$: $R^n \to R$, to minimize with respect to $m$ constraint equations; $g_i(x_1, x_2, \ldots, x_n) = 0$, $1 \leq i \leq m$; with $m < n$ (fewer restrictions than free parameters). We can visualize the $m$ constraint equations as implicitly eliminating $m$ of the variables of the object function, and our problem is then equivalent to minimizing an *unconstrained* functional of $(n - m)$ variables, for which the theory and techniques of the previous sections are applicable. For instance, in the above example we had $m = 2$ constraint equations for an object function of $n = 3$ variables, and the problem reduced to minimizing an object function of $(n - m) = (3 - 2) = 1$ variable, for which we used the familiar minimization techniques of calculus. (In order to make the above statements completely rigorous, we must call upon the Implicit Function Theorem (see Bartle 1964 p. 260). This would go far beyond the level of this text, however, and so we will pursue it no further than the above intuitive approach.)

Since we cannot usually expect to be able to explicitly use the $m$ constraint equations to eliminate $m$ variables in the object function (as we did in the above example), we must search for another procedure. An elegant, yet still very practical, alternative method was developed by Lagrange. In Lagrange's method we introduce $(m + 1)$ additional variables, $\{\lambda_0, \lambda_1, \lambda_2, \ldots, \lambda_m\}$, called *Lagrange multipliers*, and consider the new function

$$F(x_1, x_2, \ldots, x_n) \equiv \lambda_0 f(x_1, x_2, \ldots, x_n) + \sum_{i=1}^{m} \lambda_i g_i(x_1, x_2, \ldots, x_n). \qquad \textbf{(8.17)}$$

Loosely speaking, if $\mathbf{x}^* = (x_1^*, x_2^*, \ldots, x_n^*)$ is a minimum point of the object function subject to the constraint equations, then there exist values, $\lambda_0^*, \lambda_1^*, \ldots, \lambda_m^*$, for the Lagrange multipliers such that

$$\lambda_0^* \nabla f(x_1^*, x_2^*, \ldots, x_n^*) + \sum_{i=1}^{m} \lambda_i^* \nabla g_i(x_1^*, x_2^*, \ldots, x_n^*) = \boldsymbol{\theta}. \qquad \textbf{(8.18)}$$

Furthermore, if the $m$ vectors $\{\nabla g_i(\mathbf{x}^*)\}_{i=1}^{m}$ are linearly independent, then we may take $\lambda_0^* = 1$, and the remaining multipliers are unique and not all zero. In the following material we shall assume that this is true, but the reader must be aware of the possibility that it is not, in which case the variable $\lambda_0$ must be incorporated in the equations. Thus $\mathbf{x}^*$ can be found by solving the following

$(n + m)$ equations in the $(n + m)$ unknowns, $\{x_1, x_2, \ldots, x_n, \lambda_1, \lambda_2, \ldots, \lambda_m\}$:

$$\nabla f(x_1, x_2, \ldots, x_n) + \sum_{i=1}^{m} \lambda_i \nabla g_i(x_1, x_2, \ldots, x_n) = \mathbf{0},$$

*and* **(8.19)**

$$g_i(x_1, x_2, \ldots, x_n) = 0, \qquad 1 \leq i \leq m.$$

Once again we emphasize that the rigorous derivation of the equations of (8.19) or even the conditions under which they will yield a solution to our problem are beyond the scope of this text, thus the reader must be very cautious in their utilization. We also note that (8.19) usually represents a system of $(n + m)$ *nonlinear* simultaneous equations in the $(n + m)$ unknowns, and so a nonlinear method for their solution must be employed. We could possibly use Newton's method in $(n + m)$ variables to solve this system, but once again this involves computing the Hessian at each step. An alternative way to attempt to solve this system is to define the functional

$$F(\mathbf{y}) \equiv F_1(\mathbf{y})^2 + F_2(\mathbf{y})^2 + \cdots + F_{n+m}(\mathbf{y})^2,$$

where each $F_j(\mathbf{y})$ is the $j$th component of the equations (8.19) where $\mathbf{y} \equiv (x_1, x_2, \ldots, x_n, \lambda_1, \lambda_2, \ldots, \lambda_m)$. Since $F(\mathbf{y})$ is minimized at the solution vector $\mathbf{y}^*$ of (8.19), that is, $F(\mathbf{y}^*) = 0$, we can apply the method of steepest descent to $F(\mathbf{y})$ for determining $\mathbf{y}^*$. We further note that the Lagrange multipliers $\{\lambda_1, \lambda_2, \ldots, \lambda_m\}$ play the role of auxiliary variables, but their values are necessary in determining the minimizing vector, $\mathbf{x}^*$.

As motivation for the derivation of the equations in (8.19), consider the differentials of the object function and the constraint equations (recalling the definition of "differential" from calculus):

$$df = \frac{\partial f}{\partial x_1} dx_1 + \frac{\partial f}{\partial x_2} dx_2 + \cdots + \frac{\partial f}{\partial x_n} dx_n = 0,$$

and

$$dg_i = \frac{\partial g_i}{\partial x_1} dx_1 + \frac{\partial g_i}{\partial x_2} dx_2 + \cdots + \frac{\partial g_i}{\partial x_n} dx_n = 0, \qquad 1 \leq i \leq m.$$

Multiplying each $dg_i$ by $\lambda_i$ and adding yields

$$\left\{\frac{\partial f}{\partial x_1} + \sum_{i=1}^{m} \lambda_i \frac{\partial g_i}{\partial x_1}\right\} dx_1 + \left\{\frac{\partial f}{\partial x_2} + \sum_{i=1}^{m} \lambda_i \frac{\partial g_i}{\partial x_2}\right\} dx_2 + \cdots + \left\{\frac{\partial f}{\partial x_n} + \sum_{i=1}^{m} \lambda_i \frac{\partial g_i}{\partial x_n}\right\} dx_n = 0. \qquad \textbf{(8.20)}$$

If each of the quantities in brackets in (8.20) vanishes regardless of any variation $dx_j$, $1 \leq j \leq n$, then we can think of this as representing a horizontal tangent and thus a candidate for a minimum. We leave to the reader to verify that setting each bracket equal to zero *plus* satisfying $g_i(x_1, x_2, \ldots, x_n) = 0$, $1 \le i \le m$, yields precisely the equations of (8.19).

We conclude this section with two examples of the use of the Lagrange multiplier technique. In the first example we shall rework the problem of Example 8.2 by this technique and in the second example we shall illustrate the technique in terms of a data-fitting problem.

**Example 8.3.** With $f(\alpha, \beta, \gamma)$, $g_1(\alpha, \beta, \gamma)$, and $g_2(\alpha, \beta, \gamma)$ given as in Example 8.2, we have from (8.19)

$$\nabla f + \lambda_1 \nabla g_1 + \lambda_2 \nabla g_2 = \begin{bmatrix} 2\alpha(1 + \lambda_1 + \lambda_2) \\ 2\beta(1 + \lambda_1) - 4 \\ 2\gamma(1 + \lambda_2) \end{bmatrix}^T = \begin{bmatrix} 0 \\ 0 \\ 0 \end{bmatrix}^T,$$

with $\gamma = \sqrt{1 - \alpha^2}$ and $\beta = \sqrt{1 - \alpha^2}$. Algebraically solving these five equations, we obtain one solution to be: $\alpha = 0$, $\beta = 1$, $\gamma = 1$, and $\lambda_1 = 1$, $\lambda_2 = -1$. Thus $f(0, 1, 1) = 2$, as before. There are other solutions to these five equations. We leave it to the reader to find them and to interpret their significance via Fig. 8.3 (Problem 3).

**Example 8.4.** (A piece-wise discrete least-squares data-fit with continuity constraints.) We have mentioned before that polynomial data-fitting leads to an undesirable oscillatory behavior in the approximation. Cubic spline interpolation is one way of overcoming this difficulty, but for a large number of data points this process can become very cumbersome. Another approach is to partition the data points into several different sets and obtain a separate low-degree polynomial least-squares fit on each individual set. For example, let the data points $\{(x_i, y_i)\}_{i=1}^{5}$ be the first set $S_1$, let $S_2$ consist of $\{(x_i, y_i)\}_{i=6}^{10}$, and $S_3$ consist of $\{(x_i, y_i)\}_{i=11}^{20}$. Let $p_1(x)$ be the best quadratic least-squares polynomial on $S_1$, $p_2(x)$ be the best quadratic least-squares polynomial on $S_2$, and $p_3(x)$ be the best cubic least-squares polynomial on $S_3$. Then we could take $P(x)$ given by

$$P(x) = \begin{cases} p_1(x), & \text{for } x_1 \le x < x_6 \\ p_2(x), & \text{for } x_6 \le x < x_{11} \\ p_3(x), & \text{for } x_{11} \le x \le x_{20}, \end{cases}$$

as our approximating curve. The difficulty with this approach is that usually $p_1(x_6) \ne p_2(x_6)$ and $p_2(x_{11}) \ne p_3(x_{11})$, and thus $P(x)$ is discontinuous at $x_6$ and $x_{11}$, which is usually undesirable in practice. We can use the Lagrange multiplier technique to remedy this deficiency. This is done by still requiring a least-squares polynomial fit on $S_1$, $S_2$, and $S_3$ as above, but also imposing the side constraints that $p_1(x_6) = p_2(x_6)$ and $p_2(x_{11}) = p_3(x_{11})$. (Note that by altering the problem in this way we will not obtain the same polynomials $p_1(x)$, $p_2(x)$, and $p_3(x)$ as above.) Our variables are the coefficients of the three polynomials plus the two Lagrange multipliers corresponding to the side constraints. Thus we let $p_1(x) = \alpha_1 + \alpha_2 x + \alpha_3 x^2$, $p_2(x) = \alpha_4 + \alpha_5 x + \alpha_6 x^2$, and $p_3(x) = \alpha_7 + \alpha_8 x + \alpha_9 x^2 + \alpha_{10} x^3$, where $\{\alpha_i\}_{i=1}^{10}$ are the variables to be determined subject to the constraints $p_1(x_6) = p_2(x_6)$ and $p_2(x_{11}) = p_3(x_{11})$. The constraint equations expressed in these variables are

$$g_1(\alpha_1, \ldots, \alpha_{10}) \equiv (\alpha_1 + \alpha_2 x_6 + \alpha_3 x_6^2) - (\alpha_4 + \alpha_5 x_6 + \alpha_6 x_6^2) = 0,$$

$$A \equiv \begin{bmatrix} 1 & x_1 & x_1^2 & 0 & 0 & 0 & 0 & 0 & 0 & 0 \\ 1 & x_2 & x_2^2 & 0 & 0 & 0 & 0 & 0 & 0 & 0 \\ \vdots & & & & & & & & & \vdots \\ 1 & x_5 & x_5^2 & 0 & 0 & 0 & 0 & 0 & 0 & 0 \\ 0 & 0 & 0 & 1 & x_6 & x_6^2 & 0 & 0 & 0 & 0 \\ 0 & 0 & 0 & 1 & x_7 & x_7^2 & 0 & 0 & 0 & 0 \\ \vdots & & & & & & & & & \vdots \\ 0 & 0 & 0 & 1 & x_{10} & x_{10}^2 & 0 & 0 & 0 & 0 \\ 0 & 0 & 0 & 0 & 0 & 0 & 1 & x_{11} & x_{11}^2 & x_{11}^3 \\ 0 & 0 & 0 & 0 & 0 & 0 & 1 & x_{12} & x_{12}^2 & x_{12}^3 \\ \vdots & & & & & & & & & \vdots \\ 0 & 0 & 0 & 0 & 0 & 0 & 1 & x_{20} & x_{20}^2 & x_{20}^3 \end{bmatrix}$$

**Figure 8.4**

and

$$g_2(\alpha_1, \ldots, \alpha_{10}) \equiv (\alpha_4 + \alpha_5 x_{11} + \alpha_6 x_{11}^2) - (\alpha_7 + \alpha_8 x_{11} + \alpha_9 x_{11}^2 + \alpha_{10} x_{11}^3) = 0.$$

Thus

$$\nabla g_1(\alpha_1, \ldots, \alpha_{10}) = (1, x_6, x_6^2, -1, -x_6, -x_6^2, 0, 0, 0, 0)$$

and

$$\nabla g_2(\alpha_1, \ldots, \alpha_{10}) = (0, 0, 0, 1, x_{11}, x_{11}^2, -1, -x_{11}, -x_{11}^2, -x_{11}^3).$$

As in Example 8.1, we take $f(\alpha_1, \alpha_2, \ldots, \alpha_{10}) = \|A\mathbf{x} - \mathbf{b}\|_2^2$, where $\mathbf{x}^T \equiv (\alpha_1, \alpha_2, \ldots, \alpha_{10})^T$, $\mathbf{b}^T \equiv (y_1, y_2, \ldots, y_{20})^T$, and $A$ is the $(20 \times 10)$ Vandermonde-type matrix shown in Fig. 8.4. Again, as in Example 8.1, $\nabla f(\alpha_1, \alpha_2, \ldots, \alpha_{10}) = 2A^TA\mathbf{x} - 2A^T\mathbf{b}$. Thus to minimize $f(\alpha_1, \ldots, \alpha_{10})$ under the constraints $g_1(\alpha_1, \ldots, \alpha_{10}) = 0$ and $g_2(\alpha_1, \ldots, \alpha_{10}) = 0$, we must satisfy the equations of formula (8.19):

$$\begin{aligned} \nabla f(\alpha_1, \ldots, \alpha_{10}) + \lambda_1 \nabla g_1(\alpha_1, \ldots, \alpha_{10}) + \lambda_2 \nabla g_2(\alpha_1, \ldots, \alpha_{10}) &= 0 \\ g_1(\alpha_1, \ldots, \alpha_{10}) &= 0 \\ g_2(\alpha_1, \ldots, \alpha_{10}) &= 0. \end{aligned} \tag{8.21}$$

Since the bottom two equations of (8.21) are linear in the variables $\{\alpha_1, \alpha_2, \ldots, \alpha_{10}\}$, (8.21) represents a linear system of twelve equations in twelve unknowns, $\{\alpha_1, \alpha_2, \ldots, \alpha_{10}, \lambda_1, \lambda_2\}$. In matrix form these equations become:

$$\left[\begin{array}{c|cc} A^TA & \multicolumn{2}{c}{B^T} \\ \hline B & 0 & 0 \\ & 0 & 0 \end{array}\right] \begin{bmatrix} \alpha_1 \\ \alpha_2 \\ \vdots \\ \alpha_{10} \\ \tilde{\lambda}_1 \\ \tilde{\lambda}_2 \end{bmatrix} = \begin{bmatrix} A^T\mathbf{b} \\ 0 \\ 0 \end{bmatrix}, \tag{8.22}$$

where $\tilde{\lambda}_1 = \lambda_1/2, \tilde{\lambda}_2 = \lambda_2/2$ and $B$ is the $(2 \times 10)$ matrix,

$$B \equiv \begin{bmatrix} 1 & x_6 & x_6^2 & -1 & -x_6 & -x_6^2 & 0 & 0 & 0 & 0 \\ 0 & 0 & 0 & 1 & x_{11} & x_{11}^2 & -1 & -x_{11} & -x_{11}^2 & -x_{11}^3 \end{bmatrix}.$$

Using the expressions for $\nabla f$, $\nabla g_1$, and $\nabla g_2$, we leave to the reader to verify componentwise that (8.22) is precisely (8.21).

We have presented this method of data-fitting in a very specialized and simplified form to more clearly illustrate the derivation of the linear system (8.22). It should be clear to the reader that any number of data points can be used, there can be a different number of partitions and they can be separated at different points, the degrees of the polynomials on each partition can be altered, etc. We can even put further conditions on the degree of continuity between the polynomials of each partition. (For example, we could require $p_1'(x_6) = p_2'(x_6)$, etc.) For each additional constraint we have another Lagrange multiplier, however, these types of constraints are still linear with respect to the polynomial coefficients, and we still obtain a partitioned matrix form similar to (8.22) to solve for the polynomial coefficients and Lagrange multipliers. A data-fitting technique of this type is called a *smoothing* procedure, since it reduces the oscillatory behavior of the approximation.

## PROBLEMS

1. In formula (8.19) let $n = 5$ and $m = 3$. Write out (8.19) in componentwise form yielding the eight equations in eight unknowns that must be solved.
2. Show that the constraints, $g_i(\mathbf{x}) = 0, 1 \leq i \leq m$, plus setting the quantities in parentheses in (8.20) yields the equations of formula (8.19).
3. Find the other solutions to the equations in Example 8.3, and interpret their geometric significance to Fig. 8.3.
4. Verify that the matrix representation in (8.22) is equivalent to the equations of formula (8.21).
5. a) In Example 8.4, specify numerical values for $\{(x_i, y_i)\}_{i=1}^{20}$ and find $p_1(x)$, $p_2(x)$, and $p_3(x)$ by solving (8.22).

   b) Modify the problem of Example 8.4 by imposing the additional restriction that $p_2'(x_{11}) = p_3'(x_{11})$. Display the modifications this makes on the matrix system (8.22).
6. Use Lagrange's method to solve the following:
   Given the plane $\alpha + 2\beta + \gamma = 1$, find the point on this plane which is nearest the origin. (Hint: Minimize $f(\alpha, \beta, \gamma) = \alpha^2 + \beta^2 + \gamma^2$ with respect to the constraint $\alpha + 2\beta + \gamma - 1 = 0$.)
7. Find the maximum of $f(x_1, x_2, \ldots, x_n) = (x_1 x_2 \cdots x_n)^2$ subject to the constraint $x_1^2 + x_2^2 + \cdots + x_n^2 = 1$.

## 8.5 LINEAR PROGRAMMING

The minimization problem considered in Section 8.4 had *equality constraints* of the form $g_i(x_1, x_2, \ldots, x_n) = 0$, $1 \leq i \leq m$. A more complicated problem, in general, is to replace the above equality constraints by inequalities of the form

$g_i(x_1, x_2, \ldots, x_n) \leq 0$, $1 \leq i \leq m$. This general problem is beyond the level of this text, so we shall be content to consider a specialized but important problem of the latter type. The problem is called the *linear programming* problem and seeks to minimize an objective function which is *linear* in the unknowns, subject to constraints which consist of *linear* equalities and *linear* inequalities. In standard form the inequalities are changed to equalities and the problem can be expressed as:

$$\begin{aligned} \text{minimize } f(x_1, \ldots, x_n) &= c_1x_1 + c_2x_2 + \cdots + c_nx_n \\ \text{subject to } a_{11}x_1 + a_{12}x_2 + \cdots + a_{1n}x_n &= b_1 \\ a_{21}x_1 + a_{22}x_2 + \cdots + a_{2n}x_n &= b_2 \\ \vdots \qquad\qquad & \quad \vdots \\ a_{m1}x_1 + a_{m2}x_2 + \cdots + a_{mn}x_n &= b_m \end{aligned} \tag{8.23}$$

and

$$x_1 \geq 0, \qquad x_2 \geq 0, \ldots, x_n \geq 0,$$

where the $a_{ij}$'s, $c_i$'s, and $b_i$'s are known values, and the $x_i$'s are the unknowns to be determined. We shall also require that in (8.23) each $b_i \geq 0$ (since this can be accomplished by multiplication by $(-1)$ whenever necessary). Furthermore, we note that minimizing $f(x_1, x_2, \ldots, x_n)$ is equivalent to maximizing

$$(-f(x_1, x_2, \ldots, x_n)).$$

If $A = (a_{ij})$ is an $(m \times n)$ matrix, $\mathbf{x} = (x_1, x_2, \ldots, x_n)^T$, $\mathbf{c} \equiv (c_1, c_2, \ldots, c_n)^T$, and $\mathbf{b} \equiv (b_1, b_2, \ldots, b_m)^T$, then the linear programming problem, stated above, can be expressed in the more compact vector form

$$\begin{aligned} &\text{minimize } f(\mathbf{x}) \equiv \mathbf{c}^T\mathbf{x} \\ &\text{subject to } A\mathbf{x} = \mathbf{b} \quad \text{and} \quad \mathbf{x} \geq \mathbf{0}. \end{aligned} \tag{8.24}$$

Often a practical problem will not occur directly in the standard form of (8.23) or (8.24), a procedure which can be used to convert it to this form is by introducing what are known as *slack variables*. For instance, consider the following simple example where we wish to determine $x_1$ and $x_2$ such that $x_1 \geq 0$, $x_2 \geq 0$; $z = -3x_1 - 4x_2$ is minimized, and where $x_1$ and $x_2$ must satisfy the three constraints:

$$-x_1 + 2x_2 \geq -1, \qquad 5x_1 + 2x_2 \leq 7, \qquad \text{and} \qquad 2x_1 + 3x_2 \leq 8.$$

We recognize that the first constraint is equivalent to $x_1 - 2x_2 \leq 1$. We then introduce three new "slack variables"; $x_3 \geq 0$, $x_4 \geq 0$, $x_5 \geq 0$; such that the constraints become:

$$x_1 - 2x_2 + x_3 = 1, \qquad 5x_1 + 2x_2 + x_4 = 7, \qquad \text{and} \qquad 2x_1 + 3x_2 + x_5 = 8.$$

This problem is now in the standard form given by (8.23) with respect to the nonnegative variables, $x_1$, $x_2$, $x_3$, $x_4$, and $x_5$. We observe that the original problem involves only two unknowns whereas the auxiliary problem (in standard form) involves five unknowns. It can be shown, however, that a solution of the latter problem will determine a corresponding solution of the original problem, as illustrated later. The following example shows where the above linear-programming problem could occur in a hypothetical real-world setting.

**Example 8.5.** Let $x_1$ and $x_2$ represent the respective amounts of two types of pain-killing liquid that are made in one day; each type contains certain amounts of Drug $A$ and Drug $B$. Each unit of the first type of pain-killing liquid requires $\frac{5}{7}$ unit of Drug $A$, and each unit of the second type of pain-killing liquid requires $\frac{2}{7}$ unit of Drug $A$. If only one unit of Drug $A$ is available in one day, then

$$\frac{5}{7}x_1 + \frac{2}{7}x_2 \leq 1 \quad \text{or} \quad 5x_1 + 2x_2 \leq 7.$$

Each unit of the first type requires $\frac{1}{4}$ unit of Drug $B$, and each unit of the second type requires $\frac{3}{8}$ unit of Drug $B$. If only one unit of Drug $B$ is available in one day, then

$$\frac{1}{4}x_1 + \frac{3}{8}x_2 \leq 1 \quad \text{or} \quad 2x_1 + 3x_2 \leq 8.$$

Moreover FDA regulations require that the amount $x_1$ cannot exceed $1 + 2x_2$; that is

$$x_1 \leq 1 + 2x_2 \quad \text{or} \quad x_1 - 2x_2 \leq 1.$$

If three dollars and four dollars are the respective profit margins for one unit of the two pain-killing liquids, then $3x_1 + 4x_2$ represents the total profit per day. We want to determine the values of $x_1$ and $x_2$ that will maximize the profit.

In order to obtain some geometric insight into the linear programming problem, we consider a geometric approach to the above particular problem. The first constraint demands that $x_1 - 2x_2 \leq 1$. This means that our solution, $(x_1^*, x_2^*)$, must be a point in the $x_1x_2$-plane lying below the line $x_1 - 2x_2 = 1$. Similarly the other two constraints insist that $(x_1^*, x_2^*)$ lie below the lines $5x_1 + 2x_2 = 7$ and $2x_1 + 3x_2 = 8$. Since we must also have $x_1 \geq 0$ and $x_2 \geq 0$, the solution $(x_1^*, x_2^*)$ must lie in the area denoted *feasible solutions* in Fig. 8.5.

We observe that minimizing $z = -3x_1 - 4x_2$ is equivalent to maximizing $z = 3x_1 + 4x_2$. After graphing a few lines of the form $z = 3x_1 + 4x_2$ for various values of $z$, it becomes readily apparent that the largest value that $z$ can attain and still have the line intersect the area of feasible solutions comes from the line that passes through the point $C = (5/11, 26/11)$. The value of $z$ for this line is $z^* = 3(5/11) + 4(26/11) = 119/11$.

Obviously it would be very difficult to carry out a graphical procedure as above in a problem involving several variables. For example, in a problem involving three variables, the area of feasible solutions becomes a three-dimensional region bounded by planes, and in the general case of $n$ variables it becomes an $n$-dimensional solid called a *polytope* bounded by *hyperplanes*. A hyperplane in

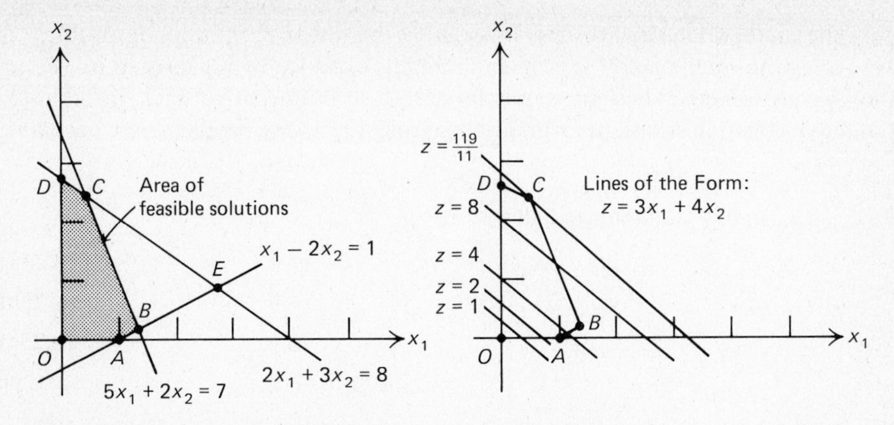


**Fig. 8.5** Area of feasible solutions and the graphical solution

$R^n$ is the set of all vectors $\mathbf{x} = (x_1, x_2, \ldots, x_n)$ whose coordinates satisfy an equation of the form: $a_1x_1 + a_2x_2 + \cdots + a_nx_n = b$. A hyperplane of this form thus divides $R^n$ into two parts or *half-spaces*. One half-space is the set of all vectors $\mathbf{x} = (x_1, x_2, \ldots, x_n)$ such that $a_1x_1 + a_2x_2 + \cdots + a_nx_n \geq b$ and the other half-space is the set of all vectors $\mathbf{x} = (x_1, x_2, \ldots, x_n)$ such that $a_1x_1 + a_2x_2 + \cdots + a_nx_n \leq b$. Clearly the hyperplane is the intersection of these two half-spaces. In Fig. 8.5, $n = 2$, and each line (say, $x_1 - 2x_2 = 1$) is a hyperplane that divides the $x_1x_2$-plane into two parts or half-spaces. If $n = 3$, a hyperplane is the set of all vectors $\mathbf{x} = (x_1, x_2, x_3)$ such that $a_1x_1 + a_2x_2 + a_3x_3 = b$. The reader should recall from calculus that this is the general equation of a plane and any plane divides $R^3$ into two parts. If $n > 3$ we lose this type of graphical perspective, but we see from above that a hyperplane in $R^n$ still divides $R^n$ into two parts. Thus we are led to suspect that some of the geometrical properties of the above two-variable problem carry over into the $n$-dimensional problem. Any vector (if any exists) which lies in the set of feasible solutions is called a *feasible point* and its coordinates satisfy all constraints of the problem. Furthermore the $n$-dimensional polytope (set of feasible points) has vertices. In Fig. 8.5, $n = 2$ and the points $O$, $A$, $B$, $C$, and $D$ are vertices. The point $E$ is not a vertex of the polytope even though it is the intersection of $x_1 - 2x_2 = 1$ and $2x_1 + 3x_2 = 8$. This is because the coordinates of $E$ do not satisfy the constraint $5x_1 + 2x_2 \leq 7$ and hence $E$ is not a feasible point. A vertex of a polytope defined by the constraints can be regarded geometrically as a *feasible* point where the boundaries of at least $n$ of the half-spaces, determined by the constraints, intersect. These vertices are called *extreme feasible points*. It can be shown that a solution point is always an extreme feasible point (such as the point $C$ in Fig. 8.5). It can happen that two extreme feasible points are solutions, in which case the entire edge connecting them consists of solution points. The same is true in higher dimensions.

The method we shall discuss for solving the linear programming problem is known as the *simplex method*, which we shall introduce by illustrating its use on the above problem. The first step is to introduce nonnegative slack variables to put the problem in standard form. For example, the above problem now becomes:

Determine $x_i \geq 0, 1 \leq i \leq 5$, such that $z \equiv f(x_1, x_2, \ldots, x_5) = 3x_1 + 4x_2 + 0x_3 + 0x_4 + 0x_5$ is maximized subject to

$$x_1 - 2x_2 + x_3 = 1 \qquad \textbf{(8.25a)}$$

$$5x_1 + 2x_2 + x_4 = 7 \qquad \textbf{(8.25b)}$$

$$2x_1 + 3x_2 + x_5 = 8, \qquad \textbf{(8.25c)}$$

$$x_i \geq 0, \qquad 1 \leq i \leq 5.$$

We recall that a solution must occur at an extreme feasible point of the polytope. Since we do not know which extreme feasible point yields the solution, however, we proceed in a systematic fashion, which will amount to choosing a vertex and moving along an edge from that vertex to another vertex in a manner such that $z$ will be increased or unchanged. This process will be repeated again and again until the solution is found. There are several complications that can arise in this procedure, such as *degeneracy*, contradictory constraint equations, multiple solutions, etc., but it is beyond the scope of this text to investigate them (see Luenberger 1973). This is primarily why we only consider the simple problem above to illustrate the technique.

Once the problem is in standard form, our first step is to set $x_1 = x_2 = 0$ and solve (8.25) for $x_3$, $x_4$, $x_5$, yielding $x_3 = 1$, $x_4 = 7$, and $x_5 = 8$. For this choice $z = 0$ and is not maximized. This follows, since $z$ can be increased by increasing either $x_1$ or $x_2$. Since $x_2$ has the larger coefficient in $z = 3x_1 + 4x_2 + 0x_3 + 0x_4 + 0x_5$ we allow it to increase and keep $x_1$ at zero. We next select one of the three remaining variables, $x_3$, $x_4$, or $x_5$, to be set equal to zero in place of $x_2$. To determine which one, we consider the equations of (8.25) with $x_1 = 0$:

$$-2x_2 + x_3 = 1 \qquad \textbf{(8.26a)}$$

$$2x_2 + x_4 = 7 \qquad \textbf{(8.26b)}$$

$$3x_2 + x_5 = 8. \qquad \textbf{(8.26c)}$$

Since $x_3, x_4, x_5$ cannot be negative, (8.26b) yields $x_2 \leq \frac{7}{2}$ and (8.26c) yields $x_2 \leq \frac{8}{3}$. Since the coefficient of $x_2$ in (8.26a) is negative, $x_3$ will be positive for any $x_2 \geq 0$, and so (8.26a) places no restriction on $x_2$. The two restrictions we do have on $x_2$ must hold simultaneously, and so we let $x_2$ equal the smaller of the two, $x_2 = \frac{8}{3}$, which requires $x_5 = 0$ due to (8.26c). Now solving equations (8.26) yields $x_1 = 0$, $x_2 = \frac{8}{3}$, $x_3 = \frac{19}{3}$, $x_4 = \frac{5}{3}$, $x_5 = 0$. With these values we have $z = 3x_1 + 4x_2 = \frac{32}{3}$, which is substantially better than our first value $z = 0$. (We note here that the above work is precisely equivalent to moving from the origin to the vertex $D$ in Fig. 8.5.) Before we can repeat this step and move to another vertex, we must reformulate equations (8.25) such that $x_2$, $x_3$, and $x_4$ play the roles of the slack variables. This means that we wish to convert the equations of (8.25) to the form $\alpha_k x_1 + \beta_k x_5 + x_k = \gamma_k$ for $k = 2$, 3, and 4, and that we wish to express $z$ as a function of $x_1$ and $x_5$ rather than $x_1$ and $x_2$. We can do this by use of (8.25c), since it can be used to solve for $x_2$ in terms of $x_1$ and $x_5$. (8.25c) becomes $\frac{2}{3}x_1 + x_2 + \frac{1}{3}x_5 = \frac{8}{3}$. Now we use it to eliminate $x_2$ from the first two equations, yielding

$$7x_1/3 \quad + x_3 \quad + 2x_5/3 = 19/3 \tag{8.27a}$$

$$11x_1/3 \quad + x_4 - 2x_5/3 = 5/3 \tag{8.27b}$$

$$2x_1/3 + x_2 \quad + x_5/3 = 8/3. \tag{8.27c}$$

Using (8.27c) to eliminate $x_2$ from $z = 3x_1 + 4x_2$, we obtain $z = \frac{32}{3} + \frac{1}{3}x_1 - \frac{4}{3}x_5$.

We have now completed one step of the simplex method and are ready to repeat the same process using equations (8.27) and $z = \frac{32}{3} + \frac{1}{3}x_1 - \frac{4}{3}x_5$. Since $x_1$ has the larger coefficient in $z(\frac{1}{3} > -\frac{4}{3})$, and its coefficient is positive, the value of $z$ can be increased by increasing the value of $x_1$. Therefore we allow $x_1$ to increase while we keep $x_5$ at zero. We now repeat the above process to select $x_2$, $x_3$, or $x_4$ to take the place of $x_1$ and be set equal to zero. From (8.27) with $x_5 = 0$ we obtain

$$7x_1/3 \quad + x_3 \quad = 19/3 \tag{8.28a}$$

$$11x_1/3 \quad + x_4 = 5/3 \tag{8.28b}$$

$$2x_1/3 + x_2 \quad = 8/3. \tag{8.28c}$$

Since $x_2$, $x_3$, $x_4$ are nonnegative, (8.28a) yields $\frac{7}{3}x_1 \leq \frac{19}{3}$, (8.28b) yields $\frac{11}{3}x_1 \leq \frac{5}{3}$, and (8.28c) yields $\frac{2}{3}x_1 \leq \frac{8}{3}$. Since these must hold simultaneously, we have from (8.28b) that $x_1 \leq \frac{5}{11}$, since this is the smallest of the three bounds on $x_1$. Setting $x_1 = \frac{5}{11}$ yields $x_4 = 0$ from (8.28b). Now solving equations (8.28) yields $x_1 = \frac{5}{11}$, $x_2 = \frac{26}{11}$, $x_3 = \frac{58}{11}$, $x_4 = 0$, $x_5 = 0$. We calculate $z = \frac{32}{3} + \frac{1}{3}x_1 - \frac{4}{3}x_5$ with these values and obtain $z = \frac{119}{11}$, which is larger than our previous value of $z = \frac{32}{3}$. (Note that this step is precisely equivalent to moving from vertex $D$ to vertex $C$ in Fig. 8.5.) Now we reformulate equations (8.27) such that $x_1$, $x_2$, and $x_3$ play the role of the slack variables. Using (8.27b) to eliminate $x_1$ in the remaining two equations yields

$$x_1 \quad + 3x_4/11 - 2x_5/11 = 5/11 \tag{8.29a}$$

$$x_2 \quad - 2x_4/11 + 5x_5/11 = 26/11 \tag{8.29b}$$

$$x_3 - 7x_4/11 + 12x_5/11 = 58/11. \tag{8.29c}$$

Using (8.29a) to eliminate $x_1$ from $z = \frac{32}{3} + \frac{1}{3}x_1 - \frac{4}{3}x_5$ yields $z = \frac{119}{11} - \frac{1}{11}x_4 - \frac{14}{11}x_5$. Since increasing either $x_4$ or $x_5$ in this expression for $z$ will decrease the value of $z$, this expression results in the maximum value for $z$. (Moving to any other vertex decreases $z$ and thus we are at the extreme feasible point where $z$ is maximized.) Thus our final answer is $z = \frac{119}{11}$, which occurs at $x_1 = \frac{5}{11}$ and $x_2 = \frac{26}{11}$, and agrees with the answer achieved by the graphical method in Fig. 8.5.

We notice that each step of the simplex method involves solutions of linear systems of equations and selection of largest *pivotal elements*, i.e., the choice of which variable to be increased to maximize the increase in $z$ in going from one vertex to another. These procedures can be organized and put in an algorithmic form through what is known as a *simplex tableau*. This primarily involves an extension of the familiar Gauss-Jordan elimination method with pivoting to solve a set of simultaneous linear equations. Since this section is intended only as a brief introduction to the linear programming problem, and, because of the many complications that must be considered to treat the problem thoroughly, we shall not pursue this further. There are numerous texts which are primarily devoted to the theory of linear programming and the efficient utilization of the simplex method. An excellent reference is the above-mentioned text by Luenberger. We also call attention to the text by Kuo (1972). In particular we reference Kuo's computer program for the simplex method for up to 34 constraints and 85 variables on pages 360–363 and his flow chart on pages 364–366.

## PROBLEMS

1. Consider the problem of maximizing $z = 3x_1 + 2x_2$ subject to the constraints: $2x_1 - x_2 \le 4$, $2x_1 + x_2 \le 8$, $x_1 + x_2 \le 6$, and $-3x_1 + x_2 \le 2$, with $x_1 \ge 0$ and $x_2 \ge 0$.
   a) Graph the region of feasible solutions in the $x_1x_2$-plane and identify the extreme feasible points.
   b) Solve this problem graphically.
   c) Solve this problem by putting it in standard form and using the simplex method. Identify your path from vertex to vertex as you proceed.
2. Consider the problem of maximizing $z = 2x_1 + 3x_2 + x_3$ subject to the constraints: $-2x_1 + x_3 \le 0$, $2x_1 + 3x_2 \le 9$, $x_1 - x_2 \le 2$, and $x_1 + x_3 \le 3$, with $x_1 \ge 0$, $x_2 \ge 0$, $x_3 \ge 0$. Repeat the steps of Parts (a), (b), and (c) of Problem 1 for this three-dimensional problem. (The solution is found after three steps of the simplex method. Note, however, that the results of the first and second steps yield the same values of the unknowns and do not increase the value of $z$. This is an example of a phenomenon known as *degeneracy*. It is theoretically possible that this situation could cycle indefinitely without reaching a maximum. This situation, however, is not common and modifications can be made to the simplex method to treat it.)
3. Mr. H. H. Schmidtke makes doll houses and toy barns as a hobby and sells them for profit. He has six days of vacation to work on his hobby and wishes to work at most eight hours each day. He furthermore wishes to spend two days cutting pieces and four days for assembly and finishing work. Each doll house requires 30 minutes to cut pieces and one hour and twenty minutes to assemble and finish. Each toy barn requires 20 minutes to cut pieces and 30 minutes to assemble and finish. His retail outlet demands that he make 13 more toy barns than doll houses. His net profit is \$20 for each doll house and \$12 for each toy barn. How many of each should he make in order to maximize his profit over the vacation, provided he is willing to leave a doll house and a toy barn only partially done?

# REFERENCES


Bartle, R. G. (1964), *The Elements of Real Analysis*, New York: John Wiley and Sons.

Cheney, E. W. (1966), *Introduction to Approximation Theory*, New York: McGraw-Hill.

Clenshaw, C. W., and A. R. Curtis (1960), "A method for numerical integration on an automatic computer," *Numerische Mathematik*, Vol. 2.

Coddington, E. A., and N. Levinson (1955), *Theory of Ordinary Differential Equations*, New York: McGraw-Hill.

Davis, P. J., and P. Rabinowitz (1967), *Numerical Integration*, Waltham, Mass.: Blaisdell.

Faddeev, D. K., and V. N. Faddeeva (1963), *Computational Methods of Linear Algebra*, San Francisco: Freeman.

Forsythe, G. E. (1967), "Today's computational methods of linear algebra," *SIAM Review*.

Forsythe, G. E., and C. Moler (1967), *Computer Solution of Linear Algebraic Systems*, Englewood Cliffs, N.J.: Prentice-Hall.

Fox, L. (1965), *Introduction to Numerical Linear Algebra*, New York: Oxford University Press.

Fröberg, C. E. (1969), *Introduction to Numerical Analysis*, Reading, Mass.: Addison-Wesley.

Givens, W. (1954), "Numerical computation of the characteristic values of a real symmetric matrix," Rep. ORNL 1574, Oak Ridge, Tenn.: Oak Ridge National Laboratory.

Henrici, P. (1962), *Discrete Variable Methods in Ordinary Differential Equations*, New York: John Wiley and Sons.

Henrici, P. (1974), *Applied and Computational Complex Analysis, Vol. 1*, New York: John Wiley and Sons.

Householder, A. S. (1970), *The Numerical Treatment of a Single Nonlinear Equation*, New York: McGraw-Hill.

Householder, A. S., and F. L. Bauer (1959), "On certain methods for expanding the characteristic polynomial," *Numerische Mathematik*, Vol. 1, pp. 29–37.

Isaacson, E., and H. B. Keller (1966), *Analysis of Numerical Methods*, New York: John Wiley and Sons.

Kuo, S. S. (1972), *Computer Applications of Numerical Methods*, Reading, Mass.: Addison-Wesley.

Lambert, J. D. (1973), *Computational Methods in Ordinary Differential Equations*, London: John Wiley and Sons.

Luenberger, D. G. (1973), *Introduction to Linear and Nonlinear Programming*, Reading, Mass.: Addison-Wesley.

Natanson, I. P. (1965), *Constructive Function Theory, Vol. III*, New York: Frederick Ungar.

Ostrowski, A. M. (1966), *Solution of Equations and Systems of Equations*, New York: Academic Press.

Ralston, A. (1965), *A First Course in Numerical Analysis*, New York: McGraw-Hill.

Rivlin, T. J. (1969), *An Introduction to the Approximation of Functions*, Waltham, Mass.: Blaisdell.

Rowland, J. H., and Y. L. Varol (1972), "Exit criteria for Simpson's compound rule," *Mathematics of Computation*, Vol. 26, No. 119.

Shampine, L. F., and R. C. Allen (1973), *Numerical Computing: An Introduction*, Philadelphia: Saunders.

Shanno, D. F. (1970), "Conditioning of quasi-Newton methods for function minimization," *Mathematics of Computation*, Vol. 24, No. 111.

Stroud, R. H., and D. Secrest (1966), *Gaussian Quadrature Formulas*, Englewood Cliffs, N.J.: Prentice-Hall.

Thomas, G. B. (1972), *Calculus and Analytic Geometry*, 4th ed., Reading, Mass.: Addison-Wesley.

Todd, J. (1962), *Survey of Numerical Analysis*, New York: McGraw-Hill.

Varga, R. S. (1962), *Matrix Iterative Analysis*, Englewood Cliffs, N.J.: Prentice-Hall.

Wilkinson, J. H. (1963), *Rounding Errors in Algebraic Processes*, Englewood Cliffs, N.J.: Prentice-Hall.

# INDEX

# INDEX